Lecture Notes in Civil Engineering

Volume 201

Lecture Notes in Civil Engineering (LNCE) publishes the latest developments in Civil Engineering - quickly, informally and in top quality. Though original research reported in proceedings and post-proceedings represents the core of LNCE, edited volumes of exceptionally high quality and interest may also be considered for publication. Volumes published in LNCE embrace all aspects and subfields of, as well as new challenges in, Civil Engineering. Topics in the series include:

- Construction and Structural Mechanics
- Building Materials
- Concrete, Steel and Timber Structures
- Geotechnical Engineering
- Earthquake Engineering
- Coastal Engineering
- Ocean and Offshore Engineering; Ships and Floating Structures
- Hydraulics, Hydrology and Water Resources Engineering
- Environmental Engineering and Sustainability
- Structural Health and Monitoring
- Surveying and Geographical Information Systems
- Indoor Environments
- Transportation and Traffic
- Risk Analysis
- Safety and Security

To submit a proposal or request further information, please contact the appropriate Springer Editor:

- Pierpaolo Riva at pierpaolo.riva@springer.com (Europe and Americas);
- Swati Meherishi at swati.meherishi@springer.com (Asia - except China, and Australia, New Zealand);
- Wayne Hu at wayne.hu@springer.com (China).

All books in the series now indexed by Scopus and EI Compendex database!

More information about this series at https://link.springer.com/bookseries/15087

Thomas Kang · Youngjin Lee
Editors

Proceedings of 2021 4th International Conference on Civil Engineering and Architecture

Springer

Editors
Thomas Kang
Department of Architecture
and Architectural Engineering
Seoul National University
Seoul, Korea (Republic of)

Youngjin Lee
Boston Architectural College
Boston, USA

Sasaki Associates, Inc.
Watertown, USA

ISSN 2366-2557 ISSN 2366-2565 (electronic)
Lecture Notes in Civil Engineering
ISBN 978-981-16-6931-6 ISBN 978-981-16-6932-3 (eBook)
https://doi.org/10.1007/978-981-16-6932-3

This Springer imprint is published by the registered company Springer Nature Singapore Pte Ltd.
The registered company address is: 152 Beach Road, #21-01/04 Gateway East, Singapore 189721, Singapore

Conference Committees

Conference Chair

Thomas Kang, Seoul National University, South Korea

Program Chairs

Alireza Afshari, Aalborg University, Denmark
Adnan Ibrahimbegovic, University of Technology of Compiègne, France
Dario Pedrabissi, University of Portsmouth, UK
Youngjin Lee, Boston Architectural College and Sasaki Associates, Inc., USA

Technical Committee

Alberto Gonzalez, University of Utah Asia Campus, South Korea
Allan Alzona, National University—Manila, Philippines
Bakhriev Nuritdin, Samarkand State Architectural and Civil Engineering Institute, Uzbekistan
Bernard Villaverde, Mapua University, Philippines
Changsoo Kim, Shandong Jianzhu University, China
Chavanont Khosakitchalert, Chulalongkorn University, Thailand
Chompoonut Kongphunphin, Thammasat University, Thailand
Dante L. Silva, Mapua University, Philippines
Deuckhang Lee, Nazarbayev University, Kazakhstan
Emina Hadzalic, University of Sarajevo, Bosnia and Herzegovina
Fadi Hage Chehade, Lebanese University, Lebanon
Gokhan Tunc, Atilim University, Turkey

Haryati Yaacob, Universiti Teknologi Malaysia, Malaysia
Hazrina Haja Bava Mohidin, University of Malaya, Malaysia
Hosein Nadepour Hosein Naderpour, Semnan University, Iran
Hyeon-Jong Hwang, Hunan University, China
Junior Didi Orihuela, Universidad Peruana de Ciencias Aplicadas, Peru
Junwon Seo, South Dakota State University, USA
Karol Slawomir Sikora, University of Wollongong in Dubai, UAE
Khan Zaib Jadoon, International Islamic University Islamabad, Pakistan
Leonardo Maximiliano Massone, University of Chile, Chile
Lyazat Aruova, Eurasian National University, Kazakhstan
Manat Srivanit, Thammasat University, Thailand
Manny Anthony Taguba, National University—Manila, Philippines
Marcelo Villacis Ormaza, Universidad Tecnológica Indoamerica, Ecuador
Mohammad Arif Kamal, Aligarh Muslim University, India
Mohammad Ichsan, Binus University, Indonesia
Muneerah Jeludin, Universiti Teknologi Brunei, Brunei Darussalam
Norhidayah Abdul Hassan, Universiti Teknologi Malaysia, Malaysia
Ryan Ahn, Texas A&M University, USA
Satawat Doungpan, Sukhothai Thammathirat Open University, Thailand
Shaoyun Pu, Southeast University, China
Shavkat Buriboev, Samarkand State Architectural and Civil Engineering Institute, Uzbekistan
Suppachai Sinthaworn, Srinakharinwirot University, Thailand
Taki Eddine Seghier, School of Architecture and Built Environment, UCSI University, Kuala Lumpur, Malaysia
Tamara Kelly Kelly, Abu Dhabi University, UAE
V. V. N. Prabhakara Rao, VR Siddhartha Engineering College, India
Wardah Fatimah Mohammad Yusoff, National University of Malaysia, Malaysia
Wonseok Chae, University of Wuppertal, Germany
Woosuk Kim, Kumoh National Institute of Technology, South Korea
Zhmagul Nuguzhinov, KazMIRR at Karaganda State Technical University, Kazakhstan

Preface

The 4th International Conference on Civil Engineering and Architecture (ICCEA 2021) was held from July 10 to 12, 2021. ICCEA aims to become a premier international conference for an in-depth discussion on the most up-to-date and innovative ideas, research projects, and practices in the field of civil engineering and architecture. Given the current global outbreak of the virus, this year's conference, which was supposed to be held in Seoul, was organized as a fully virtual conference.

This year, 4 keynote speeches have been delivered by myself, Prof. Atsuko K. Yamazaki (Japan), Prof. Youngjin Lee (USA), and Prof. Jorn Altmann (South Korea). More than 60 presentations were given from about 20 countries/regions, making this conference truly international. The whole conference was held online in Zoom, and we also had tests for each presenter in advance to ensure the success of delivery of the conference. For those who had Internet problems, a pre-recorded video presentation is accepted as an alternative. Meanwhile, the whole conference had been recorded for conference backup only.

ICCEA 2021 proceedings span over created 8 topic tracks including *geological survey and road engineering, earthquake engineering, concrete structure design and performance analysis, building materials and structural engineering, architecture and environmental engineering, urban space planning and development, project construction and management, and ancient architecture and architectural history,* which are well balanced in content and manageable in terms of many contributions, and create a discussion space for trendy topics.

Efforts taken by peer reviewers contributed to improving the quality of papers provided constructive critical comments; improvements and corrections to the authors are gratefully appreciated. We are very grateful to the international/national advisory committee, session chairs, and administrative assistants who selflessly contributed to the success of this conference. Also, we are thankful to all the authors who submitted papers, because of which the conference became a story of success. It was the quality of their presentation and passion to communicate with the other participants that made this conference a grand success.

We sincerely hope that the academic community and industrial practitioners will continue to support us in our attempts to provide even more meaningful conferences with numerous critical idea exchanges, diverse opportunities for fruitful networking, and future collaborations between the delegates.

Warmest regards,
Dr. Thomas Kang
Conference Chair, ICCEA 2021
Professor
Seoul National University
Seoul, South Korea

Contents

Geological Survey and Road Engineering

Stability Analysis Using the Finite Element Method of a Slope in the Virgen de Fátima Sector of the San Juan de Lurigancho District of Lima, Peru

N. Carrizales, R. Rodriguez, and J. Vasquez

Abstract The geometric alteration of slopes is a reality that can be observed in several districts of the Peruvian capital. The construction of houses, roads, and other infrastructure can produce some slope instability and cause tragic events. Thus, a stability analysis was carried out for a slope located in the San Juan de Lurigancho district, specifically in the Virgen de Fátima sector. Therefore, for the present investigation, two models were made with the help of Phase2 software to identify the resistance reduction factors (SFR), which is a finite element-based program. To start the analysis, possible failures were identified, demonstrating that the slope does not present any possibility of failure. For the first model, we worked with a dip of 30° and identified an SFR = 19.26 for static conditions and an SFR = 9.66 for pseudo-static conditions. For the second model, we worked with a dip of 55°, this change in slope geometry shows a possible wedge failure, according to the kinematic analysis. Also, an SFR = 0.89 was identified for static conditions and an SFR = 0.48 for pseudo-static conditions. The results show very considerable changes and are due to the geometry of the slope, the presence of discontinuities, and the participation of seismic forces.

Keywords Finite elements method · Kinematic analysis · Rock slope · Slope stability

N. Carrizales · R. Rodriguez · J. Vasquez (✉)
Universidad Peruana de Ciencias Aplicadas, Lima, Perú
e-mail: pccijiva@upc.edu.pe

N. Carrizales
e-mail: u201417184@upc.edu.pe

R. Rodriguez
e-mail: u201414727@upc.edu.pe

T. Kang and Y. Lee (eds.), *Proceedings of 2021 4th International Conference on Civil Engineering and Architecture*, Lecture Notes in Civil Engineering 201,
https://doi.org/10.1007/978-981-16-6932-3_1

1 Introduction

A current problem in one of the most populated districts of Lima is the construction of housing on the slopes of the hillsides by families in need of a home of their own. This context translates into the alteration of the slope, through cuts or the intervention of machinery, intending to create flat spaces for the construction of housing, access roads, or common areas such as the construction of a sports slab. For this reason, the way of cutting the slope, without any engineering criteria, often generates instability, leading to a latent risk such as rockfall and thus causing very important damages such as human and material losses.

In this study, although an analysis will be made on a slope that is potentially stable, it is necessary to know the behavior of stability with the change of slope, due to the geometric variation caused by the cuts in the slope. In addition, it is important to mention that most of the constructions in this type of places lack a technical study due to the lack of advice from a specialist and the absence of State institutions to improve the quality of life of the population.

The study area is located in the Virgen de Fátima sector, in the district of San Juan de Lurigancho, Lima, Peru. To carry out the investigation, an identification of the fractures present is made, and then the analysis is carried out to determine the types of rupture that can be generated, and which are the intervening fractures. Previously, a geotechnical study was carried out by INGEMMET [1], in which a geomechanical classification of the slope under study was performed. On this occasion, in general, what is intended to be done is a model of the slope before and after cutting. For this purpose, a slope stability analysis will be performed using the finite element method, where the representation of the models will be done with Phase2 software. The objective is to identify the vulnerability of the slope after making cuts or geometric changes in its surface.

2 Current State of Knowledge

Mithresh et al. focused on the rock falls that may occur along the hill slopes on a busy road, as these problems cause both as loss of money, inconvenience to travelers and sometimes there are human losses due to these natural phenomena. Thus, the authors investigated the mechanism of such faults to control such problem, for this the respective analysis of a rock slope located in the Himalayas of Sikkim are carried out. These analyses consist of 3 steps, which were performed through a kinematic analysis, a limit equilibrium analysis, and a numerical analysis that focuses on the finite element method. The purpose of the study was to obtain the safety factor which was obtained by the shear strength reduction (SRF) method, therefore, the results obtained show that the numerical analysis was the one that gave a result more related to the kinematic analysis and showing that the slope is not stable, which ends up failing due to its respective explicit discontinuities [2].

Anangsha et al. evaluated the fault mechanisms in a basalt rock slope which is located on the Deccan State Highway in western India. For their respective analysis, they were made through a kinematic analysis which focuses on being able to visualize the possible failures that occur in the basalt rock slope and numerically based on the finite element method with the technique of reducing shear strength parameters (SSR). The shape was also applied to represent the three-dimensional orientation to a two-dimensional orientation of said slope in order to represent an efficient analysis in a 2D model thanks to the "Phase 2" program. Finally, it was announced that the slope presented stability risk because of the safety factor (SFR) that was obtained [3].

Fumming used the resistance reduction method associated with the finite element method with the purpose of not having error and low analysis efficiency of a rock slope, the case study is of a rock slope of the Danjiangkou reservoir with which a simulation must be obtained to know if it is more precise in the results and better satisfy the needs of the geotechnical analysis of slope stability. Given the final results of the research, it was identified that, compared to the traditional method, the safety factor (SFR) is more related to the real data with less analysis error and greater efficiency in less time [4].

Ahmed and Seshagiri focused on how to improve the behavior of rock mass stability under static and dynamic conditions through analytical analysis. The stability of the slope of the slope is generally affected by the material property of the rock in conjunction with the discontinuities in the rock mass. It also depends on the angle of the slope, the height of the slope, the surcharge, the groundwater conditions, the rain, and the dynamic forces such as an earthquake. Block and core samples were collected from each category of rocks and their respective geotechnical properties were experimentally measured and presented. The rock slope stability analysis was performed using numerical modeling techniques, two-dimensional analysis based on the fine element method (FEM) (RS2), or Phase2 9.0 software from Rocscience. The slope of the rock having four different qualities of rock mass was analyzed under static and pseudostatic conditions. Therefore, the optimal support measurements were provided using shotcrete (fck = 35 MPa and 100 mm thick) and fully bonded bolts (32 mm diameter, 10 m long with 3 3 m spacing) because the factor safety for the most critical section which was increased from 0.48 to 1.2 under dynamic load conditions [5].

3 Tools

To carry out the research, a geotechnical study of the slope and three computational tools were used.

3.1 Geotechnical Study

One of the main characteristics in any research process is the field work to collect the corresponding and necessary data. Therefore, for this document, the geotechnical study carried out by INGEMMET [1], is taken as the starting point, where it makes a geological description of the place and the geomechanical classifications of the slopes that were considered for said report, where the characteristics of the rocky matrix and discontinuities.

3.2 Software Dips

Within the analysis it is necessary to identify the possible failure mechanisms and to achieve this characteristic a kinematic analysis is carried out. For this reason, the Dips software is a tool that helps to represent the orientations of the planes based on their dips, dip direction and heading. It also helps to identify the type of failure whether it be planar, wedge or overturning, with the data of the slope and the friction angle.

3.3 Rocdata V3

This software was used to determine the input parameters for the application of the finite element method in the Phase2 software. From the compressive strength of the intact rock (σci), the parameter (mi), geological resistance index (GSI), the disturbance factor (D) and the consideration of the Generalized Hoek Brown criterion, a constitutive method suitable for an analysis in a rocky massif [6], it identifies the cohesion and friction angle values, as well as parameters of the rock mass tensile strength, global compressive strength, deformation modulus and the uniaxial compressive strength of the massif rocky.

3.4 Phase2

It is a finite element program that allows slope stability analysis, whose main characteristic is the shear strength reduction method to determine the safety factor [7]. Phase2 allows evaluating the risk of slope failure and for this purpose, the geometry, geological conditions of the study site and the results or identified parameters of the constitutive model found with Rocdata were identified. In addition, within the modeling, the most representative fracture families or systems are considered.

Table 1 Orientation characteristics of discontinuities

Kind structure	Orientation	
	Dip	Dip direction
J1	59	299
J2	67	85
J3	86	162
Slope	30	242

4 Geological Description of the Study Area

The study area is in the "Cerrito Rico" human settlement, in the Virgen de Fátima Sector, San Juan de Lurigancho, Lima, Peru. The geological aspect of the place is represented by intrusive rocks covered by deposits of different physical and mechanical properties. Within this category of rocks is the presence of gray granodiorite, in slightly weathered outcrop with xenolites, with a high degree of fracturing [1]. On the other hand, the slope has two main joint families and a secondary family as detailed in Table 1.

5 Methodology

5.1 Kinematic Analysis

For the present analysis, it is important to know the orientation of the discontinuities since it is one of the most influential aspects in determining stability. The slope understudy presents two main families J1, J2, and a secondary family J3 with the orientation of their discontinuities indicated in Table 1 and a friction angle of 32°. In addition, the stereographic projections of the DIPS software [3], show that there is no presence of wedge, planar and direct toppling, with a dip of the slope equal to 30° as shown in Fig. 1, characteristics of the natural slope. On the other hand, the geometric alteration of the slope, specifically the change of slope, where at first, we worked with the slope dip of a value of 30°, which did not present possible failures. For a second kinematic analysis where the slope dip has a value of 55°, added to the characteristics of the discontinuities present in the slope, a possible wedge failure is observed due to the intersection of the main plane and a secondary plane, J1 and J3 respectively, as shown in Fig. 2.

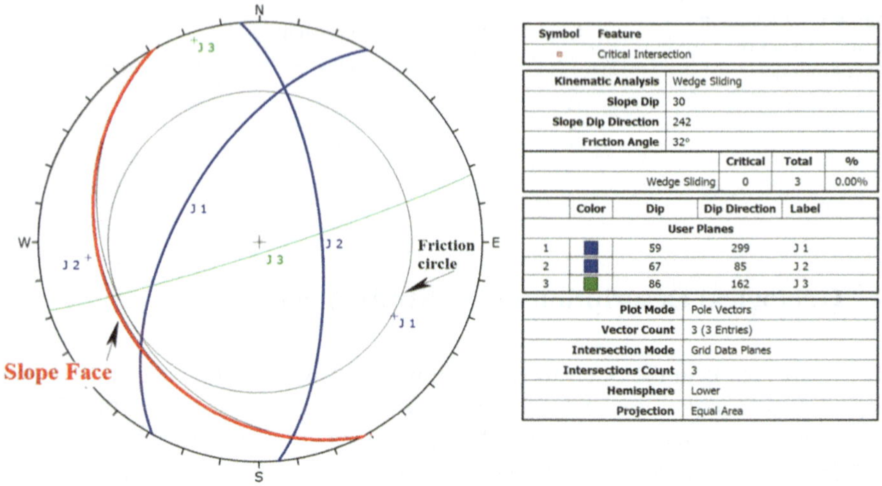

Fig. 1 Kinematic analysis model 1, natural slope. *Source* DIPS

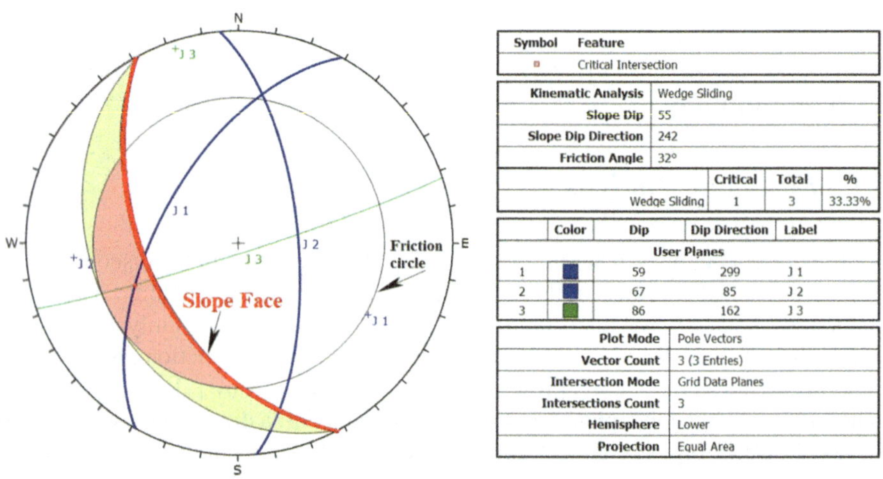

Fig. 2 Kinematic analysis model 2, wedge failure. *Source* DIPS

5.2 Numerical Analysis for the Rock Slope

In this present investigation, the finite element method was used to be able to analyze the stability of the ET-002 slope by calculating the safety factor by means of the shear strength parameter reduction (SRF) method, which represents a linkage between the resistance of the system at failure and the resistance found.

With the help of the Phase2 program, it was possible to represent the two-dimensional model of said slope with its respective geometry, this is observed in

Table 2 Slope input parameter ET-002

Parameters	Values
Uniaxial compressive strength (σ_{ci}, MPa)	157
Geological resistance index (GSI)	54
Intact rock (mi)	30
Disturbance factor (D)	0.7

Table 3 Properties of the ET-002 slope both intact rock and the discontinuities

Properties	
Intact rock	Specific weight = 0.026 MN/m^3
	Young's modulus (E, MPa) = 8183.02
	Poisson's ratio = 0.25
	Friction angle (°) = 61.56
	Angle of dilatation (°) = 0
Discontinuities	Apparent dip/direction (Set 1, °) = 42.19/299
	Apparent dip/direction (Set 2, °) = 65.24/85
	Apparent dip/direction (Set 3, °) = 68.07/162
	Friction angle (°) = 61.56
	Cohesion (C, MPa) = 0
	Average apparent separation (Set 1, m) = 1.10
	Average apparent separation (Set 2, m) = 0.65
	Average apparent separation (Set 3, m) = 3.46

Fig. 2. To determine the properties of said rocky material found in the slope under study (ET-002), they were found using the ROCDATA V3 program, in which the following initial parameters were introduced, which are observed in Table 2.

Once the data has been entered into "ROCDATA", the following output parameters are obtained, which are shown in Table 4.

Also, the intact rock properties and discontinuities of slope ET-002 were identified, as detailed in Tables 3 and 4.

Once having the necessary properties and parameters, the slope is designed with its respective geometry, which is shown in Table 5.

Table 4 Slope output parameter ET-002

Parameters	Values
mb	2.396
s	0.0013
a	0.504
Cohesion (C, MPa)	0.821
Friction angle (°)	61.56
Young's modulus (E, MPa)	8183.02

Table 5 Geometry for the design of model 2 of slope ET-002

Geometry for the design of the slope ET-002	
Slope height	10 m
Slope toe length	10 m
Slope head length	10 m
Horizontal length of slope	17.32 m
Slope angle	30°

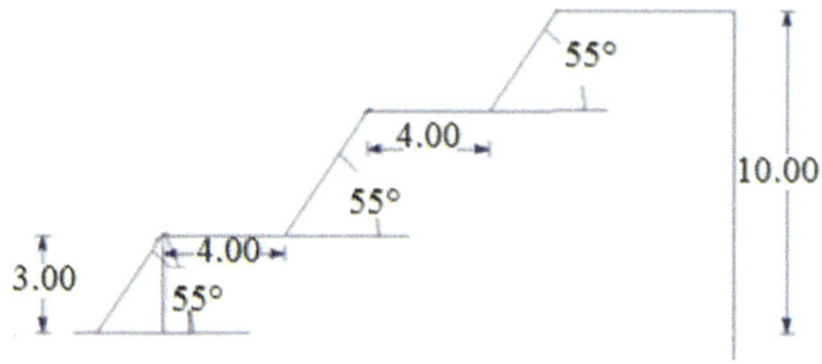

Fig. 3 Design of the ET-002 slope model with benches. *Source* Phase 2

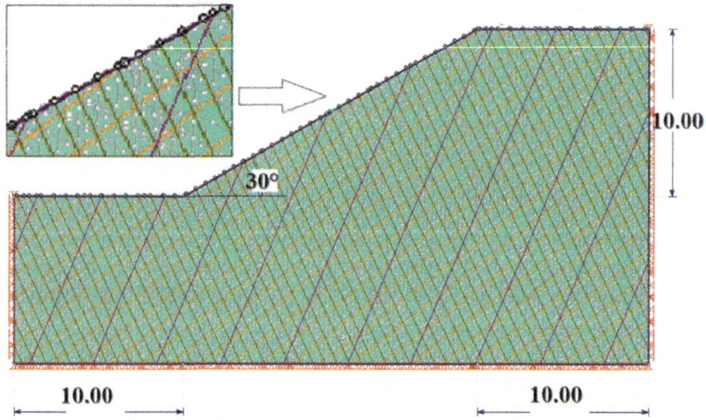

Fig. 4 Slope ET-002 (model 1) meshed and discretized. *Source* Phase 2

For this investigation, 2 cases were considered in relation to the ET-002 slope. The first model consists of a slope without the presence of banks while, in the second model, the slope is presented with the presence of 2 banks.

In the slope under study, 2 banks were considered which were produced by cuts, these banks have a height of 3 m and a length of 4 m, the respective angle of cut is 55°. It is necessary to emphasize that the banks implemented in the slope represent areas for the construction of houses and roads in certain cases.

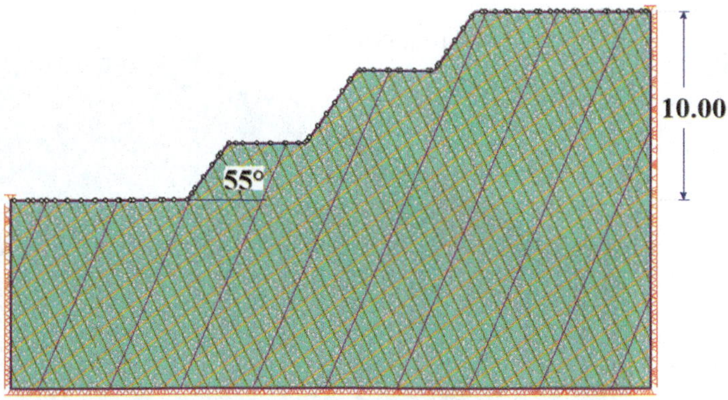

Fig. 5 Slope ET-002 (model 2) meshed and discretized. *Source* Phase 2

Figures 3 and 4 show the design of the two-dimensional models with their corresponding mesh made by the using the Phase 2 program (Fig. 5).

6 Results and Discussion

In this section, the results of the cases proposed in this investigation are shown in Table 6. For model 1, the safety factor (SFR) was statically calculated which resulted in "19.26" and it was also calculated pseudostatically giving as a result a safety factor of "9.66", using the Phase2 software which applies the numerical method of finite elements. For both cases, the system is stable, this can be seen in Figs. 6 and 7.

In this model 2, the safety factor (SFR) was statically calculated which resulted in "0.89" and it was also calculated pseudostatically resulting in a safety factor of "0.48", using Phase2 software which applies the finite element numerical method. For both cases the system is unstable, this can be seen in Figs. 8 and 9.

Table 6 shows the results of the two models under study of the ET-002 slope that have been analyzed statically and pseudostatically.

The result in the kinematic analysis due to the change in geometry in the dip due to the implementation of banks within the system generates a possible wedge failure due to the participation of the families of discontinuities J1 and J3. Due to the more representative participation of the main families, J1 and J2 compared to the discontinuities J3, it becomes a second family, which would not have a greater influence on

Table 6 Results of the safety factor of the slope ET-002

Slope ET-002 cases	SRF (static)	SRF (pseudostatic)
Model 1	19.26	0.89
Model 2	9.66	0.48

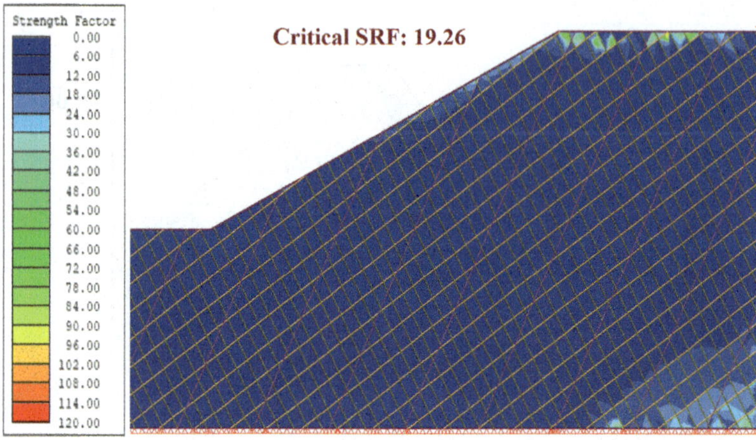

Fig. 6 Critical SFR of the ET-002 slope for statically analyzed model 1. *Source* Phase 2

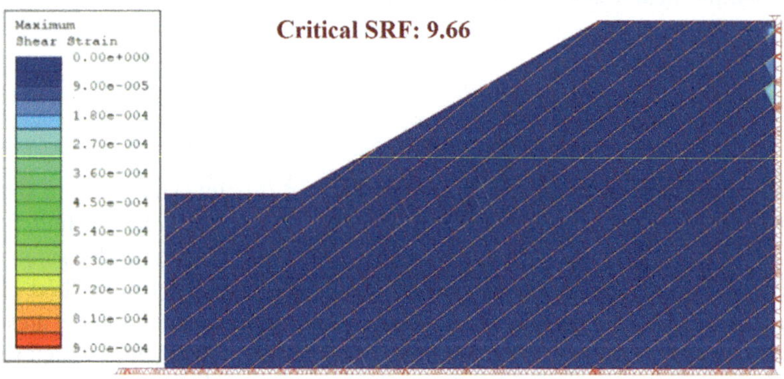

Fig. 7 Critical SFR of the ET-002 slope for model 1 analyzed pseudo statically. *Source* Phase 2

the occurrence of a wedge fault. On the other hand, the representation of said discontinuities by Phase2 shows a tendency to planar failure due to the participation of J1 in the slope understudy, but according to the conditions according to the kinematic analysis, the slope has a dip of 55° which is less to the dip of discontinuity J 1 (59°), where it would be failing as a planar fault. One of the aspects that were noticed was the direct influence of the sudden change in geometry to determine possible failures and thus obtain slope stability.

One of the criteria by which the investigation was carried out was to witness how much difference there could be between a slope without banks and with banks since these cases can be a bit like our problem.

On the other hand, the study area is in the Pacific Ring of Fire, an area of seismic activity. Therefore, static, and pseudo-static analysis was performed for both models under study, considering a seismic coefficient of 0.3 because the Phase2 program

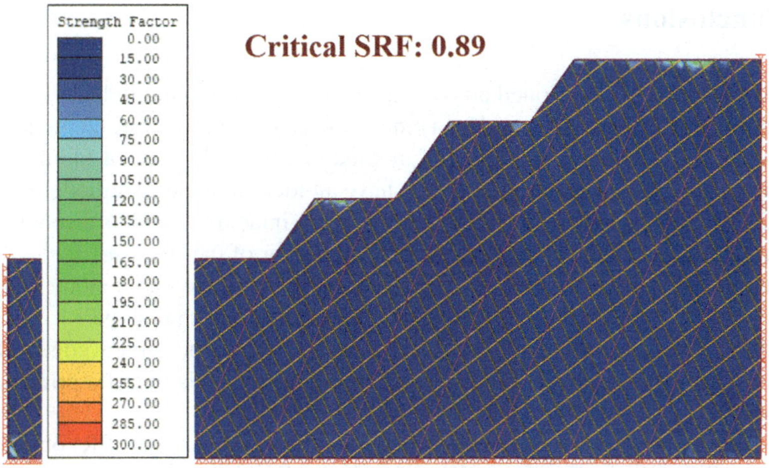

Fig. 8 Critical SFR of the ET-002 slope for statically analyzed model 2. *Source* Phase 2

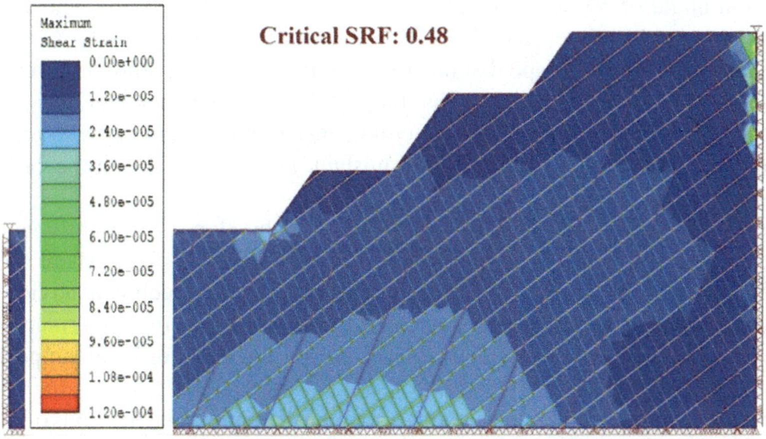

Fig. 9 Critical SFR of the ET-002 slope for model 2 analyzed pseudo statically. *Source* Phase 2

only accepts this coefficient as the maximum value. Therefore, an approximate value was used according to Technical Standard E.030 Seismic-resistant Design of the National Building Regulations, where Lima is in zone 4 with a seismic coefficient of 0.45 [8].

According to the results of the ET-002 slope for static analysis, an SFR = 19.26 (model 1, inclination angle 30°) and an SFR = 0.89 (model 2, inclination angle 55°) is obtained, Figs. 6 and 7, respectively. On the other hand, for a pseudo-static analysis of the same slope understudy, an SFR = 9.66 (model 1, inclination angle 30°) and an SFR = 0.48 (model 2, inclination angle 55°) are obtained, Figs. 8 and 9, respectively.

7 Conclusions

Based on the results mentioned above, the following can be concluded:

The research base is supported by a kinematic analysis and a finite element numerical method. On the one hand, the first method identified the type of failure and the behavior of the discontinuities, and thus have an idea in numerical modeling of the Phase2 software, considering the parameters of the intact rock and its discontinuities present on the slope, to later identify the safety factor of the study area.

In model 1, by means of a kinematic analysis of the slope under study "ET-002" it was identified that there are no possible faults that affect said slope.

In the kinematic analysis of model 2 with a dip of 55°, a possible wedge failure is observed due to the intersection of two discontinuities consisting of a principal one (J1 and J3).

Given the results of the safety factors with reduction to shear (SFR) obtained by the Phase 2 program, the decrease in "SFR" can be evidenced. For an inclination angle of 30° with the participation of the seismic coefficient, it has an SFR = 9.66, a value less than the SFR = 19.26 in static conditions. On the other hand, for an inclination angle of 55° with the participation of the seismic coefficient, it has an SFR = 0.48, a value lower than the SFR = 0.89 in static conditions.

The abrupt change in slope that occurs for both cases under study is given by the inclusion of a bank. Having a considerable change in the geometry of the slope and the presence of discontinuities can generate considerable variations with respect to the stability levels (going from stable to unstable) observed through the calculated safety factors.

One of the causes that lead to the alteration of the slopes of the hills is due to the intervention of families or people who are looking for a place where they can build their own home. In addition, the lack of knowledge and technical advice for an evaluation with respect to stability can cause future risks, thus the importance of this type of analysis since many of these places do not have institutional support from the state.

For future research, it would be good to compare the numerical methods of discrete and finite elements to see how much difference one method can influence to model the fractures and the other for the rest of the domain.

References

1. Ingemmet (2011) Geological geotechnical study and susceptibility to mass movements in the Virgen de Fátima sector, energy and mining sector. Lima
2. Mithresh KP, Amalesh J, Arindam D, Murali A (2017) Evaluation of the stability of a rock slope using finite element modeling. Civil Engineering Department, India
3. Anangsha A, Seema P, Amalesh J (2017) A methodology for modeling an effective two-dimensional plane deformation analysis of articulated rock slope. Civil Engineering Department, India

4. Fuming L (2020) Stability analysis of geotechnical slope based on strength reduction method. Geotech Geol Eng
5. Ahmed S, Seshagiri R (2020) Rock mass slope stability analysis under static and dynamic conditions in Mumbai, India. J Civ Eng
6. Hoek E, Carranza-Torres C, Corkum B (2001) The Hoek-Brown failure criterion. ResearchGate
7. Rocscience (2020) Finite element analysis for excavations and slopes. Rocscience Inc., Toronto
8. Ministerial Resolution N°. 043-2019-HOUSING (2019) Technical standard E.030 earthquake-resistant design of the national building regulations. Lima

The Environmental Performance of Earth Construction Systems in Jordan's Mediterranean Climatic Zone

Zahra Al-Shatnawi

Abstract The aim of this paper was to investigate the thermal and energy performance of different earth construction methods, particularly in Jordan's Mediterranean climatic zone. An IES-VE simulation was conducted to compare five proposed earth construction methods against historic and conventional construction methods used in Jordan, in terms of indoor air temperature performance and operational heating and cooling demand. The findings from the simulation analysis show that a double compressed earth block (CEB) wall constructed model and a hybrid wall constructed model consume the smallest amount of operational energy and have the best thermal winter performance. Moreover, the historic earth wall construction methods perform slightly better than other options during summer. It is concluded that if the appropriate earthen construction method is adopted and combined with suitable passive strategies, earth construction can perform better than conventional building methods used in Jordan's Mediterranean climatic zone, resulting in energy-efficient and comfortable structures.

Keywords Earth construction · Thermal and energy performance · Jordan

1 Introduction

Earth been used as a natural building material for thousands of years. More than half of the population of the developing countries currently lives in earth-constructed dwellings [1]. This method is now attracting attention in industrialized countries, due to its low carbon footprint and low thermal conductivity, as combining earth as a natural building material with advanced construction techniques can result in modern ecological buildings [1–3].

Earth has long been used as a construction material in Jordan. The two construction methods primarily used in the past were adobe and mud with stone [4–6]. Dwellings in the Jordan valley were built using sun dried mud-bricks, while some in the mountainous regions of Jordan were constructed using stone and mud [7].

Z. Al-Shatnawi (✉)
Oxford Brookes University, Oxford OX3 0BP, UK

T. Kang and Y. Lee (eds.), *Proceedings of 2021 4th International Conference on Civil Engineering and Architecture*, Lecture Notes in Civil Engineering 201,
https://doi.org/10.1007/978-981-16-6932-3_2

However, it would be difficult to find complete structures built of earth in Jordan today. Most surviving examples are in a dilapidated state, abandoned and on their way to extinction [8].

The manufacture of cement, the main building material in Jordan, is associated with primary emissions from chemical reactions [9]. More sustainable building materials should thus be considered; and it may be possible to revive earth construction in Jordan with the application of new technologies and methods.

Compressed earth blocks (CEBs) represent an evolution of adobe bricks, with a device used to compress the earth inside a mould [10]. Binders can be added to the earth-block mixture to enhance its mechanical properties. Moreover, CEB-constructed buildings can be rendered using a thin layer of cement-based render to prevent erosion and to cover uneven block laying [11]. Earth-bag construction may also be considered a potential future earth-building method in Jordan. Although this method is not widely known, it is sustainable and affordable [12]. Earth bags are soil-filled sacks used to construct both straight walls and dome structures. The method is quick, easy, and inexpensive, and it may provide a solution for temporary or disaster relief housing, especially with the rapid influx of refugees into Jordan [13]. Several experimental earth-bag constructed projects have been completed in the country [14], but their environmental performance in this climate has not previously been studied.

The aim of this paper is to investigate the thermal and operational energy performance of multiple earth construction methods and compare them to the conventional building methods currently used in Jordan.

2 Methodology

IES-VE, a modern dynamic building energy simulation software program, was used to explore the thermal and operational energy performance of five practical earth construction techniques for external walls in Jordan. The five techniques were then compared with one another. They were also compared against two historic earth-construction methods and two conventional construction methods for external walls. Several authors have used simulation software such as IES-VE to assess the performance of building envelope wall materials [15–17]. Indoor air temperature provides the thermal performance parameter, whereas heating and cooling energy consumption represents the operational energy parameter.

Indoor air temperature simulation was conducted in winter (10–20 January) and summer (1–10 August). A simulation was also performed on 14 January and 1 August to test the thermal performance of the construction methods during different hours of the day. An annual heating and cooling demand simulation was conducted for the construction methods. The heating and cooling systems were only turned on when the annual energy demand was calculated. The simulation dates were chosen to reflect the typical coldest and hottest climates in Jordan.

A single storey dwelling (Dar) with an area of 165 m^2 was used as the simulation model. Invariable parameters were used for all options. The five potential systems for earth wall construction are described in Table 1, along with the historical and conventional earth wall construction methods used in the country. The table is followed by illustrative sections for the tested construction external wall methods (Fig. 1).

There are four climatic zones in Jordan: the Mediterranean, Irano-Turanian, Saharo-Arabian, and Sudanian. However, 90% of Jordan's inhabitants occupy the Mediterranean climatic zone, which is essentially the mountainous plateau [18]. Therefore, this climatic zone was considered for the simulation.

Table 1 External wall-construction options to be tested

Historic earth construction methods used in Jordan	
Option 1: adobe (outside to inside)	20 mm mud plaster with straw—500 mm adobe blocks—20 mm mud plaster (U value = 0.91 W/m^2 K)
Option 2: mud and stone (outside to inside)	150 mm local stone—350 mm mud and straw (cob) with integrated small stones—20 mm mud plaster (U value = 1.04 W/m^2 K)
Conventional construction methods used in Jordan	
Option 3: concrete blocks (outside to inside)	20 mm cement plaster—200 mm concrete block—20 mm gypsum plaster (U value = 1.39 W/m^2 K)
Option 4: concrete blocks, cement, and stone (outside to inside)	50 mm local stone—100 mm Portland cement—50 mm rock wool insulation—100 mm concrete block—20 mm gypsum plaster (U value = 0.53 W/m^2 K)
Potential systems for earth construction in Jordan	
Option 5: simple CEB	20 mm stabilized earth plaster—140 mm CEB—20 mm mud plaster (U value = 2.2 W/m^2 K)
Option 6: CEB with external insulation (outside to inside)	20 mm stabilized earth plaster—50 mm cork board insulation—140 mm CEB—20 mm mud plaster (U value = 0.61 W/m^2 K)
Option 7: CEB double wall system (outside to inside)	20 mm stabilized earth plaster—140 mm CEB—50 mm rock wool insulation—140 mm CEB—20 mm mud plaster (U value = 0.55 W/m^2 K)
Option 8: CEB hybrid wall system (outside to inside)	20 mm cement plaster—200 mm concrete blocks (existing)—50 mm rock wool insulation—140 mm CEB—20 mm mud plaster (U value = 0.47 W/m^2 K)
Option 9: earth bag (outside to inside)	20 mm stabilized earth plaster—450 mm earth bag—20 mm mud plaster (U value = 1.13 W/m^2 K)

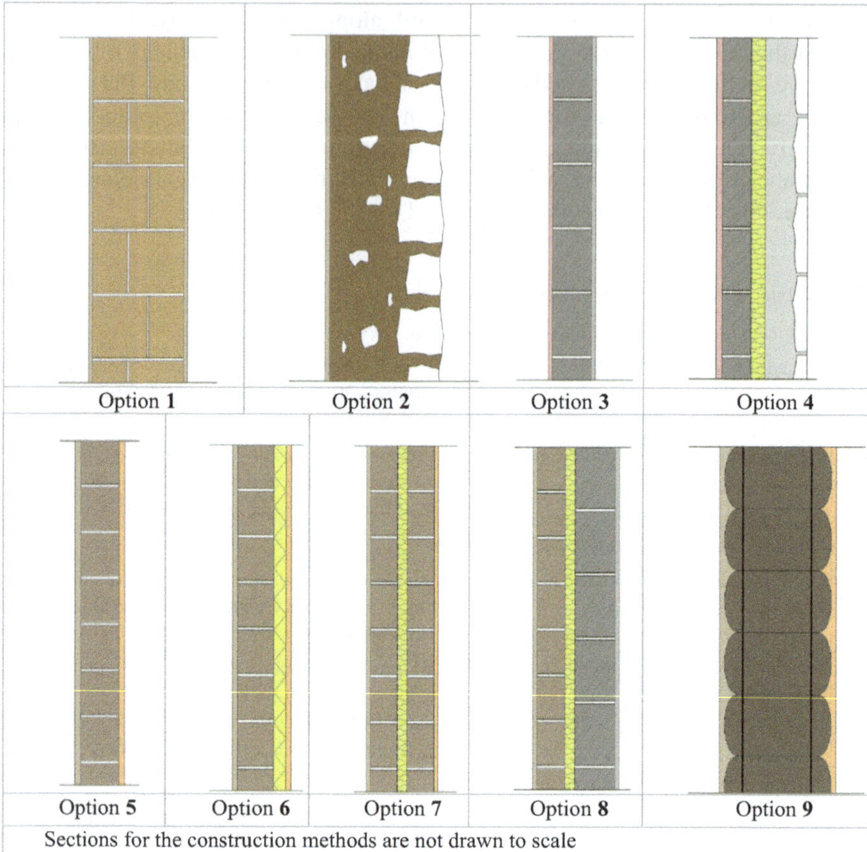

Fig. 1 Illustrative sections for the nine options

3 Results and Analysis

Figures 2, 3, 4 and 5 represent the indoor air temperature simulation results for the nine options during January (winter) and August (summer).

It is noted in Figs. 2 and 3 that all insulated options performed better thermally in winter than the non-insulated options did in terms of the model's internal air temperatures. However, the analysis of the proposed insulated earth wall construction methods reveals that a double CEB wall system (Option 7) performed slightly better than a CEB wall with external insulation (Option 6) in January. This might be because the internal CEB (Option 7) was providing thermal mass for heat storage, while the insulated cavity prevented heat loss to the outside. At the same time, the external CEB acted as a moderating layer against the cold. Furthermore, a hybrid wall system (Option 8) performed slightly better than both Options 6 and 7 in January. This might be explained by the thicker external layer in Option 8.

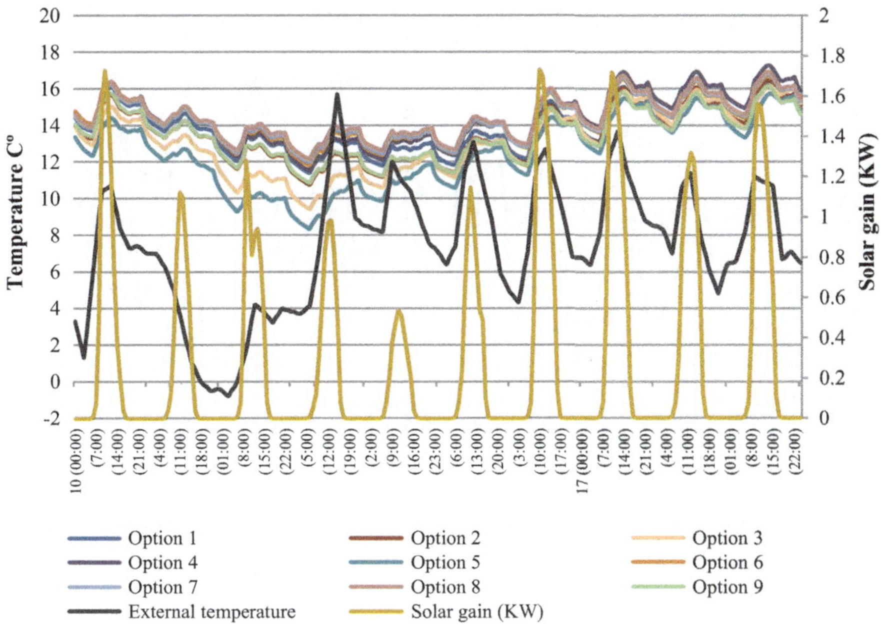

Fig. 2 Indoor air temperature simulation results (10 Jan–20 Jan)

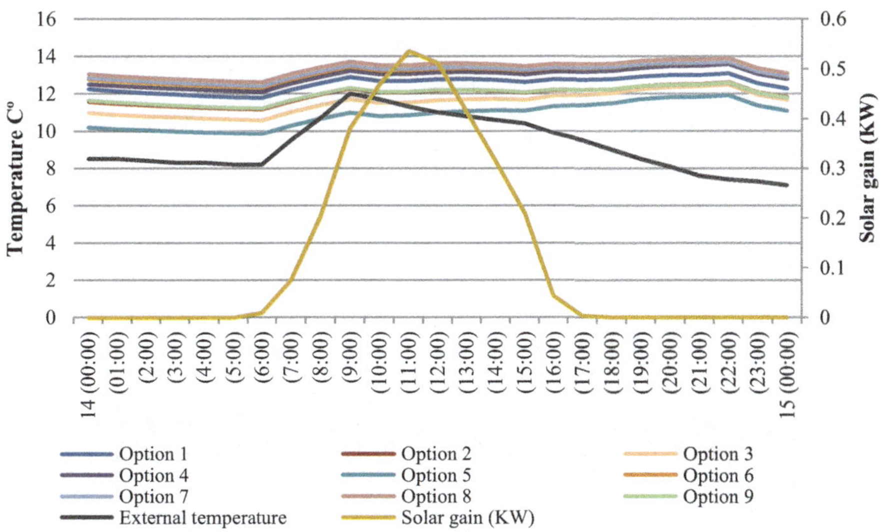

Fig. 3 Indoor air temperature simulation results (14 Jan)

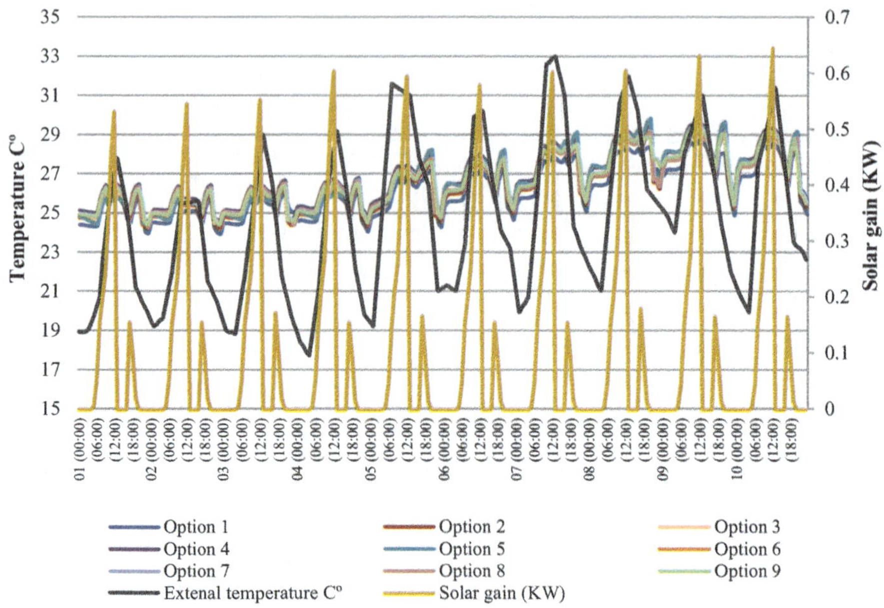

Fig. 4 Indoor air temperature simulation results (1 Aug–10 Aug)

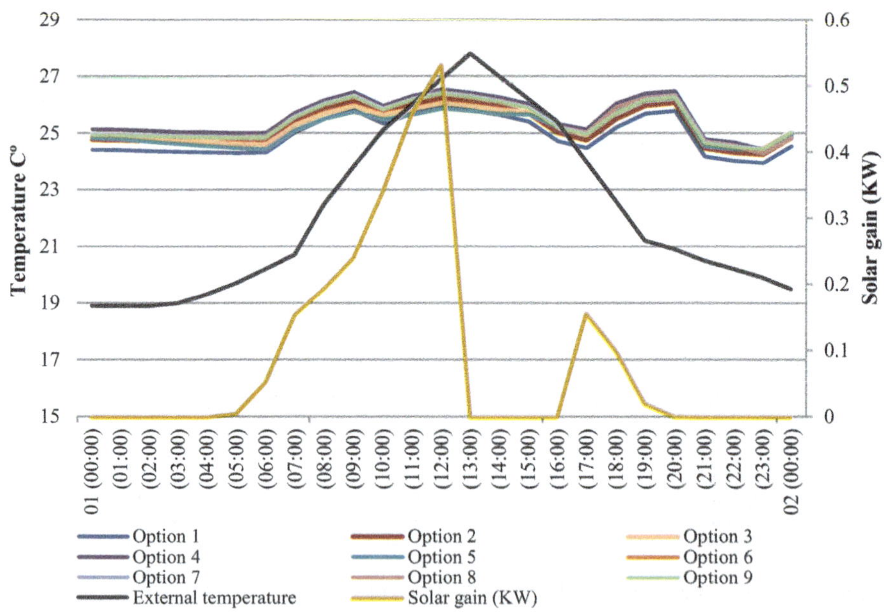

Fig. 5 Indoor air temperature simulation results (1 Aug)

The earth-bag wall-construction method (Option 9) and the two historic earth-wall construction methods used in Jordan (Options 1 and 2) performed better than the conventional concrete-block wall-construction method (Option 3) and a simple CEB wall-construction method (Option 5) in January. This can be explained by the lower U-values for Options 1, 2, and 9.

Turning to the model's summer internal air temperatures analysis, Figs. 4 and 5 show that all nine options had very similar performances. This may be because all options were considered within the high thermal mass category, and no option was considered a lightweight construction method. Nonetheless, both historic earth wall-construction methods used in Jordan performed slightly better than other options in August. This can be explained by the thickness of these historic earth-construction options. Moreover, for the proposed earth wall-construction methods, Options 6, 7, 8, and 9 have slightly more stable temperatures than Option 5.

The addition of insulation to the external walls had an impact on winter performance only. This supports the findings of previous research that showed adding insulation to rammed earth walls in a hot and arid climate improved winter performance and had minimum impact on summer performance [14].

The thermal performance of the model influenced its energy performance. To illustrate this, Fig. 6 presents the simulation results of the annual heating and cooling load for the nine options. The heating and cooling set points are 19 and 25 °C, respectively.

The model's operational energy consists of the space heating and cooling energy load, in addition to the appliances and lighting energy load. In this simulation, the appliances and lighting energy consumption was assumed to be the same for all options.

All options in Fig. 6 show that the energy required to heat the building exceeded that needed to cool the building in Jordan's Mediterranean climatic zone. Figure 6

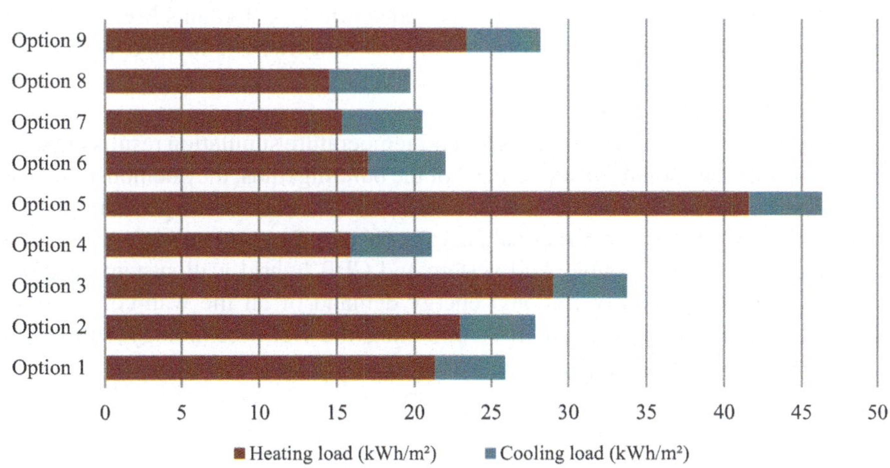

Fig. 6 Annual heating and cooling loads for the nine options

indicates that a model constructed using simple CEB walls (Option 5) consumed the largest amount of operational energy compared to all other options. A double CEB wall constructed model and a hybrid wall-constructed model (Options 7 and 8), on the other hand, had the smallest operational energy load. According to [17], the position of a CEB in a hybrid model depends on the space occupancy period, as the CEB layer must be placed on the interior side of the wall when the room is occupied primarily during the day and on the exterior side when the room is occupied primarily at night. A conventional concrete, cement, and stone wall and an externally insulated CEB wall (Options 4 and 6) performed slightly less well than Options 7 and 8 in terms of the model's operational energy consumption. Furthermore, the earth-bag wall constructed model (Option 9) required more operational energy than all the insulated CEB options and the two historic earth wall constructed models (Options 1 and 2).

4 Conclusions

This assessment of the IES-VE winter indoor air temperature simulation results for the five potential systems of earth wall-construction has shown that those options that incorporate insulation between two layers of thermal mass perform the best in Jordan's Mediterranean climatic zone. These options embrace a CEB double-wall system and a CEB hybrid wall system. An external insulated CEB wall-construction method also has remarkably good winter performance. These methods performed better than the two historic earth wall-construction methods. Moreover, they performed much better than a conventional concrete-block wall-construction method.

According to the summer indoor air temperature simulation results, the five potential earth wall-construction methods performed similarly in Jordan's Mediterranean climatic zone. However, the historic earth wall-construction method performed slightly better than the conventional wall-construction method and the potential earth wall-construction method during this season.

The IES-VE winter and summer indoor air temperature simulation results indicate the associated operational energy demand of the building; thus, they demonstrate that wall-construction options with the best winter and summer air temperature performances lead to the lowest operational energy consumption. Accordingly, it is noted that the proposed CEB double-wall system and CEB hybrid wall system have the overall lowest associated operational energy demand of all the wall-construction options. Hence, it is concluded that combining the right technology with earth construction results in energy-efficient structures.

References

1. Hoff E (2016) Appraisal of the sustainability of compressed stabilized earthen masonry. Dissertation architectural engineering. University of Nebraska, Lincoln
2. Hall M, Cassey S (2012) Hygrothermal behavior and occupant comfort in modern earth buildings. In: Modern earth buildings, pp 17–40
3. Love S (2013) Architecture as material culture: building form and materiality in the Pre-Pottery Neolithic of Anatolia and Levant. J Anthropol Archaeol 32:746–758
4. Shiyyab A (2014) Recent discoveries in city center of Petra, Jordan: a preliminary excavation report. J Mediterr Archaeol Archaeom 14:103–118
5. Tarrad M, Al-Omari O, Alrawashdeh M (2012) Natural stone in Jordan: characteristics and specifications and its importance in interior architecture. J Environ Sci Eng B1:720–727
6. Alami A (2010) Design and construction of a green, controlled-environment dwelling in Jordan. Int J Therm Environ Eng 3
7. Abu-Hammad N (2011) Architectural mud brick prototypes as efficient and sustainable shelters for the low-income group in Jordan. Jordan J Civ Eng 5:82–88
8. IUCN-The World Conservation Union (1991) National environment strategy for Jordan: a resource book of information and guidelines for action. Amman-Jordan, IUCN
9. Torgal F, Jalali S (2011) Eco-efficient construction and building materials. Springer, London
10. Minke G (2006) Building with earth, design and technology of a sustainable architecture. Birkhäuser–Publishers for Architecture, Basel-Switzerland
11. Barnes B, Kang M, Cao H (2015) Sustainable characteristics of earthbag housing. Hous Soc 33:21–32
12. Makind O (2012) Ecological and sustainability issues in earth construction. IOSR J Environ Sci Toxicol Food Technol 1:20–28
13. Jannat N, Hussien A, Abdullah B, Cotgrave A (2020) A comparative simulation study of the thermal performances of the building envelope wall materials in the tropics. Sustainability 12:4892
14. RBG (Royal Botanic Garden) (2012) Eco-domes at RBG, sustainable living (Online). Available at: http://www.royalbotanicgarden.org/. Accessed 11 May 2021
15. Teixeira E, Machado G, de Paula C, Junior A, Guarnier C, Fernandes J, Silva S, Mateus R (2020) Mechanical and thermal performance characterisation of compressed earth blocks. Energies 13:02978
16. Hema C, Messan A, Lawane G, Van Moeseke G (2020) Impact of the design of walls made of compressed earth blocks on the thermal comfort of housing in hot climate. Buildings 10:157
17. Al-Jaloudy M (2001) Country pasture/forage resource Jordan. Food and Agriculture Organization of the United Nations (Online). Available at: http://www.fao.org/. Accessed 30 Mar 2021
18. Cheikhi W, Baba K, Nounah A, Cherradi C (2020) Effect of adding insulation on the energy performances of rammed earth buildings in hot and arid climates. E3S Web Conf 150:01011

Measuring the Impact of Pedestrian Flow on Vehicle Delays for a Proposed DLT Intersection with Two Displaced Left Turns in an Urban Commercial Area

B. E. Acuña, L. C. Amaya, and A. M. Sanchez

Abstract This study evaluates the impact of pedestrian flow through vehicular delays in a proposed DLT intersection located in an urban area with a high vehicular and pedestrian traffic volume. In addition, it evaluates the pedestrian–vehicle conflict points, the effect on the other accesses after the implementation of the displaced left turns, and the consequences of reorganizing pedestrian flow. Three scenarios are evaluated through microsimulation; the existing one, which is a conventional cross-shaped intersection; a proposed DLT intersection with two displaced left turns on adjacent accesses; and a third one, similar to the last one, but with the inclusion of a pedestrian underpass. Results indicate that a proposed conventional DLT increases the vehicle delays of the intersection by 6.0% compared to the existing condition, mainly due to the high volume of pedestrians and the conflicts it generates. However, with the inclusion of a pedestrian underpass, vehicle delays are reduced significantly by up to 34.0%.

Keywords DLT intersection · Unconventional intersection · Displaced left turn

1 Introduction

The high traffic volume, inadequate design and operation are generating high congestion at intersections, and it is more severe in developing countries where the design and traffic control seem to be inadequate. Currently in Lima-Peru has a total of 1250 intersections which are affected by high congestion due to the increase of pedestrian and vehicle traffic volume, the absence or poor traffic light synchronization, and lack of connectivity to a traffic control center (TCC). Less than 20% of traffic lights are connected to a TCC because of the unrelated communication protocol uses in each district of Lima [1]. During the last years, various solutions have been proposed to mitigate traffic congestion at intersections in 17 districts of Lima, such as left-turn restriction, do not block intersection box marks, police enforcement, and bypasses,

B. E. Acuña (✉) · L. C. Amaya · A. M. Sanchez
Department of Civil Engineering, Universidad Peruana de Ciencias Aplicadas (UPC), Santiago de Surco, Lima 15023, Peru
e-mail: u201410003@upc.edu.pe

T. Kang and Y. Lee (eds.), *Proceedings of 2021 4th International Conference on Civil Engineering and Architecture*, Lecture Notes in Civil Engineering 201,
https://doi.org/10.1007/978-981-16-6932-3_3

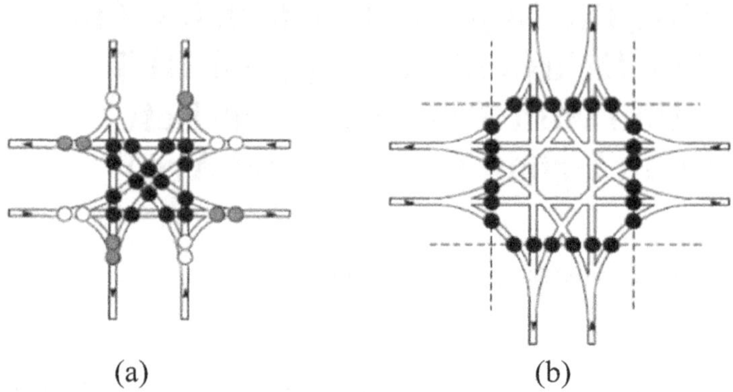

Fig. 1 Cross-shaped intersection. **a** 32 points of vehicular conflict; **b** 24 points of pedestrian–vehicle conflict. Black circles indicate crossing, gray circles indicate merging and white circles indicate diverging. Adapted from [3]

but unfortunately, any of them have solved the problem or they only translated the problem to other intersection [2]. According to [3, 4], conventional intersections have 32 vehicular conflict points and 24 pedestrian–vehicle conflict points as seen in Fig. 1.

Conflict points at intersections are one of the main factors that generate traffic congestion since they create delays and mobility barriers to pedestrians and vehicular flow. Traffic engineers have been working on innovative solutions to solve the problems that conventional intersections can no longer master. One of them is the Displaced Left Turn (DLT) intersection which can reduce vehicle delays by up to 30% [3]. This innovative design has fewer points of conflict than the conventional cross-shaped intersection. Depending on the number of shifted left turns, the number of conflict points varies. According to [3]; including two displaced left turns located on opposite sides causes 30 vehicle conflict points and 18 or 20 vehicle–pedestrian conflict points depending on whether the crossing is one-stage or two-stage, as shown in Fig. 2.

The DLT intersection has some disadvantages in terms of pedestrian management, such as more stages for crossing, longer pedestrian crossing, and indirect movements that may be required to cross streets. These disadvantages complicate the vehicle and pedestrian movements affecting the operation of a DLT intersection [3]. The main objective of this study is to measure the impact of pedestrian flow on DLT intersection by proposing an underpass pedestrian crossing which allows for measuring the impact of pedestrians on the DLT intersection. The study area is a conventional cross-shaped intersection, selected due to it is affected by heavy pedestrians and vehicle traffic during various hours in a typical day, as shown in Fig. 3. The intersection is located between the Panamericana Norte (main avenue) and Proceres de Huandoy (minor avenue), in Lima, Peru. This study evaluates the selected intersection in three scenarios to measure the pedestrian impact on the intersection. The three scenarios are: The current situation, a DLT proposal with two displaced left turns on adjacent

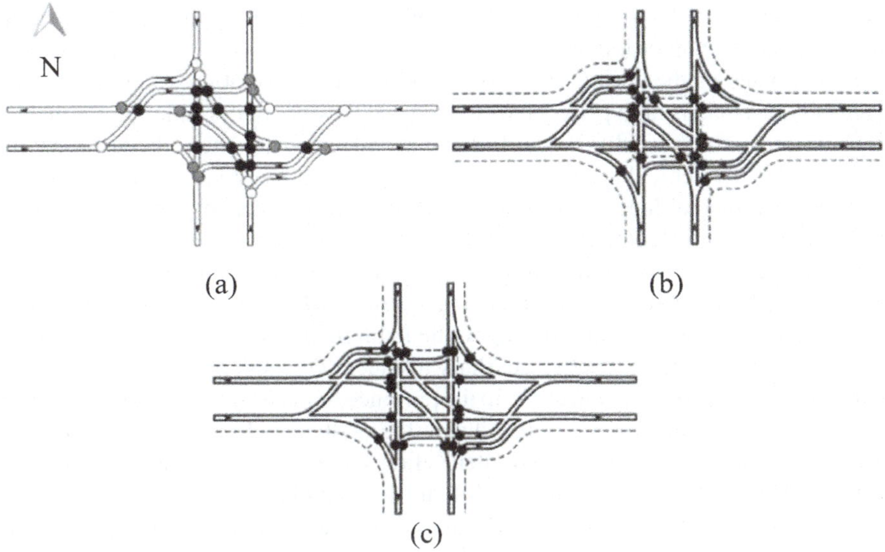

Fig. 2 DLT intersection with two displaced left turns (west and east). **a** 30 points of vehicular conflict; **b** 18 points of pedestrian–vehicle conflict in a two-stage crossing; **c** 20 points of pedestrian–vehicle conflict in a one-stage crossing. Black circles indicate a crossing, gray circles indicate merging and white circles indicate diverging. Adapted from [3]

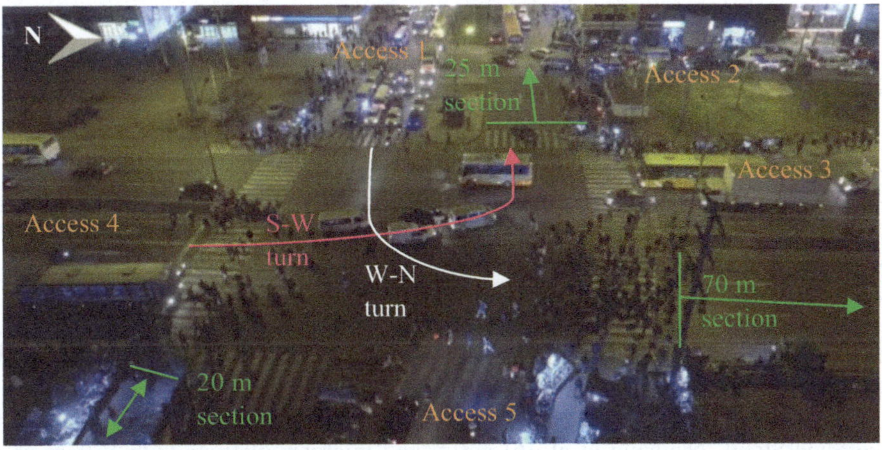

Fig. 3 Current situation of the intersection. Left S–W and W–N turns are the ones that increase congestion

accesses, and the DLT proposed in the previous scenario with the inclusion of an underpass pedestrian crossing.

Figure 3 shows the current congestion and the difficulty of crossing the studied intersection. There is high pedestrian demand and how uncomfortable it is for pedestrians to cross the road due to the little space allocated for them. Pedestrians are forced to use the vehicular lanes. The presence of street vendors, in large numbers, invades the vehicular lanes, thus reducing the designated capacity for vehicles. Due to the mess at the intersection; traffic police officers try to control and mitigate traffic congestion as the current traffic light fails to control traffic conflicts. The crossing has 6 vehicular accesses. Accesses 1 and 5 correspond to the minor road (Proceres Avenue), and accesses 3 and 4 belong to the main road (Panamericana Norte), with accesses 2 and 6, their respective auxiliary avenues. The high pedestrian and vehicular demand at the intersection is due to the presence of local businesses and shopping centers in the surroundings. It is also due to the existence of urban districts located near the intersection, so the studied intersection serves as the main access for these districts. This high demand makes it difficult for vehicles to make left turns to the west and north of the intersection, thus requiring the traffic police to organize left turns from accesses 4 and 1, rendering the traffic light without effect.

This document is organized into four sections: Sect. 2 describes the literature review, Sect. 3 provides the methodology used, Sect. 4 includes the results obtained and the analysis, Sect. 5 contains the conclusions and suggests future research directions.

2 Literature Review

Unconventional intersections have been studied to reduce traffic congestion. Autey et al. [5] evaluated the performance of four unconventional road designs using microsimulations with Vissim. The intersections evaluated, as shown in Fig. 4, were: Up-stream signalized crossover (USC), Double crossover intersection (DXI), Crossover displaced left-turn (XDL), and Median U-Turn (MUT).

Autey et al. [5] discovered that USC, DXI, and XDL intersections had 99%, 50%, and 50%, respectively, extra capacity compared to a conventional one and showed that the biggest problem in conventional intersections is the left turn.

Another unconventional intersection evaluated against a conventional one was the SuperStreet. Naghawi et al. [6] performed microsimulations with Vissim and Synchro and obtained a 70% decrease in vehicular delays and 73% in the maximum queue length formed by vehicles.

Finally, Shu et al. [7], prioritizing the left turn of public transport buses, proposed to use a two-way lane for the displaced left-turn (DLT), since there was not much space at the intersection. Figure 5 shows the design proposed.

Delays were reduced by 5% for every 30 vehicles/h that traveled straight on the roads and by 5% for every 40 vehicles/h that turned left. When the vehicle/h ratio exceeded 108, delays were reduced by more than 15%.

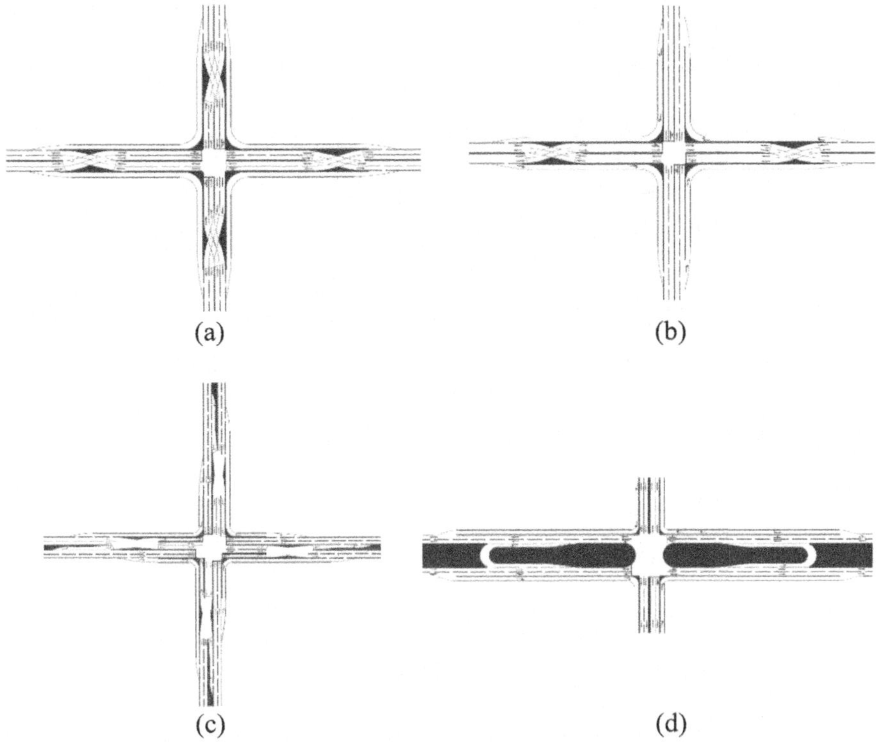

Fig. 4 **a** Up-stream signalized crossover (USC); **b** double crossover intersection (DXI); **c** crossover displaced left-turn (XDL); **d** median U-turn (MUT). Adapted from [5]

Therefore, due to the problems observed at the studied intersection; it was necessary to solve the left-turn problem in two directions, westbound and northbound of the intersection. Consequently, the Displaced Left Turn (DLT) intersection was chosen for the present study. The methodology used to evaluate the established scenarios is shown below.

3 Methodology and Procedures

The study followed the methodology suggested by the FHWA with the use of microsimulation software; Vissim 9.00 [8]. The procedure was as follow:

- A field investigation was conducted during two typical days using a drone. It was on Thursday, September 03 and 10, 2020. At each visit, the travel time of 31 vehicles in each section (25 and 70 m) and of 31 pedestrians in the 20 m section

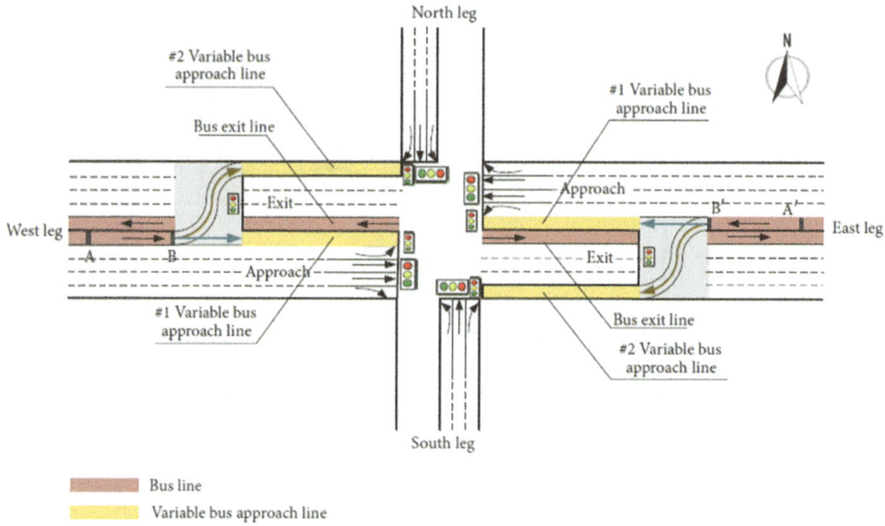

Fig. 5 Variation of the conventional design of a DLT for a smaller working area. Taken from [7]

was recorded (see Figs. 1 and 6). Information on vehicular and pedestrian volume and geometric information were also obtained.

- The current situation (first scenario) was modeled using data collected in the first field visit. It was considered the behavior of informal vehicles, which, although they were light vehicles, behaved as public transportation buses. The software included the average entry and exit time of formal and informal public transport vehicles at the buses stop, speed areas (customized for each type of vehicle), priority rules which helped to represent the behavior of the vehicles as their paths

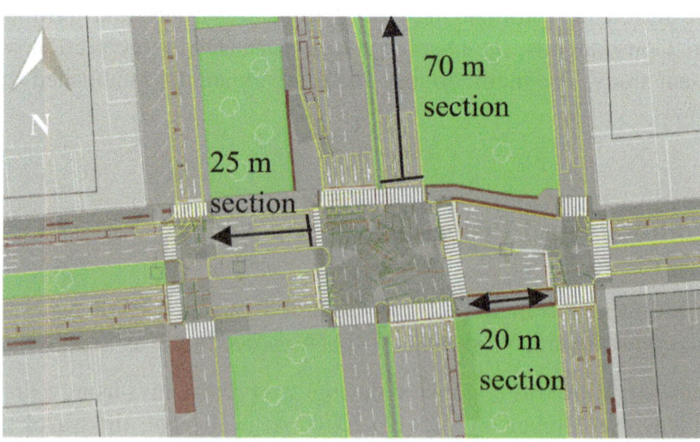

Fig. 6 First scenario—current intersection modeled in Vissim

cross, bus stops and objects on the sidewalks to simulate the presence of street vendors. Figure 6 shows a part of the current modeled intersection.

- The first scenario was calibrated and validated considering, as evaluation parameters, vehicle travel times (s) and pedestrian speeds (km/h). To find the vehicle travel times (s) and pedestrian speed, the 31 travel times (s) obtained from the vehicular section (25 and 70 m) and the pedestrian section (20 m) were used as shown in Figs. 1 and 6. Each section was strategically chosen to help represent the model more realistically. The number of seeds and runs was 31. Each run lasted 70 min, the first 10 min corresponded to the warm-up of the model.

For calibrating, data collected in the first field visit was used to obtain the speed curves for each section. The speed curves were plotted as a function of their speeds (km/h) (X-axis) and their cumulative relative frequencies (H) (Y-axis). These curves, whose coordinates were entered into Vissim in their respective sections, represented the speeds of the vehicles and pedestrians recorded in the first field visit. Then, the values of the Wiedemann 74 and Social Force parameters were determined iteratively until these values represent, in the model, the vehicular and pedestrian behavior observed in the first field visit.

For validating, the vehicle and pedestrian volumes obtained in the second field visit were used, but the speed curves and parameters established in the calibration are maintained.

By completing the 31 runs for each process, the data documented in the first and second field visit were statistically compared to those obtained from Vissim. The mean (M), the standard deviation (S), the error (E = 10% M), the minimum number of runs (N_{min}) (Eq. 1), and the null hypothesis analysis (H0: u1 = u2) were obtained. For the null hypothesis analysis, the "t-student" table was used with N − 1 degrees of freedom (N = 31) and 95% reliability for the evaluation parameters. The virtual platform Statkey was used.

$$N_{\min} = \left(t_{0.025,N-1} \times \frac{S}{E}\right)^2 \qquad (1)$$

- For modeling, the second (DLT) and third scenario (DLT + pedestrian underpass), the manuals of the FHWA, MTC, and A.120 Technical Standard for Universal Accessibility were considered in the design [3, 9, 10]. A DLT intersection with two displaced left turns located at the south and west accesses was considered, and not on the opposite sides as proposed by the FHWA (Fig. 2). As a result, an "X" shaped crossing was proposed on the west side of the intersection (Fig. 9). Bicycle lanes (red roads in Figs. 7 and 8), a road called "Exit N–W" were added to the north of the intersection (see Fig. 8) and traffic lights timing was optimized by using the Webster method (Figs. 8 and 9).

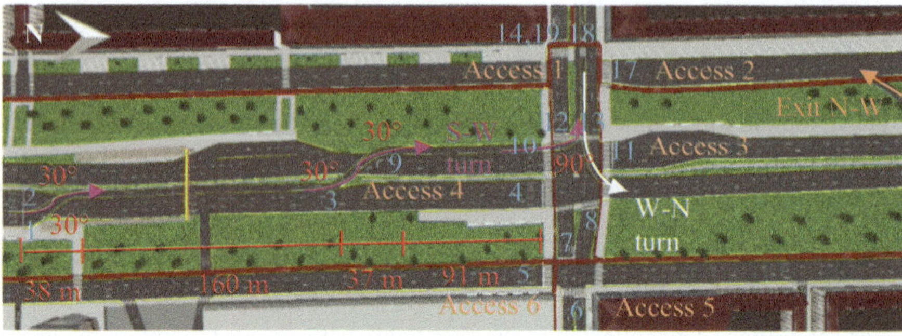

Fig. 7 Second scenario—DLT with two displaced left turns with no pedestrian underpass modeled in Vissim

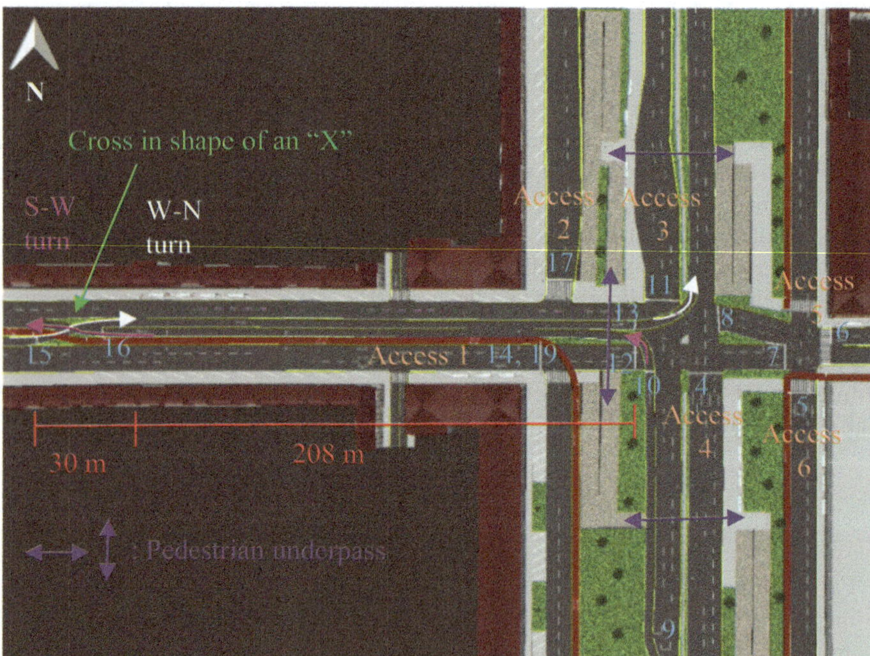

Fig. 8 Third scenario—DLT with two displaced left turns with pedestrian underpass modeled in Vissim

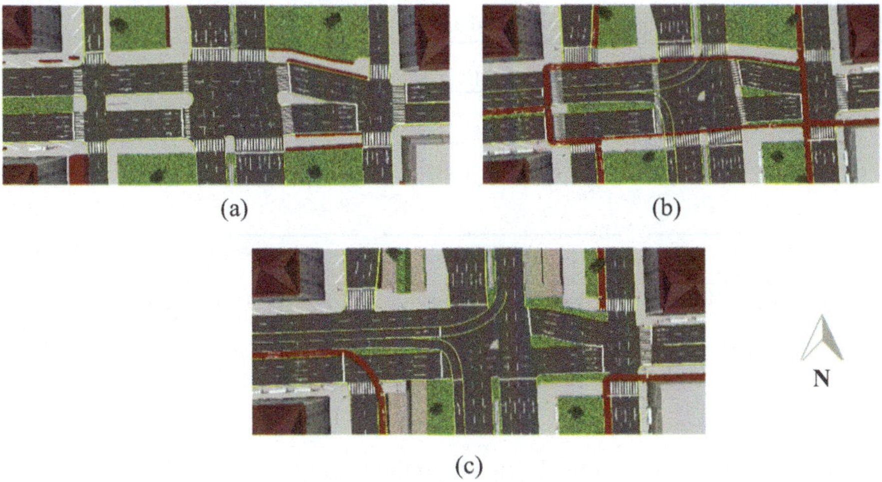

Fig. 9 **a** First scenario—current intersection; **b** second scenario—DLT proposed without pedestrian underpass; **c** third scenario—DLT proposed with pedestrian underpass

4 Results and Discussion

4.1 Field Information

The studied intersection is under formal and informal activities that generate high vehicle traffic congestion. There are 6 vehicular accesses with a lane width of 3.30 m. The traffic light cycle is 123 s with two phases; however, it is not effective, then, police enforcement is required to control the intersection. From the first field visit, it was obtained that the peak hour (PH) is from 06:30 to 07:30 p.m., the peak hour factor is 0.976, and the pedestrian and vehicle volume at this hour is 1649 pph and 8047 vph, respectively. The classification of vehicles is 81.7, 7.3, and 11%, which correspond to personal cars, heavy vehicles, and buses (Table 1).

It is important to indicate that during the second field visit; similar traffic volumes were recorded during the PH. Vehicle and pedestrian volumes were 8251 vph and 1802 pph, respectively.

4.2 Calibration and Validation of the Model (First Scenario)

The vehicle sections were chosen because in those sections, after the intersection, vehicles showed similar behavior. Also, both sections were the receiving roadways for the left turn that the DLT intersections attempted to resolve. The pedestrian section was chosen because this sidewalk is one of those with the highest pedestrian traffic

Table 1 Vehicle movements at each access during peak hour

Access 1 (vph = 463)		Access 2 (vph = 346)		Access 3 (vph = 2320)	
Movement	**% vph**	**Movement**	**% vph**	**Movement**	**% vph**
W-N	33%				
W-E	44%			N-S	98%
W-S	23%	N-W	100%	N-W	2%

Access 4 (vph = 4326)		Access 5 (vph = 281)		Access 6 (vph = 311)	
Movement	**% vph**	**Movement**	**% vph**	**Movement**	**% vph**
S-W	12%	E-S	34%	S-W	19%
S-N	86%	E-W	63%	S-N	57%
S-E	2%	E-N	3%	S-E	24%

and the highest number of street vendors in the vicinity, as shown in Figs. 3 and 6. The data recorded in the field were travel times (s) Therefore, since the distance of the sections was known, the speeds (km/h) could be found.

Calibration of the First Scenario. The average vehicle speeds, recorded during the first field visit, in the 70 m and 25 m sections were 39,600 km/h and 30,847 km/h, respectively. And the average pedestrian speed was 4,745 km/h. The speed curves were then entered in Vissim for each section. The values of the Wiedemann 74 parameters were: Average Standstill Distance = 1.00 m, Additive Part of Safety Distance = 2.00 m, and Multiplic. Part of Safety Distance = 2.00 m. The values of the Social Force parameters were: Tau (t) = 1.00, Lambda (λ) = 0.010, and Noise = 0.010. These last three values referred to the relaxation time, anisotropy factor, and randomness factor, respectively. Finally, the 31 runs were executed, and the following was obtained (Table 2).

The evaluation parameters were analyzed using the null hypothesis analysis. Vehicle travel times complied with the null hypothesis analysis (0.04 and −0.05 for the 70 m and 25 m sections, respectively). Pedestrian speeds did not comply with the null hypothesis analysis, since the mean difference (0.369 km/h) was 10 cm/s. However, the value was accepted due to the disorganized large pedestrian flow observed in the field.

Validation of the First Scenario. When comparing the data recorded in the second field visit versus those obtained from Vissim after the 31 runs, was obtained (Table 3).

It was obtained that the vehicle travel times complied with the null hypothesis analysis (0.03 and −0.07 for the 70 m and 25 m sections, respectively), and the pedestrian speeds also complied with the analysis of the null hypothesis (−0.07).

In this way, the model designed for the current intersection was calibrated and validated, since, if it were decided to make changes in the model, after the microsimulations, the model would show the real behavior of the intersection.

Table 2 Calibration results of the first scenario

	Vehicle travel times (s)				Pedestrian speeds (km/h)	
	70 m		25 m		20 m	
	Obtained from the field	Obtained from Vissim	Obtained from the field	Obtained from Vissim	Obtained from the field	Obtained from Vissim
Mean (M)	6.376	6.335	2.937	2.983	4.745	4.376
Standard deviation (S)	0.287	0.021	0.243	0.058	0.069	0.721
Error (E = 10%M)	0.638	0.633	0.294	0.298	0.475	0.438
t0.025, n − 1	2.042	2.042	2.042	2.042	2.042	2.042
N_{min}	0.845	0.004	2.857	0.156	0.088	11.320
Mean difference	0.04		−0.05		0.369	
H0: u1 = u2	Complies		Complies		Does not comply	

Table 3 Validation results of the first scenario

	Vehicle travel times (s)				Pedestrian speeds (km/h)	
	70 m		25 m		20 m	
	Obtained from the field	Obtained from Vissim	Obtained from the field	Obtained from Vissim	Obtained from the field	Obtained from Vissim
Mean (M)	6.395	6.367	3.22	3.294	4.579	4.651
Mean difference	0.03		−0.07		−0.07	
H0: u1 = u2	*Complies*		*Complies*		*Complies*	

4.3 Modeling of the Second and Third Scenarios

The scenarios were modeled considering an 18.30 m long articulated bus as the design vehicle. The 30° and 90° curves had a turning radius of 50 m and 15 m, respectively. The width of the lanes was 3.30 m. The width of the pedestrian overpass was 6.5 m and the slope of the ramps; 2%. In the second scenario, 19 traffic lights were considered; and in the third one; 18 since the traffic light No. 18 was not necessary.

It should be noted that traffic light No. 14 only controls the turn to the right. In both scenarios, for the cross in the shape of an "X"; Cycle Time (CT) = 30 s. Numbers, in Figs. 3 and 4, indicate the location of the traffic lights (Table 4).

Table 4 Traffic light timing plan for the proposed scenarios

			Phase 1				Phase 2	Phase 3	
	Interval	Time (Sec.)	Traffic light No. 2, 4, 9 y 11	Traffic light No. 5 y 17	Traffic light No. 14	Traffic light No. 10	Traffic light No. 12, 19, 7, 8 and 6	Traffic light No. 1 and 3	Traffic light No. 13 and 18 (not required for third scenario)
Second scenario CT = 114 sec.	1	30	G	G	G	G	R	R	R
	2	4	G	G	G	Y	R	R	R
	3	15	G	G	G	R	R	R	R
	4	4	Y	G	G	R	R	R	R
	5	4	R	G	G	R	R	G	G
	6	4	R	G	Y	R	R	G	G
	7	2	R	G	R	R	R	G	G
	8	4	R	Y	R	R	R	G	G
	9	16	R	R	R	R	G	G	G
	10	4	R	R	R	R	G	G	Y
	11	3	R	R	R	R	G	G	R
	12	4	R	R	R	R	Y	G	R
	13	16	R	R	R	R	R	G	R
	14	4	R	R	R	R	R	Y	R
Third Scenario CT = 104 sec.	1	50	G	G	G	G	R	R	R
	2	4	Y	G	G	G	R	R	R
	3	22	R	G	G	G	R	G	G
	4	4	R	Y	G	Y	R	G	Y
	5	6	R	R	G	R	G	G	R
	6	4	R	R	Y	R	G	G	R
	7	10	R	R	R	R	G	G	R
	8	4	R	R	R	R	Y	Y	R

4.4 Vehicle Delays

The vehicular delays of the three scenarios were evaluated considering 372 m on the W–E axis and 682 m. on the N–S axis. The results for each access are shown in Table 5. The percentages indicate an increase or decrease from the current intersection. Improvements are found in both scenarios. However, the third was more beneficial due to the absence of pedestrian flow through the intersection (Fig. 9).

The second scenario was more beneficial for accesses 2 and 5, since, in a certain way, it dissipated the pedestrian capacity. In the third scenario, both entrances were filled as they served as a path to the pedestrian underpass. Access 1, although it was more orderly, is a similar case to accesses 2 and 5 for this reason, the results of the second and third scenarios in access 1 did not have much difference. Access 2 also showed a large increase in vehicles, due to the "N–W exit" implemented. This new access prevents vehicles from turning to the right from access 3. The turn to the right "N–W", from access 3, refers to the "N–W exit" for the second and third scenarios. Accesses 1 and 4 demonstrated that a DLT reduces vehicle delays for left turns by up to 83%, especially if there is no pedestrian flow through the intersection. Finally, access 6 was damaged by the second scenario, since there was no good traffic flow through the access.

Finally, vehicle delays throughout the intersection were 137.76 s, 146.65 s, 91.34 s for the first, second, and third scenarios, respectively. The results indicate that the conventional DLT scenario increases delays by +6%. In addition, if a pedestrian underpass is considered, it decreases by −34%, compared in both cases with the current situation scenario.

Table 5 Vehicle delays, for accesses and movements, obtained from the three studied scenarios

Acc.	Mov.	Scenario	Vehicular Delays (Sec.)	Vehicular Stop Delays (Sec.)
1	↰ W-N	First	195.78	172.01
		Second	52.93 (-73%)	37.88 (-78%)
		Third	43.94 (-78%)	32.19 (-81%)
	↑ W-E	First	165.45	139.14
		Second	48.57 (-71%)	31.83 (-77%)
		Third	45.86 (-72%)	32.68 (-77%)
	↱ W-S	First	138.39	114.01
		Second	23.81 (-83%)	11.92 (-90%)
		Third	6.27 (-95%)	1.03 (-99%)
2	↱ N-W	First	3.18	0.66
		Second	17.61 (+454%)	9.89 (+1399%)
		Third	26.27 (+726%)	4.39 (+565%)
3	↑ N-S	First	127.33	78.59
		Second	122.88 (-3%)	86.81 (+10%)
		Third	57.84 (-55%)	34.74 (-56%)
	↱ N-W	First	106.15	66.62
		Second	37.50 (-65%)	26.73 (-60%)
		Third	47.88 (-55%)	28.25 (-58%)
4	↰ S-W	First	367.89	242.40
		Second	141.09 (-62%)	103.03 (-57%)
		Third	63.62 (-83%)	39.30 (-84%)
	↑ S-N	First	271.66	161.21
		Second	206.12 (-24%)	135.10 (-16%)
		Third	179.98 (-34%)	110.82 (-31%)
	↱ S-E	First	374.45	259.60
		Second	153.60 (-59%)	100.19 (-61%)
		Third	132.18 (-65%)	79.28 (-69%)
5	↰ E-S	First	151.23	124.18
		Second	64.93 (-57%)	43.63 (-65%)
		Third	65.61 (-57%)	43.23 (-65%)
	↑ E-W	First	68.85	54.52
		Second	48.47 (-30%)	36.28 (-33%)
		Third	63.72 (-7%)	44.11 (-19%)
	↱ E-N	First	75.67	65.55
		Second	36.55 (-52%)	28.94 (-56%)
		Third	54.16 (-28%)	42.44 (-35%)
6	↰ S-W	First	77.13	63.55
		Second	71.22 (-8%)	57.50 (-10%)
		Third	63.74 (-17%)	46.28 (-27%)
	↑ S-N	First	22.59	14.02
		Second	17.72 (-22%)	10.69 (-24%)
		Third	11.69 (-48%)	5.48 (-61%)
	↱ S-E	First	17.75	11.69
		Second	25.48 (+44%)	19.38 (+66%)
		Third	6.17 (-65%)	2.87 (-75%)

5 Conclusions and Future Research

Results indicate that replacing a conventional intersection with a DLT intersection with two displaced left turns on adjacent access in an urban area is not recommended, since its increases the vehicle delays, mainly due to the high pedestrian traffic volume. Informal commercial activities on streets like street vendors, informal bus stops greatly influences traffic congestion. The proposed scenarios in this study included the informality observed in the field. The inclusion of a pedestrian underpass allows for quantifying the effect of pedestrian flow over the DLT intersection. Results indicate that pedestrian flow has an impact of +34% on the average delay at the intersection. It is recommended to restrict heavy traffic during peak hours and to organize street vendors who invade pedestrian and vehicular lanes. Thus, if the informality is resolved, the inclusion of a pedestrian underpass would be even more beneficial.

Future studies should be focused on evaluating the effect of informality, especially in the labor market on the congestion of the intersection. It is observed that informality creates conflict points not distinguished yet, between pedestrians and vehicles that are not considered in conventional scenarios. These studies should also be focused on quantifying the effects of the social conditions presented in this study affects a DLT intersection with two displaced left turns located on opposite. Assess whether the social conditions presented in this study equally affect a DLT intersection with the two displaced left turns located on opposite accesses.

References

1. El Comercio. https://elcomercio.pe/lima/transporte/semaforos-mal-coordinados-quitan-30-minutos-dia-noticia-596006-noticia/. Accessed 14 May 2021
2. Municipalidad de Lima (Municipality of Lima). http://www.gmu.munlima.gob.pe/index.php/es-es/noticias-gtu/item/239-municipalidad-de-lima-mejora-la-circulacion-vehicular-y-peatonal-con-100-senales-de-transito. Accessed 14 May 2021
3. Steyn H, Bugg Z, Ray B, Daleiden A, Jenior P, Knudsen J, Kittelson & Associates, Inc. (2014) Displaced left turn informational guide. U.S. Department of Transportation Federal Highway Administration
4. Hughes W, Jagannathan R, Sengupta D, Hummer J (2010) Alternative intersections/interchanges: informational report (AIIR). U.S. Department of Transportation Federal Highway Administration
5. Autey J, Sayed T, El Esawey M (2012) Operational performance comparison of four unconventional intersection designs using micro-simulation. J Adv Transp 47(5):536–552
6. Naghawi H, AlSoud A, AlHadidi T (2018) The possibility for implementing the superstreet unconventional intersection design in Jordan. Period Polytech Transp Eng 46(3):122–128
7. Shu S, Zhao J, Han Y (2019) Novel design method for bus approach lanes with bus guidance and priority controls for prioritizing through and left-turn buses. J Adv Transp 2019:1–15
8. Wunderlich K, Vasudevan M, Wang P (2019) TAT volume III: guidelines for applying traffic microsimulation modeling software 2019 update to the 2004 version. U.S. Department of Transportation Federal Highway Administration
9. Dirección General de Caminos y Ferrocarriles (Directorate General of Roads and Railways) (2018) Manual de carreteras: diseño geométrico DG—2018 (Manual of highways: geometric design DG—2018). Ministerio de Transportes y Comunicaciones (Ministry of Transport and Communications)
10. Ministerio de Vivienda, Construcción y Saneamiento (Ministry of Housing, Construction and Sanitation) (2017) Norma A.120 Accesibilidad para personas con discapacidad y de las personas adultas mayores 8 (2017) (Standard A.120 Accessibility for people with disabilities and the elderly. Article 8)

Identification of Black Spots on Bekasi City Roads

W. B. Dermawan and E. A. Wardani

Abstract A traffic accident is one of the traffic problems that often occur in Bekasi. Accidents that occurred were caused by several factors such as human factors, vehicle factors, and road environment factors. One of the efforts that can be done to overcome the traffic accident problems is analyzing black spots on the road and analyzing its causative factors so that we can determine appropriate solutions to overcome this matter. The analysis of black spots on the roads carried out using the Frequency and Equivalent Property Damage Only Method (EPDO). The result of the analysis showed that roads that are considered prone to traffic accidents (black spots) are Jalan Raya Narogong-Siliwangi, Jenderal Achmad Yani, and Sultan Agung. The result of secondary data analysis shows that the causative factors of accidents were caused by 70% of human factors, 22% of vehicle factors, and 8% of road environment factors, whereas the primary data analysis result shows that the causative factors of accidents were caused by the average vehicle's speed that exceeds the speed limit standard and poor road infrastructure conditions.

Keywords Black spot · Frequency · EPDO

1 Introduction

In Transportation, a traffic accident is one of the biggest causes of death in the world. Based on the data from The World Human Organization (WHO) in The Global Status Report on Road Safety in 2018, the number of deaths due to traffic accidents in the world reached 1.35 million people annually [1]. Indonesia is one of the countries in the world with a high mortality rate death due to traffic accident matters. According to The Indonesia National Police Data in 2019, there has been an increase in traffic accidents by 3% from the previous year. The accidents that occurred in 2019 caused 23,530 fatalities [2]. It also caused the victim to suffer serious injuries, minor injuries, and material losses. Human factors, vehicle factors, and road

W. B. Dermawan (✉) · E. A. Wardani
University of Mercubuana, Jakarta Barat, Indonesia
e-mail: widodo.budi@mercubuana.ac.id

T. Kang and Y. Lee (eds.), *Proceedings of 2021 4th International Conference on Civil Engineering and Architecture*, Lecture Notes in Civil Engineering 201,
https://doi.org/10.1007/978-981-16-6932-3_4

environment factors are the three main factors that can cause traffic accidents. Bekasi is one of the cities in Indonesia that have high traffic accident cases. According to Traffic Police Corps in Bekasi, there has been an increase in traffic accidents throughout the year. There have been 686 traffic accidents cases in 2019, and 485 cases in 2018. Most accidents were involved with two-wheel vehicles [2]. According to the report from Directorate General of Land Transportation in The Blueprint of Road Safety and Road Transport, traffic accidents were not only causing fatalities, injury, or material losses, but also reducing the victim's productivity level, causing stress, and prolonged trauma [3]. One of many ways to minimize the rate of traffic accidents occur in Bekasi is conducting a research entitled Identification of Black Spots on Bekasi City Roads. The purpose of this research is to analyze the black spot area, analyze factors that cause traffic accidents, and to determine appropriate solutions to minimize the number of accidents in Bekasi.

2 Literature Review

2.1 The Definition of Traffic Accident-Causing Factors

According to the Law of The Republic of Indonesia number 22 the year 2009, concerning road traffic and transportation. A traffic accident is an unexpected event and accidental road involving vehicles with or without other road users resulting in human casualties and/or property loss [4]. A traffic accident is divided into 3 (three) categories. There are Minor Traffic Accident which caused property damage, Serious Traffic Accident which caused a minor injury to the victim and property damage, and Fatal Traffic Accident which caused the victim seriously injured or died. In general, 3 (three) accident-causing factors are human factors as road users, vehicle factors, and road environment factors. Road environment factors consist of surface conditions, sign conditions, lighting conditions, and geometry conditions [5].

2.2 The Definition of Black Spot

Based on Construction and Building Guidelines Number Pd T-09-2004-B from The Department of Settlements and Infrastructure in 2004, Black Spot is an area with a high rate of traffic accidents with repeated accidents and caused by certain causes [3].

2.3 *Collision Diagram*

Collision diagrams used to show important road accident patterns. The use of collision diagrams provides a broader understanding of the number of accidents and common contributory factors at analyzed locations with graphic symbols [6].

2.4 *Accident Rate and Fatality Rate*

Accident Rate of Bekasi can be calculated by dividing the number of accidents by the number of residents per 100,000. The Fatality Rate can be calculated by dividing the number of death by the number of residents per 100,000 [7].

3 Method

3.1 *The Frequency Method*

Frequency Method is the method that will be used to analyze the Black Spot by calculating the frequency of accidents that had caused the victim died and/or seriously injured [8].

3.2 *The Equivalent Property Damage Only (EPDO) Method*

The Equivalent Property Damage Only (EPDO) is the method that will be used to convert the value of accident fatality to property damage cost. The EPDO method is carried out by giving weight factors on each accident based on the severity of the accident (Died, Injured, Property Damage Only) to obtain a combination of the frequency and severity of the accident. The number of dead victims will be multiplied by Rp 406.520.934, the number of the victim with seriously injured will be multiplied by Rp 111.726.411, the number of the victim with minor injury will be multiplied by Rp 31.098.486, and the number of loss in the material will be multiplied by Rp 40.000.000 [9]. If simplified, each category weights approximately 10:3:1:1 (Died: Serious Injured: Minor Injured: Property Damage). After that, to convert the value of fatality to property damage cost, all total costs are divided by Rp 40 million to get the weights of accident value.

4 Data Collection

There are 2 (two) types of data that will be used in this study, named primary data and secondary data.

4.1 Primary Data

Primary data obtained through the survey in black spot areas. The survey conducted when Bekasi was implementing Large-Scale Social Restrictions due to a Covid-19 pandemic. Primary data consist of actual vehicle speed data, interview data, and road infrastructure data such as surface conditions, road signs condition, lighting condition, and geometry condition.

4.2 Secondary Data

Secondary data consist of 2016–2019 traffic accident data from The Indonesian National Traffic Police Corps and 2016–2019 total population data obtained from Bekasi Central Bureau of Statistics.

5 Discussion

Based on Fig. 1, it can be seen that the accident rate of Bekasi increased from 2017 to 2019. The increase of accident rate in contrast to the fatality rate. From the chart presented above, the fatality rate of Bekasi in 2019 decreased by 2.92% per 100,000 residents and became the lowest fatality rate compared to previous years. This indicates that the number of death of traffic accidents in Bekasi has decreased. In general, 3 (three) accident-causing factors in Bekasi were caused by 70% of human factors with 1492 cases, 22% of vehicle factors with 457 cases, and 8% of road environment factors with 172 cases [2]. The types of vehicles that are often involved in accidents on Jalan Narogong-Siliwangi were motorcycles 70%, minibus 12% and trucks 11% (Fig. 2; Tables 1 and 2).

Three roads with a high rate of the accident based on the analysis using the frequency method and EPDO method are Jalan Narogong Siliwangi, Jalan Sultan Agung, and Jalan Jenderal Achmad Yani. The result showed that Jalan Narogong-Siliwangi is the black spot location for traffic accidents. The evidence was shown by the result of data analysis in 2016–2019. 71 accidents were causing the death of victims and 23 accidents causing seriously injured, so there were 94 accidents. With a total loss of material Rp 70,811,364,509 and accident weight 177,028. Jalan

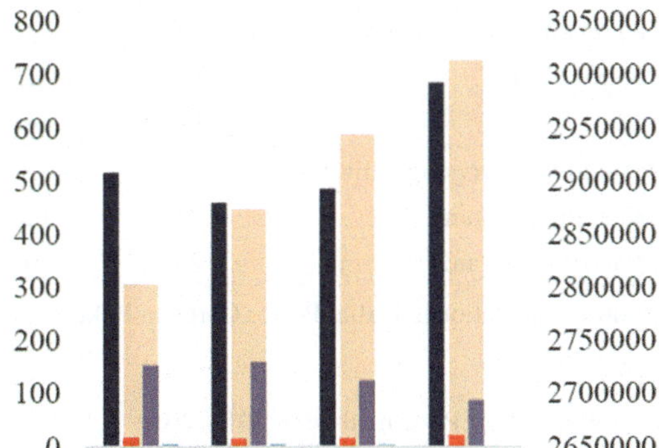

	2016	2017	2018	2019
Population (Million)	2803283	2873484	2943859	3013851
Number of Accidents	517	459	485	686
Accident Rate (%)	18.443	15.974	16.475	22.762
Number of Deaths	154	160	125	88
Fatality Rate (%)	5.494	5.568	4.246	2.920

Fig. 1 Graph of accident rate and fatality rate in Bekasi

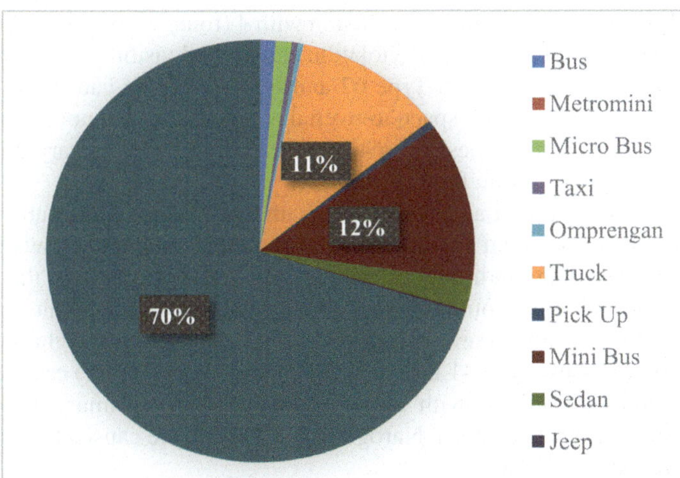

Fig. 2 Graph of vehicle types involved in traffic accidents

Narogong-Siliwangi is a residential and a shopping area with a length of road ± 10.55 km, road width 30 m, and road type 2/2 UD [10]. The highest number of accidents occurred in Segment 9 on GPS Coordinates (−6.272747, 106.992975)–(−6.281369, 106.990236) with a length of road ± 1 km. cases, Type 62 through with right-angle collision with 5 (five) cases, Type 41 Left-Angle Collision with 4 (four)

Table 1 Road ranking by frequency method in 2016–2019

No.	Name of road	Frequency of accidents		Total
		Death	Seriously injured	
1	Jalan Raya Narogong-Siliwangi	71	23	94
2	Jalan Sultan Agung	25	12	37
3	Jalan Ahmad Yani	26	9	35

Source The Indonesian National Traffic Police Corps in Bekasi [2]

Table 2 Road ranking by EPDO method in 2016–2019

No.	Name of road	Death	Seriously injured	Minor injured	Property damage	Cost	Weight
1	Jalan Raya Narogong-Siliwangi	76	25	194	138,700,000	70,811,364,509	1770.28
2	Jalan Jenderal Ahmad Yani	28	12	75	247,450,000	27,288,188,910	682.20
3	Jalan Sultan Agung	26	13	77	82,200,000	26,281,836,927	657.05

Source Processed Author

cases, and Type 61 through with left-collision with 4 (four) cases. Type 63 Rear-End Collision can be caused by speeding violation and limited visibility so the driver can not see the vehicle in front of it. Type 62 and type 61 can be caused by a limited view and not available enough space to overtake, type 62 is also not recommended because of violating traffic rules. Type 41 can be caused by an unwary driver and a road that does not have traffic lights (Fig. 3).

Above is one example of a Collision Diagram that illustrates the pattern of incidents in the type of accident with the highest number in Segment 9 with a length of road 300 m. The diagram used to understand the types of accidents that occurred in that segment. The most common types of accidents on this segment are Type 63 Rear-End Collision with 7 (seven) cases, Type 62 through with right-angle collision with 5 (five) cases, Type 41 Left-Angle Collision with 4 (four) cases, and Type 61 through with left-collision with 4 (four) cases. Based on primary data analysis, those Accidents occurred on Jalan Narogong-Siliwangi were caused by several road infrastructure factors (Table 3).

To support both primary data and secondary data, it is necessary to have interviews and questionnaires. The respondents were four residents who live in Jalan Narogong-Siliwangi and fourteen friends who resided around that road or passed the road. The interview was conducted in May 2020, when Bekasi implemented Large-Scale Social Restrictions due to the Covid-19 pandemic. The results that the types of vehicles that are often involved in accidents were motorcycles and trucks, the time of accident often happened during the night (18.00–00.00 WIB) when in the cloudy condition, with the most victims suffering a serious injury, due to poor road surface conditions

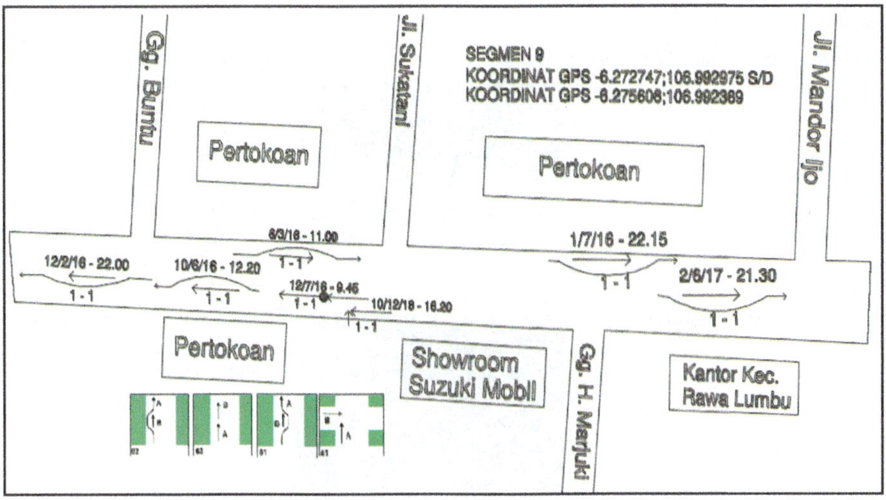

Fig. 3 Collision diagram I segment 9

such as potholes, cracked, and slippery roads, vehicle's speed, and the frequency of accidents is 1–3 cases in 1 (one) month.

The average speed that exceeds the standard was one of the causative factors of accidents. Therefore, the average vehicle's speed was observed to see the potential of accidents. The survey is conducted by calculating the travel time of the vehicle that crosses the distance of 81 m using a stopwatch. The average vehicle's speed did not reflect the real condition when the accident happened, but it was being calculated to compare with the vehicle's speed standard on that road. If the average vehicle's speed exceeds the vehicle's speed standard, then the potential of accidents will be greater. According to the results of the survey, the average actual speed of the vehicles that pass through Jalan Narogong-Siliwangi is 54.81 km/h. The actual speed exceeds the predefined standards of 40 km/h.

6 Conclusion

Based on the description above, the appropriate solutions that can be done to overcome and minimize traffic accidents that occur in Bekasi are Installing Speed Limit Signs. Jalan Narogong-Siliwangi is a region of central activity, industry, and residential. There are also several schools. The maximum speed limit allowed is 40 km/h, and in the school environment, the highest speed limit is 30 km/h [11], Installing Rumble Strip to make drivers increase more alertness, reduce vehicle speed, and remind drivers of the black spots roads. The thickness of the rumble strip is 40 mm. The installation distance between the strips is the closest to 500 mm and the farthest to 5000 mm. The slope of the edge of the strip is at most 15%. It can be placed

Table 3 Primary data analysis

No	Road infrastructure factors	Figure	Description
1	Surface condition		Poor quality because most of the road has potholes and cracked roads, and at some points, uneven asphalt patches
2	Road sign condition		Some signs are not obvious because of obstructed trees and the number of signs is inadequate
3	Lighting condition		The lighting condition is good because every 200 m, the road is equipped with the lamp and adequate lighting
4	Geometry condition		Most of the road has no road markings and not equipped with road median. So the driver drove in another lane when overtaking other vehicles

Source Processed Author

and installed before school area or dangerous places [12], Installing Signposts and Warning Signs. Signs are placed on the left according to the direction of traffic. The distance between the closest signposts to the outer edge of the vehicle traffic lane is at least 0.6 m. The installation of signposts and Warning signs are placed on the side of the road before dangerous places or sections of road, with a minimum distance of 50 m for speeds $\leq$ 60 km/h [12], Repainting of Road Markers, Road Surface Repairs and Regular Road Maintenance, Taking care of trees so they would not be blocking the traffic signs and Providing Pedestrian Facilities such as crossing facilities in school environments.

References

1. World Health Organization (2018) Global status report on road safety 2018. World Health Organization, Geneva, p 2
2. The Indonesia National Police (2020) Road traffic accident data 2020. The Indonesia National Police. Author F, Author S, Author T (1999) Book title, 2nd edn. Publisher, Location
3. Directorate of Traffic and Road Transport (2004) The blueprint of road safety and road transport. Directorate General of Land Transportation, Jakarta, p 1
4. President of the Republic of Indonesia (2009) The law of the Republic of Indonesia number 22 year 2009: traffic and road transport. Republic of Indonesia, Jakarta, p 4
5. Dermawan WB, Tjahjono T (2020) A conceptual framework for development of safety performance functions for rural roads in Indonesia. IOP Conf Ser Earth Environ Sci 498:3–4
6. Integrated Road Safety Management System (2013) Traffic accident data collection manual. Integrated Road Safety Management System, Jakarta, p 21
7. Pfundt K (1969) Three difficulties in the comparison of accident rates. Accid Anal Prev 1(3):253–259
8. Ghadi M, Torok A (2017) Comparison different black spot identification methods. Transp Res Procedia 27:1105–1112
9. Bekasi Central Bureau of Statistics (2020) Kota Bekasi Dalam Angka 2020. BPS-Statistics Indonesia
10. Transportation Minister of Republic of Indonesia (2015) Regulation of the minister of transportation number PM 111 year 2015: procedures for setting speed limits. Ministry of Transportation, Jakarta, pp 23–24
11. Transportation Minister of Republic of Indonesia (2018) Regulation of the minister of transportation number PM 82 year 2018: road user controllers and safeguards. Ministry of Transportation, Jakarta, p 19
12. Directorate of Urban Transportation System (2006) Guidelines for placement of road equipment facilities. Directorate General of Land Transportation, Jakarta, pp 19–57

The Vibration Attenuation Properties of Locally Resonant Track Bed for Subway Engineering

Lijian Lei, Linchang Miao, Chao Li, Junjie Wang, and Maolin Zhang

Abstract Due to the high speed, large capacity, safety and reliability of subway traffic, it is of great significance to alleviate urban congestion. However, the ensuing vibration problem has caused a major adverse impact on building safety, precision instrument operation and human health. Although there are many kinds of vibration reduction measures, it is difficult to attenuate low frequency vibration and achieve satisfactory results between vibration attenuation effect and cost control. Based on the locally resonant phononic crystal theory, this paper proposes three-component locally resonant track bed (LRTB) and investigates the effects of geometrical parameters on the band gap features in detail. The band structures are calculated using improved plane wave expansion (IPWE) to determine the frequency attenuation range, and the transmission spectra are obtained by finite element method (FEM) to represent the attenuation efficiency. The results indicate LRTB designed based on locally resonant theory is able to give low frequency band gap to cover the main frequency of subway environment, and the band gap can be regulated by properly configurating the geometrical parameters to actualize low frequency broadband attenuation. With the increasing of the primitive cell number, the band gap attenuation efficiency continues to improve and eventually remain stable. To facilitate the application, the detailed flow chart of LRTB is proposed for actual engineering vibration environment. The investigations can provide inspiration for the design of subway vibration isolation structure.

Keywords Phononic crystal · Locally resonant · Track bed · Band gap tunability · Subway vibration mitigation

---

L. Lei · L. Miao (✉) · C. Li · J. Wang · M. Zhang
Institute of Geotechnical Engineering, Southeast University, Nanjing 211189, Jiangsu, China
e-mail: Lc.miao@seu.edu.cn

L. Lei
e-mail: lljseu@163.com

C. Li
e-mail: lcseu@foxmail.com

T. Kang and Y. Lee (eds.), *Proceedings of 2021 4th International Conference on Civil Engineering and Architecture*, Lecture Notes in Civil Engineering 201,
https://doi.org/10.1007/978-981-16-6932-3_5

1 Introduction

The environmental vibration problem induced by urban subway operation has caused significant negative impact on building safety [1, 2], residents health [3, 4] and precision instrument operation [5]. Therefore, the vibration control of subway transportation has become an important research orientation. The propagation path of subway vibration is generally as follows. The vibration generated by wheel-rail interaction during the subway operation is transmitted along the track to the fasteners, sleepers and further to the track bed, and then the soil is used as the transmission medium to cause the vibration and damage of the surrounding buildings. During the process of vibration propagation, the high-frequency waves decay rapidly as the distance increases, while the low-frequency waves continue to propagate because of slower attenuation. Consequently, the attenuation of low-frequency vibration has always been the crucial research content of subway vibration control.

In the field of civil engineering, the common vibration mitigation measures are to arrange wave barriers in the wave transmission path. Thereinto, open and infilled trench have good performance in reducing environmental vibration, but they require wider site space and better soil stability [6–8]. Besides, the larger excavation depth needs to be guaranteed in both types of trenches to ensure considerable protective effectiveness [9]. In contrast, the discontinuous barriers represented by rows of piles has lower requirements for soil stability and have the characteristics of flexible layout, simple operation and low cost. However, in the arrangement and design of multirow piles system, the larger pile length and tighter pile spacing are necessary to obtain enough low-frequency shielding effects [10, 11], that is, the densely-rowed long piles have better vibration reduction performance. It can be concluded that the wave barriers mentioned above have high requirements for urban underground space, which is sometimes difficult to achieve in the crowded urban environment. Furthermore, there is no vibration isolation protection between the vibration source and wave barriers.

The application of track bed can avoid the above problems, which is placed directly under the track to isolate train vibration. Existing studies have indicated that the track bed possesses better vibration reduction effect than fasteners, sleepers and wave barriers [12]. Among them, the steel spring floating slab track bed is the most efficient. It is to set the concrete track bed slab with specific mass and rigidity on the steel spring vibration isolator, which can offset dynamic load generated by the vehicle through sufficient inertial mass. Although the steel spring floating slab track bed has excellent vibration damping effect, it has complicated structure, complex procedures, difficult maintenance, long construction period and expensive cost. Therefore, it is generally applied in high-level vibration damping areas such as hospital, laboratories and research institutes [13]. In summary, although the existing vibration isolation measures can produce considerable effects, it is difficult to achieve a balance between low-frequency damping, application environment and cost control. Therefore, it is still of great significance to research the new vibration reduction structures.

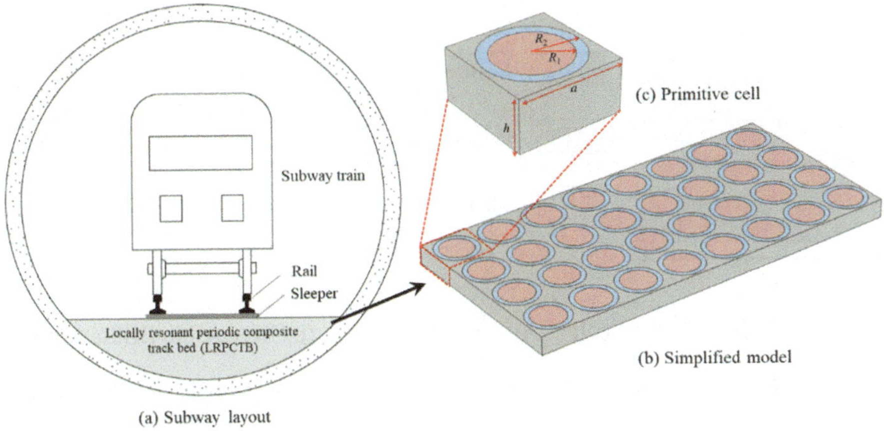

Fig. 1 The schematic illustration of LRTB in subway engineering

Phononic crystals (PCs) are the periodic structures which is able to prohibit the propagation of elastic wave in a specific band gap region [14, 15]. Early research mainly focused on the Bragg PCs, which has a high operating frequency and is not convenient for low-frequency engineering application. Fortunately, the local resonance mechanism presented by Liu et al. [16] makes it possible to obtain low-frequency band gap for small size structure. The application of local resonance theory in subway track beds has broad prospects, but it is rarely involved in the current research.

Consequently, the locally resonant track bed (LRTB) is proposed in this study to attenuate the low-frequency subway vibration. The main structure of this paper is as follows: the simplified calculation model is demonstrated and theoretical derivation of LRTB is carried out based on improved plane wave expansion (IPWE). The band structure is analyzed by IPWE and transmission spectra is calculated by finite element (FEM), and the time domain analysis is investigated to verify the shielding effects. The parameter analysis is executed to regulate the band gap, and the design flow chart is presented to customize the LRTB structure for specific engineering environment.

2 Model and Method

2.1 Calculation Model

The schematic illustration of subway operation is shown in Fig. 1. Thereinto, the LRTB can be simplified as three-component locally resonant plate structure shown in Fig. 1b, which is able to mitigate the vibration to protect surrounding building and constructions. Its basic structure can be extracted as a primitive cell, which includes

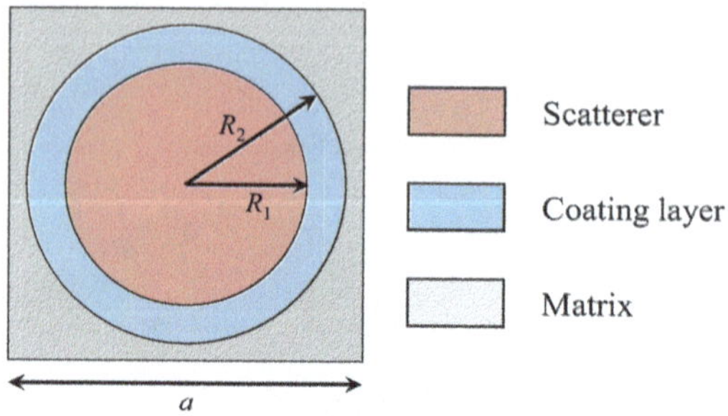

(a) Primitive cell

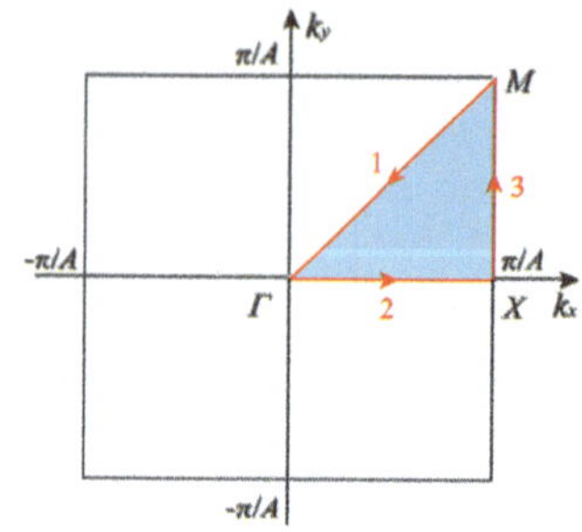

(b) Irreducible Brillouin zone

Fig. 2 The schematic illustration of **a** the primitive cell of LRTB and **b** irreducible Brillouin zone (blue shaded area)

Table 1 The material constant of LRTB

	Mass density (kg/m^3)	Lamé's constants (GPa)	
	ρ	λ	G
Scatterer	7780	122	81
Coating layer	1300	6e−4	4e−5
Matrix	2000	9.58	14.4

steel scatterer, rubber coating layer and cement matrix. The configuration of primitive cell is shown in Fig. 1c, the scatterer radius $R_1 = 0.07$ m, the coating layer radius $R_2 = 0.09$ m, the lattice constant $a = 0.2$ m and $h = 0.1$ m. Based on Bloch-Floquet theory, the band structure can be calculated by dispersion curve analysis for the single cell, which can indicate the vibration attenuation ability of the LRTB.

The plane projection of primitive cell shown in Fig. 1c is presented in Fig. 2a. The corresponding irreducible Brillouin zone is revealed in Fig. 2b, and the red line indicate the high-symmetry boundary and the arrows indicate the wave vector scanning sequence. The material constants of each component are shown in Table 1.

2.2 *Improved Plane Wave Expansion*

The dispersion curve analysis of LRTB can be executed by IPWE method. The differential governing equation for free vibration of heterogeneous thick plate can be expressed as:

$$\begin{cases} \frac{\partial}{\partial x}\left[m_1(\boldsymbol{r})\left(\frac{\partial w}{\partial x}-\psi_x\right)\right]+\frac{\partial}{\partial y}\left[m_1(\boldsymbol{r})\left(\frac{\partial w}{\partial y}-\psi_y\right)\right]=m_2(\boldsymbol{r})\frac{\partial^2 w}{\partial t^2} \\ \frac{\partial}{\partial x}\left[m_3(\boldsymbol{r})\frac{\partial \psi_x}{\partial x}+m_4(\boldsymbol{r})\frac{\partial \psi_y}{\partial y}\right]+\frac{\partial}{\partial y}\left[m_5(\boldsymbol{r})\left(\frac{\partial \psi_x}{\partial y}+\frac{\partial \psi_y}{\partial x}\right)\right]+m_1(\boldsymbol{r})\left(\frac{\partial w}{\partial x}-\psi_x\right)=m_6(\boldsymbol{r})\frac{\partial^2 \psi_x}{\partial t^2} \\ \frac{\partial}{\partial x}\left[m_5(\boldsymbol{r})\left(\frac{\partial \psi_x}{\partial y}+\frac{\partial \psi_y}{\partial x}\right)\right]+\frac{\partial}{\partial y}\left[m_4(\boldsymbol{r})\frac{\partial \psi_x}{\partial x}+m_3(\boldsymbol{r})\frac{\partial \psi_y}{\partial y}\right]+m_1(\boldsymbol{r})\left(\frac{\partial w}{\partial y}-\psi_y\right)=m_6(\boldsymbol{r})\frac{\partial^2 \psi_y}{\partial t^2} \end{cases} \tag{1}$$

where, the material parameters $m_1(\mathbf{r}), m_2(\mathbf{r}), m_3(\mathbf{r}), m_4(\mathbf{r}), m_5(\mathbf{r}), m_6(\mathbf{r})$ are the function of position coordinates, $w(x, y)$, $\psi_x(x, y)$ and $\psi_y(x, y)$ are generalized displacement function.

$$\begin{cases} m_1(\mathbf{r}) = K_s Eh/2(1+\nu) \\ m_2(\mathbf{r}) = \rho h \\ m_3(\mathbf{r}) = D \\ m_4(\mathbf{r}) = D\nu \\ m_5(\mathbf{r}) = D(1-\nu)/2 \\ m_6(\mathbf{r}) = \rho J \end{cases} \tag{2}$$

Expanding the generalized displacement function into Fourier series:

$$\begin{cases} w(\boldsymbol{r}, t) = e^{i(\mathbf{k}\cdot\mathbf{r}-\omega t)} \sum_{\mathbf{G}_1} W_k(\mathbf{G}_1) e^{i\mathbf{G}_1\cdot\mathbf{r}} \\ \psi_x(\mathbf{r}, t) = e^{i(\mathbf{k}\cdot\mathbf{r}-\omega t)} \sum_{\mathbf{G}_1} \theta_{xk}(\mathbf{G}_1) e^{i\mathbf{G}_1\cdot\mathbf{r}} \\ \psi_y(\mathbf{r}, t) = e^{i(\mathbf{k}\cdot\mathbf{r}-\omega t)} \sum_{\mathbf{G}_1} \theta_{yk}(\mathbf{G}_1) e^{i\mathbf{G}_1\cdot\mathbf{r}} \end{cases} \tag{3}$$

Besides, expanding the material parameters into Fourier series:

$$m_i(\mathbf{r}) = \sum_{\mathbf{G}_2} M_i(\mathbf{G}_2) e^{i\mathbf{G}_2\cdot\mathbf{r}}, \quad i = 1, 2, \ldots, 5, 6 \tag{4}$$

Substituting Eqs. (2) and (3) into Eq. (1) and simplifying:

$$\begin{cases} \omega^2 \sum_{\mathbf{G}_1} M_2(\mathbf{G}_3 - \mathbf{G}_1) W_k(\mathbf{G}_1) = \sum_{\mathbf{G}_1} \left[(\mathbf{k}+\mathbf{G}_1)_x (\mathbf{k}+\mathbf{G}_3)_x + (\mathbf{k}+\mathbf{G}_1)_y (\mathbf{k}+\mathbf{G}_3)_y \right] \\ \quad \cdot M_{1(\mathbf{G}_3-\mathbf{G}_1)} W_k(\mathbf{G}_1) \\ \quad + \sum_{\mathbf{G}_1} i(\mathbf{k}+\mathbf{G}_3)_x M_{1(\mathbf{G}_3-\mathbf{G}_1)} \theta_{x\mathbf{k}}(\mathbf{G}_1) \\ \quad + \sum_{\mathbf{G}_1} i(\mathbf{k}+\mathbf{G}_3)_y M_{1(\mathbf{G}_3-\mathbf{G}_1)} \theta_{y\mathbf{k}}(\mathbf{G}_1) \\ \omega^2 \sum_{\mathbf{G}_1} M_6(\mathbf{G}_3 - \mathbf{G}_1) \theta_{x\mathbf{k}}(\mathbf{G}_1) = -\sum_{\mathbf{G}_1} i(\mathbf{k}+\mathbf{G}_1)_x M_{1(\mathbf{G}_3-\mathbf{G}_1)} W_k(\mathbf{G}_1) \\ \quad + \sum_{\mathbf{G}_1} \left[(\mathbf{k}+\mathbf{G}_1)_x (\mathbf{k}+\mathbf{G}_3)_x M_3(\mathbf{G}_3 - \mathbf{G}_1) \right. \\ \quad \left. + (\mathbf{k}+\mathbf{G}_1)_y (\mathbf{k}+\mathbf{G}_3)_y M_{5(\mathbf{G}_3-\mathbf{G}_1)} + M_{1(\mathbf{G}_3-\mathbf{G}_1)} \right] \theta_{x\mathbf{k}}(\mathbf{G}_1) \\ \quad + \sum_{\mathbf{G}_1} \left[(\mathbf{k}+\mathbf{G}_1)_y (\mathbf{k}+\mathbf{G}_3)_x M_{4(\mathbf{G}_3-\mathbf{G}_1)} \right. \\ \quad \left. + (\mathbf{k}+\mathbf{G}_1)_x (\mathbf{k}+\mathbf{G}_3)_y M_{5(\mathbf{G}_3-\mathbf{G}_1)} \right] \theta_{y\mathbf{k}}(\mathbf{G}_1) \\ \omega^2 \sum_{\mathbf{G}_1} M_6(\mathbf{G}_3 - \mathbf{G}_1) \theta_{x\mathbf{k}}(\mathbf{G}_1) = -\sum_{\mathbf{G}_1} i(\mathbf{k}+\mathbf{G}_1)_y M_{1(\mathbf{G}_3-\mathbf{G}_1)} W_k(\mathbf{G}_1) \\ \quad + \sum_{\mathbf{G}_1} \left[(\mathbf{k}+\mathbf{G}_1)_x (\mathbf{k}+\mathbf{G}_3)_y M_{4(\mathbf{G}_3-\mathbf{G}_1)} \right. \\ \quad \left. + (\mathbf{k}+\mathbf{G}_1)_y (\mathbf{k}+\mathbf{G}_3)_x M_{5(\mathbf{G}_3-\mathbf{G}_1)} \right] \theta_{x\mathbf{k}}(\mathbf{G}_1) \\ \quad + \sum_{\mathbf{G}_1} \left[(\mathbf{k}+\mathbf{G}_1)_x (\mathbf{k}+\mathbf{G}_3)_x M_{5(\mathbf{G}_3-\mathbf{G}_1)} \right. \\ \quad \left. + (\mathbf{k}+\mathbf{G}_1)_y (\mathbf{k}+\mathbf{G}_3)_y M_{3(\mathbf{G}_3-\mathbf{G}_1)} + M_{1(\mathbf{G}_3-\mathbf{G}_1)} \right] \theta_{y\mathbf{k}}(\mathbf{G}_1) \end{cases} \tag{5}$$

Solving the eigen equation by sweptwing the wave vector k along the high-symmetry boundary of irreducible Brillouin zone shown in Fig. 2b, the relationship of $k - \omega$ and band structure can be obtained to demonstrate the band gap. The factors involved in above governing equation include the material parameters and structural parameters, and subsequent parameter analysis is carried out on these arguments.

2.3 Frequency Response Calculation

To investigate the shielding performance of LRTB, the frequency response analysis (FRA) is conducted using FEM, which has good convergence and high precision. As shown in Fig. 3, a prescribed acceleration is applied to the left excitation port of LRTB model in the y direction, and the displacement response is extracted on the receiving boundary, then the frequency response function can be expressed as:

$$T = 20 \log \frac{U_{\text{output}}}{U_{\text{input}}} \tag{6}$$

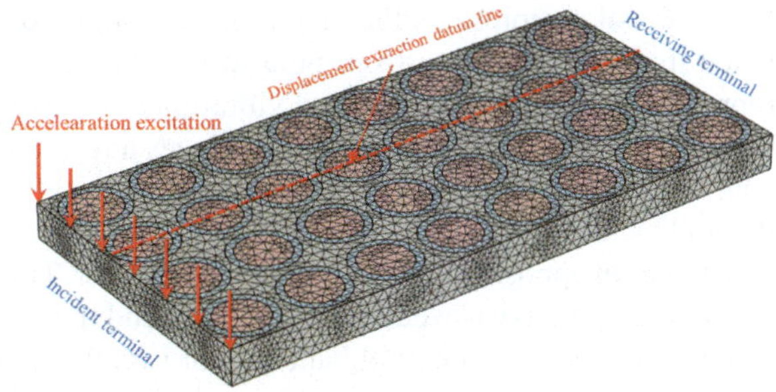

Fig. 3 Schematic representation of LRTB used for frequency response analysis

where T is frequency response function, dB; U_{input} and U_{output} are the acceleration amplitude of the receiving and excitation terminal, respectively.

During the subway operation, the track bed is mainly subjected by vertical flexural wave vibration [17], then the energy band calculation and transmission analysis are mainly carried out for the flexural waves.

3 Band Gap Property

3.1 Band Structure

The band structure of LRTB is represented in Fig. 4, and the band gap range is marked by blue shade. It can be seen that the flexural band gap range is 36.3–79.2 Hz, and the band gap width is 42.9 Hz. Furthermore, there is an obvious flat line in 65.4 Hz. This shows obvious local resonance characteristics, which is similar to the property of two-dimensional three-component PCs [18].

To explain the band gap mechanism, the displacement fields of the band gap boundaries (point A and B) shown in Fig. 4 is presented in Fig. 5. In the displacement

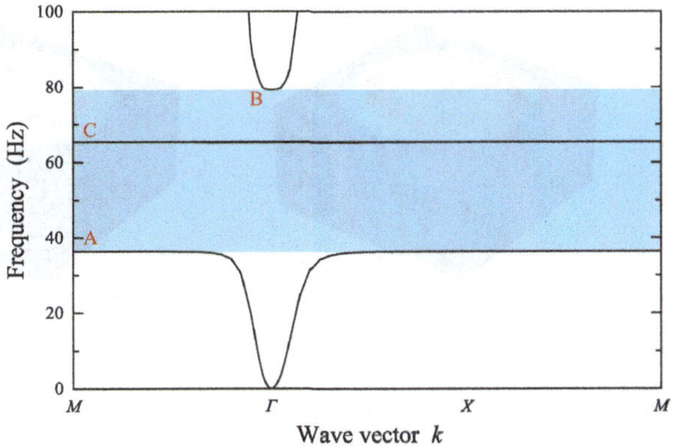

Fig. 4 The band structure of LRTB

field, the color represents the displacement amplitude, and the amplitude increases when the color changers from bule to red. In the starting frequency, the scatterer represents prominent translational vibration along z direction but the matrix remains stationary. The motion of scatterer produces a resultant force in z direction, and the coupling effect between the resultant force and traveling wave in the matrix opens the flexural band gap. For the ending frequency, the matrix maintains the larger vibration and moves in opposite directions with the scatterer. Then, the coupling effect between scatterer and matrix is weakened relatively and the band gap is closed. Therefore, it can be concluded the flexural band gap of LRTB depends on the coupling vibration between scatterer and matrix in the z direction, and the strength of the coupling effect determines the generation of band gap. For the flat line (point C) shown in Fig. 4, the scatterer makes a flip motion along z axis, but the matrix keeps stable. During the process, there is no resultant force generated in the z direction, then the coupling between the scatterer and matrix cannot be produced. Therefore, the correlative dispersion curve appears as a straight flat line across the entire Brillouin zone without creating the new band gap (Fig. 6).

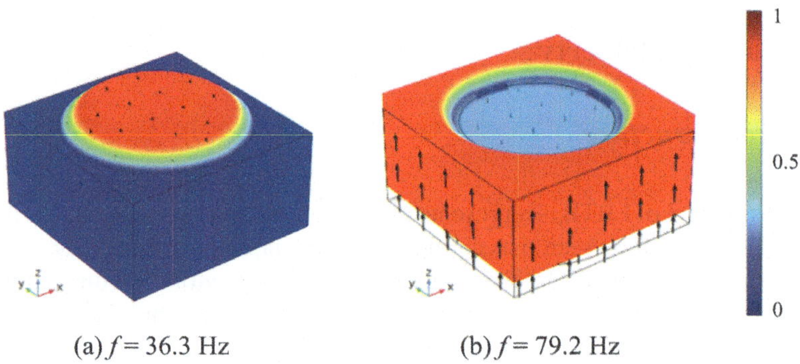

Fig. 5 The displacement field of band gap boundaries

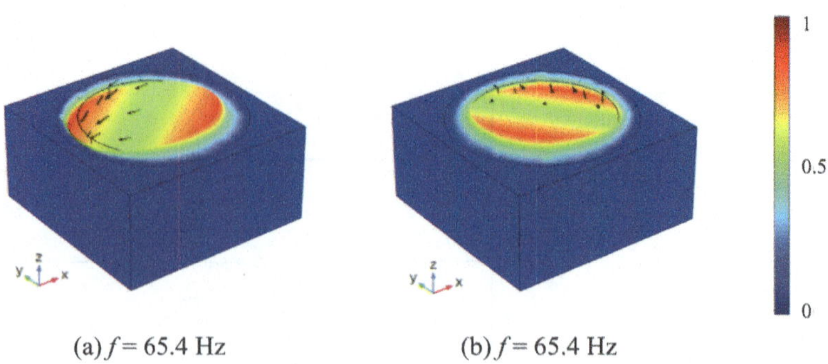

Fig. 6 The displacement field of flat line

3.2 Transmission Spectra

The transmission spectra of LRTB with different cell number are demonstrated in Fig. 7, they have similar development trends, that is, the transmission decreases rapidly at the starting frequency and then gradually rises as the frequency increases. Certainly, the band gap attenuation effect becomes more significant as the number increases, and LRTB exhibits more uniform effect in the entire frequency range when there are 10 cells. Generally, the representative size of single track bed plate is 4.6 m × 2.4 m × 0.26 m, which has sufficient space to arrange enough primitive cells for desired effect.

In order to analyze the attenuation efficiency in quantification, we integrate the red filled area of the transmission curve below the zero line in the band gap interval, as shown in Fig. 8. As the number of primitive cells increases, the total attenuation

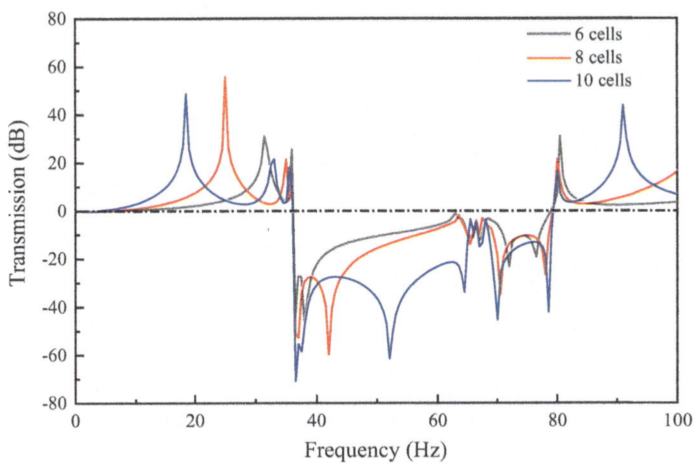

Fig. 7 The transmission spectra of LRTB

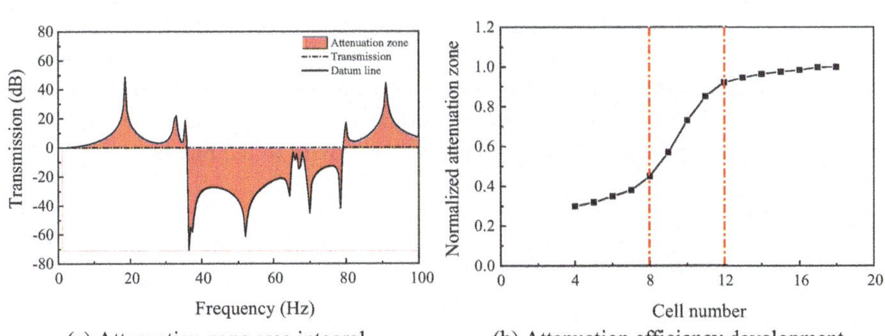

Fig. 8 The quantitative analysis of attenuation efficiency

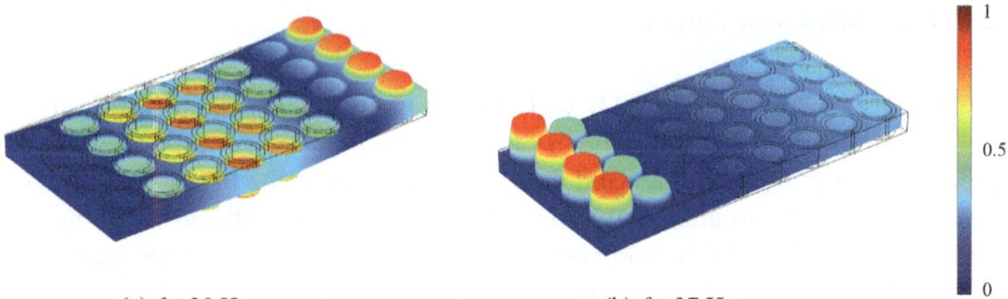

(a) $f = 30$ Hz (b) $f = 37$ Hz

Fig. 9 The kinetic energy distribution of LRTB

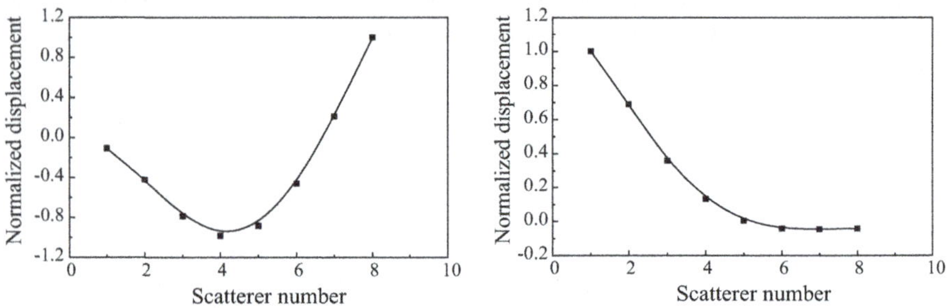

(a) $f = 30$ Hz (b) $f = 37$ Hz

Fig. 10 The displacement distribution of LRTB

area increases, that is, the attenuation efficiency gradually increases. After the cell number increased to 12, the growth rate of the attenuation area is basically stable. Therefore, the designer can reasonably choose the cell number to save costs in the actual project.

The kinetic energy distribution corresponding to the passband ($f = 30$ Hz) is shown in Fig. 9a, and the scatterer displacement along the extraction datum line shown in Fig. 3 is displayed in Fig. 10a. Obviously, the displacement amplitude of the receiving end is significantly greater than that of the excitation, which indicates there is no attenuation in passband. The difference is that the kinetic energy distribution and normalized displacement corresponding to band gap ($f = 37$ Hz) decreases exponentially in Figs. 9b and 10b, which demonstrates the elastic wave energy has been significantly attenuated along the propagation path in the band gap.

3.3 *Time Domain Analysis*

The measured vertical acceleration excitation of subway from the literature [19] shown in Fig. 11 is applied in the LRTB shown in Fig. 3 to conduct the time domain

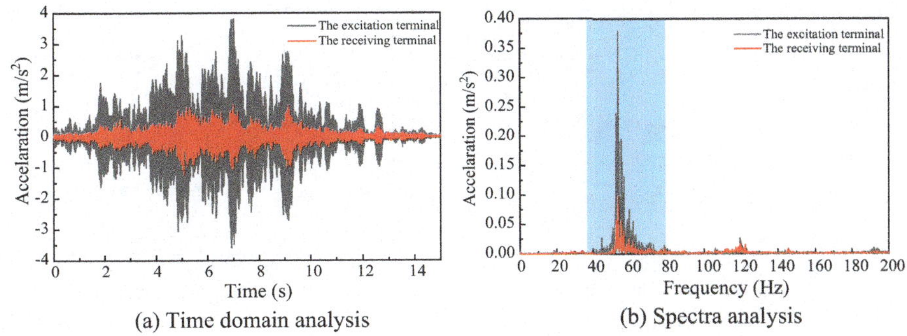

Fig. 11 The time transient and spectra analysis of LRTB under the measured subway vibration

analysis. Its frequency spectra are presented in Fig. 11b, the main frequency of subway is 40–80 Hz. The signal extracted on the receiving end indicate the vibration amplitude is significantly decreased, and the spectrum analysis manifests the results consistent with the band gap range.

4 LRTB Design and Application

The main frequency of the actual subway system is varied because of vehicle types, track and operation speed. Thus, it is necessary to design corresponding vibration reduction measures for specific vibration environments. The relevant parameters have been given in Sect. 2.2. Due to the space limitations, this section only lists limited structure parameters to discuss the influence law.

4.1 Parameter Analysis

After selecting the material combination, the band gap can be customized by adjusting the structure form. The development of band gap can be explained by equivalent spring-mass system [20], which can be expressed as:

$$f_{\text{start}} = \frac{1}{2\pi}\sqrt{K/m_1} \tag{7}$$

$$f_{\text{end}} = \frac{1}{2\pi}\sqrt{K/m_1+K/m_2} \tag{8}$$

where, f_{start} and f_{end} are starting and ending frequency of band gap, respectively. K is equivalent stiffness, which is provided by coating layer. m_1 and m_2 are scatterer and matrix equivalent mass, respectively.

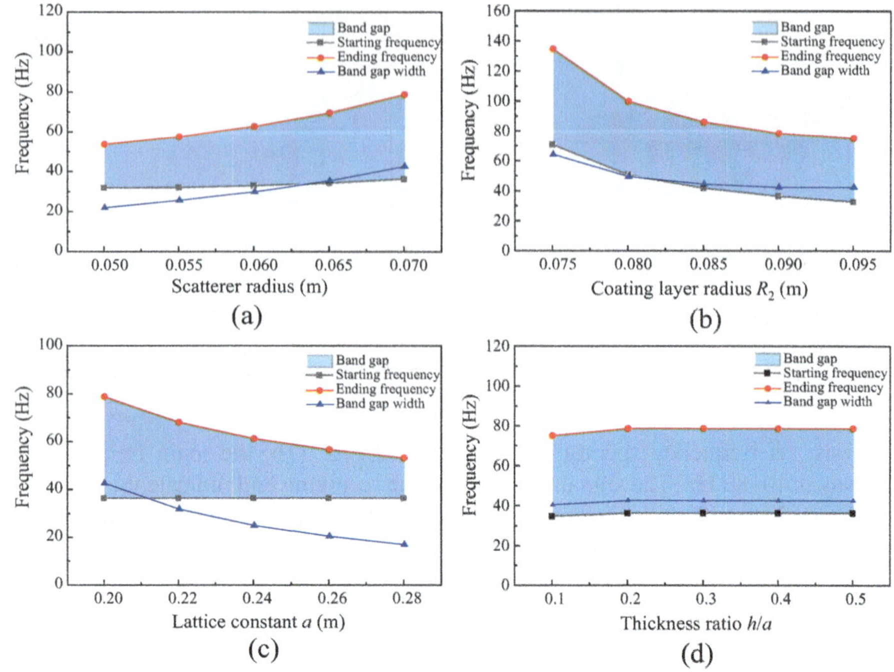

Fig. 12 The influence of structural parameters on the band gap

With the increases of scatterer radius R_1, the starting and ending frequency increase synchronously, and the band gap width increases accordingly. Instead, both the band gap boundary and the band gap width decrease as the coating layer radius R_2 increases, because the thicker coating layer provides a smaller equivalent stiffness K but the equivalent mass m_1 remain unchanged, which induces the lower band gap. Only increasing the lattice constant a, the initial frequency does not change but the band gap width decreases. The reason is equivalent stiffness K and equivalent mass m_1 remain unchanged, but the equivalent mass m_2 is decreased. It can be seen from Fig. 12d that the band gap is not sensitive to plate thickness, thus the thickness of LRTB can be appropriately reduced to save costs in actual application.

4.2 Design Flow Chart

To customize the attenuation structure for specific engineering conditions, the design flow chart is presented in Fig. 13. First of all, the designer needs to perform frequency spectrum analysis of the target environment vibration to determine the main frequency and the desired vibration reduction effect. After selecting materials and structural parameters, the band structure calculation can be performed for

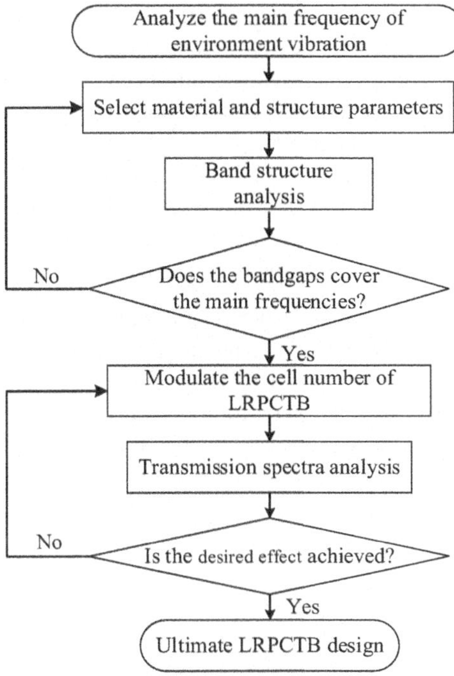

Fig. 13 The design flow chart of LRTB

LRTB. When the attenuation interval meets the main frequency, the primitive cells is adjusted to achieve the attenuation efficiency target, and the ultimate LRTB design is obtained.

Furthermore, the structure of LRTB is easy to realize, and its main materials are commonly used in engineering. Therefore, it is easier for engineers to prepare mold and manufacture LRTB, which can be customized for specific engineering environments to present superior vibration isolation performance.

5 Conclusions

In conclusion, this paper proposed LRTB by introducing locally resonant PCs in track bed. The band structure is calculated by IPWE method to obtain the band gap range, and the FEM is applied to achieve the transmission spectra to represent the attenuation efficiency. Besides, the parameter analysis is carried out to discuss band gap development in detail. The major conclusions are as follows:

(1) The proposed LRTB is able to obtain low-frequency broad band gap, which is proved by time domain analysis of measured subway vibration. Mechanism analysis indicates that the band gap generation depends on the coupling effect between scatterer and matrix in the z direction.

(2) Transmission spectra analysis demonstrates that the frequency within the band gap can be significantly attenuated, and the attenuation efficiency is gradually significant as the number of primitive cells increases, the most economical cell number can be obtained by attenuation quantification analysis to save engineering costs.
(3) The proposed LRTB is easy to construct, and the attenuation effect can be customized by regulating reasonably the geometric parameters. The corresponding design flow chart has been provided to facilitate the application in specific subway engineering.

This article theoretically analyzes the band gap characteristics of LRTB, and the band gap mechanism and tunability law are obtained to guide the design optimization. Furthermore, the LRTB specimen can be prepared and subjected to shaker tests to verify its real attenuation performance. Considering the engineering application, the compressive strength of LRTB needs to meet the requirements of the subway, which is an unavoidable challenge from the laboratory to the field application.

References

1. Zou C, Wang YM, Wang P, Guo JX (2015) Measurement of ground and nearby building vibration and noise induced by trains in a metro depot. Sci Total Environ 536:761–773
2. Vogiatzis K, Mouzakis H (2017) Ground-borne noise and vibration transmitted from subway networks to multi-storey reinforced concrete buildings. Transport 33(2):446–453
3. Zhu SY, Wang JW, Cai CB, Wang KY, Zhai WM (2017) Development of a vibration attenuation track at low frequencies for urban rail transit. Comput-Aided Civ Inf 32:713–726
4. Zhang XA, Zhai WM, Chen ZW, Yang JJ (2018) Characteristic and mechanism of structural acoustic radiation for box girder bridge in urban rail transit. Sci Total Environ 627:1303–1314
5. Ma XF, Cao MY, Gu XQ, Zhang BM, Yang ZH, Guan PF (2020) Vibration-isolation performance of a pile barrier in an area of soft soil in shanghai. Shock Vib 2020:8813476
6. Adam M, von Estorff O (2005) Reduction of train-induced building vibrations by using open and filled trenches. Comput Struct 83:11–24
7. Çelebi E, Fırat S, Beyhan G, Çankaya İ, Vural İ, Kırtel O (2009) Field experiments on wave propagation and vibration isolation by using wave barriers. Soil Dyn Earthq Eng 29(5):824–833
8. Ekanayake SD, Liyanapathirana DS, Leo CJ (2014) Attenuation of ground vibrations using in-filled wave barriers. Soil Dyn Earthq Eng 67:290–300
9. Alzawi A, Hesham El Naggar M (2011) Full scale experimental study on vibration scattering using open and in-filled (GeoFoam) wave barriers. Soil Dyn Earthq Eng 31:306–317
10. Tsai PH, Feng ZY, Jen TL (2008) Three-dimensional analysis of the screening effectiveness of hollow pile barriers for foundation-induced vertical vibration. Comput Geotech 35:489–499
11. Xia TD, Sun MM, Chen C, Chen WY, Ping X (2011) Analysis on multiple scattering by an arbitrary configuration of piles as barriers for vibration isolation. Soil Dyn Earthq Eng 31(3):535–545
12. Huang XD, Li P, Luo XW, Ling L, Shuaibu AA, Zeng ZP (2020) Performance analysis of prefabricated steel-spring floating-slab track and its application to urban express rail transit. Adv Civ Eng 2020:4515319
13. Shi L, Ye XQ, Zhang PF, Guo L (2019) A comparative study on vertical dynamic responses of three types of elevated railway tracks subjected to a moving train. Math Probl Eng 2019:3290958

14. Kushwaha MS, Halevi P, Dobrzynski L, Djafari-Rouhani B (1993) Acoustic band structure of periodic elastic composites. Phys Rev Lett 71(13):2022–2025
15. Sigalas MM, Economou EN (1992) Elastic and acoustic wave band structure. J Sound Vib 158:377–382
16. Liu ZY, Zhang XX, Mao YW, Zhu YY, Yang ZY, Chan CT, Sheng P (2000) Locally resonant sonic materials. Science 289(5485):1734–1736
17. Aggestam E, Nielsen JCO (2020) Simulation of vertical dynamic vehicle-track interaction using a three-dimensional slab track model. Eng Struct 222:110972
18. Lei LJ, Miao LC, Li C, Liang XD, Wang JJ (2021) Locally resonant periodic wave barriers for vibration isolation in subway engineering. KSCE J Civ Eng 25(4):1239–1251
19. Gao M, Shi ZF (2019) A wave guided barrier to isolate antiplane elastic waves. J Sound Vib 443:155–166
20. Wang G, Shao LH, Liu YZ, Wen JH (2006) Accurate evaluation of lowest band gaps in ternary locally resonant phononic crystals. Chinese Phys 15(8):1843–1848

Vibration Analysis on a Railway Bridge Structure

R. Kusumawardani, N. Qudus, A. B. Sabba, N. C. Nugroho, N. H. A. Hamid, S. A. Kudus, A. Alisibramulisi, R. Hassan, A. Kusbiantoro, N. M. Ermansyah, and P. P. Pangestu

Abstract The railway bridge would be vibrating when a series of train passed on it. The movement a train triggers a noise to the bridge structure which is generated by the combination of small-scales undulation between wheel of train, railway and bridge railway contact surfaces. Nowadays, railway researches focused on dynamic behaviour of railway track when the train passed on it. This study discusses the dynamic behaviour of railway bridges vibrations caused by train traffic load in terms of three axis direction (x, y and z direction). A field investigation was conducted by applied a series of accelerometers to find out the value of frequency natural and amplitude. The tools are placed on the beam at a half span and at a third span. The results were compared between passengers train and carriage train. During unloaded condition, the range of frequency natural is 0.78–3.73 Hz and amplitude is in between 1.82 and 1.89. When the train passes by, the range of frequency natural is in between 1.03 and 5.77 Hz and amplitude 1.70–4.

Keywords Vibration analysis · Railway bridge · HVSR

1 Introduction

The occurrence of earth noise ambient could be triggering through many ways by several independent sources. Since it caused the noise to environmental then probably could affect the frequency band of natural frequency itself. The oceanic and large-scale meteorological could generated the frequencies below 1 Hz. For frequency above 1 Hz are dominated by traffic vibration and environmental natural sources [1].

By using the ambient noise with frequency band below 1 Hz could help to get better understanding about relationship between the earth, oceans and atmosphere.

R. Kusumawardani (✉) · N. Qudus · A. B. Sabba · N. C. Nugroho · A. Kusbiantoro · N. M. Ermansyah · P. P. Pangestu
Department of Civil Engineering, Universitas Negeri Semarang, Semarang, Indonesia
e-mail: rini.kusumawardani@mail.unnes.ac.id

N. H. A. Hamid · S. A. Kudus · A. Alisibramulisi · R. Hassan
Faculty of Civil Engineering, Universiti Teknologi MARA, Shah Alam, Malaysia

T. Kang and Y. Lee (eds.), *Proceedings of 2021 4th International Conference on Civil Engineering and Architecture*, Lecture Notes in Civil Engineering 201,
https://doi.org/10.1007/978-981-16-6932-3_6

Normally, the range of ambient noise frequency is in between 1 and 20 Hz and it commonly known by microtremor analysis. This method is quite powerful in order to provide a low-cost and non-invasive exploration when the geotechnical information data is difficult to obtain [2]. Many studies were applied of microtremor measurements results to determine several parameter vibrations such as frequency of fundamental resonant, unconsolidated shallow sediment thickness and shear wave velocity. This parameter could be used to predict the ground motion local amplification during the earthquake and could be said it has an important role for seismic hazard assessment [3, 4]. A seismic microzonation maps could be built by referring to this parameter. Nowdays, the ground motion amplification could be reach by comparing the observed data during earthquake and amplification pattern modelled by microtremor measurements and this method widely used by many researchers in the world [5–8].

The responses of the railway bridge when the train passed on it is one of crucial aspect in the point of view bridge structure railway capacity support the dynamic loading. Lorieux [9] revealed the train speed induced the dynamic behavior of bridge particularly the amplitude and frequency. They are having a positive correlation; the increasing of train speed will be increase the amplitude of dynamic behavior of a railway bridge and rails. Rigid railway bridge and tracks have non-elastic behaviour, therefore when vibrations occur, the ballast particle displacement is often irreversible and the accumulation of these minor displacements results in settlements and degradation of the track itself. The complexity of the structural response of train load and speed based on the interaction of parts of the track-railway bridge system which is consist of rails, sleepers, ballast, girders and foundations. Each elements of the bridge has its own behavior with different characteristics. Due to the non-linear differences of each element, it is very difficult to predict the distribution of stresses and vibrations of each element. The aims of this article to revealed the dynamics behavior of railway bridge by using microtremor methods to understand the dynamic properties of railway bridge.

2 Materials and Methods

In this work is using the HSVR (Horizontal to Vertical Spectral Ratio) which could be used as indicator the subsurface structure which shows the relationship between the Fourier spectrum ratios of the horizontal component microtemor signal to its vertical component. This method utilized three direction components which consist of two horizontal components and one vertical component. The results of the HSVR method are natural frequency and amplification [10] could be used to determine the maximum value of vibration acceleration, amplitude and natural frequency.

The type of railway bridges is closed frame bridge which could be seen in Fig. 1. A railway bridge in Semarang city is chosen as appropriate model for this study. This railway bridge is fully composed by steel material and constructed over the river which cross cut the east of Semarang city, Central Java province, Indonesia.

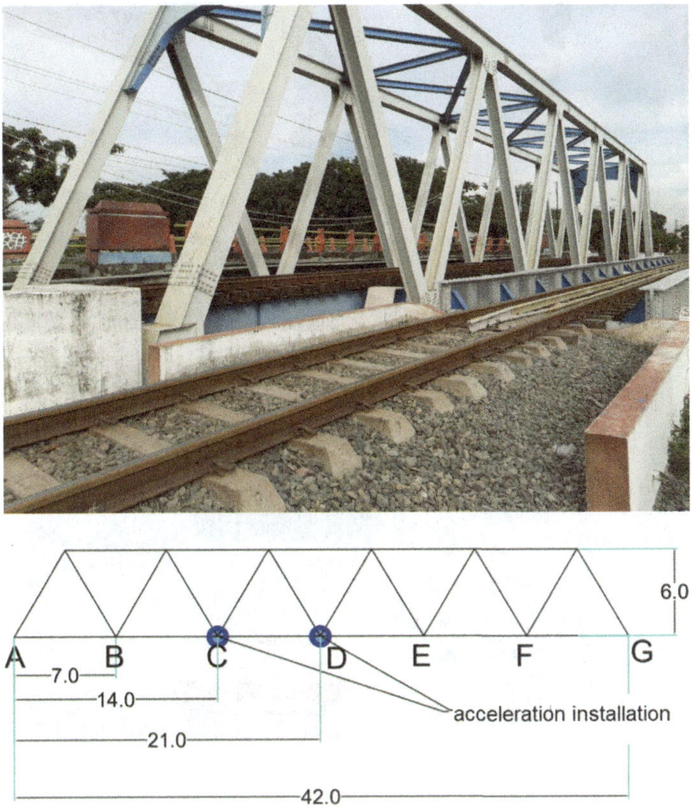

Fig. 1 Railway bridge and acceleration installation

The measurements length and width of railway bridge during the field investigation are 42 m and 5.13 m respectively and the support is identified as roll at both side.

A set of sensor was installed in several parts of the span bridge frame which could be seen in Fig. 2. The vibration data due to passing train known as a microtremor. The sensor collected the data of frequency and time during unloaded bridge and when the train passed. Data collections were conducted in two techniques, firstly, the data was taken when the bridge was passed by train and secondly, the data was collected when the bridge was unloaded state. The two sensors were located in the mid of span and the third of span. The total of data collection was 12 on which 7 were taken when the train passing by and the other when the bridge was unloaded. Data collection was performed by applying by sensor and using GEOPSY software to analyse signals from the acceleration data recording [11].

Fig. 2 Sensor installation and a set of measurements

3 Results and Conclusion

According to the technical field and the data analysis using Geopsy program, it obtained natural frequencies of the bridge, vibration maximum, vertical displacement, and amplitude. Further, the data of result analysis obtained from the program was presented in the form of figure and table and then analyzed.

3.1 Train Characteristic as a Load

The natural frequency of railway was obtained from HVSR curve for the data of the bridge taken while train passing by with the assumption that loads passing it were at maximum. Figure 3 illustrated HVSR curve of the bridge while Agro Anggrek train passed. There were two HVSR curves obtained during field investigation, while the train passed and while the bridge was unloaded. Based on Fig. 3, it was obtained that the natural frequency value of bridge was 5.5 Hz.

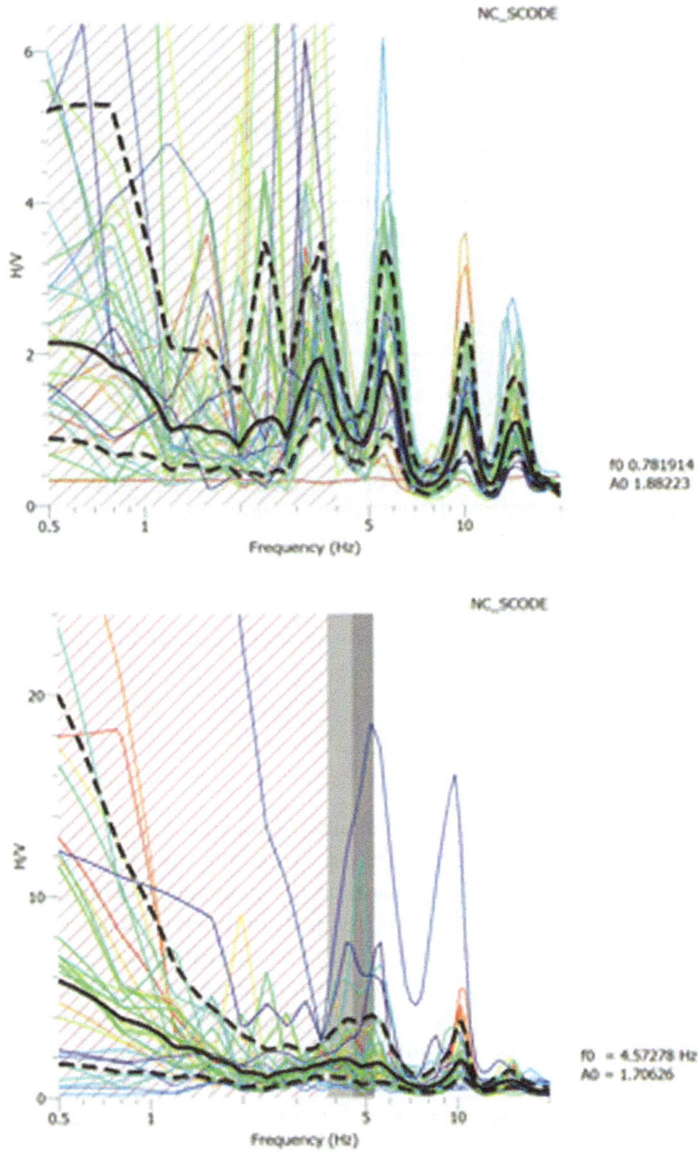

Fig. 3 H/V graphic when unloaded condition (**a**) and Maharani train passed (**b**)

3.2 *The Natural Frequency*

The natural frequency of railway was obtained from HVSR curve for the data of the bridge taken while train passing by with the assumption that loads passing it were at maximum. When the railway in the condition of unloaded, the measurement recorded

Table 1 Measurement result

Train name	Weight (tons)	Acceleration (g)			Velocity (km/h)	f (Hz)	A
		X	Y	Z			
No train		0.003751	0.00145	0.00385		0.78	1.88
Maharani	902	0.57	0.827	1.896	72.42	4.572	1.706
Freight train	2250	0.754	**1.017**	1.787	51.5	1.03	2.6
Kedung Sepur	264	0.345	0.427	0.838	56.33	5.66	3.88
Argo Anggrek	612	**0.815**	0.724	**1.903**	78.86	5.5	4.14

For the maximum value for each direction is indicated by bold number

the natural frequency and amplitude are 0.78 Hz and 1.88. Figure 3 illustrated the comparison of HVSR curve when the bridge conditions are unloaded and with a train passed on it. Based on Fig. 3, it was obtained that the natural frequency value of bridge was 5.5 Hz. Moreover, we could be seen the curve while the bridge is unloaded, the initial frequency natural (f_n) is 0.78 Hz. The detail of natural frequency (f) and amplification (A) of each train in the study site could be seen in Table 1.

3.3 Acceleration and Velocity

The data related to acceleration was obtained by using sensor equipment which connected to a seismic monitoring recorder. Thus, the raw data from sensor was processed by Geopsy software to determine the parameters of natural frequency, amplification and frequency with two difference condition (unloaded and loaded by dynamic load of train passed). When the bridge structure does not have loads passing on it, the vibrations that occur come from earth/microseismic vibrations, vibrations due to wind and surroundings. The maximum acceleration of vibration when there is no train passing is very small, namely 0.003751 g in the x direction (east–west), 0.001450 g in the y direction (north–south) and 0.00835 g in the z direction. The natural frequency and an amplitude value is 0.78 Hz and 1.88, respectively.

While the Maharani train passes at a speed of 72.42 km/h, it causes a maximum acceleration of 0.570 g in the x direction, 0.827 g in the y direction and 1.896 g in the z direction, with a natural frequency value of 4.572 Hz and an amplification of 1.706. At moment the freight train get through at a speed of 51.50 km/h causes vibrations with a maximum acceleration of 0.754 g in the x direction, 1.017 g in the y direction and 1.787 g in the z direction.

The natural frequency is 1.03 Hz and the amplification is 2.60. In time the Kedung Sepur passenger train passing at a speed of 56.33 km/h causes vibrations with a maximum acceleration of 0.345 g in the x direction, 0.427 in the y direction and

0.838 in the z direction, has the natural frequency 5.66 Hz and the amplification 3.88. Furthermore the Argo Anggrek passenger train passes with the highest speed of 78.86 km/h causing vibrations with a maximum acceleration of 0.815 g in the x direction, 0.724 y direction and 1.903 z direction, possesses the natural frequency 5.5 Hz and the amplification 4.14.

3.4 Conclusion

Based on the research results, the acceleration of vibration is influenced by the load and speed of the train. The highest vibration acceleration in Y direction occurred when the freight train with a load of 2250 tons and with a speed of 51.5 km/h passed, which was 0.754 m/s^2 in the x direction, 1.017 m/s^2 in the y direction and 1.787 m/s^2 in the z direction. Meanwhile, the highest vibration acceleration in X and Z direction occurred when the Argo Anggrek train with a load of 612 tons and with a speed of 75.64 km/h passed, which was 1.35 m/s^2 in the x direction, 1.63 m/s^2 in the y direction and 1.31 m/s^2 in the z direction.

From data processing using Geopsy software, the natural frequency of the structure is 0.78 Hz and an amplitude of 1.88. Meanwhile, the value of the natural frequency when the train passes is 1.03–5.77 Hz and the amplitude is 1.70–4.14. The greatest frequency value occurs when the Argo Anggrek train passes at a speed of 78.86 km/h which produces a natural frequency of 5.50 Hz and an amplitude of 4.14.

For further research, data collection samples should be reproduced so that more accurate results are representative of the bridge conditions. Especially if you want to get the bridge shape mode data, which is really needed if you want to do a finite update of the model.

Acknowledgements This research was funded by Joint-Research scheme between Universitas Negeri Semarang (Indonesia) and Universiti Teknologi MARA (Malaysia) fiscal year 2020–2021.

References

1. Kusumawardani R, Nugroho U, Fansuri MH, Mindiastiwi T, Yuniarti W, Hilmi AS (2018) The impact of vehicle load inducing vibrations on the subgrade soil particle acceleration. J Eng Sci Technol 13(6):1440–1450
2. Kusumawardani R, Suryolelono KB, Suhendro B, Rifa'i A (2016) The dynamic response of unsaturated clean sand at a very low frequency. Int J Technol 7(1):123–131. https://doi.org/10.14716/ijtech.v7i1.1163
3. Hidayat S, Warnana DD, Koesuma S, Cari C (2017) Local site effects evaluation using microtremor measurements at north side of Pandan Mountain. J Phys Theor Appl 1(2):89–96. https://doi.org/10.20961/jphystheor-appl.v1i2.19120
4. Stanko D, Markušić S, Gazdek M, Sanković V, Slukan I, Ivančić I (2019) Assessment of the seismic site amplification in the city of Ivanec (NW part of Croatia) using the microtremor

HVSR method and equivalent-linear site response analysis. Geosciences 9(7):312. https://doi.org/10.3390/geosciences9070312
5. Nur AD, Citra DSP, Rini TL, Dwa DW (2016) Earthquake microzonation and VS, 30 mapping based on microtremor measurement (case study in Kaliwates and Sumbersari sub-district, Jember Regency). Procedia Soc Behav Sci 227:354–360
6. Rezaei S, Choobbasti AJ (2017) Application of the microtremor measurements to a site effect study. Earthq Sci 30:157–164. https://doi.org/10.1007/s11589-017-0187-2
7. Kusumawardani R, Nugroho U, Handayani S, Fananda MA (2019) The analysis of liquefaction phenomenon of the flexible pavement using seismic monitoring equipment. IIUM Eng J 20(1):70–78. https://doi.org/10.31436/iiumej.v20i1.1031
8. Kusumawardani R (2017) Investigation of subgrade particles acceleration due to dynamic loading. Aceh Int J Sci Technol 6(3):97–103. https://doi.org/10.13170/aijst.6.3.8427
9. Lorieux L (2008) Analysisi of train-induced vibrations on a single-span composite bridge. Royal Institute of Technology (KTH), Sweden
10. Sunkgono WEJ (2013) Estimasi Indeks Kerentanan Tanah menggunakan Metode HVSR (Horizontal to vertical spectral ratio) (in Indonesia language). J Sains Seni Pomits 2(1):2337–3520
11. Kusumawardani R, Zelin MA, Kusbiantoro A (2019) Ground vibration analysis of railroad dynamic loads on rail structure. J Teknik Sipil dan Perencanaan 21(2):62–70. https://doi.org/10.15294/jtsp.v21i2.20957

Investigation on Effective Charging Time of Electro-Osmotic Consolidation for Marine Clayey Soils

Lei Zhang, Wenbo Zhang, Binghui Wang, and Chen Fang

Abstract This paper presented an investigation on the effective charging time of electro-osmotic consolidation (EC) for marine clayey soils. According to the previous experimental data, the theoretical analysis and the relevant formulae are deduced for the soil parameters, including water discharge, water discharge rate, normalized current, effective charging time, and so forth. Based on the different experiments, the water discharge-time curves can be approximated as a polynomial curve. The water discharge rate is defined as the water discharge in a unit time, it characterized the drainage velocity during the electro-osmosis process. The normalized current $I_0(t)$ which is defined as the ratio of the current at a certain time to the initial current is employed to analyze the termination time and current. The analysis results provide the reasonable theories for the related researches and also demonstrate the feasibility of EC for the clayey soils.

Keywords Marine clayey soil · Elctro-osmosis consolidation · Water discharge · Water discharge rate · Normalized current · Effective charging time

1 Introduction

Electro-osmotic consolidation (EC) is a useful method for improving soil strength, it is a consolidation treatment on the soft soils owing to the mechanism of dewatering from the anode to the cathode caused by an applied direct current. The dewatering

L. Zhang (✉) · W. Zhang · B. Wang
School of Architecture and Civil Engineering, Jiangsu University of Science and Technology, Zhenjiang 212100, China
e-mail: lei.zhang@just.edu.cn

B. Wang
e-mail: jkdwbh@just.edu.cn

L. Zhang
Institute of Geotechnical Engineering, Nanjing Tech University, Nanjing 211816, China

C. Fang
Department of Civil Engineering, University of Nebraska-Lincoln, Lincoln, NE 68588, USA

T. Kang and Y. Lee (eds.), *Proceedings of 2021 4th International Conference on Civil Engineering and Architecture*, Lecture Notes in Civil Engineering 201,
https://doi.org/10.1007/978-981-16-6932-3_7

process results in the consolidation and the increase in the mechanical strength of the clayey soil. Meanwhile, EC can promote the aggregation of soil particles and produce the cementation of soil. The strength of the soil receives a substantial improvement to enhance the efficiency of EC method.

Electro-osmosis was proposed and reported by Reuss in 1809, and Casagrande first utilized EC for the clayey soil foundation. Many researchers have investigated the electro-osmosis and demonstrated its feasibility using various theoretical analysis, laboratory experiments, numerical simulation, and field studies [1–9]. Although many scholars have proven the effectiveness of EC, but the serious erosion of electrodes imposes restrictions on the application of the EC method to the foundation projects [10, 11]. Many types of research have been focused on the theory of EC since the electro-osmosis have been proposed [4, 6, 12, 13]. Casagrande [2] developed a formula of the relation between the electro-osmosis rate and the electric field intensity. Based on the theories, Esrig [14] presented one-dimensional consolidation theory, and also described a definition equation of the pore pressure of the soil during the electro-osmosis process. Burnotte [15] used EC method to strengthen a soft soil foundation in Montreal, Canada. The ground settlement of the consolidated foundation was 0.5 m and a substantial improvement of the strength was obtained for the foundation. Rittirong and Shang [16] built a two-dimensional EC model to analyze this model using the finite difference method. Yuan and Hicks [17, 18] conducted the finite-element-analysis on the EC with the large displacements.

Many types of research on EC have been reported, however, few studies on the theoretical equation of the method are performed. Based on the results of related researches [19, 20], this paper proposed the theoretical formula of the water discharge, discharge rate, current, energy consumption, and so forth. Then we received the effective charging time during EC on the basis of the formula of the water discharge. To validate the feasibility of the proposed formula, a series of laboratory experiments were conducted using self-designed devices. In the experiment, several soil parameters including the water discharge, discharge rate, current, normalized current, and energy consumption were considered to analyze the effective charging time during EC process. The experimental results demonstrated that the theoretical results matched well with the experimental results. And the results of this research also provide some references for the application of EC for marine clayey soils in future engineering projects.

2 Theoretical Analysis

The reaction mechanism and process during the electro-osmosis are complicated. Few researches had been performed on the theoretical analysis of marine clayey soils during EC method. In this paper, several basic soil parameters in terms of the water discharge, discharge rate, current, normalized current and energy consumption were analyzed to obtain the theoretical formulae. These formulae presented the effects of these parameters during the electro-osmosis process.

2.1 Water Discharge

The water discharge is defined as the amount of the water discharged from the soils, and the unit is mL. Based on the different experiments, the water discharge-time curves can be approximated as a polynomial curve. Hence, the water discharge in this paper is assumed as a polynomial equation related to the charging time (t), shown in Eq. (1).

$$Q_j = m_{nj}t^n + m_{(n-1)j}t^{(n-1)} + \cdots + m_{2j}t^2 + m_{1j}t + c_j$$
$$= \sum_{n=1}^{n\to\infty} m_{nj}t^n + c_j = (m_{1j}\ m_{2j}\ \cdots\ m_{(n-1)j}\ m_{nj}) \begin{pmatrix} t \\ t^2 \\ \vdots \\ t^{(n-1)} \\ t^n \end{pmatrix} + c_j \quad (1)$$

In the equation, though the polynomial fitting, the maximum value of $m_{nj} \sim m_{3j}$ was 10-5 which were much less than m_{1j} and m_{2j}, so $m_{nj} \sim m_{3j}$ were defined as zero in the equation. Therefore, the fitting equation of the water discharge is formulated as a quadratic polynomial equation related to the water discharge, as shown in Eq. (2). While Q_j reached the maximum value, the corresponding tj was the effective charging time.

$$Q_j = m_{2j}t_j^2 + m_{1j}t_j + c_j \quad (2)$$

where Q_j is the water discharge, and j is the number of the test group. m_{2j}, m_{1j} are the coefficients, and $m_{1j} < 0$. t_j is the charging time. c_j is a constant.

According to the quadratic polynomial equation, the corresponding maximum value of the water discharge Q_{jmax} and the effective charging time t_{ect} can be obtained, as shown in Eqs. (3) and (4).

$$Q_{j\max} = \frac{4m_{2j} \cdot c_j - m_{1j}^2}{4m_{2j}} \quad (3)$$

$$t_{ect} = -\frac{m_{1j}}{2m_{2j}} \quad (4)$$

2.2 Water Discharge Rate

The water discharge rate (mL/h) is defined as the water discharge in a unit time, it characterized the drainage velocity during the electro-osmosis process, which is formulated as Eq. (5).

$$v = \frac{dQ}{dt} \tag{5}$$

The formula of the discharge rate was obtained inserting Eq. (1) into Eq. (5), as shown in Eq. (6).

$$\begin{aligned} v_j &= nm_{nj}t_j^{(n-1)} + (n-1)m_{(n-1)j}t_j^{(n-2)} + \cdots + 2m_{2j}t_j + m_{1j} \\ &= \begin{pmatrix} n & n-1 & \cdots & 2 & 1 \end{pmatrix} \begin{bmatrix} m_{nj} & 0 & 0 & 0 & 0 \\ 0 & m_{(n-1)j} & 0 & 0 & 0 \\ \vdots & \vdots & \ddots & \vdots & \vdots \\ 0 & 0 & 0 & m_{2j} & 0 \\ 0 & 0 & 0 & 0 & m_{1j} \end{bmatrix} \begin{pmatrix} t_j^{(n-1)} \\ t_j^{(n-2)} \\ \vdots \\ t_j \\ 1 \end{pmatrix} \end{aligned} \tag{6}$$

where v_j is the fitted discharge rate.

With the polynomial fitting, $m_{nj} \sim m_{3j}$ were determined as zero in Eq. (6) which were same as the definition in Eq. (1). A simplified equation of the discharge rate is formulated as Eq. (7).

$$v_j = 2m_{2j}t_j + m_{1j} + q(t_j) \tag{7}$$

where $q(t_j)$ is a coefficient to be determined with the relation of the charging time.

When the discharge rate is set as zero, the increment of the water discharge is obtained as zero and the accumulated water discharge does not increase. The termination charging time is defined as the time at which the discharge rate is zero. The termination charging time is formulated as Eq. (8).

$$t_{vj} = -\frac{m_{1j} + q(t_j)}{2m_{2j}} \tag{8}$$

where t_{vj} is the termination charging time. Owing to the complication of the EC process, the undetermined coefficient $q(t_j)$ is not considered for the equation and assumed as $q(t_j) = 0$. While the undetermined coefficient $q(t_j)$ is assumed as zero, $t_{vj} = -m_{1j}/2m_{2j}$, the value of the termination charging time t_{vj} is equal to the effective charging time t_{ect}.

2.3 Current and Normalized Current

Since the charging time in the project is much different from the laboratory experiment, the current corresponding to the effective charging time is used to determine the termination time. It is difficult to determine the currents which are corresponding to the effective charging time. The normalized current is employed to analyze the termination time and current. The normalized current $I_0(t)$ is defined as the ratio of the current at a certain time to the initial current, formulated as Eq. (9).

$$I_0(t) = \frac{I(t)}{I_0} \tag{9}$$

where $I(t)$ is current at a certain time, and I_0 is the initial current.

The normalized current–time curve is assumed as a straight line, and hence the linear fitting was used to obtain the normalized current–time curve, as shown in Eqs. (10) and (11).

$$I_0(t_j) = a_j t_{Ij} + b_j \tag{10}$$

And

$$t_{Ij} = \frac{I_0(t_j) - b_j}{a_j} \tag{11}$$

where $I_0(t_j)$ is the normalized current. a_j, b_j are the related coefficients of the normalized current. t_{Ij} is the charging time.

The normalized current was set as zero in Eq. (11), and also the current at this time was zero based on Eq. (9). The charging time obtained by Eq. (11) is the time to terminate the test, and the value of the charging time was $-b_j/a_j$. The value of charging time based on Eq. (11) might be equal to the value based on Eq. (8), but it is larger than the effective charging time, namely, $t_{Ij} > t_{ect}$. Since the water discharge stopped increasing, some water molecules can be found in the soils and the current did not decrease to zero. If the time corresponding to the current decrease to zero is used as the termination time, the energy consumption will be too large, meanwhile, the water discharge will not increase significantly. Therefore, it is a better choice to use the effective charging time to terminate the test. But the effective charging time in the laboratory and the field is not easy to measure, and hence the reference current and the normalized current are used to determine the termination time of the test.

The current in which the water discharge does not increase, or the water discharge has reached the maximum value calculated on the basis of Eq. (3) is assumed as the current corresponding to the effective charging time. The range of the normalized current corresponding to the effective charging time is formulated as Eq. (12).

$$\varepsilon_j \geq \frac{I(t_{ect})}{I_0} \tag{12}$$

where ε_j and $I(t_{ect})$ is the normalized current and current at the certain time t_{ect}, respectively. According to Eq. (12), the real current in the test was received as $I(t_{ect}) \leq \varepsilon_j \cdot I_0$, and the range of the real current corresponding to the effective charging time was determined.

2.4 Undetermined Coefficient $q(t_j)$

Through the theoretical analysis and formula derivation of the water discharge rate, current and normalized current, in the process of determining the effective charging time, the undetermined coefficient $q(t_j)$ is assumed as zero. However, the $q(t_j)$ cannot be ignored as zero. The $q(t_j)$ is important to determine the effective charging time and its corresponding current. According to the analysis on the water discharge rate and the normalized current in Eqs. (4) and (8), the coefficient $q(t_j)$ is formulated as Eq. (13).

$$q(t_j) = 2m_{2j} \cdot (t_{ect} - t_{vj}) \tag{13}$$

With Eq. (13), the value of the undetermined coefficient $q(t_j)$ can be received for the research.

3 Conclusions

The paper presented the theoretical analysis on the water discharge, water discharge rate, current, normalized current, and also proposed equations for these parameters. Based on the laboratory tests [19, 20], the law of dewatering, effective charging time were investigated, and several conclusions were developed.

(1) The theoretical equations in terms of the water discharge, water discharge rate, current, normalized current, and energy consumption were proposed. The results obtained using these equations were validated against the experimental results to demonstrate the feasibility and accuracy of these equations.
(2) The formula of effective charging time was developed. According to the formula, the calculated charging time was determined and verified with the experimental time. The formula was proved to be accurate without considering the effect of the coefficient $q(t_j)$. The fitted and experimental water discharge rate were compared and investigated to determine the range of the effective charging time.
(3) With the same electrode spacing, the water discharge and water discharge rate were proportional to the voltage. The water content had a significant effect on the dewatering during EC. Under the identical voltage, the water discharge with the small electrode spacing was larger than that with the large electrode spacing at the initial stage of testing. With the increase in the time, the difference of the water discharge decreased.

Acknowledgements This project was supported by Natural Science Foundation of Jiangsu Province (Grant NO. BK20200996), China Postdoctoral Science Foundation (Grant NO. 2020M681566), and the National Natural Science Foundation of China (Grant NO. 51978317).

References

1. Bejrrum L, Moum J, Eide O (1967) Application of electro-osmosis to a foundation problem in a Norwegian quick clay. Géotechnique 17(3):214–235
2. Casagrande L (1983) Stabilization of soils by means of electroosmotic state-of-art. J Boston Soc Civ Eng 69(3):255–302
3. Lo KY, Inculet II, Ho KS (1991) Electroosmotic strengthening of soft sensitive clays. Can Geotech J 28(1):62–73
4. Micic S, Shang JQ, Lo KY et al (2001) Electrokinetic strengthening of a marine sediment using intermittent current. Can Geotech J 38(2):287–302
5. Fu HT, Fang ZQ, Wang J et al (2018) Experimental comparison of electroosmotic consolidation of Wenzhou dredged clay sediment using intermittent current and polarity reversal. Mar Georesour Geotechnol 36(1):131–138
6. Zhou YD, Deng A (2019) Modelling combined electroosmosis–vacuum–surcharge preloading consolidation considering large-scale deformation. Comput Geotech 109:46–57
7. Shen Y, Feng JT, Ma YH et al (2019) Two-dimensional electroosmotic consolidation theory of nonlinear soil voltage distribution characteristics. Adv Civ Eng 2019(2):1–10
8. Shen Y, Feng JT, Qiu CC et al (2020) Two-dimensional consolidation theory of vacuum preloading combined with electroosmosis considering the distribution of soil voltage. Soil Mech Found Eng 57(1):25–34
9. Joshua HN, Kara F (2020) Numerical investigation of dynamic pipe-soil interaction on electrokinetic-treated soft clay soil. J Waterway Port Coast Ocean Eng 146(1):04019033
10. Ivliev EA (2008) Electro-osmotic drainage and stabilization of soils. Soil Mech Found Eng 45(6):211–218
11. Lin YS (2014) Strength improvement and cementation mechanism of kaolinite in electroosmotic chemical treatment. National Taiwan University of Science and Technology
12. Xue ZJ, Tang XW, Yang Q et al (2018) Mechanism of electro-osmotic chemical for clay improvement: process analysis and clay property evolution. Appl Clay Sci 166:18–26
13. Hu LM, Zhang L, Wu H (2019) Experimental study of the effects of soil pH and ionic species on the electroosmotic consolidation of kaolin. J Hazard Mater 368:885–893
14. Esrig MI (1968) Pore pressure consolidation and electrokinetics. J Soil Mech Found Div 94(SM4):899–921
15. Burnotte F, Lefebvre G, Grondin G (2004) A case record of electroosmotic consolidation of soft clay with improved soil electrode contact. Can Geotech J 41(6):1038–1053
16. Rittirong A, Shang JQ (2008) Numerical analysis for electro-osmotic consolidation in two-dimensional electric field. In: Proceedings of 18th international offshore and polar engineering conference, Vancouver, Canada, pp 566–572
17. Yuan J, Hicks MA (2013) Large deformation elastic electroosmosis consolidation of clays. Comput Geotech 54:60–68
18. Yuan J, Hicks MA (2015) Numerical simulation of elasto-plastic electro-osmosis consolidation at large strain. Acta Geotech 1:1–17
19. Zhang L, Wang NW, Jing LP et al (2016) Experimental studies and engineering application of electro-osmotic consolidation of clay soil. J Nat Disasters 25(3):78–86
20. Zhang L, Wang NW, Jing LP et al (2017) Electro-osmotic chemical treatment for marine clayey soils: a laboratory experiment and a field study. Geotech Test J 40(1):1–12

Earthquake Engineering

Cost Study of Construction of a 10-Storey Building in Different Seismic Zones According to Modern Seismic Codes

Theodoros Chrysanidis and Maria Stergiou

Abstract The purpose of this study is the investigation of the construction of a multi-storey building in the three seismic zones of Greece and its dimensioning according to modern regulations. The building has ten floors with a conventional floor plan and solid reinforced concrete slabs. The simulation has been done in the software program SAP2000 according to the modal response spectrum analysis. Then follows the dimensioning according to EC2 and EC8 and finally, the measurements of the required concrete for the whole building and the required steel for the ground floor in each of the three seismic zones. The purpose of this study is to estimate the differences arising from the construction of the building in the three seismic hazard zones, in terms of costs arising from the required amount of construction materials.

Keywords Cost study · Seismic zones · Reinforced concrete

1 Introduction

It is world-wide accepted that a multi-storey reinforced concrete (R/C) building should be adequate against all types of loads [1, 2]. Special provisions are considered for seismic actions in the relevant codes and regulations. These provisions have to do, among other things, with the seismic hazard zone that each structure should be designed in. Usually, the relevant regulations like, e.g., EC8 [3] or the Greek Seismic Code 2000 [4], etc. include specific maps showing each region in which seismic zone belongs to. According to these maps, the consulting engineer chooses the seismic zone that the building belongs to and designs it accordingly to the earthquake action provided for the seismic zone in question. Having said that, it is noteworthy to be said that the engineer can choose a higher seismic zone if he wishes to for reasons of extra safety. Of course, such an action by the engineer has an effect to the construction cost, too apart from the safety [5–12]. In that case, other phenomena have to be taken into account to, e.g., the lateral buckling of R/C walls, etc. [13–25]. The present study investigates the influence to the construction cost of the load-bearing structure

T. Chrysanidis (✉) · M. Stergiou
Hellenic Open University, 26335 Patras, Greece

T. Kang and Y. Lee (eds.), *Proceedings of 2021 4th International Conference on Civil Engineering and Architecture*, Lecture Notes in Civil Engineering 201,
https://doi.org/10.1007/978-981-16-6932-3_8

Table 1 Geometrical characteristics of structural elements

Floor	Height (m)	Beam dimensions (cm)		Wall thickness (cm)	Columns (cm)
		Perimetric	Internal		
1st	4.5	25 × 70	25 × 60	37.5	75 × 75
2nd	3.0	25 × 70	25 × 60	37.5	75 × 75
3rd	3.0	25 × 70	25 × 60	37.5	70 × 70
4th	3.0	25 × 70	25 × 60	37.5	65 × 65
5th	3.0	25 × 70	25 × 60	37.5	60 × 60
6th	3.0	25 × 70	25 × 60	25	55 × 55
7th	3.0	25 × 70	25 × 60	25	50 × 50
8th	3.0	25 × 70	25 × 60	25	45 × 45
9th	3.0	25 × 70	25 × 60	25	40 × 40
10th	3.0	25 × 70	25 × 60	25	35 × 35

of a ten-storey building when moving from one seismic zone to a higher one. Useful conclusions arise from the present research which could be proven helpful to practice engineers world-wide. The codes used are Eurocode 2 (EC2) [26] for the design of concrete and Eurocode 8 (EC8) [3] for the earthquake loads.

2 Structure Description

This study concerns a ten-storey building without a basement. It has a rectangular plan 25 × 25 m with a total floor plan area of E = 625 m^2. There are two walls at the perimeter of the floor plan and in the centre, there is the staircase and the elevator surrounded by eight walls, forming two symmetrical cores. The height of each floor is 3.0 m except for the ground floor which is 4.5 m and the dimensions of the columns change in height and are reduced per floor by 5 cm. Also, the thickness of the walls changes in height and reduces per five floors by 12.5 cm. Table 1 displays the geometrical characteristics for all structural elements and for all floors of the building, while Fig. 1 displays the floor plan. The concrete used is quality C30/37 and according to EC2 its modulus of elasticity is equal to Ecm = 33 GPa while the steel quality is B500C. SAP2000 software is used for the building modelling. Figure 2 displays the model of the building.

3 Analysis of Results

Tables 2 and 3 display the measurement results for concrete and reinforcement steel materials.

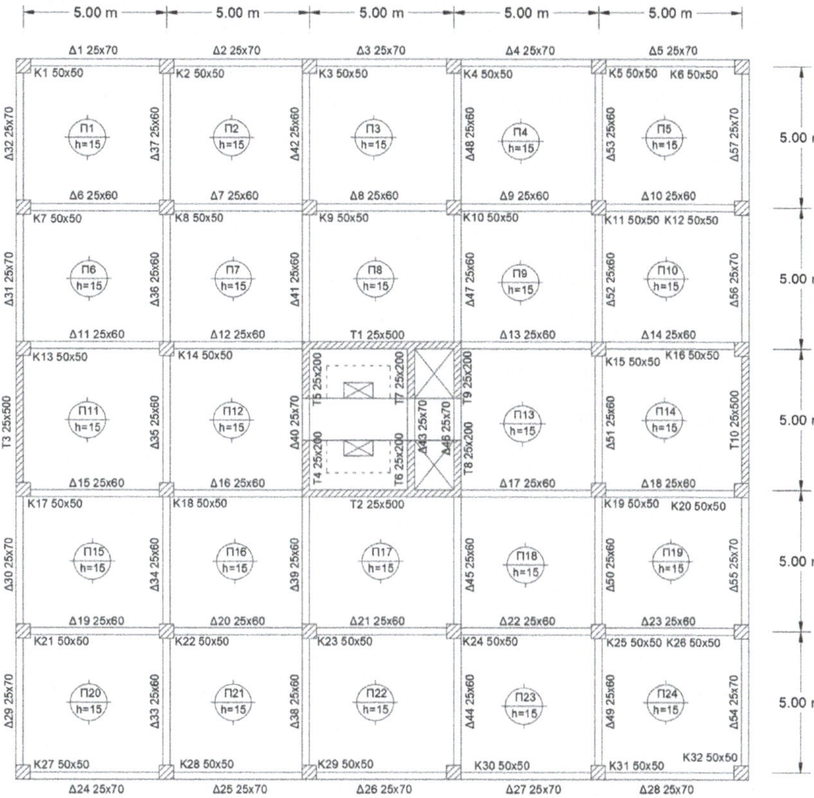

Fig. 1 Building's floor plan

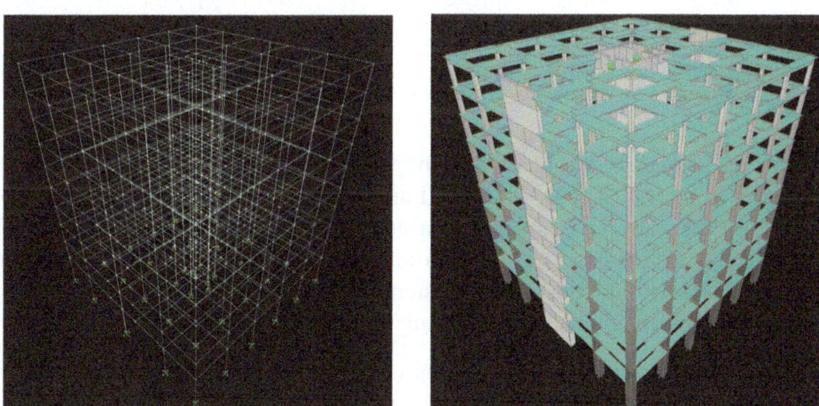

(a) (b)

Fig. 2 Views of the 3D building model: **a** linear finite elements, **b** sections

Table 2 Concrete measurements for each structural element per floor

Concrete measurements per floor					
Storey	Slabs (m^3)	Beams (m^3)	Columns (m^3)	Walls (m^3)	Total (m^3)
1	105	37	69	57	267
2	105	37	45	39	226
3	105	37	39	37	218
4	105	38	34	36	213
5	106	38	29	36	208
6	106	39	24	25	193
7	106	39	20	24	189
8	106	39	16	24	186
9	107	40	13	23	183
10	107	40	10	23	180
Parapet					14
Total	1057	384	298	324	2077

Table 3 Measurements of concrete and reinforcement steel for ground floor

Concrete and steel measurements						
Zone	**Slabs (kg)**	**%**	**Columns (kg)**	**%**	**Beams (kg)**	**%**
I	3292	17.06	5260	27.26	4844	25.11
II	3292	14.65	5093	22.67	5242	23.33
III	3292	13.59	5060	19.35	6605	25.26
Zone	**Walls (kg)**	**%**	**Total steel (kg)**		**Total concrete (m^3)**	
I	5898	30.57	19,294		2077	
II	8844	39.36	22,471		2077	
III	11,188	42.79	26,145		2077	

From Figs. 3 and 4, it appears that the steel required for the whole building is 19294 kg in Zone I, 22,471 kg in Zone II and 26,145 kg in Zone III. Also, it seems that while for the slabs the requirement in steel remains the same from zone to zone, on beams and walls increases, while on columns decreases (Table 3). This is because the building's structural load-bearing system is a predominantly wall system.

From the above diagrams, the following remarks can be stated:

- The largest percentage of steel is occupied by the walls in all three seismic zones. For Zone I 31%, for Zone II 39% and for Zone III 43% (Fig. 4).
- The beams for Zone I occupy 25%, for Zone II 23% and for Zone III 25% (Fig. 4).
- The columns for Zone I occupy 27%, for Zone II 23% and for Zone III 19% (Fig. 4).
- The slabs for Zone I occupy 17%, for Zone II 15% and for Zone III 13% (Fig. 4).

Fig. 3 Percentage change of reinforcement steel weight per change of seismic zone

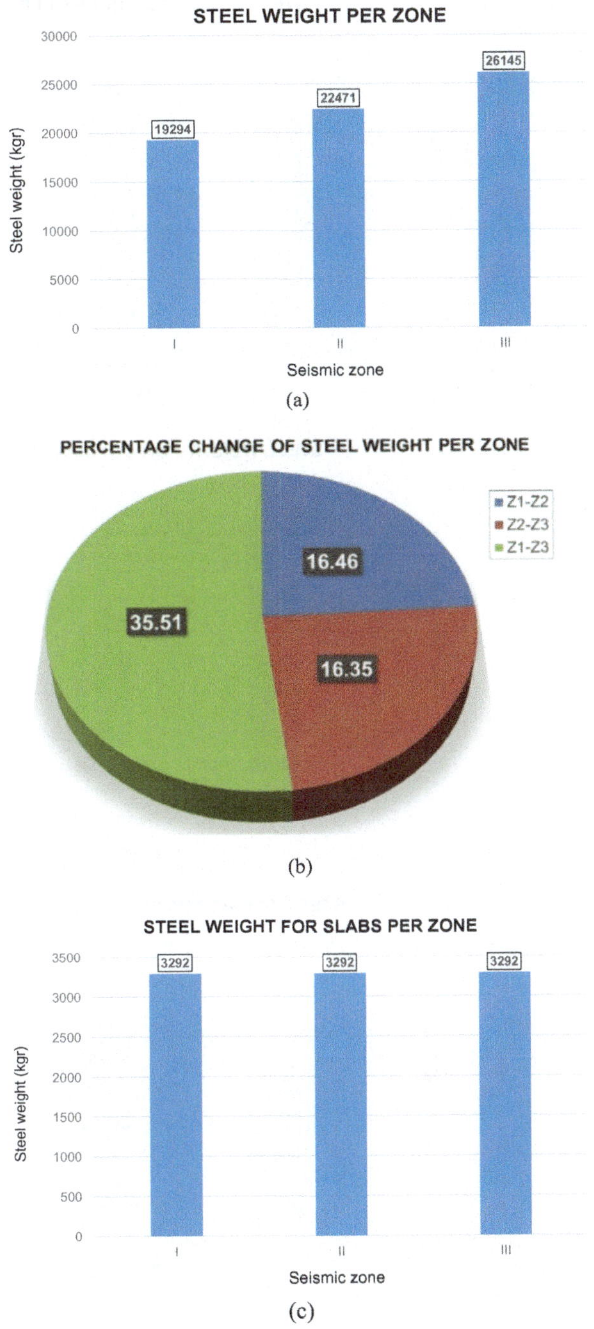

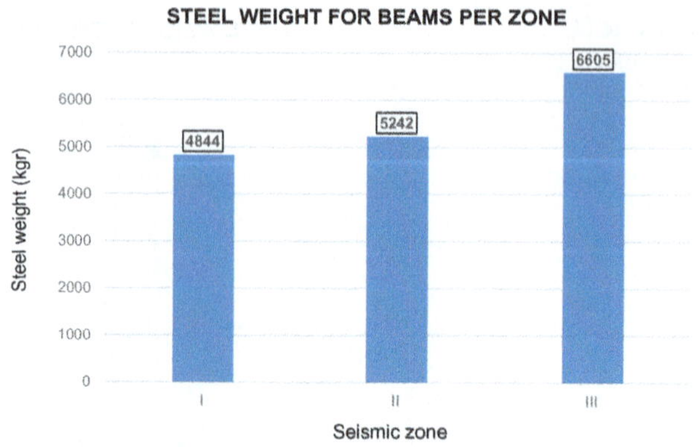

(d)

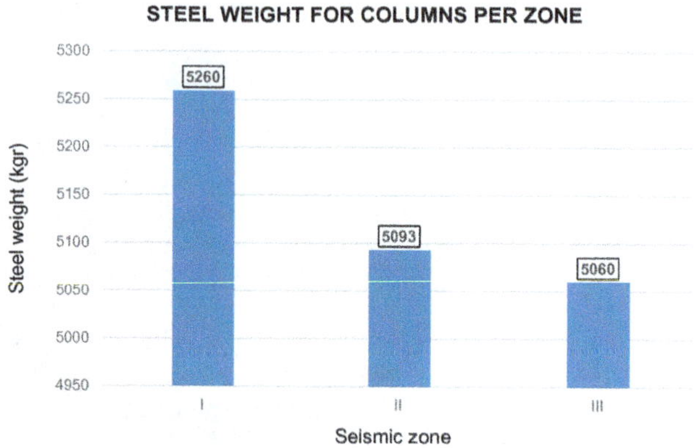

(e)

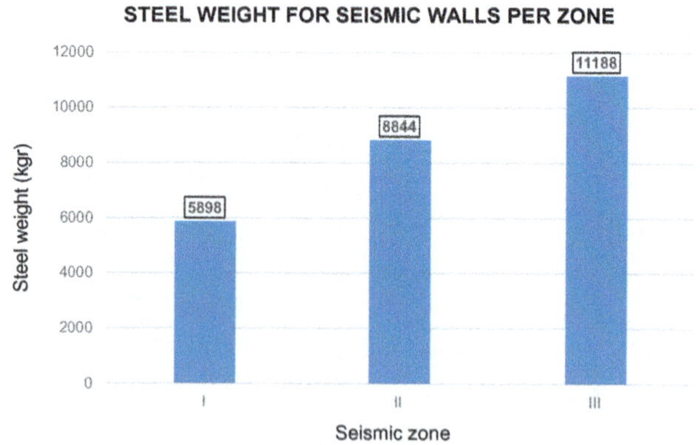

(f)

Fig. 3 (continued)

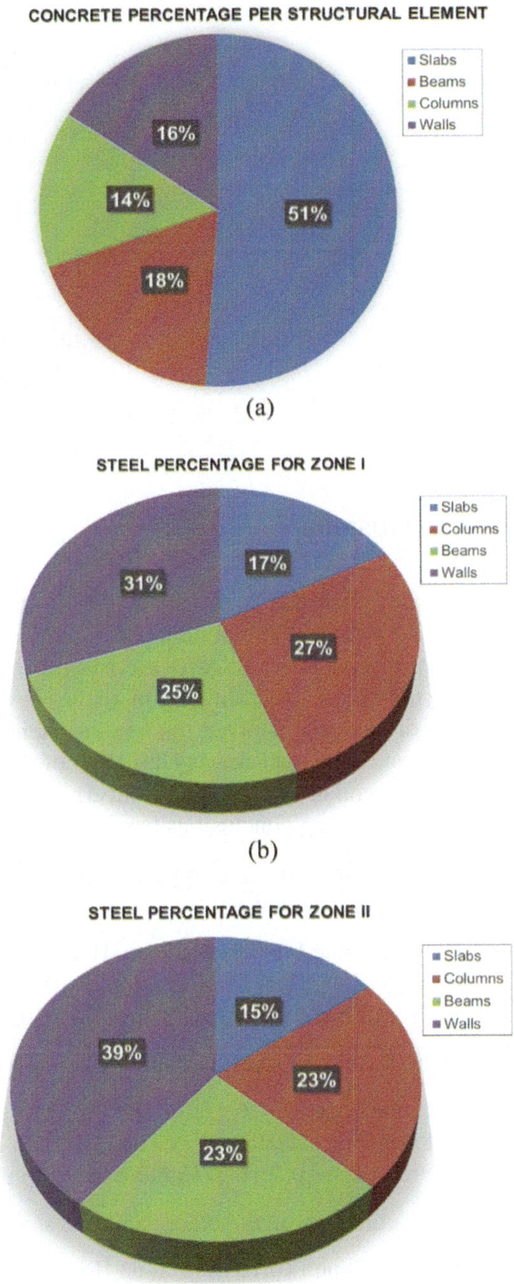

(c)

Fig. 4 Percentage change of concrete and steel weight per change of seismic zone

Fig. 4 (continued)

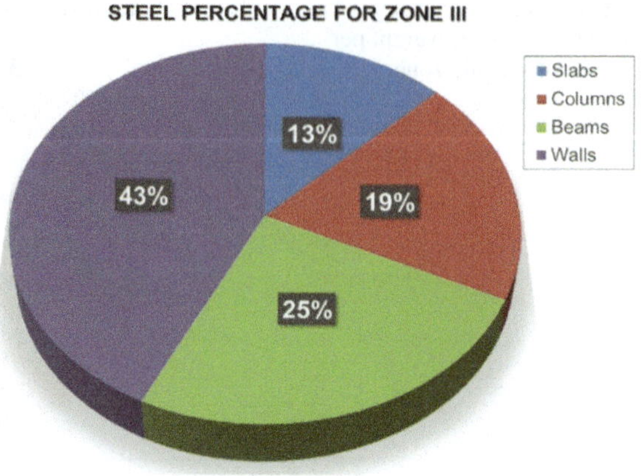

(d)

4 Conclusions

The building analysed is a ten-storey building of square conventional floor plan without a basement. The building was analysed using the spectral dynamic analysis according to EC8 [3]. After the analysis has finished, the dimensioning of the building took place according, again, to the provisions of EC2 [26] and EC8 [3]. The last step of analysis involved the measurement of the quantities demanded for the construction of the load-bearing structure of the building, as far as the concrete and steel reinforcement materials are concerned. During its design in the three seismic zones, the following conclusions have risen:

1. The walls show the largest percentage increase in steel weight with the increase of seismic acceleration.
   - From Zone I (0.16 g) to Zone II (0.24 g): Increase 49.94%
   - From Zone II (0.24 g) to Zone III (0.36 g): Increase 26.5%
   - From Zone I (0.16 g) to Zone III (0.36 g): Increase 89.67%
2. In the beams the steel rises much less than the walls:
   - From Zone I (0.16 g) to Zone II (0.24 g): Increase 8.2%
   - From Zone II (0.24 g) to Zone III (0.36 g): Increase 26%
   - From Zone I (0.16 g) to Zone III (0.36 g): Increase 36.36%
3. In the columns, on the other hand, the steel is reduced as follows:
   - From Zone I (0.16 g) to Zone II (0.24 g): Reduction of 3.16%
   - From Zone II (0.24 g) in Zone III (0.36 g): Reduction 0.65%
   - From Zone I (0.16 g) to Zone III (0.36 g): Reduction of 3.79%
   - This reduction is justified by the building's load-bearing system which is a predominantly wall system.

4. The increase of seismic acceleration:
   - From Zone I to Zone II, it is of the order of 50%, while the percentage increase in the total demand of steel is 16.46%, i.e., much smaller.
   - From Zone II to Zone III, it is of the order of 50%, while the percentage increase in the total demand in steel is 16.35%, i.e., much smaller.
   - From Zone I to Zone III, it is of the order of 125%, while the percentage increase in the total demand in steel is 35.51%, i.e., much smaller.
5. The conclusion is that for a ten-storey building with strong walls and a strong core of walls, the percentage increase of materials required is proportional to the seismic risk, but not to the extent someone would expect, thus the cost increase can be considered as not too much, especially compared to the corresponding increase of the seismic acceleration.
6. The challenges still existing is the analysis, dimensioning and material measurement for more buildings with different parameters. For example, different number of storeys can be used and different floor plans should be used, too. Furthermore, the influence of the foundation and the soil could be taken into account to, instead of using only rigid supports for all vertical elements at the base of the ground floor. These challenges could be confronted in a future research.

References

1. Penelis G, Stylianidis K, Kappos A, Ignatakis C (1995) Reinforced concrete structures. A.U. Th. Press, Thessaloniki, Greece
2. Penelis GG, Kappos AJ (1996) Earthquake-resistant concrete structures. E & F N SPON (Chapman & Hall), London, UK
3. European Committee for Standardization: EN 1998-1:2004, Presented at the Eurocode 8: Design of structures for earthquake resistance—Part 1: General rules, seismic actions and rules for buildings
4. Ministry of Environment Planning and Public Works (2003): Presented at the Greek Earthquake Resistant Design Code
5. Chrysanidis T, Panoskaltsis V, Tegos I (2016) Parametrical cost analysis of an ultra high-rise building: detailed design. Int J Appl Eng Res 11:9644–9650
6. Chrysanidis T, Panoskaltsis V, Tegos I (2016) Preliminary design and analysis of cost parameters of a high-rise building: Braced shear wall core system. Int J Civil Eng Technol 7:137–152
7. Chrysanidis T, Tegos I (2016) Cost comparison and parametrical investigation of the R/C shear wall core of a tall building. Int J Eng Res Technol 5:592–595
8. Papageorgiou G, Papadimitriou E, Alamanis N, Xafoulis N, Chouliaras I, Lazogiannis K (2019) Construction cost comparative analysis of highways in Greece. PRIME Int J Pract Res Innovative Manag Entrepreneurship 12:68–85
9. Tsiknas A, Athanasopoulou A, Papageorgiou G (2020) Evaluation of flexible pavement construction cost according to the design method. Proceedings of the institution of civil engineers (ICE)—transport, vol 173, pp 3–12

10. Papageorgiou G, Alamanis N, Chouliaras I, Kapsali P (2019) Decision making for designing infrastructure projects: the case of the city of Larissa, Greece. J Eng Architect 7, 115–125. https://doi.org/10.15640/jea.v7n1a13
11. Srivastava M, Kumar V (2018) The methods of using low cost housing techniques in India. J Build Eng 15:102–108. https://doi.org/10.1016/j.jobe.2017.11.001
12. Jasvi A, Bera D (2015) Sustainable use of low cost building materials in the rural India. Int J Res Eng Technol 4, 534–547. https://doi.org/10.15623/ijret.2015.0425077
13. Chrysanidis T (2021) The effect of longitudinal reinforcement ratio on the lateral buckling behavior of R/C walls modelled using prism elements. J Build Eng 42. https://doi.org/10.1016/j.jobe.2021.102456
14. Chrysanidis T (2020) Evaluation of out-of-plane response of R/C structural wall boundary edges detailed with maximum code-prescribed longitudinal reinforcement ratio. Int J Concr Struct Mater 14:3. https://doi.org/10.1186/s40069-019-0378-4
15. Chrysanidis T (2019) Influence of elongation degree on transverse buckling of confined boundary regions of R/C seismic walls. Constr Build Mater 211:703–720. https://doi.org/10.1016/J.CONBUILDMAT.2019.03.271
16. Chrysanidis T, Tegos I (2016) Size of seismic tensile strain and its influence on the displacements due to transverse buckling of ultra-highly reinforced structural walls. ARPN J Eng Appl Sci 11:13884–13890
17. Haro A, Kowalsky M, Chai Y, Lucier G (2018) Boundary elements of special reinforced concrete walls tested under different loading paths. Earthq Spectra 34:1267–1288. https://doi.org/10.1193/081617EQS160M
18. Welt T, Massone L, LaFave J, Lehman D, McCabe S, Polanco P (2017) Confinement behavior of rectangular RC prisms simulating wall boundary elements. J Struct Eng 143:1–12. https://doi.org/10.1061/(asce)st.1943-541x.0001682
19. Herrick K, Kowalsky M (2016) Out-of-plane buckling of ductile reinforced structural walls due to in-plane loads. J Struct Eng 143:1–15. https://doi.org/10.1061/(asce)st.1943-541x.0001660
20. Chai Y, Kunnath S (2005) Minimum thickness for ductile RC structural walls. Eng Struct 27:1052–1063. https://doi.org/10.1016/j.engstruct.2005.02.004
21. Chrysanidis T, Tegos I (2020) Axial and transverse strengthening of R/C circular columns: conventional and new type of steel and hybrid jackets using high-strength mortar. J Build Eng 30. https://doi.org/10.1016/j.jobe.2020.101236
22. Terzioglu T, Orakcal K, Massone LM (2018) Cyclic lateral load behavior of squat reinforced concrete walls. Eng Struct 160. https://doi.org/10.1016/j.engstruct.2018.01.024
23. Christidis K, Trezos K (2017) Experimental investigation of existing non-conforming RC shear walls. Eng Struct 140:26–38. https://doi.org/10.1016/j.engstruct.2017.02.063
24. Ho NM, Doh JH, Fragomeni S (2017) Instability analysis of reinforced concrete walls with various support conditions. Struct Design Tall Special Build 26. https://doi.org/10.1002/tal.1353
25. Lim J-J, Park H-G, Eom T-S (2017) Cyclic load tests of reinforced concrete columns with high-strength bundled bars. ACI Struct J 114:197–207. https://doi.org/10.14359/51689254
26. European Committee for Standardization: EN 1992-1-1:2004, Presented at the Eurocode 2: Design of concrete structures—Part 1-1: General rules and rules for buildings

Importance of Shear Wall to Control Deflection and Drift of RC Buildings for Different Earthquake Zones with Respect to Aspect Ratio

Muhammad Rafiul Mahdi, Md. Rabib Masnun, Mahedi Hasan Shuvo, and Tarek Uddin Mohammed

Abstract Bangladesh is situated in a geographically vulnerable location making it susceptible to major earthquakes and disastrous cyclones in the near future. The situation is exacerbated due to the ever-increasing colossal population density of the country. In this study, the impact of shear wall in controlling lateral loads has been investigated with respect to aspect ratios (Height/Breadth) for different Breadth/Length ratios (0.5, 0.75, 1) to realize the effect of narrow and symmetrical buildings. Shear wall is an efficacious alternative for controlling lateral loads of a structure. The building models have been developed in Etabs for four different seismic zones ($Z = 0.12, 0.2, 0.28, 0.36$) to explore the ramifications of Earthquakes for each building type. The study concludes that shear wall can reduce story drift and lateral deflection by almost 90% irrespective of seismic zones and building shapes. Furthermore, the necessity of shear walls in buildings with a lower Breadth/Length ratio is higher as they are more vulnerable to lateral loads, which can be visualized from the data obtained from the study. The study also recommends the aspect ratios at which shear walls will be necessary to restrict the deflection and drift within the limit for buildings in different seismic zones mentioned in BNBC 2020.

Keywords Earthquake zones · Shear wall · Lateral loads · Drift · Aspect ratio

1 Introduction

Southeast Asia is one of the densely populated areas of the world facing the risk of high seismic activity due to Neo-tectonic depressions and the presence of active geographical fault lines [1]. Furthermore, history dictates that several major earthquakes occurred along the margin of the Indian and Eurasian plate boundaries [2] as a result of subduction tectonics in the northern and north-eastern part of the subcontinent [3]. Bangladesh, one of the major countries of Southeast Asia with a population

M. R. Mahdi (✉) · Md. R. Masnun · M. H. Shuvo · T. U. Mohammed
Department of Civil and Environmental Engineering, Islamic University of Technology, K B Bazar Rd, Gazipur 1704, Bangladesh
e-mail: rafiulmahdi@iut-dhaka.edu

T. Kang and Y. Lee (eds.), *Proceedings of 2021 4th International Conference on Civil Engineering and Architecture*, Lecture Notes in Civil Engineering 201,
https://doi.org/10.1007/978-981-16-6932-3_9

density of 1103 per sq. km. [4], is more vulnerable to the predicted seismic activity along the region due to rapid and unplanned urbanization [5].

The capital of Bangladesh, Dhaka city has an increasing trend of urban population density and is situated in the vicinity of Madhupur Fault, [6] making it more susceptible to the damages of an earthquake. Studies have forecasted that if in any circumstance an earthquake of Richter scale magnitude of 6.0 originates with the epicenter being Dhaka, will enumerate a total of US$1,075 million financial loss and decimate a staggering 78,323 urban structures [6, 7]. And so, the proper consideration of Earthquake Loads during modelling of structures in Bangladesh is of paramount importance [8]. The Bangladesh National Building Code (BNBC) of 1993 did not overly emphasize on the seismic design criteria of a building due to a considerably lower population density then. But as of now, due to the swift increase in urban population density, the latest edition of BNBC 2020 stressed on the stringent seismic design criteria as per ASCE 7-05.

Wind Load is another one of the strong concerns for a lower riparian, deltaic country like Bangladesh. Bangladesh has been ranked as one of the world's most disaster-prone country with 97.7 percent of its total population at risk of numerous hazards including cyclones [9]. The wind speed of the cyclones can go up to 223 km/hr (SIDR, 2007) [4] resulting in catastrophic loss of life and devastation of structures. Hence, the wind load and the earthquake load are the prime lateral loads for concern in the case of building structures in Bangladesh.

Lateral loads create sway movement and vibration in structures which makes the structure unsuitable for any kind of service [10]. Effects of lateral loads on RC buildings are measured with the help of story drift and deflection. Story drift is defined as the relative difference between the displacement of two consecutive floors [10]. Story drifts act as the assessment point for structural damages and provide ample information in case of rigid body displacement. Story deflection is defined as the total displacement of any story with respect to ground. Both story drift and deflection have a maximum limit prescribed in BNBC 2020 and ASCE 7-05. Shear wall is one of the most effective vertical structural elements designed to resist in-plane lateral forces, typically wind and seismic loads by influencing the center of mass and center of rigidity. Studies have proven that shear wall reduces maximum story drift efficiently and provides better seismic performance [11].

This study investigates the effect of shear wall in controlling the lateral loads for four different seismic zones and three different Breadth/Length (B/L) ratios with increasing aspect ratio (Fig. 1). It also explores the impacts of lateral loads on the structural stability of a narrow building (B/L = 0.5). The stability of the building has been checked with the help of permissible story drift and deflection limit mentioned in BNBC 2020 and ASCE 7-05. This study will aid future structural designers to create efficient and sustainable earthquake-resistant structures in the Southeast Asia region.

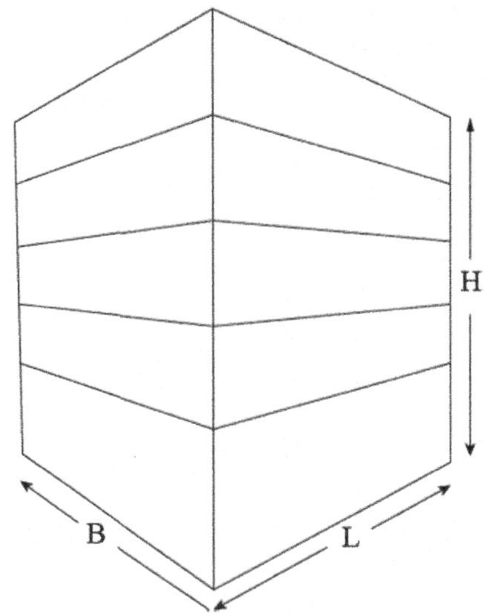

Fig. 1 Aspect ratio is defined as the ratio of height to the breadth of building. Here, aspect ratio = H/B

2 Methodology

The investigation is going to be carried out by performing equivalent static analysis of buildings in separate regions involving different aspect ratios as well as B/L ratios. Analysis of the building models has been performed in Etabs. The vulnerability of the structure to lateral load will be determined in comparison with the sway limitation and the story drift limitation mentioned in BNBC 2020 [12] and ASCE 7-05.

Collecting the results of the models of the highest aspect ratio at which the building frames are susceptible to failure, the aspect ratio of the structures is increased with the implementation of shear walls. Shear walls will be oriented along the periphery of the building for higher efficiency [13].

2.1 Sample Selection

As per the BNBC 2020 [12], Bangladesh is divided into four earthquake regions which have been duly considered while modeling the buildings. Barisal, Dhaka, Chittagong, and Sylhet have been selected as the representatives of the four earthquake zones. The zone co-efficient and wind speed of the respective locations have been mentioned in Table 1. Three breadth to length ratios is considered which are 0.5, 0.75, 1 for proper evaluation of the impact of lateral loads on narrow and symmetrical buildings. In this case, the breadth of the building models has been taken as 50 feet and the lengths are considered 100 feet, 66.67 feet, and 50 feet respectively for the aforementioned B/L ratios which are displayed in Figs. 2, 3 and 4.

Table 1 Wind speed and seismic zones according to BNBC 2020

Location	Basic wind speed, V (m/s)	Seismic zone and zone coefficient (Z)
Barisal	78.7	Zone 1: 0.12
Dhaka	65.7	Zone 2: 0.20
Chittagong	80.0	Zone 3: 0.28
Sylhet	61.1	Zone 4: 0.36

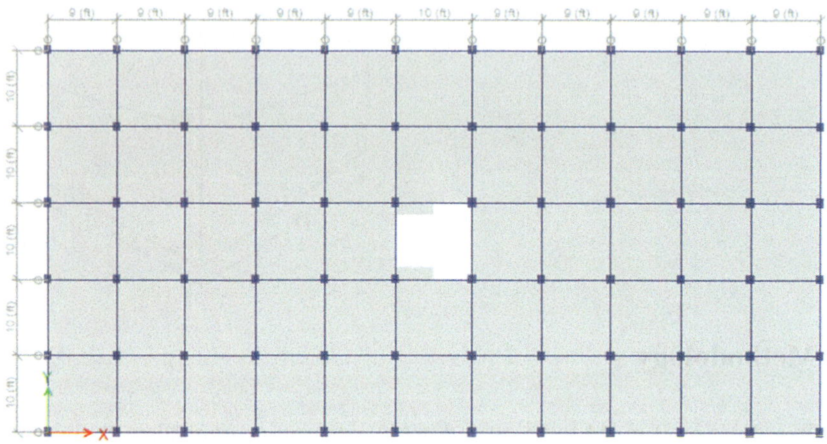

Fig. 2 B/L ratio 0.5

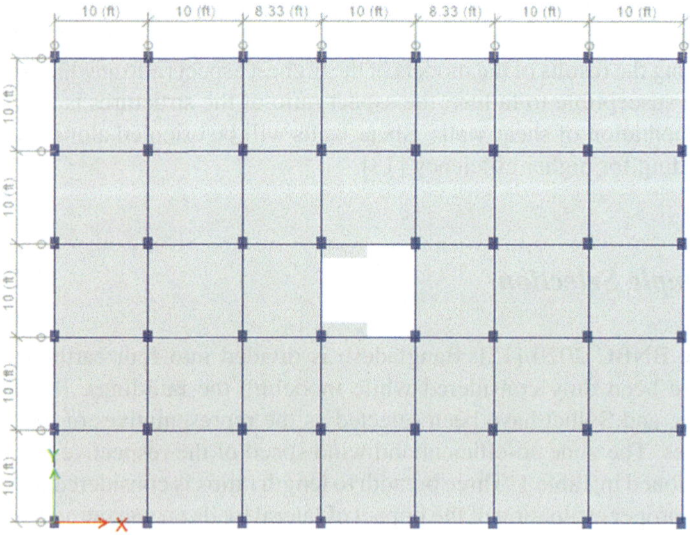

Fig. 3 B/L ratio 0.75

Fig. 4 B/L ratio 1

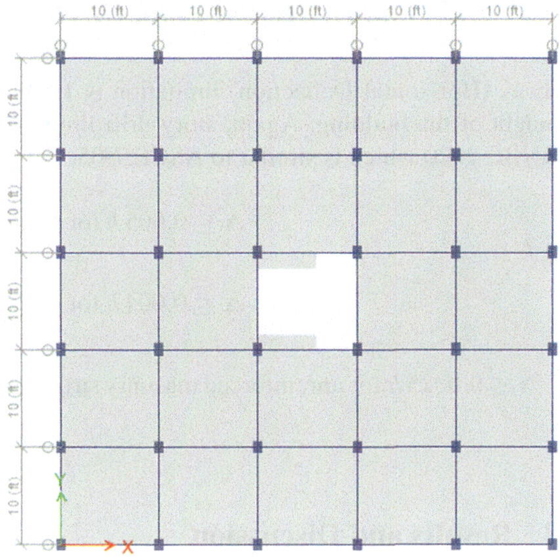

2.2 *Structural Modeling*

A typical residential structure made of reinforced cement concrete (RCC) of 3ksi design strength with a special moment resisting frame (SMRF) is constructed as per ACI 318-14. Lateral loads will be considered in the X and Y direction. Serviceability limit state has not been considered for this case. Reinforcement bar has been selected as A615 60 grade. Story height is 10 feet and shear wall thickness has been taken as 10 in. The earthquake load is calculated using Eq. 1 where site class has been considered as SC.

$$S_a = (2ZIC_s)/(3R) \tag{1}$$

where S_a, Z, I, R and C_s represent the design spectral acceleration, seismic zone coefficient, structure importance factor, response reduction factor and normalized acceleration response spectrum respectively. The time period (T) of the building is approximated by using

$$T = C_t(h_n)^{\mathrm{m}} \tag{2}$$

where h_n is the height of the building in meters from the foundation or from the top of the rigid basement.

2.3 Drift and Deflection Limit (BNBC 2020 Art 1.5.6)

Sway (Horizontal Deflection) limitation is 1/500 times of H, where H is the total height of the building. Again, story drift limitation is mentioned in Art 1.5.6.1 of BNBC 2020 which is similar to ASCE 7-05.

$$\Delta \le 0.005\,h \text{ for } T < 0.7\text{ s} \tag{3}$$

$$\Delta \le 0.004\,h \text{ for } T \ge 0.7\text{ s} \tag{4}$$

$$\Delta \le 0.0025\,h \text{ for unreinforced masonry structures, where h is the story height} \tag{5}$$

3 Results and Discussion

3.1 Shear Wall Orientation

Shear walls have been provided along the periphery of the building models as the reduction of deflection and drift is maximum in this type of orientation. This result resembles other researches as this type of orientation reduces the deflection up to 33.31 and 32.03% compared to L type and cross-type shear wall orientations [12] (Tables 2 and 3).

Table 2 Building model specifications

Serial no	Building properties	Description
1	Importance factor (I)	1
2	Response modification factor, R	8
3	Site coefficient, S	1.15
4	Loadings as per BNBC 2020 (Art 2.2-2.3)	

Table 3 Load combinations used (BNBC 2020 Art. 2.7.3.1)

Serial no.	Load combinations used
1	1.4D
2	1.2D + 1.6L
3	1.2D + L + 0.8W
4	1.2D + 1.6W + L
5	1.2D + 1.0E + 1.0L
6	0.9D + 1.6W
7	0.9D + 1.0E

3.2 *Control of Deflection and Drift for Different Seismic Zones and B/L Ratios*

As per BNBC 2020, comparisons have been made for four seismic zones taking Barisal, Dhaka, Chittagong and Sylhet as the representatives of zone-1 (z = 0.12), zone-2 (z = 0.20), zone-3 (z = 0.28) and zone-4 (z = 0.36) respectively.

The models with B/L ratio 0.5 for zone-1, zone-2, zone-3 and zone-4 exceed either the deflection limit or the drift limit as per BNBC 2020 at aspect ratios 1.32, 1.52, 0.92 and 0.72 respectively. For zone-1, at an aspect ratio 1.32, the deflection and drift values are 1.83 in. and 0.0149 m. respectively whereas the maximum allowable deflection and drift limits are 1.58 inches and 0.0153 m. respectively. Providing shear wall at the periphery of the model controls both the deflection and drift as the values go down to 0.16 in. and 0.00072 m. respectively, which is within the maximum allowable deflection and drift limits. Similar results are obtained for zone-2, zone-3 and zone-4 which are shown in Table 4.

The models with B/L ratio 0.75 for zone-1, zone-2, zone-3 and zone-4 exceed either the deflection limit or the drift limit as per BNBC 2020 at aspect ratios 1.92, 1.72, 0.92 and 0.92 respectively. For zone-1, at an aspect ratio 1.92, the deflection and drift values are 2.43 inches and 0.0141 m. respectively whereas the maximum allowable deflection and drift limits are 2.30 inches and 0.0122 m. respectively. Providing shear wall at the periphery of the model controls both the deflection and drift as the values go down to 0.37 inches and 0.00117 m. respectively, which is within the maximum allowable deflection and drift limits. Similar results are obtained for zone-2, zone-3 and zone-4 which are shown in Table 5.

The models with B/L ratio 1.00 for zone-1, zone-2, zone-3 and zone-4 exceed either the deflection limit or the drift limit as per BNBC 2020 at aspect ratios 2.12, 1.92, 1.12 and 0.92 respectively. For zone-1, at an aspect ratio 2.12, the deflection and drift values are 2.38 in. and 0.0127 m. respectively whereas the maximum allowable deflection and drift limits are 2.54 in. and 0.0122 m. respectively. Providing shear wall at the periphery of the model controls both the deflection and drift as the values go down to 0.40 in. and 0.00115 m. respectively, which is within the maximum

Table 4 Analysis results of building models of B/L ratio 0.5

B/L ratio	Seismic zone	Unsafe aspect Ratio	Deflection			Story drift		
			Limit (in.)	No shear Wall (in.)	With shear Wall (in.)	Limit (m.)	No shear Wall (m.)	With shear Wall (m.)
0.50	1	1.32	1.58	1.83	0.16	0.0153	0.0149	0.00072
	2	1.52	1.82	1.85	0.30	0.0122	0.0119	0.00123
	3	0.92	1.10	1.35	0.20	0.0153	0.0147	0.00059
	4	0.72	0.86	1.09	0.06	0.0153	0.0146	0.00065

Table 5 Analysis results of building models of B/L ratio 0.75

B/L ratio	Seismic Zone	Unsafe Aspect Ratio	Deflection			Story drift		
			Limit (in.)	No Shear Wall (in.)	With Shear Wall (in.)	Limit (m.)	No Shear Wall (m.)	With Shear Wall (m.)
0.75	1	1.92	2.30	2.43	0.37	0.0122	0.0141	0.00117
	2	1.72	2.06	2.07	0.31	0.0122	0.0122	0.00110
	3	0.92	1.10	1.11	0.11	0.0153	0.0131	0.00069
	4	0.92	0.86	1.25	0.07	0.0153	0.0146	0.00046

Table 6 Analysis results of building models of B/L ratio 1.00

B/L ratio	Seismic zone	Unsafe Aspect Ratio	Deflection			Story drift		
			Limit (in.)	No shear Wall (in.)	With shear Wall (in.)	Limit (m.)	No shear Wall (m.)	With shear Wall (m.)
1.00	1	2.12	2.54	2.38	0.40	0.0122	0.0127	0.00115
	2	1.92	2.30	2.35	0.26	0.0122	0.0126	0.00084
	3	1.12	1.34	1.40	0.09	0.0153	0.0128	0.00047
	4	0.92	1.10	1.20	0.06	0.0153	0.0132	0.00037

allowable deflection and drift limits. Similar results are obtained for zone-2, zone-3 and zone-4 which are shown in Table 6.

3.3 *Failure Pattern with Respect to Aspect Ratio and Different Seismic Zones*

For B/L ratio 0.5, the aspect ratios at which the maximum deflection or drift limits are exceeded are 1.32, 1.52, 0.92 and 0.72 for zone-1, zone-2, zone-3 and zone-4 respectively. For B/L ratio 0.75, the aspect ratios at which the maximum deflection or drift limits are exceeded are 1.92, 1.72, 0.92 and 0.92 for zone-1, zone-2, zone-3 and zone-4 respectively. Similarly, for B/L ratio 1.00, the aspect ratios are 2.12, 1.92, 1.12 and 0.92 respectively.

From the analysis, a descending pattern can be observed among the aspect ratios from zone-1 to zone-4 models (considering the B/L ratio of the buildings does not go below 0.5), at which the maximum allowable deflection and drift limits are exceeded for a particular B/L ratio. An exception can be noticed for the models with B/L ratio 0.5 among zone-1 and zone-2 as the aspect ratio at which the maximum deflection or drift limits are exceeded in the zone-1 model is less than that of the zone-2 model. The

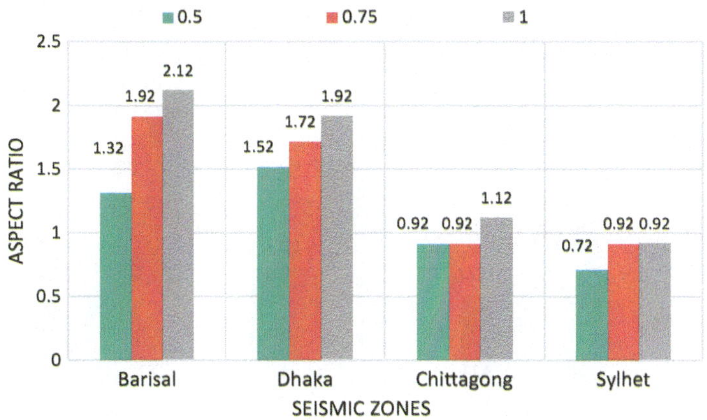

Fig. 5 Unsafe aspect ratio for different zones

reason for the exception is the difference in wind speed between zone-1 (78.7 m/s) and zone-2 (65.7 m/s) which affects only the models with B/L ratio 0.5 as the building models are slender (Fig. 5).

3.4 Failure Pattern with Respect to Aspect Ratio and Different B/L Ratios

For zone-1, aspect ratios at which the maximum deflection or drift limits are exceeded are 1.32, 1.92, 2.12 for B/L ratios 0.5, 0.75 and 1.00 respectively. For zone-2, aspect ratios at which the maximum deflection or drift limits are exceeded are 1.52, 1.72, 1.92 for B/L ratios 0.5, 0.75 and 1.00 respectively. Similarly, for zone-3 and 4, aspect ratios at which the maximum deflection or drift limits are exceeded are 0.92, 0.92, 1.12 and 0.72, 0.92, 0.92 for B/L ratios 0.5, 0.75 and 1.00 respectively.

From the analysis, an ascending pattern can be observed among the aspect ratios from models with B/L ratio 0.5 to 1.00, at which the maximum allowable deflection and drift limits are exceeded for a particular B/L ratio (Fig. 6).

4 Future Scopes

The study has been conducted based on 4 representative locations of the 4 seismic zones defined in BNBC 2020 which can be expanded and turned into a database for all the districts of Bangladesh. This database will include data of more than 60 districts and will serve as a guideline for engineers in local RC building projects to find out the optimum aspect ratios at which shear walls will be necessary. The database

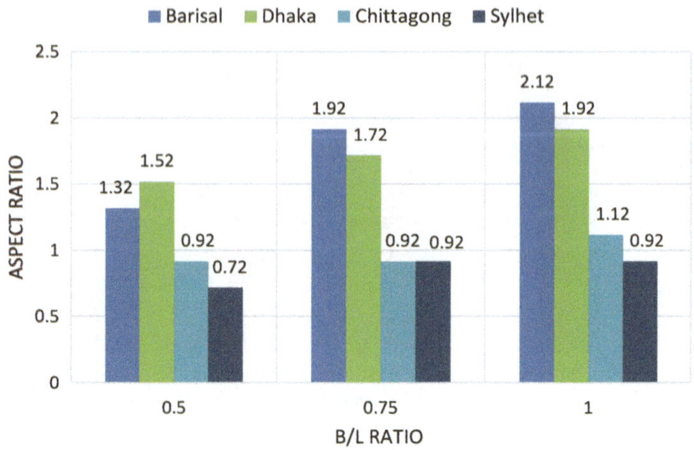

Fig. 6 Unsafe aspect ratio for different B/L ratio

will not only provide guidance to the local engineers but will also help structural engineers of different international projects with almost similar soil, seismic and wind characteristics to develop an understanding of the necessity of shear wall with respect to aspect ratio. Using the same methodology, the optimum aspect ratios at which shear walls will be necessary for RC buildings can be obtained for any place around the globe if the soil, seismic and wind data are provided.

5 Conclusion

(i) As the value of the seismic zone coefficient (z) increases, the buildings become more vulnerable towards lateral deflection and story drift. To counter the lateral deflection and story drift, buildings in higher seismic zones need shear walls at a lower aspect ratio.

(ii) As the B/L ratio decreases, that is the buildings become narrower, it is more vulnerable towards lateral deflection and story drift. As a result, buildings in lower B/L ratios need shear walls at a lower aspect ratio.

(iii) Shear walls are provided to add stiffness to the structures which decreases the lateral deflection and story drift. From the obtained data, providing shear walls in an RC building can reduce the lateral deflection and story drift by up to 90%.

(iv) The buildings located on seismic zone-1, zone-2, zone-3, and zone-4 (as per BNBC-2020) need a shear wall if the aspect ratios of the buildings exceed 1.32, 1.52, 0.92, and 0.72 respectively, irrespective of their B/L ratios (considering the B/L ratio does not go below 0.5).

(v) The buildings with B/L ratios 1.00, 0.75, and 0.50 need a shear wall if the aspect ratio of the buildings exceeds 0.92, 0.92, and 0.72 respectively, without considering the seismic zones they are located on. Therefore, as the B/L ratio of a building moves closer to 1, the building becomes more resistant towards lateral loads like wind load and earthquake load.

References

1. Khan AA, Hoque MA (2002) Quaternary paleo-geography and geohazard scenario of the Bengal delta of Bangladesh. Orient Geogr 46(2):1–16
2. Paul BK, Bhuiyan RH (2010) Urban earthquake hazard: perceived seismic risk a preparedness in Dhaka City, Bangladesh. Disasters 34(2):337–359
3. Nath SK, Mandal S, Adhikari MD, Maiti SK (2017) A unified earthquake catalogue for South Asia covering the period 1900–2014. Nat Hazards 85(3):1787–1810
4. Statistics BBS (2019) Statistical yearbook of Bangladesh. Statistics & Informatics Division, Ministry of Planning, Dhaka, Government of the People's Republic of Bangladesh
5. Hossain N (2014) 'Street'as accessible open space network in earthquake recovery planning in unplanned urban areas. Asian J Humanit Soc Sci (AJHSS) 2(4):103–115
6. Rahman N, Ansary MA, Islam I (2015) GIS based mapping of vulnerability to earthquake and fire hazard in Dhaka city, Bangladesh. Int J Disaster Risk Red 13:291–300
7. CDMP (2010) Report on earthquake risk assessment of Dhaka, Chittagong and Sylhet City Corporation Area. Govt. of Bangladesh, Dhaka
8. Barua K, Rahman SH, Das S (2013) Performance based analysis of seismic capacity of Mid Rise building. Int J Emerg Technol Adv Eng 3(11):44–52
9. Dilley M (2005) Natural disaster hotspots: a global risk analysis, vol 5. World Bank Publications
10. Hosseini M, Rao NR (2018) Study the impact of the drift (lateral deflection) of the tall buildings due to seismic load in concrete frame structures with different type of RC shear walls. Global J Res Eng
11. Ali J, Bhatti AQ, Khalid M, Waheed J, Zuberi S (2015) A comparative study to analyze the effectiveness of shear walls in controlling lateral drift for medium to high rise structures (10–25 Storys). In: Proc. 2015 2nd international conference on geological and civil engineering, IPCBEE, pp 31–36
12. BNBC-2020, 2021, February. Bangladesh National Building Code 2020, Ministry of Housing and Public Works, Government of the People's Republic of Bangladesh
13. Ali SE, Aquil MMU Study of strength of RC shear wall at different location on multi-storied residential building. J Eng Res Appl. ISSN: 2248-9622

Investigations of the Microstructure of the Contact Zone Between Bricks and Masonry Mixtures in Seismic Regions

N. F. Bakhriev, Sh. A. Buriboev, and G. Sh. Odilova

Abstract The article presents the results of studies on the study of the contact zone of masonry mixtures, with the aim of developing compositions for the construction of walls and partitions made of bricks and blocks in seismically active regions. Analyzes of the developed compositions, theoretical and practical substantiation of high adhesive capacity in conditions of high seismic activity are presented. Methods of using non-fired binders based on high-modulus silicate components are proposed. The technological scheme and methods of obtaining universal masonry mortars providing high adhesive capacity and adhesion strength are presented. When studying the "brick–mortar" contact zone by X-ray thermal electron probe microanalysis, a regular relationship between the chemical compounds of the system was established and the correlation dependence of the micro hardness of neoplasm's in the contact zone was determined. It has been established that an increased concentration of sodium, aluminum, silicon is observed in the contact zone, which contribute to the intensification of the formation of a dense microstructure due to the presence in it of water-resistant neoplasm's such as low-basic hydro silicates and alkaline hydro aluminates. These new formations constitute the structural matrix (frame) of the contact zone and play the role of a damper at the time of the action of alternating seismic loads. The developed compositions have found application as universal masonry mortars in the construction of brick walls with the provision of category I adhesion strength.

Keywords Bricks and masonry mixtures · Alkaline component · Masonry mortars · Contact zone

N. F. Bakhriev · Sh. A. Buriboev (✉) · G. Sh. Odilova
Department of Construction Materials, Samarkand State Architectural and Civil Engineering Institute (SSACEI), Samarkand, Uzbekistan

N. F. Bakhriev
e-mail: bakhriyev.nf@mail.ru

T. Kang and Y. Lee (eds.), *Proceedings of 2021 4th International Conference on Civil Engineering and Architecture*, Lecture Notes in Civil Engineering 201,
https://doi.org/10.1007/978-981-16-6932-3_10

1 Introduction

Slag-alkaline binders are hydraulic binders obtained by fine grinding of granular slag together with a low-hygroscopic alkaline component or mixing of ground slag with aqueous solutions of alkali metal compounds (sodium, lithium, potassium), the scientific basis of which was developed by professor Glukhovsky V.D. and are currently being developed by the researchers of his school under the guidance of professor P.V. Krivenko [1, 2].

The purpose of these studies was to develop compositions of universal masonry mixes with the use of slag-alkali binders. In accordance with the task, a series of experiments was carried out to optimize the composition of masonry mixtures, which were used to prepare masonry mortars used for laying walls and partitions from dense materials (concrete blocks, ceramic, or silicate bricks).

2 Proposed Methodology

The masonry mixture was produced in the following way, all the constituent components (slags, alkaline components, and additives in dry form with a moisture content of not more than 1 wt%) in a calculated amount were ground in a ball drum mill to the required dispersion. The finished ground powder was used as a binder for the preparation of masonry mortars (Fig. 1). A series of experiments on the selection of the composition [3–6] of masonry mortars was carried out, "samples of two-brick masonry mortar-brick" were made. The adhesion properties of which were determined in a special tensile testing machine (Fig. 2) [3].

For a more detailed study of the physic and chemical processes occurring in the contact zone "brick—masonry mixture" we carried out a set of studies.

Based on the studies carried out, it has been established that slag-alkaline masonry mortars are characterized by high adhesion to piece materials. Tests to determine the category of masonry showed rather high results in comparison with the normative ones and at the same time ensuring the solidity of the contact zone (Table 1).

One of the important characteristics that determine the adhesion strength and solidity of the contact zone is the change in the micro hardness of the formed micro-conglomerates in this zone.

To determine the kinetics of changes in the micro hardness of conglomerates in the direction of "masonry mortar—contact zone—brick" studies were carried out on the device PMT-3. For this purpose, from specially prepared samples of "twos" of experimental and control compositions, "polished sections" were prepared from the contact zone and the micro hardness of conglomerates formed in the contact zone was determined according to the method presented in the source [7]. According to the test data, a graph of changes in the micro hardness in the contact zone was constructed (Fig. 3).

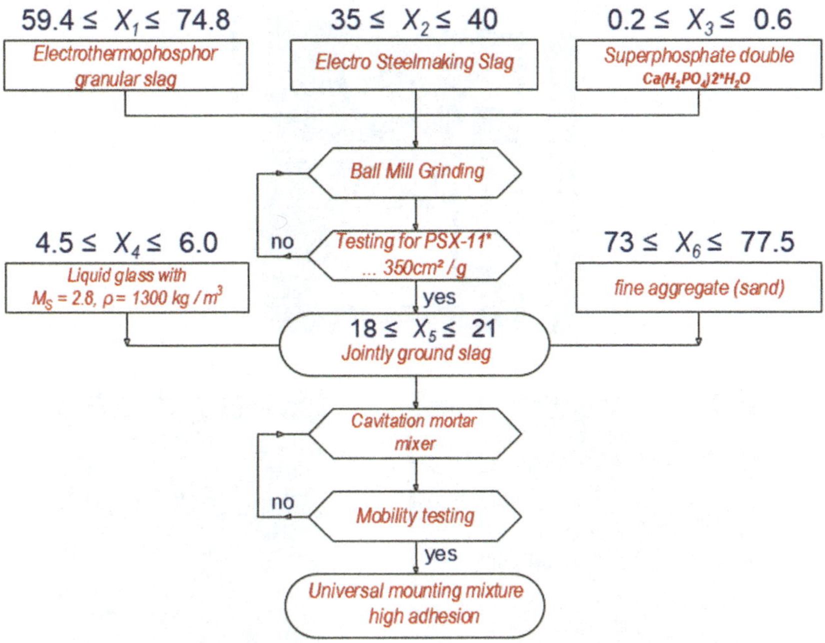

Fig. 1 Technological scheme of production of building mixture of high adhesion: X_1 … X_n—constituent components, wt.%

Thus, tests have shown that the micro hardness in the contact zone "masonry mortar-brick" depends on the composition of the masonry mortar. The maximum value of micro hardness is observed in the contact zone, where slag-alkaline binders with silicate alkaline components are used as a masonry mortar (Table 2). For masonry mortars 1 composition, 2 composition and 3 composition (Table 1) micro hardness in the contact zone ranges from 227.2 to 410.6 MPa (Table 2).

Observation of the micro hardness of the conglomerates in a certain area from the middle of the mortar joint (Fig. 3) showed that the (very) micro hardness in the contact zone increases. This indicates the formation of highly hard crystal-like minerals. At the same time, when using cement-lime mortars, the value of micro hardness in the contact zone decreases. The increase in micro hardness and density in the contact layer between the masonry mortar and the silicate brick is explained by the fact that on the surface of the silicate brick it intensively proceeds in the process of chemical interaction of the components with the appearance of reaction borders and neoplasm's such as alkaline-alkaline earth minerals, as well as the formation of a crystal lattice of the products of hydration of the binder and brick that can be observed under an electron microscope or using petro graphic analysis. In addition, another reason for the formation of a dense contact zone when using slag-alkaline masonry mortar is the high water-retention capacity and increased viscosity of the latter due to the significant viscosity of the grout (aqueous solutions of alkaline

Fig. 2 The nature of fracture of samples "twos" for rupture: **a** universal tensile testing machine UMM-5; **b** a gap along the mortar joint; **c** brick break; **d** break in the contact area; **e** monolithic masonry seam

Table 1 Composition and properties of masonry mixtures

Conditional designation	Components	Content, wt. %	Ultimate tensile strength of masonry elements under tension, MPa			Masonry efficiency coefficient	Nature of destruction under axial tension, %		
			Bricks	Mortar	Masonry "two"		By brick	By mortar	By contact area
1 composition	Ground electro thermo phosphoric slag	18–21	0.87 0.65 0.74 0.8	0.54 0.95 0.83 0.57	0.6 0.72 0.67 0.5	1.11 0.75 0.8 0.75	0 60 40 20	70 20 40 50	30 20 20 30
	Ground slags of non-ferrous metallurgy	2–6							
	Sodium silicate (in terms of R_2O)	4.5–6							
	Additive, hardening retarder (double superphosphate)	0.1–0.2							
	Aggregate (sand)	Rest							
2 composition	Slag-alkali binder (SAB-2 according to Table 3.6 [3])	18–21	0.72 0.84 0.382 0.75	0.35 0.32 0.29 0.37	0.22 0.27 0.25 0.31	0.63 0.27 0.86 0.83	0 0 0 0	40 60 85 70	60 40 15 30
	Plasticizer	0.2–0.6							
	Aggregate (sand)	Rest							
3 composition	Slag-alkali binder (SAB-3 according to Table 3.6 [3])	18–21	0.66 0.70 0.8	0.34 0.35 0.32	0.28 0.35 0.32	0.82 1.00 1.00	0 0 0	70 60 40	30 40 60
	Plasticizer	0.2–0.6							
	Aggregate (sand)	Rest							

(continued)

Table 1 (continued)

Conditional designation	Components	Content, wt. %	Ultimate tensile strength of masonry elements under tension, MPa			Masonry efficiency coefficient	Nature of destruction under axial tension, %		
			Bricks	Mortar	Masonry "two"		By brick	By mortar	By contact area
traditional	Portland cement M400	10–12	0.75	0.28	0.07	0.25	0	0	100
			0.84	0.32	0.08	0.26	0	5	95
	Lime dough	8–9	0.77	0.26	0.1	0.38	0	20	80
	Aggregate (sand)	Rest							

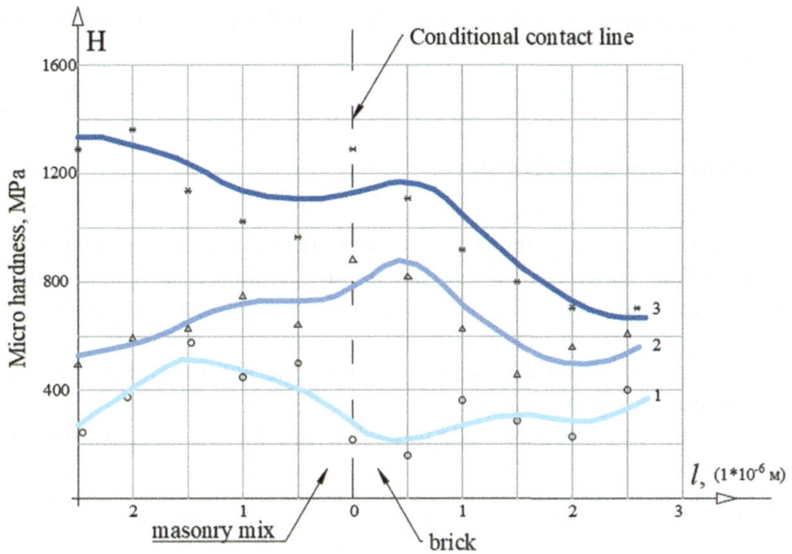

Fig. 3 Study of the micro hardness value of neoplasms in the contact zone: 1—control composition (traditional cement-lime masonry mortar, Table 1, pos. 4); 2—compositions 2 and 3 (Table 1, pos. 2 and 3); 3—the same composition 1, pos. 1

Table 2 Micro hardness in the contact zone "masonry mortar—silicate brick"

Name of the masonry mortar	Contact zone width, microns	Micro hardness, MPa	
		Of the mortar part	Of the contact zone
Cement-lime mortar (control)	70–100	410.6 227.2 240.3	298.7 297.2 315.1
Slag-alkaline based on non-silicate alkaline components (2–3 composition)	105–150	648.1 531.4 537.2	624.6 804.6 658.4
Slag-alkaline based on silicate alkaline components (1 composition)	120–170	634.0 720.5 968.0	1328.0 1132.0 1115.0

components of water glass or sodosulfate mixture). Due to the high density and viscosity of the alkaline medium, during the adsorption of the masonry mortar inside the pores and capillaries of the brick, not only the liquid medium is involved, but also the dispersed phases of the binder, which interact with the surface of the pore space walls of the constituent low-basic calcium hydro silicates or amorphous silica acid minerals, activate this surface, react, and seal contact zones. The contact zone between the slag-alkaline masonry mortar and the silicate brick is characterized by a dense structure with an increase in the layer width in the range of 105–150 or 120–170 μm (Table 2).

3 Experiments and Results

Based on the results of the studies carried out to determine the micro hardness, we studied the contact zone of the system under consideration also by optical methods to establish the physicochemical processes of the appearance of neoplasms in this zone.

Masonry mortars were made with distinguishable aluminosilicate components based on alkaline components (Table 1).

The results of optical studies made it possible to establish the following patterns in the contact zone [3]:

1. The contact zone between the mortar and the surface of the silicate brick is not pronounced, indistinct. Diffusion of two substances is noticeable, especially intensive penetration into the pores of brick cracks. No reaction edges are observed. The contact is fragile. Discontinuities were noted along the entire contact plane of the solution due to its low strength.

   The hydration process of cement in the solution is normal; the residual clinker grains are 50–55%. Neoplasms are presented in the form of a gel-like brown phase, presumably calcium hydro silicates. In the region of the contact zone, sub microcrystalline formations of portlandite $Ca(OH)_2$ with a grain size of 0.01 mm are noted, (Fig. 4, item 1 and 2).
2. The sections were made of a mortar seam, where a dry-milled slag-alkaline binder was used (compositions according to Table 1, item 2.3).

   The contact is tight; the edge of the neoplasms is noticeable, arising from the reaction of the substances of the binder and silicate bricks. The diffusion of the binder paste into the brick through pores and micro cracks was noted.

   There are very thin (0.01–0.02 μm) reaction edges around the grains of sand and in the contact zone itself, which consist of sub microcrystalline formations no larger than 0.01 μm.

   In the contact zone, in the mortar seam, a strip (0.03–0.08 mm) of opaque dense neoplasms with a gel-like structure is noticeable. There are also many subs microcrystalline unconventional neoplasm in the contact zone and in the solution part itself, (Fig. 3, item 3 and 4).
3. The sections are made of a mortar seam, where a slag-alkaline binder is used on a silicate alkaline component (composition according to table 1, item 1). The contact is expressed by a sharp line 0.02–0.06 mm wide, where gel-like products of a chemical reaction of brown color are concentrated. Particles of ground slag are covered, as in other thin sections, with dark globules or micelles, (Fig. 4, item 5 and 6).

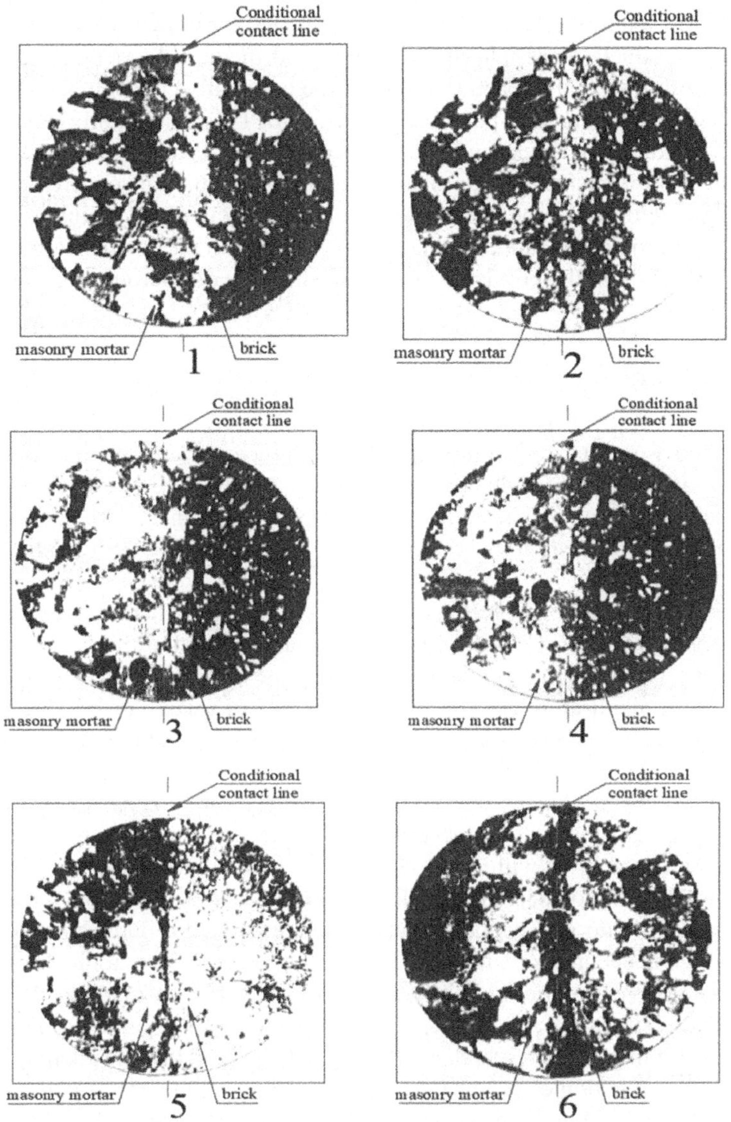

Fig. 4 Microstructure of contact zones "masonry mortar—brick": 1—cement-lime mortar (control), magnification ×300; 2—the same, mag. X400; 3—slag-alkaline solution with sulphate mixture, mag. ×300; 4—the same, mag. ×400; 5—slag-alkaline solution on liquid glass, mag. × 300; 6—the same, mag. ×400

In the solution part, isotropic substances of dark brown color can be traced, likely, products of neoplasms with high and medium birefringence of 0.02–0.03 mm in size; the reaction products are especially concentrated around the sand grains, i.e., around particles with a size of 1–1.5 mm, while destruction in the form of micro cracks and dullness is noticeable. Probably, fine-grained sand particles also interact with an alkaline medium since they are covered with dense shells of hydration products. Up to 5% of non-hydrated slag grains are noted, so the hydration of the binder in the mortar joint and in the contact, zone proceeds intensively with the provision of a dense structure of micro-conglomerates.

4 Discussion

With an increase in the concentration of Na_2O in the contact zone, the solubility of calcium hydroxide significantly decreases due to the presence of identical hydroxyl ions, which, accordingly, is reflected in the graphs. This can explain the decrease in the concentration of CaO in the contact zone, as it approaches the conditional line of contact.

An analysis of the concentration curves in Fig. 2 shows that with an increase in the concentration of Na_2O in the contact zone, an increase in the concentration of Al_2O_3 is observed, which is probably a consequence of the formation of compounds of the type of alkali aluminosilicates. Calculations of quantitative values made it possible to express the opinion that there is a certain correlation between chemical compounds.

The ratio between oxides over the width of the contact zone relative to Na_2O showed that its greatest value is observed with distance from the conditional line of contact towards the brick within 450–1600 μm, and immediately upon approaching this line, the correlation decreases to a minimum.

Balanced correlation dependence between the compounds in the contact zone, as well as a high concentration of Na_2O in the latter, can contribute to the formation of hardly soluble hydrated minerals of alkaline-alkaline earth formations or low-basic calcium hydro silicates.

Thus, according to the research results, it can be judged that the alkaline medium promotes the formation of a dense structure and creates favorable conditions in the contact zone for the active interaction of the solution with the brick surface. As a result of this interaction, a layer of new formations of sub microcrystalline size and gel-like dispersed phases are formed, which form a matrix of interweaving in the contact zone and act as a damper now of action of alternating seismic loads.

References

1. Krivenko PV (1986) Synthesis of cementitious materials in a system $R_2O–Al_2O_3–SiO_2–H_2O$ with required properties. DSC (Eng) Thesis, Kiev Civil
2. Krivenko PV (2013) Alkali cements. In: Proceedings of the first international conference on alkaline cements and concretes, Kiev, Ukraine, Curriers
3. Bakhriev NF (1991) Highly adhesive slag-alkaline mortars for masonry walls of silicate brick. Abstract of dissertation for the degree of candidate of technical sciences Kiev 1991
4. Bakhriev NF, Buribaev SA, Buribaev A (2020) Recommendations for the processing of phosphogypsum waste. In: 3rd International conference on civil engineering and architecture (ICCEA 2020). France. 28.06.20
5. Bakhriev NF, Karimov GU (2021) Modeling and optimization of compositions of knitting low water requirements for producing cellular concrete. http://www.jcreview.com/?mno=125412. Access: 29 Janu 2021. https://doi.org/10.31838/jcr.07.16.380
6. Fakhritdinovich BN, Shairovna OG (2020) Universal building mixtures on high-module silicate components. Int J Orange Technol 2(10):144–148. https://doi.org/10.31149/ijot.v2i10.770
7. Larionova ZM, Vinogradov BN (2016) Petrography of cements and concretes. Number of pages, 348 s, 1974. Moscow 2(5):99–110

Numerical Study of the Tensile Experiments Modelling the Elongation Degree of Highly-Reinforced Extreme Edges of Seismic Walls for Studying the Transverse Buckling Phenomenon

Theodoros Chrysanidis and Vassilis Panoskaltsis

Abstract In the framework of the present study, the influence of the elongation degree on the phenomenon of out-of-plane buckling of reinforced concrete walls is examined. The present investigation is mainly analytical but it contains experimental results of 4 test specimens published by the first author in the past. These specimens simulate the extreme boundary edges of structural walls. All columns simulate only the extreme reinforced areas of the walls, in order to study the basic mechanism of the phenomenon. The detailing of the specimens consists of 4 rebars with a diameter of 12 mm for each bar. The geometric dimensions are the same for all specimens. What differentiates the specimens from each other is the degree of tension they have sustained. More specifically, the tensile degrees used are 10, 20, 30, 50‰. The loading stages of each specimen for all specimens are as follows: (a) Uniaxial central tensile loading on each test specimen, (b) Uniaxial central compression loading on each specimen till its failure due to buckling or due to an excess of its cross-section compressive strength. The present study focuses on the tensile loading stage only. Extreme tensile strains are also used, e.g., 30 and 50‰, to take into account the cases of extreme seismic excitations. First, the experimental results from a previous publication are presented and afterwards they are followed by the numerical investigation of these 4 specimens using an appropriate finite element framework. Useful conclusions are drawn regarding the precision of the experimental tests investigating the influence of the degree of elongation on the phenomenon of transverse buckling. These conclusions are substantiated both experimentally and analytically, since the results of the tensile experiments are compared with the corresponding results of the analytical investigation.

Keywords Numerical study · Elongation degree · Transverse buckling

T. Chrysanidis (✉) · V. Panoskaltsis
Demokritos University of Thrace, 67100 Xanthi, Greece
e-mail: chrysanid@uth.gr

V. Panoskaltsis
e-mail: vpanoska@civil.duth.gr

T. Kang and Y. Lee (eds.), *Proceedings of 2021 4th International Conference on Civil Engineering and Architecture*, Lecture Notes in Civil Engineering 201,
https://doi.org/10.1007/978-981-16-6932-3_11

1 Introduction

One crucial type of failure of reinforced concrete seismic walls is the lateral buckling [1, 2]. In international bibliography, it can be found either as lateral buckling or transverse buckling or out-of-plane buckling. The terminology instability is used sometimes, too instead of buckling. Several researchers worldwide have investigated this particular phenomenon and the mechanical parameters influencing it [3–21]. The basic parameters examined were degree of elongation, longitudinal reinforcement ratio and slenderness [22–28].

The present study uses experimental tests performed in the past by the first author trying to investigate the mechanical parameters affecting the transverse instability [4]. These specimens were subjected to low, medium and high tensile strains equal to 10, 20, 30 and 50‰. In the framework of the present work, the four specimens subjected to tensile loading are modelled using finite element analysis and the results of this analytical investigation are compared to the existing experimental results concerning the tensile loading stage. It is noted that the experiments have taken place at the Laboratory of Strength of Materials of the Aristotle University of Thessaloniki and the analysis of the results has taken place at the Demokritos University of Thrace.

2 Experimental Research

2.1 Test Specimen Characteristics

The experimental investigation for the four test specimens has been described in detail by the first author in the past [4]. Figure 1a shows the geometrical characteristics of the four test specimens, while Fig. 1b displays the load test setup used for the application of the tensile loading. It is noted that the tensile loading is the first stage of the two loading stages. In the framework of the present study, only the experimental results of the tensile loading stage are compared to the analytical ones. Table 1 shows the test specimens' characteristics.

2.2 Materials

For all test specimens, materials used for their construction and their characteristics were also described in the past [4]. Table 2 displays the concrete resistance for all test specimens at 28 days and at the day the compression test has taken place while Table 3 shows the mechanical properties for the longitudinal steel and the transverse ties.

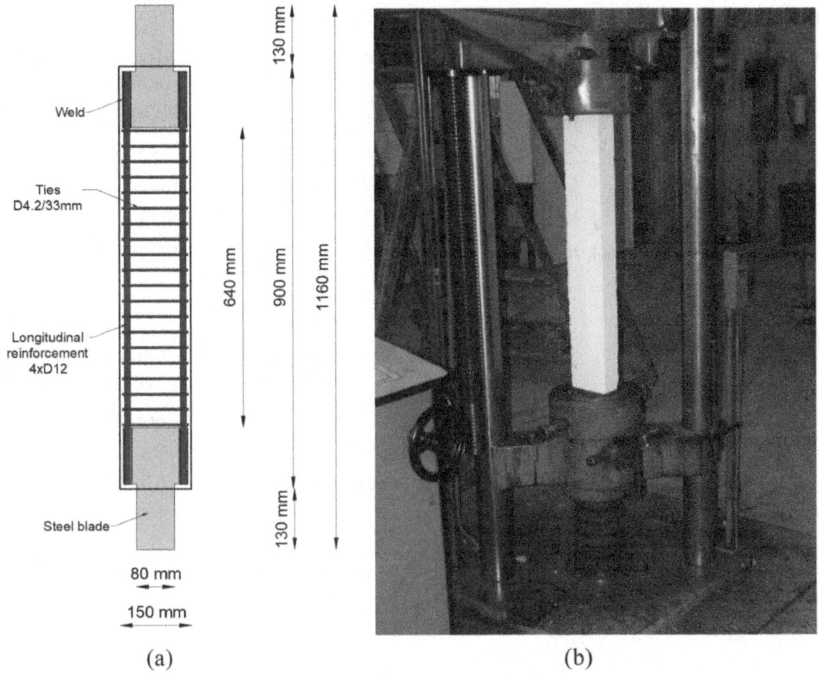

Fig. 1 **a** Vertical reinforcement layout, **b** Test setup for tensile loading

Table 1 Dimensions of the test specimens

N/A	Specimen	Dimensions (cm)	Longitudinal reinforcement
1	C-10	15 × 7.5 × 76	4xD12
2	C-20	15 × 7.5 × 76	4xD12
3	C-30	15 × 7.5 × 76	4xD12
4	C-50	15 × 7.5 × 76	4xD12
N/A	Transverse reinforcement	Longitudinal rebar ratio (%)	Elongation degree (‰)
1	D4.2@33 mm	4.02	10.00
2	D4.2@33 mm	4.02	20.00
3	D4.2@33 mm	4.02	30.00
4	D4.2@33 mm	4.02	50.00

2.3 *Experimental Results*

Each one of the four test specimens has been subjected to a different degree of elongation [4]. Figure 2 displays the shape of specimens after the uniaxial tensile

Table 2 Concrete mechanical properties

N/A	Specimen	Cube resistance (28 days) (MPa)	Cube resistance (compression test) (MPa)	Cylinder resistance (compression test) (MPa)
1	C-10	34.96	34.52	28.23
2	C-20	34.96	34.52	28.23
3	C-30	34.96	34.52	28.23
4	C-50	34.96	34.52	28.23

Table 3 Reinforcement mechanical properties

Reinforcement	Yield strength (MPa)	Ultimate strength (MPa)
D12 (longitudinal reinforcement)	560.27	666.43
D4.2 (transverse ties)	674.01	674.01

test has taken place. It is obvious that several cracks of different width have formed as it has happened in other similar experiments [29].

3 Analytical Research

3.1 *Modeling of Test Specimens*

The analytical research has taken place using a finite element software. 3D elements were used to model all four test specimens subjected to tensile loading. It is noted, as it has been mentioned before, that the present work focuses on modeling only the first stage of loading; meaning the tensile loading path till certain preselected and different degrees of elongation. For the concrete material, the inelastic concrete model of isotropic plasticity from the software library has been chosen. For the reinforcement bar material, the properties derived from experiments are implemented in the software in order to model the inelastic behavior of reinforcement steel. A bilinear isotropic model has been chosen for the behavior of rebar steel. The same inelastic model has been selected to model the behavior of the steel used for the transverse ties. 3D finite elements having an edge of 2 cm are used for the modeling of the concrete column section. Both the longitudinal reinforcement and the transverse ties are modelled using 3D finite elements having a length equal to 1 cm. Figure 3 shows the 3D model of the column both for the whole column section and the reinforcement steel. The column model is considered fixed at its base.

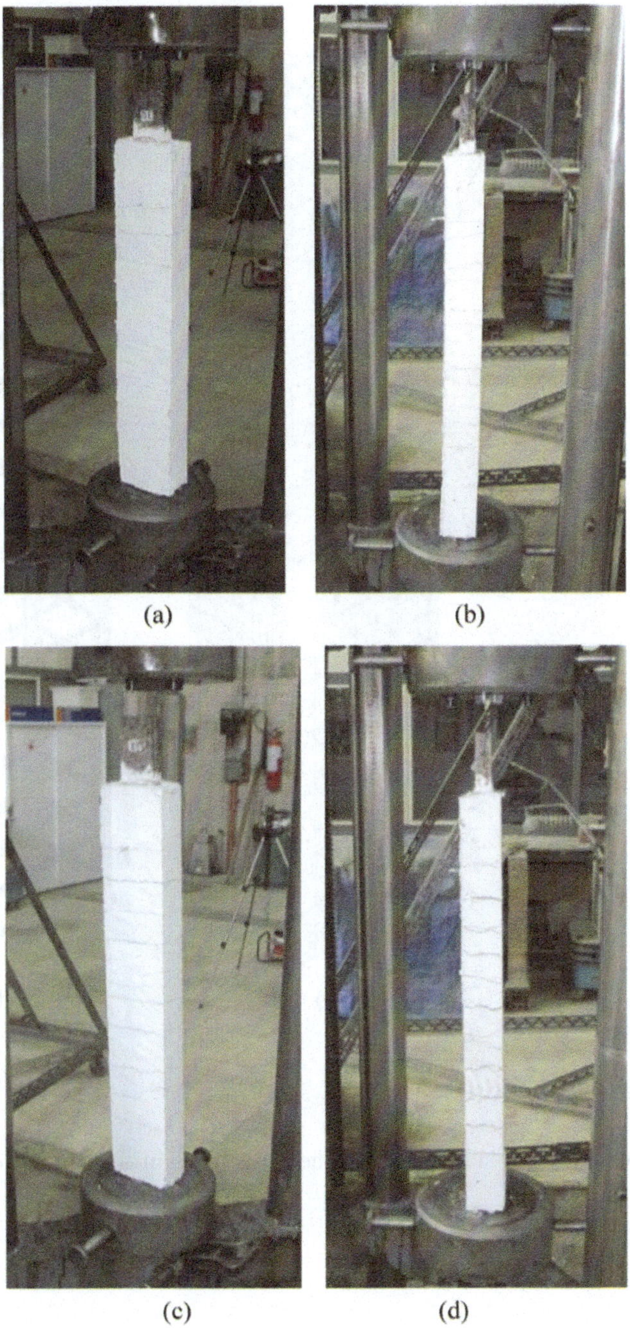

Fig. 2 Specimens with high reinforcement ratio after the uniaxial tensile test: **a** C-10, **b** C-20, **c** C-30, **d** C-50

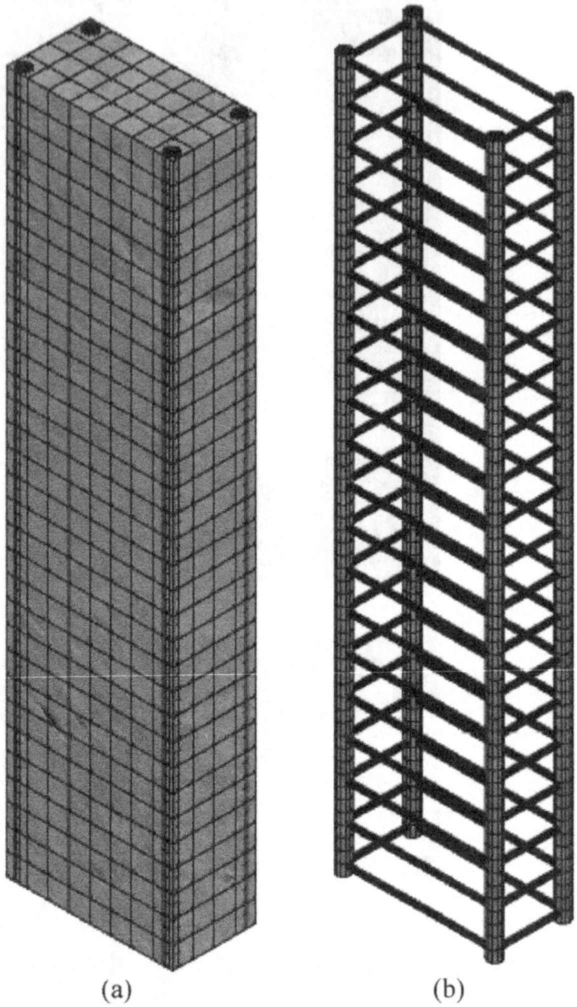

Fig. 3 a 3D model for column, b 3D model for the column reinforcement

3.2 Analytical Results

Figure 4 displays the displacement along the column height after the end of the tensile loading test.

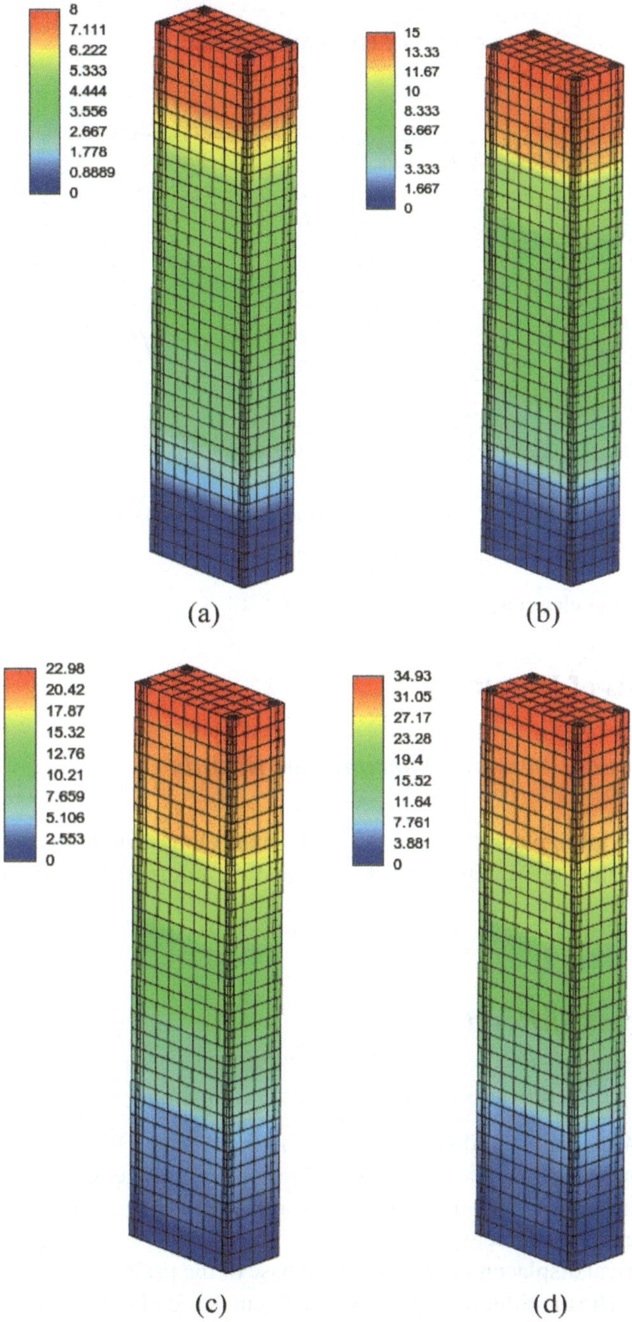

Fig. 4 Percentage change of concrete and steel weight per change of seismic zone

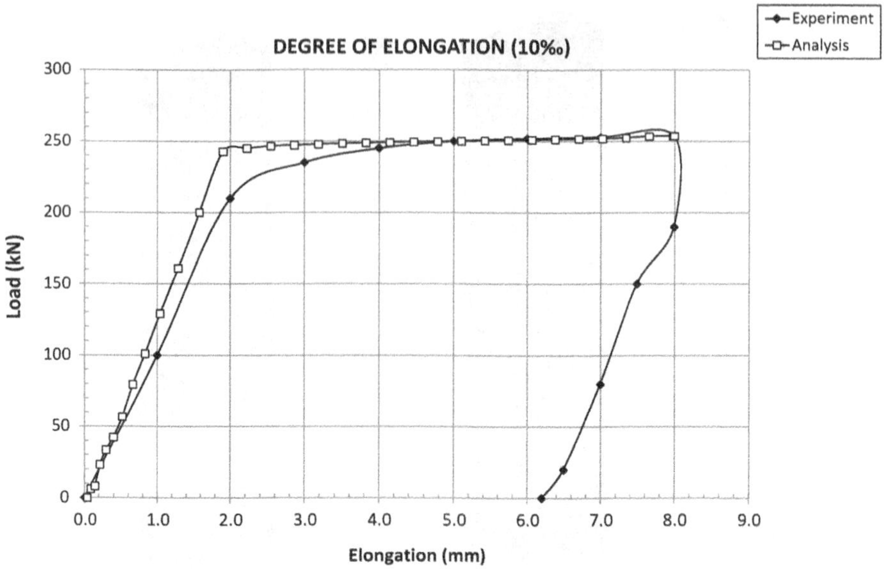

Fig. 5 Load versus elongation diagram for specimen C-10

4 Analysis of Results

4.1 Analytical Versus Experimental Results

A comparison takes place between the load versus elongation diagrams which have resulted from the experimental tensile tests and the numerical tensile tests (Figs. 5, 6, 7 and 8).

4.2 Analysis of Results

The analysis of the previous results leads to the following:

1. Figure 4 displays the displacement along the Z-axis which coincides with the height of the prism specimens modeling the extreme edges of seismic walls. All specimens are strained till the preselected displacement according to the preselected tensile degree.
2. The vertical displacement is zero at the base of the prisms since the base is fixed (Fig. 4). The maximum vertical displacement is found towards the upper part of the specimens where the tensile load is applied. The same phenomenon takes place in the tensile experiments where the tensile load is applied at the upper

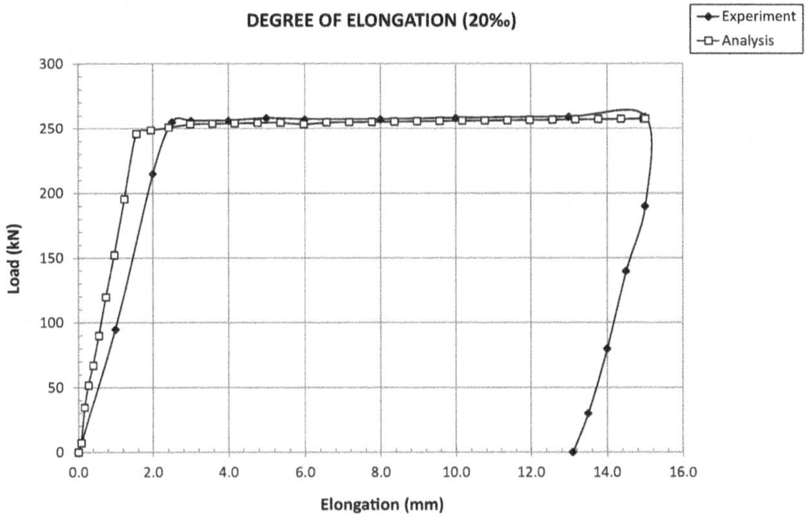

Fig. 6 Load versus elongation diagram for specimen C-20

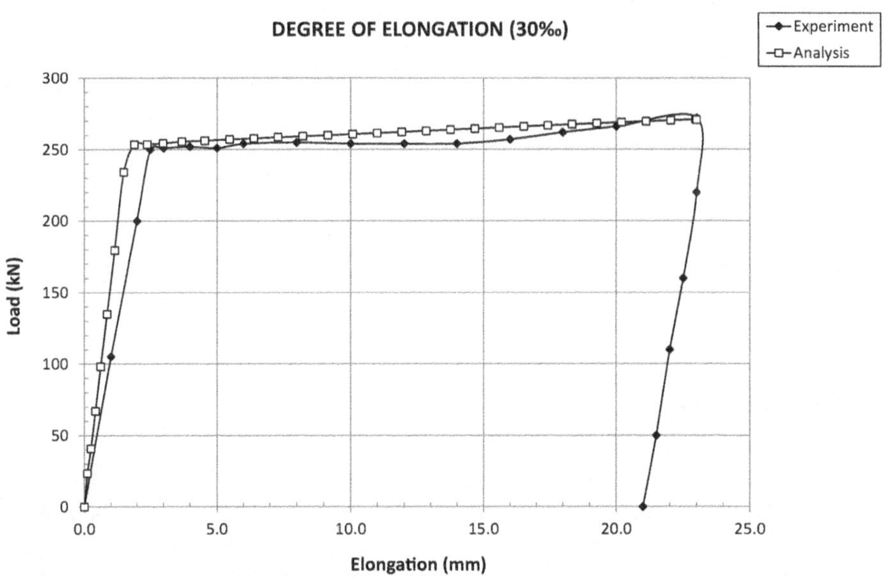

Fig. 7 Load versus elongation diagram for specimen C-30

part of the test specimens while at the bottom part the test specimens are held rigidly by the grapples of the tensile machine.

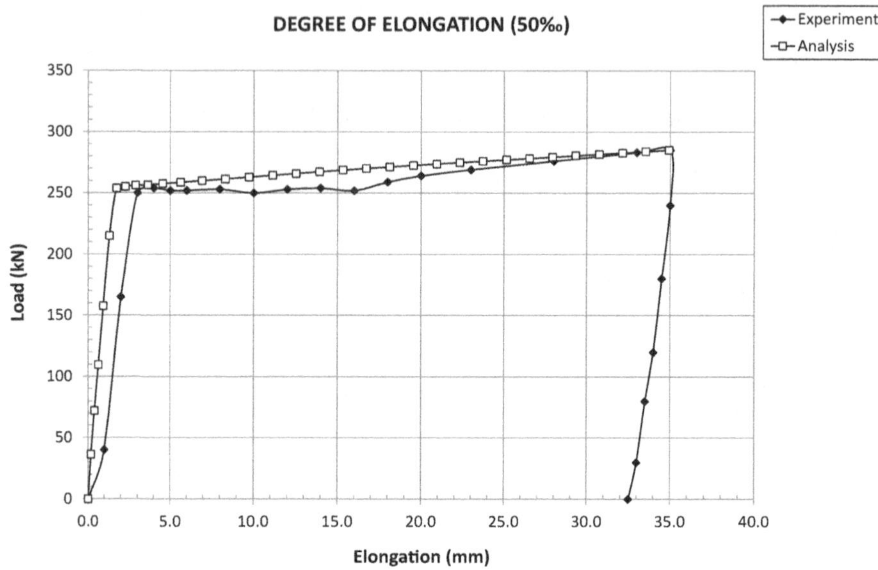

Fig. 8 Load versus elongation diagram for specimen C-50

3. It is obvious that there is an almost perfect correlation between the experimental and the analytical results for all test specimens (Figs. 5, 6, 7 and 8). Both the elastic and the plastic branch coincide between the experiments and the analyses.
4. It is noteworthy that yielding takes place almost at the same load both for the tests and the analyses for all test specimens.
5. Only the effective length of the specimens has been modelled, since this is the length within which the tensile elongation appears (Fig. 1). The effective length is equal to 640 mm (Fig. 1).

5 Conclusions

In the framework of the current research, an analytical investigation has taken place including numerical analyses of four test specimens modeling the extreme ends of R/C seismic walls. The numerical analyses have taken place using 3D finite elements. The results from the numerical analyses are compared to the relevant experimental results from the same test specimens found in a previous publication of the first author. The following conclusions can be drawn:

1. There is an almost excellent correlation between the experimental results and the analytical results concerning the load–displacement diagrams of the tensile loading.
2. The elastic branch, the yielding point and the plastic branch coincide almost perfectly for all four test specimens.

3. The degree of elongation is a very crucial mechanical parameter that affects tremendously the behavior of the boundary edges of structural walls and its investigation has to be applied in the proper and right way following a correct procedure. The convergence between the experimental and the numerical results proves that the procedure applied, in the present work, follows the right path.
4. A future research could and should model the whole test specimen and not only the effective length in order to simulate even more precisely the experimental behavior. Then this more precise analytical behavior could be compared again with the relevant experimental results.

References

1. Penelis GG, Kappos AJ (1996) Earthquake-resistant Concrete Structures. E & F N SPON (Chapman & Hall), London
2. Penelis G, Stylianidis K, Kappos A, Ignatakis C (1995) Reinforced concrete structures. A.U. Th. Press, Thessaloniki
3. Chrysanidis T (2021) The effect of longitudinal reinforcement ratio on the lateral buckling behavior of R/C walls modelled using prism elements. J Build Eng 42. https://doi.org/10.1016/j.jobe.2021.102456
4. Chrysanidis T (2020) Evaluation of out-of-plane response of R/C structural wall boundary edges detailed with maximum code-prescribed longitudinal reinforcement. Ratio. https://doi.org/10.1186/s40069-019-0378-4
5. Chrysanidis T (2019) Influence of elongation degree on transverse buckling of confined boundary regions of R/C seismic walls. Constr Build Mater 211:703–720. https://doi.org/10.1016/J.CONBUILDMAT.2019.03.271
6. Chrysanidis T, Tegos I (2017) Does reinforcement ratio affect displacements due to lateral buckling behavior of concrete walls? Int J Appl Eng Res 12:382–388
7. Chrysanidis T (2016) Low reinforced shear walls: displacements and failure modes due to lateral buckling. Int J Sci Eng Invest 5:143–148
8. Chrysanidis T (2016) Degree of elongation of maximum code-prescribed reinforced walls: modes of failure and displacements of lateral buckling phenomenon. Int J Trend Res Dev 3:435–440
9. Chrysanidis T, Tegos I (2016) Displacements and mode of failure of medium high reinforced walls due to transverse buckling. Int J Eng Dev Res 4:628–634
10. Chrysanidis T, Tegos I (2016) Size of seismic tensile strain and its influence on the displacements due to transverse buckling of ultra-highly reinforced structural walls. ARPN J Eng Appl Sci 11:13884–13890
11. Chrysanidis T, Tegos I (2016) Influence of elongation degree on out-of-plane buckling of R/C structural walls with a medium high reinforcement ratio. Int J Civil Eng Technol 7:358–364
12. Chrysanidis T (2014) Size of seismic tensile strain and its influence on the lateral buckling of highly reinforced concrete walls. IOSR J Mech Civil Eng 11:18–22. https://doi.org/10.9790/1684-11121822
13. Chrysanidis T, Tegos I (2014) Out-of-plane buckling of highly reinforced seismic walls: displacements and mode of failure. Int J Civil Eng Technol 5:101–107
14. Rosso A, Jiménez-Roa L, De Almeida JP, Zuniga AP, Guerrero, Blandón C, Bonett R, Beyer K (2018) Cyclic tensile-compressive tests on thin concrete boundary elements with a single layer of reinforcement prone to out-of-plane instability. Bull Earthquake Eng 16:859–. https://doi.org/10.1007/s10518-017-0228-1

15. Dashti F, Dhakal R, Pampanin S (2018) Evolution of out-of-plane deformation and subsequent instability in rectangular RC walls under in-plane cyclic loading: Experimental observation. Earthquake Eng Struct Dynam 47:2944–2964. https://doi.org/10.1002/eqe.3115
16. Rosso A, Jiménez-Roa L, Almeida J, Beyer K (2017) Experimental campaign on thin RC columns prone to out-of-plane instability: numerical simulation using shell element models. In: VIII Congreso Nacional de Ingeniería Sísmica. Barranquilla, Colombia
17. Almeida J, Prodan O, Rosso A, Beyer K (2017) Tests on thin reinforced concrete walls subjected to in-plane and out-of-plane cyclic loading. Earthq Spectra 33:323–345. https://doi.org/10.1193/101915EQS154DP
18. Rosso A, Almeida JP, Beyer K (2016) Stability of thin reinforced concrete walls under cyclic loads: state-of-the-art and new experimental findings. Bull Earthq Eng 14:455–484. https://doi.org/10.1007/s10518-015-9827-x
19. Taleb R, Tani M, Kono S (2016) Performance of confined boundary regions of RC walls under cyclic reversal loadings. J Adv Concr Technol 14:108–124. https://doi.org/10.3151/jact.14.108
20. Rosso A, Almeida J, Beyer K (2016) Out-of-plane behaviour of reinforced concrete members with single reinforcement layer subjected to cyclic axial loading: beam-column element simulation. In: Proceedings of the 2016 NZSEE conference. Christchurch, New Zealand, pp 1–9
21. Rosso A, Almeida JP, Jimenez L, Guerrero ZP, Blandon C, Bonett-Diáz R, Beyer K (2016) Experimental tests on the out-of-plane response of RC columns subjected to cyclic tensile-compressive loading. In: 16th World conference on earthquake engineering (2016)
22. Chai Y, Kunnath S (2005) Minimum thickness for ductile RC structural walls. Eng Struct 27:1052–1063. https://doi.org/10.1016/j.engstruct.2005.02.004
23. Chai Y, Elayer D (1999) Lateral stability of reinforced concrete columns under axial reversed cyclic tension and compression. ACI Struct J 96:780–789
24. Chrysanidis T, Tegos I (2015) The influence of the degree of elongation to the displacements of seismic walls with maximum code-prescribed reinforcement ratio. Am Acad Scholar Res J 7:273–279
25. Chrysanidis T, Tegos I (2015) How does degree of elongation affect lateral buckling behavior of seismic walls? Am Acad Scholarly Res J 7:266–272
26. Chrysanidis T, Tegos I (2021) Does the web of seismic walls play a role to their out-of-plane stability? In: Gnana S (ed) Advanced aspects of engineering research, vol 2. Book Publisher International, London, pp 15–22. https://doi.org/10.9734/bpi/aaer/v2/7485D
27. Chrysanidis T (2020) Evaluation of lateral buckling phenomenon using R/C prisms detailed with an ultra high reinforcement ratio. In: Salisu L (ed) Emerging trends in engineering research and technology, vol 3. Book Publisher International, London, pp 55–64. https://doi.org/10.9734/bpi/etert/v3
28. Chrysanidis T, Tegos I (2020) Can the diameter of longitudinal bars of extreme edges of R/C walls halt transverse buckling? In: Salisu L (ed) Emerging trends in engineering research and technology, vol 3. Book Publisher International, London, pp 65–72. https://doi.org/10.9734/bpi/etert/v3
29. Chrysanidis T, Panoskaltsis V (2020) Experimental investigation of the influence of tensile strain on the cracking of R/C vertical structural elements. In: Papadrakakis M, Fragiadakis M, Papadimitriou C (eds) Proceedings of the XI international conference on structural dynamics (EURODYN 2020), Athens, Greece, pp 3174–3185. https://doi.org/10.47964/1120.9259.18582

Concrete Structure Design and Performance Analysis

Comparative Study on Dynamic Response of Square Section RC-Members and CFRP-Reinforced Members Under Unequal Transverse Impact Loading

Liu Yanhui, Khalil AL-Bukhaiti, Zhao Shichun, Hussein Abas, Khaled Zaidi, Nan Xu, and Dong Aoran

Abstract In recent years, fiber-reinforced polymer (FRP) use to enhance existing structural members' impact resistance has become a research hotspot. This paper investigates Carbon Fiber Reinforced Polymer (CFRP) effectiveness to increase Impact Resistance of Reinforced Concrete Square Member under lateral impact loads. This article clearly said that increased strength of concrete members using CFRP layers leads to increased CFRP layers' strain. To investigate the effectiveness of Carbon Fiber Reinforced Polymer's use to increase the impact resistance of reinforced concrete square members under lateral impact loads is studied; falling-weight impact tests were conducted. The impact load was applied to the 2L/9 from the length of the RC member's span with a free-falling hammer. All of the specimens with both ends fixed support with short span members were used for these experiments. In this study, the strengthening method and material properties of the CFRP layers and the number of CFRP layers were varied. The results show that using CFRP to strengthen the square RC member specimen in shear effectively prevents shear failure and leads the specimen's response to be flexural dominated. The short span of the square RC member turns to be shear failure mode dominated by diagonal cracks. The square RC member strengthens with CFRP is shown as bending failure with vertical cracks longitudinal bars rupture. Compared with the square RC member, the impact peak value of the CFRPRC members changes little.

Keywords Impact test · Failure mechanism · Flexural performance · Unequal trans-lateral impact

1 Introduction

The application of fiber-reinforced polymer (FRP) materials in construction have increased in the past 20 years. Carbon fiber reinforced polymer (CFRP) is a type of FRPs, which is a polymer matrix composite material reinforced by carbon fibers. The major advantages of CFRP sheets are pointed out as high strength, corrosion

---

L. Yanhui · K. AL-Bukhaiti (✉) · Z. Shichun · H. Abas · K. Zaidi · N. Xu · D. Aoran
Southwest Jiaotong University, Chengdu 610031, Sichuan, China

T. Kang and Y. Lee (eds.), *Proceedings of 2021 4th International Conference on Civil Engineering and Architecture*, Lecture Notes in Civil Engineering 201,
https://doi.org/10.1007/978-981-16-6932-3_12

resistance, ease of implementation, and less impact on the original geometry. CFRP sheets are bonded to external faces of RC structures for multipurpose strengthening. This method can be carried out quickly, without using heavy equipment compared to the other conventional methods (strand splices, external post-tensioning tendons, or steel plate jackets) [1]. There have been numerous cases in which some large-scale civil engineering projects have suffered from shocks in recent years. For example, in the September 11 attacks in the United States, terrorists hijacked a passenger plane, crashing into the twin towers of the World Trade Center in New York, causing a fire, resulting in structural loss of capacity and eventually collapse. In 2005, a train derailment accident occurred on the JR Fukuchiyama Line in Hyogo Prefecture, Japan [2]. On a highway in Hunan, China, in 2009, trucks lost control and hit the pier, and caused the collapse of the upper structure of the bridge (Fig. 1a) [3]. Also, in the USA, Texas, in Corpus Christi, on May 14, 2014, truck Impact on the Piers of Tancahua Street Bridge over IH-37 as (Fig. 1b) [4].

The absence of transverse reinforcement can result in shear failure, which is both fragile and catastrophic. The shear capacity of lacking members can be essentially improved by bonded remotely fortified CFRP laminates with fibers in the hoop direction, as appeared in Fig. 2. Researches have demonstrated that an increase in the thickness of CFRP and AFRP laminates relatively builds the shear strength of the upgraded member or pier [4–7].

Therefore, accidents such as rockfall impact and station derailment trains that may be encountered in the civil engineering structure in daily life belong to the category of low-speed shock, which is also the type of impact load studied in this paper. The accuracy of the experimental results will be verified based on the results of combined experiments. The influence of CFRP reinforcement on the failure mechanism of RC members be analyzed. In this paper, the factor is sought to better represent the impact-resistant characteristics for RC members. For this purpose, 12 specimens were carried out on the DHR-9401 drop hammer impact tester of the Taiyuan University of Technology, China, on RC members that theoretically exhibit flexural failure under the lateral impact CFRP shear reinforcement. The most important thing that can be

China Hunan Vehicle Impact Pier [3]

The pedestrian bridge at Qingchun Road, Hangzhou, China, collapsed by a truck.[4]

Fig. 1 Vehicles impact pier

Fig. 2 Application of FRP for seismic retrofitting of RC members [5]

considered as a new and innovative point in this paper is that it has advantages that differ from the rest of the previous scientific research, namely:

The point of the falling hammer for impact force is with unequal transverse span.

The cross-section of the specimens is minimal compared to previous studies.

All of the specimens have both ends fixed supports.

1.1 Research Significance

This research aims to develop an impact-resistant design method for the square concrete member when strengthened by wrapping them with carbon fiber reinforced polymer (CFRP) sheets following the performance of a based design concept based on experimental results. The work described in this paper investigates the dynamic response characteristics of CFRP square concrete specimens under drop hammer impact force loading with different input impact energies.

2 Experimental Overview

2.1 Dimensions and Static Design Values for Test Specimens

This study investigates the failure mode, force, and deformation of the RC and CFRP members under the lateral impact of CFRP shear reinforcement. The specimen numbers in the table below are defined as shown in Table 1 for un-strength (RC) Specimen and Table 2 for strength (CFRP) Specimen.

Where:

Table 1 RC test specimen information

No.	Long. Reinf. ratio	Volume stirrup ratio	Impact height	No. of test specimens	factor
FH1	2.01% (4 ϕ 8)	0.62% (ϕ 4@50)	1	2	height
FH2	2.01% (4 ϕ 8)	0.62% (ϕ 4@50)	2	2	
FH3	6.16% (4 ϕ 14)	0.64% (ϕ 4@50)	2	2	Longitudinal reinforcement ratio
FH4	2.01% (4 ϕ 8)	0.31% (ϕ 4@100)	2	2	stirrup ratio
FH5	2.01% (4 ϕ 8)	0.62% (ϕ 4@50)	2	2	Axial compression ratio

Table 2 Information on CFRP specimens

No.	Wrapping Method	Impact height	Number of layers	Number of test specimens
FH2-Z1	Vertical	2	1	2

L: is the length of specimen = 1500 mm; B: is the width of specimen = 120 mm;

Both ends fixed drop hammer mass 270 kg; impact kinetic energy 1905 J for FH2 and 5292 J for CFRP specimens.

As shown in Fig. 3 for the Schematics of the cross-section and dimensions for the specimens, it is clear to see the direction and positions of wrapping for CFRP layers.

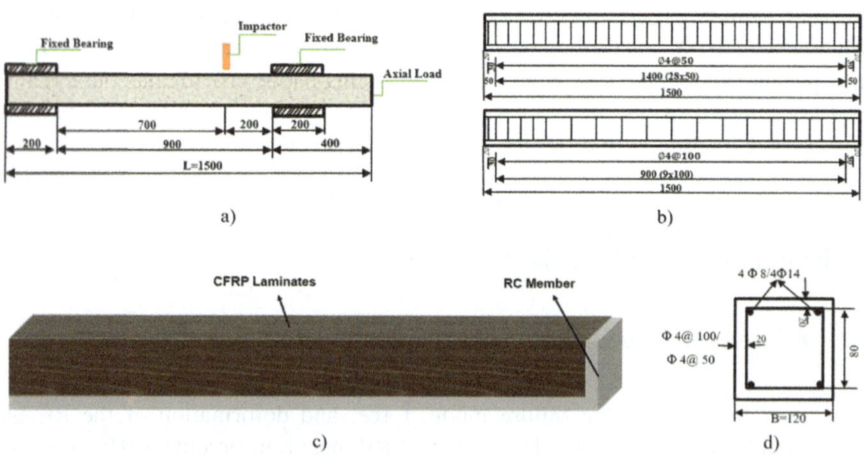

Fig. 3 The schematic diagram for **a** specimens dimensions, **b** specimen reinforced details, **c** CFRP laminate wrapping, **d** specimen cross-section

2.2 Test Devise, Loading, and Measurement Scheme

The maximum height of the drop hammer guide is 13.47 m, and the maximum impact speed can reach 15.7 m/s. The cross-section of the drop hammer is rectangular, 80 mm long, and 30 mm wide. Made of Cr with high hardness and little deformation under the impact, the weight of a single mass is 65 kg, as shown in Fig. 4. In this test, the total impact head and mass are 270 kg. The impact time history curve measured by a force sensor between the impact head and the mass, and the high-speed camera placed in front of the specimen capture the entire process of impact deformation. The shooting frequency is 2500 frames/s. The deflection time curve of the impact point is obtained by deforming the image of the specimen captured by the high-speed camera, the deformation of the member can be captured once every 0.4 ms.

Drop hammer impact tests were conducted by dropping the weight from a prescribed height onto the point of 2 L/9 of the member using the impact test apparatus, as shown in Fig. 4. CFRP members are placed on the supports equipped with impact hammer load for measuring the reaction forces and are clamped at their ends using cross members to prevent lifting off. The supports are Fixed and unable to rotate freely while restraining in any direction of the members. After each test, the residual deflection was measured, and crack patterns observed on the side surface of the specimens were sketched for documentation.

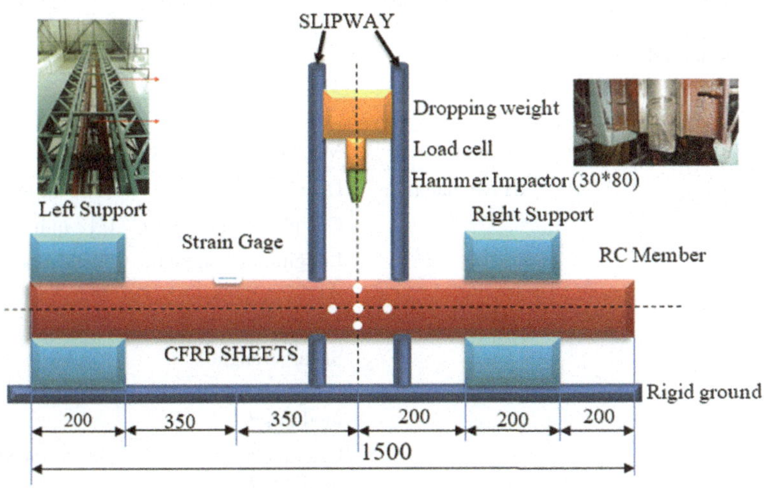

Fig. 4 Falling-weight impact tests apparatus

2.3 Material Properties

- CFRP: To measure the strength of the CFRP longitudinal (Xt), transverse (Yt), and shear (S) directions, the test is in according to the American Society for Testing and Materials (ASTM D3039-08) [7]. A pull-up tester Fig. 5 tested the CFRP sheet after the epoxy resin was impregnated and cured in three directions. The final measured average strength, elastic modulus, and ultimate failure strain are listed in Table 3.
- Concrete: Concrete strength grade is C30. Several cubic concrete test blocks (150 × 150 × 150 mm) were fabricated simultaneously as the casting members. The compressive strength test was carried out (Fig. 7c), measured the average compressive strength of concrete for 28 days $f_{cu} = 43.5$ MPa.

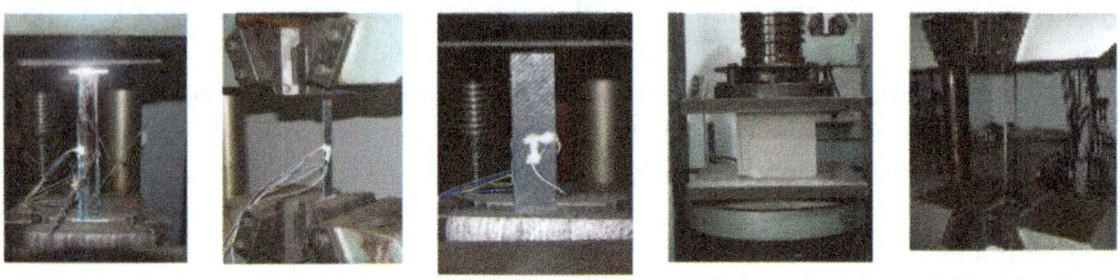

a) Longitudinal tensile test b) Lateral tensile test c) Oblique 45° tensile test d) Cube uniaxial compression test e) Steel bar tensile test

Fig. 5 Material test

Table 3 CFRP performance indicators

Material	Mechanical parameters	
CFRP material properties	Density ρ	1800 kg/m^3
	Calculated thickness t	0.17 mm
	Longitudinal strength Xt	3950 MPa
	Lateral strength Yt	74 MPa
	Shear strength S	108 MPa
	Longitudinal elastic modulus E_{aa}	2.1×105 MPa
	Lateral elastic modulus E_{bb}	1.8×104 MPa
	Shear modulus G_{ab}	4657 MPa
	Longitudinal failure strain ε_{aa}	1.6%
	Lateral failure strain ε_{bb}	0.8%

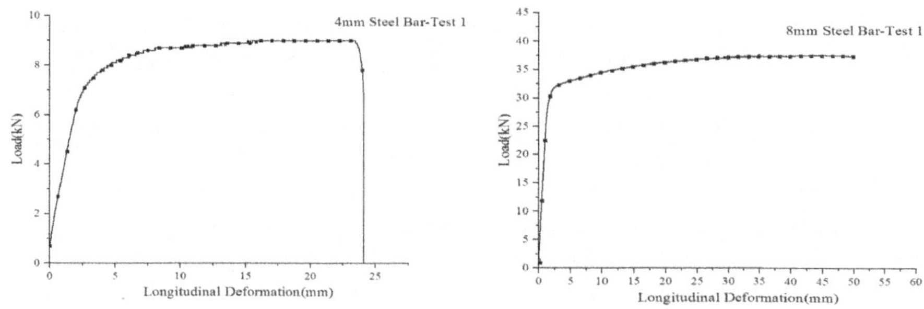

a) Tensile Test of Type Φ4 b) Tensile Test of Type Φ8

Fig. 6 Test curves of steel bars

Table 4 Rebar performance indicators

Material	fu/MPa	E/MPa	δ
Longitudinal reinforcement	320	2.1 × 105	0.21
Stirrup	520	2.0 × 105	0.22

- Steel Bar: The steel bars Type Φ4 were used as transverse bars, and Type Φ8 steel bars were used as longitudinal bars. The test curves of steel bars are presented below, as shown in Fig. 6.

The strength of the steel bar is measured according to the "Metal Material Tensile Test Method" (GB/T228-2010) (Fig. 6b) [8]. The average ultimate tensile strength (fu) of the steel bar is measured. Parameters such as modulus (E) and elongation (δ) are detailed in Table 4.

3 Details of Test

To clarify the dynamic response characteristics of the structural member under the lateral impact, the structural member is placed in the illustrated (Fig. 3a). The impact body under the design speed hits the structural column at 200 mm from the edge of the right support.

3.1 A Crack of Specimens Development

Due to the limitation of the imaging distance, the imaging range is only concentrated in the right half of the specimen (i.e., the short-span region); since the crack is difficult to observe in the early stage of the impact, marked with arrows, the direction of the arrow is perpendicular to the direction of the crack, also by the schematic diagram of the crack development. It can be observed that from FH1 to FH5 in Fig. 7, when the drop hammer is in contact with the member (0.4 ms), in FH2, two cracks can be observed near the impact point sections, which are the vertical cracks below the sections, and a 45° downward deflection crack to the right of the section. Then, at 1.6 ms, two nearly parallel oblique cracks appearing in the upper part of the right span of the member are observed and connected to the impact point. The two cracks then widened quickly (4 ms). They extended toward the top surface the right support, eventually leading to the destruction of FH1 to FH5. In addition, it can be observed that the occurrence of any cracks is not observed in the long span of the members (i.e., the left span). For specimen FH2-Z1 in Fig. 8, it can be found that similar to RC specimens. When the drop hammer is in contact with FH2-Z1 (0.4 ms), the vertical

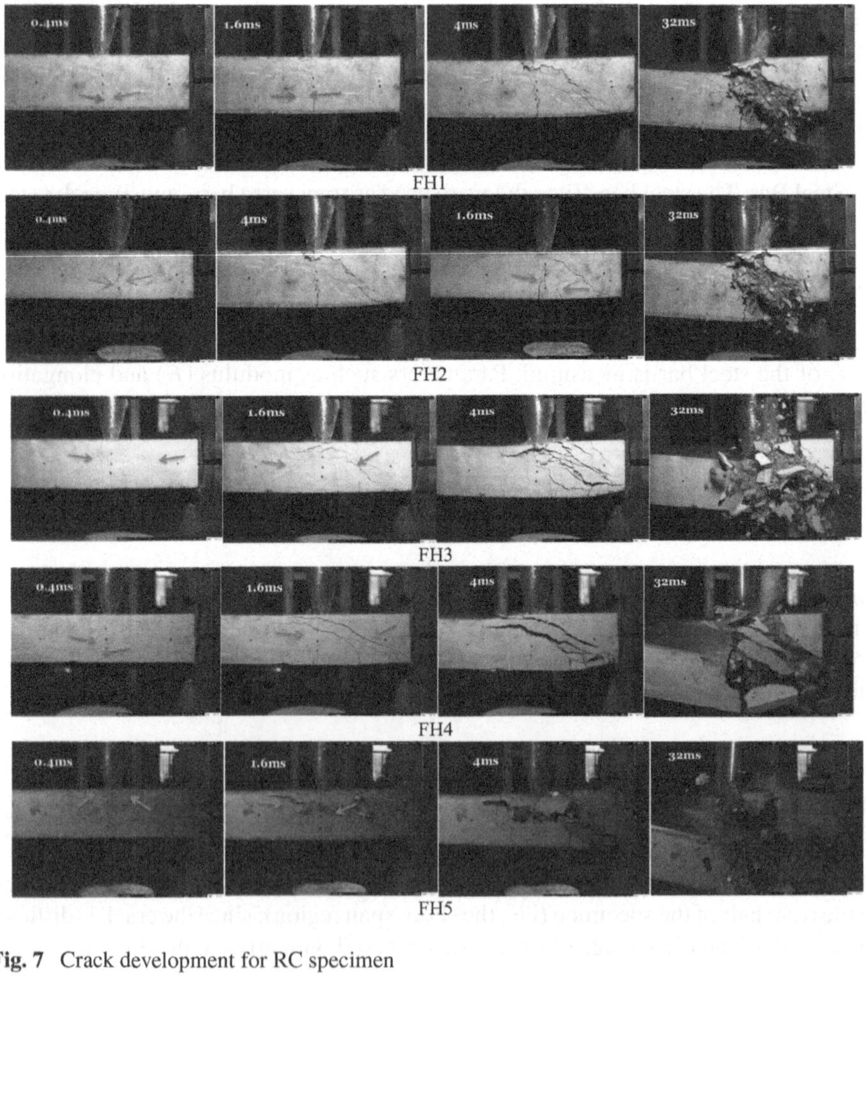

Fig. 7 Crack development for RC specimen

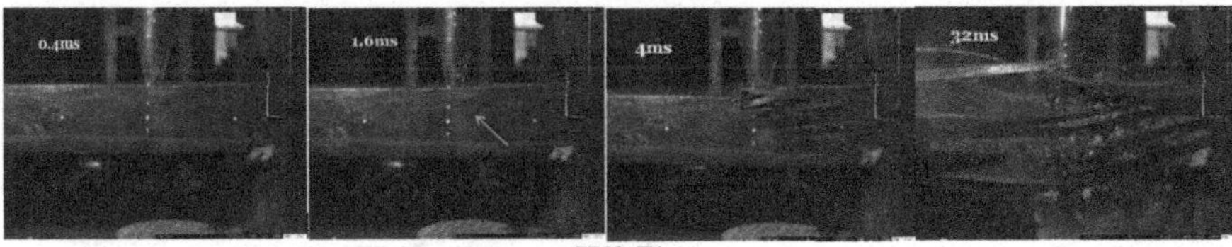

FH2-Z1

Fig. 8 Crack development for CFRPRC specimen

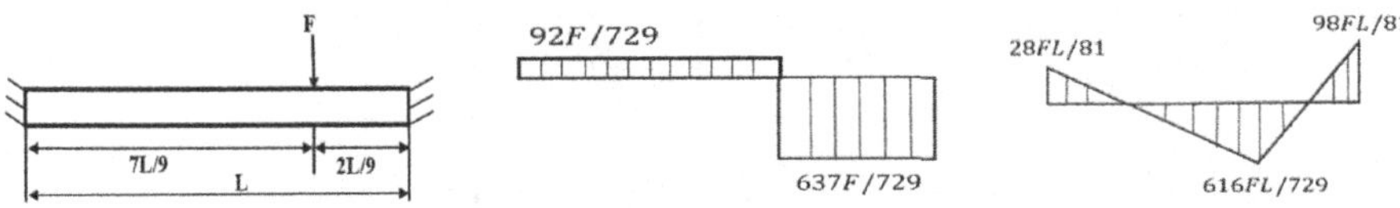

(a) Force diagram

(b) Schematic diagram of shear distribution

(c) Diagram of bending moment distribution

Fig. 9 Force diagram of specimens under the static force

cracks at the impact point cross-section also occur at 4 ms. It can be seen that there are multiple vertical cracks in the impact point section. Then similar vertical cracks appear in the upper part of the right bearing section. The cracks of the two sections gradually develop upward and downward, and the crack width gradually expands. In addition to vertical tensile cracking throughout the process, CFRP did not exhibit other forms of damage. It is worth noting that although the concrete inside the short span may have a certain degree of damage under the impact, the whole member is still dominated by the partial bending deformation due to the CFRP wrapping.

In contrast, vertical cracks do not play a leading role in the destruction of specimens, also found similar phenomena in recent research Pham and Hao [9]. From the static point of view, it is easy to know that the shear force of the member in the short span is much larger than that of the long span (Fig. 9), so the damage of the short span is more obvious.

3.2 Failure States

The article shows in Fig. 10 the failure state of specimens after impact force. It can be seen that for all RC specimens, massive cracking of concrete with the exposure of the steel reinforcement bars was observed in the diagonal crack region. For perception, the slope up line in Fig. 10 is the crack appearance. Within the rectangular dotted line is the shear failure. It can be seen that FH1 to FH5 has a large lateral deformation, the concrete fraction damage in the right span area, and the bearing capacity is lost. Only the steel bars in this area can continue to afford, indicating that RC specimens have lost the overall force mechanism during the impact process. For FH1 with the

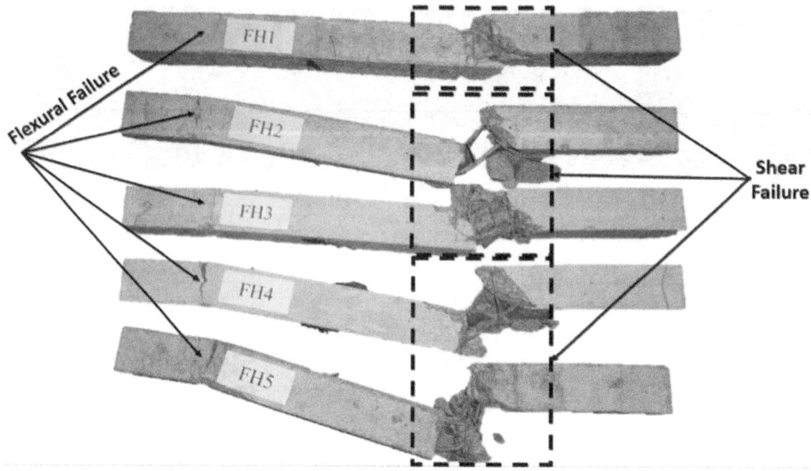

Fig. 10 RC specimens failure modes after the end of the impact

impactor, the height was 1 m. There is a small vertical crack at the left support that could be identified.

Furthermore, when the drop height increases to 2 m, shear failure at the impact point with vertical cracks at the upper end of the left support were increased as in specimen FH2, FH3, FH4, and FH5 Indicating that the bending damage is mainly subjected close to the specimens left support. The results revealed that the shear capacity of the strengthened RC specimen was significantly larger than its flexural capacity, leading to a flexural failure mode that occurred during the impact. For FH3, the increase of longitudinal and transverse reinforcement ratio supports the enhancement of RC specimen resistance against impact load. Importantly, the increases of longitudinal reinforcement from 2.01% in specimen FH2 to 6.16% in FH3 greatly affect the specimen damages under impact loading. The reduction of transverse reinforcement from 0.62% in FH2 to 0.31% in FH4 leads to increased local and overall failure modes. FH5 regarding the axial compression ratio factor leads to a decrease in flexural failure during the impact on the specimen left support compared with FH2 with the same design condition.

FH2-Z1 presents a completely different failure mode shown in Fig. 11. The damaged areas are concentrated on the impact point section and the right support section. Both sections have vertical tensile cracks and are divided at the impact point. Except for the uppermost row of steel bars, all the other parts are broken. On the contrary, the steel bars of the right support section are all broken except the lowermost row. From the comparison point of view, the damage of the left bearing section is similar to that of FH2 as a flexural failure.

Furthermore, the rectangular dotted line is defined as a shear failure on the impact position. Although the right span inner concrete may be damaged under the strong impact, and there is no severe cracking, which also enables FH2-Z1 to the whole body is deformed by bending. The above perceptions demonstrated in Fig. 11 that CFRP

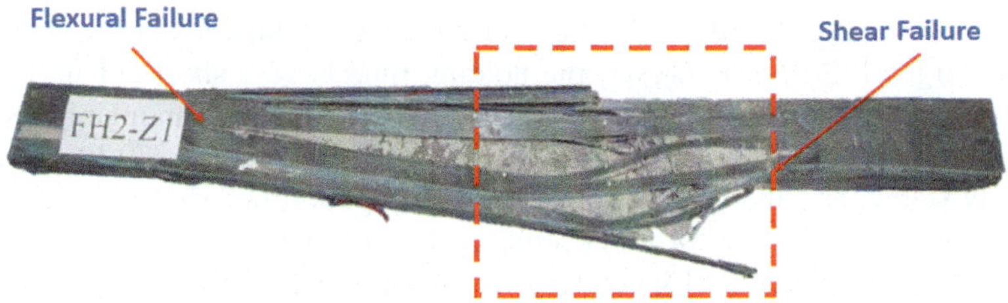

Fig. 11 FH2-Z1 failure modes after the end of the impact

strengthening could change the failure instrument of the RC component from shear failure mode to flexural failure mode. CFRP strengthening hence could successfully keep the showing up of shear failure. This paper's direction is confirmed with the available experimental works in the previous studies, which field the same conclusion [10–12].

4 Test Results and Discussion

4.1 Impact Time History Analysis

According to the impact force trend, the impact force curve can divide into three stages: the peak stage, the plateau phase, and the unloading stage. Among them, the peak point occurs when the drop hammer is in contact with the member. At this stage, the deformation of the member is very small, the inertial force provides the resistance, the support is not completely provided, and the reaction force is provided, the force of the member enters a relatively stable state. The point is from the end of the impact failure value to the phase of the force gradually decreasing to zero. Figure 13 for FH1 shows that the peak impact force is (294 kN) compared with FH2 (493 kN), the peak impact force for FH1 is lower than in FH2, which is because of the decrease of drop hammer height. In this stage, the impactor fluctuated at the same speed as the RC member. Due to a decrease in their speed, a large amount of kinetic energy is transferred and dissipated through the deformation of the specimen. It can also recognize that the increase in the longitudinal reinforcement ratio for FH3 has a limited effect on the peak impact force.

Nevertheless, it has more bunchy in the plateau stage Fh3 (100 kN) compared with other specimens (80 kN, 48 kN, and 50 kN) for (FH1, FH4, and FH5) respectively. The impact force of FH2-Z1 was significantly higher than FH2, and experienced during the ascent to descent process, the force climbed from 50 to 130 kN (3 ms), then began to decline slowly, and suddenly saw a sudden drop at (4 ms). This is because the two components' different deformation and failure modes directly affect the respective stress states. The internal force of the section was redistributed; the steel bars of other rows continued to bear, and the impact force entered the falling section. In general, FH2-Z1, due to the fracture of longitudinal steel bars, the plateau value is large (112 kN). However, the holding time is very short (11 ms). At the same time, the concrete damage of FH2 is serious. The right span mainly relies on the continued force of the steel bars. The failure value is small (70 kN), but it is relatively stable, and the time is up to 27.5 ms. Figure down shows the compared Impact time history curve for all specimens. It is clear to see the effect of CFRP layers on impact force for all specimens, as shown in Fig. 12.

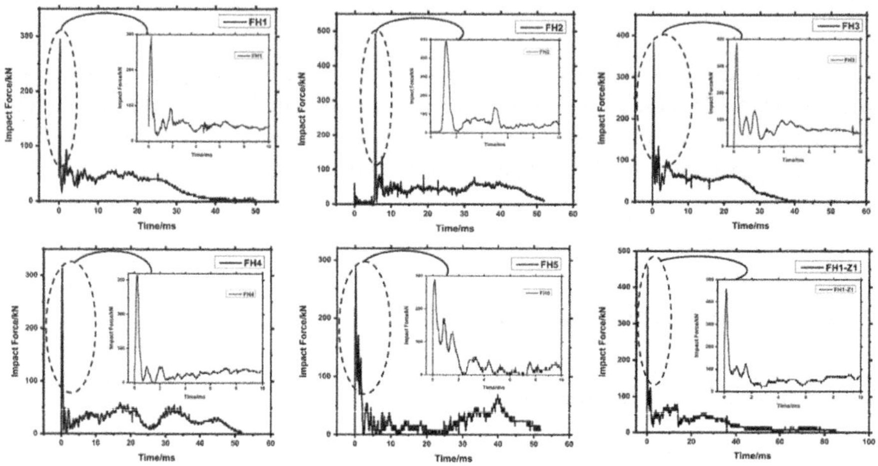

Fig. 12 Impact time history curve

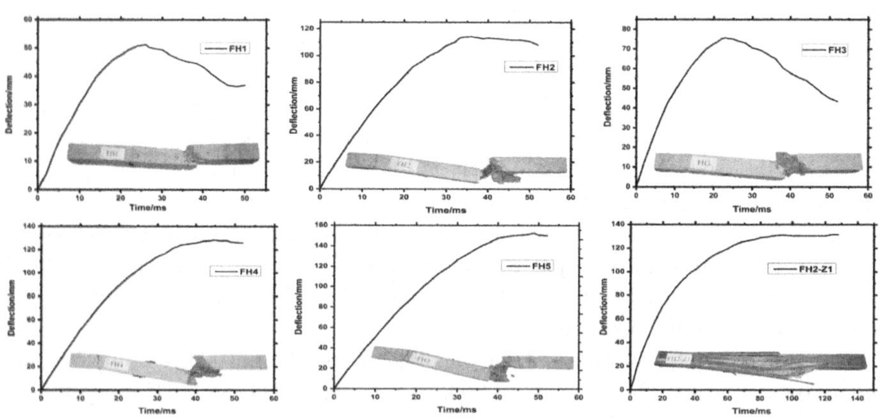

Fig. 13 Lateral deflection time history curve

4.2 Lateral Deflection Time History

Figure 13 shows the deflection time history curve of each component. The deflection of all types of components shows a slight rebound after reaching the maximum value, indicating that although all types of components have undergone different types of severe damage, they still have residual lateral directions Carrying capacity. This is indicating that the FH2-Z1 with bending failure has better resistance to deformation than the shear-destroyed FH2. However, FH2-Z1 reaches the maximum deflection for a longer period, about 37 ms, because FH2-Z1 weakens its impact resistance with the gradual fracture of the steel in the second stage of the impact. This shows how strong the CFRP layers resist deflection and how effective and demonstrates their high resistance to such force effects.

4.3 Relationship Between Impact Force - Lateral Deflection Curve

As shown in Fig. 14, the impact-deflection curves were obtained by combining the impact force and deflection time-history curves. The area enclosed by the curve is the energy consumption of the entire component during the impact process. As mentioned above, the deflection in this test is obtained by processing the picture of the high-speed camera. Therefore, the interval between the points of the deflection time-history curve is 0.4 ms, and the interval between the impact and the time history is different. Therefore, the deflection and the impact time are not synchronized happening. To synthesize the impact-deflection curve, this section utilizes

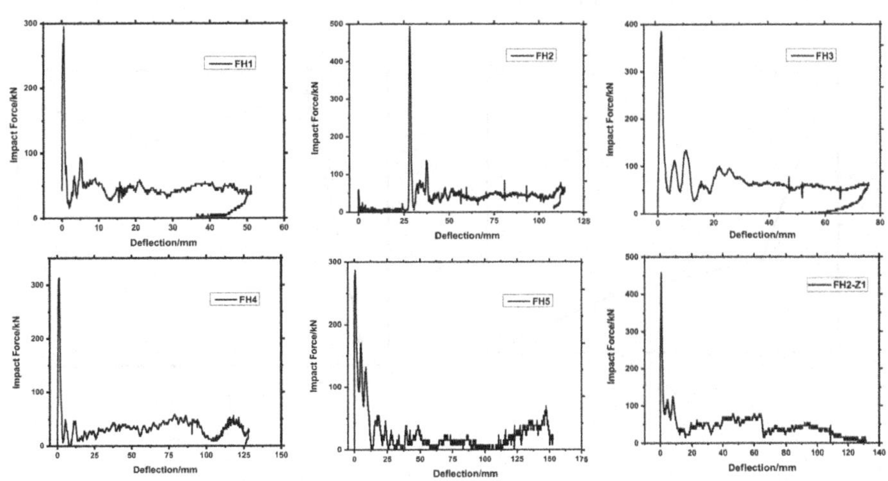

Fig. 14 Impact force-lateral deflection curve

the Interpolate module in Origin 9.0 software to supplement the missing points in the deflection time-history curve by linear interpolation to deflect time history and impact time history in the same time the number of coordinate points of the curve is the same. It can be found that when the impact force peaks, the deflection of all members is very small (<3 mm). After the end of the impact force, the peak point enters the failure point; the deflection of the specimen begins to develop significantly. The area enclosed by the curve and the abscissa is the energy consumption of the entries specimens during the impact process.

4.4 *Effect of Number CFRP Layers*

This section considers the influence of multi-layers of the wrap member. Based on the RC specimens with a speed of 4.5 m/s, the effects of zero, one, four, and six layers of CFRP on the response of the specimens were considered. Figures 15 compare the impact and maximum lateral deflection of unreinforced specimens with zero, one, four, and six-layer CFRP specimens. It can be found that when the number of layers changes from zero to one, the impact platform values, duration, and displacement of the members is all very obvious, this is because the reinforcement of the one layer CFRP not only provides the shear resistance of the original RC specimens; it also directly changes its deformation mode, which improves the overall deformation ability of the specimen, resulting in a significant increase in impact resistance. However, when the number of layers changes from one to four layers and four to six layers, there is no significant change in the indicators of the specimens, which may be due to the impact energy and specimen performance of this paper. The one-layer CFRP has provided sufficient lateral bearing capacity to maintain the specimen in an overall flexed state. At this time, the resistance of the member is controlled by

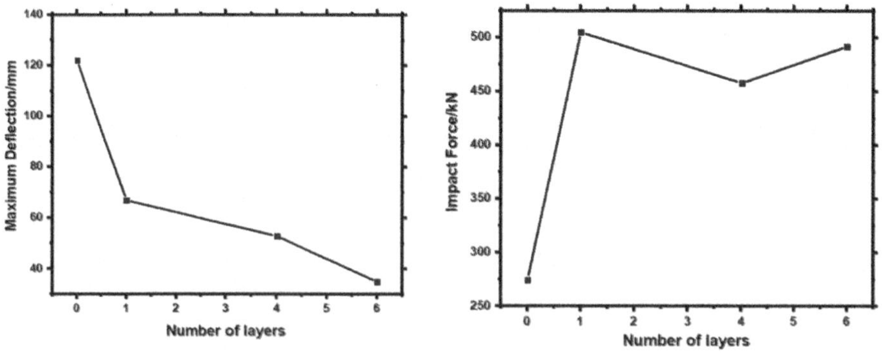

(a) (b)

Fig. 15 Effect of CFRP layers on the **a** maximum lateral deflection and **b** impact force of specimens

its bending strength, and the excessive number of layers does not provide greater rigidity to the member [13–15].

4.5 Discussion

To study the effectiveness of the use of CFRP shear-reinforced to increase the impact resistance of RC square members under the lateral impact, loads are studied. An unequal lateral impact test is carried out in this paper, RC specimen and multi-layers CFRP specimen. This paper mainly carried out aspects of work: the experimental study on the response of RC and CFRP specimens under lateral impact; the failure mechanism of RC specimens and CFRP specimens are analyzed; a formula that used to assess the deflection; the effects of the dynamic response of the piece and the CFRP specimen were analyzed.

Most of the research Wang and Belarbi [16], Yang et al. [17], Issa et al. [18] said the results showed that increasing the reinforcement ratio and adding steel fibers decreased the strain of the FRP bars, but in this article, increasing the strength of concrete members by using CFRP layers leads to increase the strain of the CFRP layers.

5 Conclusion

This technical paper establishes the impact time history, lateral deflection time history, number of CFRP layers for square concrete members. The accuracy of the proposed models was investigated based on experimental results. The main conclusions obtained are as follows:

1. The typical failure modes of RC and CFRP members under lateral impact are obtained. Several oblique cracks occur in the concrete of the right span of the RC member, and shear failure occurs. CFRP member impact point and right bearing section produce vertical crack, section steel bar fracture, showing bending failure; the failure strength of the member is controlled by bending strength. Before the steel bar breaks, the CFRP specimen has a higher impact force than the RC specimens. The CFRP specimen consumes more energy than the RC specimens.
2. Both members are severely damaged under the impact. Still, the maximum lateral deflection of the CFRP specimen is more significant than that of the RC specimens but with a more extended period of resistance for it.
3. There is little difference in the peak impact force of all specimens. However, the CFRP specimen has a higher impact failure value than the RC specimens. The energy consumption is stronger before entering the falling section. Still, due to the longitudinal reinforcement fracture, the failure section has a short

duration. The impact performance drops rapidly after entering the descending section. The impact force is reduced, and the energy consumption capacity is weakened.

The test results presented do not enable any decisive conclusions regarding the effect of specimen size (cross-sectional area) on the strength increase. Further tests, together with nonlinear finite element modeling aimed at duplicating the experimental observations, would greatly enhance the information provided by the current tests. These preliminary tests indicate that CFRP wrapping is effective as a technique for increasing the impact resistance of reinforced concrete square members, particularly if a Horizontal concrete layer is the first wrap around the member.

Acknowledgements The National Natural Science Foundation of China (Grant no. 51378427). The National Key Research and Development Program of China (Grant no. 2016YFC0802205-9).

References

1. Godat A, Qu Z, Lu XZ, Labossiere P, Ye LP, Neale KW (2010) Size effects for reinforced concrete beams strengthened in shear with CFRP strips. J Compos Constr 14(3):260–271. https://doi.org/10.1061/(asce)cc.1943-5614.0000072
2. Fuse A, Yokota H (2010) An analysis of Japan disaster medical assistance team (J-DMAT) deployments in comparison with those of J-DMAT's counterpart in the United States (US-DMAT). J Nippon Med Sch 77(6):318–324. https://doi.org/10.1272/jnms.77.318
3. Gu Z, Liu Q, Lu Y, Shi Z, Su G, Luan D (2016) Analysis and prevention of sinkhole collapses during the reconstruction and extension of Guang-Qing freeway, china. Environ Earth Sci 75(9). https://doi.org/10.1007/s12665-016-5625-2
4. Agrawal AK (2020) Scholars' mine scholars' mine performance based design for bridge piers impacted by heavy performance based design for bridge piers impacted by heavy trucks trucks. Accessed 28 Jan 2020. Available: https://scholarsmine.mst.edu/inspire_webinars
5. Dias SJE, Barros JAO (2010) Performance of reinforced concrete T beams strengthened in shear with NSM CFRP laminates. Eng Struct. https://doi.org/10.1016/j.engstruct.2009.10.001
6. El-Tawil S, Severino E, Fonseca P (2005) Vehicle collision with Bridge Piers. J Bridg Eng 10(3):345–353. https://doi.org/10.1061/(ASCE)1084-0702(2005)10:3(345)
7. D 3039/D3039M ASTM (2000) Standard test method for tensile properties of polymer matrix composite materials. Annu B ASTM Stand 15(03):105–116. https://doi.org/10.1520/D3039_D3039M-08
8. GB/T 228.1-2010: Translated English of Chinese Standard
9. Pham TM, Hao H (2017) Effect of the plastic hinge and boundary conditions on the impact behavior of reinforced concrete beams. Int J Impact Eng 102:74–85. https://doi.org/10.1016/j.ijimpeng.2016.12.005
10. Liu B, Fan W, Guo W, Chen B, Liu R (2017) Experimental investigation and improved FE modeling of axially-loaded circular RC columns under lateral impact loading. Eng Struct. https://doi.org/10.1016/j.engstruct.2017.09.009
11. Kishi N, Mikami H, Matsuoka KG, Ando T (2002) Impact behavior of shear-failure-type RC beams without shear rebar. https://doi.org/10.1016/S0734-743X(01)00149-X
12. Yao SJ, Zhang D, Lu FY, Wang W, Chen XG (2016) Damage features and dynamic response of RC beams under blast. Eng Fail Anal. https://doi.org/10.1016/j.engfailanal.2015.12.001

13. Liu Y, Al-Bukhaiti K, Abas H, Shichun Z (2020) Effect of CFRP shear strengthening on the flexural performance of the RC specimen under unequal impact loading. Adv Mater Sci Eng 2020:1–18. https://doi.org/10.1155/2020/5403835
14. Mashrei MA, Makki JS, Sultan AA (2019) Flexural strengthening of reinforced concrete beams using carbon fiber reinforced polymer (CFRP) sheets with grooves. Lat Am J Solids Struct 16(4):176. https://doi.org/10.1590/1679-78255514
15. Tavakkolizadeh M, Saadatmanesh H (2003) Strengthening of steel-concrete composite girders using carbon fiber reinforced polymers sheets. J Struct Eng 129(1):30–40. https://doi.org/10.1061/(ASCE)0733-9445(2003)129:1(30)
16. Wang H, Belarbi A (2020) Flexural behavior of fiber-reinforced-concrete beams reinforced with FRP Rebars. Accessed 29 Jan 2020. Available: https://www.researchgate.net/publication/237520601
17. Liu X, Sun Y, Wu T (2019) Flexural capacity and deflection of fiber-reinforced lightweight aggregate concrete beams reinforced with GFRP bars. Sensors (Switzerland) 19(4). https://doi.org/10.3390/s19040873
18. Issa MS, Metwally IM, Elzeiny SM (2011) Influence of fibers on flexural behavior and ductility of concrete beams reinforced with GFRP rebars. Eng Struct 33(5):1754–1763. https://doi.org/10.1016/j.engstruct.2011.02.014

Influence of Fiber on the Strength Characteristics of Fine-Grained Concrete

M. N. Nurbayeva, L. B. Aruova, S. S. Kalym, Sh. B. Toleubayeva, Zh. I. Urkinbayeva, Zh. M. Aukazhieva, Zh. N. Ospanova, A. Budikova, and A. N. Zhakanov

Abstract This paper reviews findings of research on the dependence fiber concrete with steel, polypropylene and glass fiber for fine-grained concrete, as fiber concrete with improved strength and deformation characteristics is increasingly used in construction. At the same time, fibers of different composition and origin, geometric characteristics and physical and mechanical properties are used as dispersed reinforcement. The characteristics of the fiber types that can be used to manufacture fiber concrete are given. The article presents the main characteristics of the considered fiber types. The authors studied the effect of different fiber types on the strength characteristics in the fine-grained mix concrete. The article's authors determined as a whole that the addition of steel fibers reduces concrete shrinkage and significantly raises the concrete's strength characteristics compared to polypropylene fibers. The experimental results studies in the studied articles indicated the productivity using fibers as comparison with fine-grained concrete without fiber. The article compares the main types of fiber concrete, and notes the high strength indicators of this actively developing material. The specialists can also be interested in the fact that each type of fiber gives certain properties to concrete with its advantages and disadvantages.

Keywords Fibers · Fine-grained concrete · Short AR-glass fibers · Strength

1 Introduction

For technological progress in construction is the widespread use of new effective materials, ways to increase the performance characteristics of concrete.

In recent years, new types of concrete have been increasingly used in construction: especially high-strength, fine-grained, self-compacting. The need for the use

M. N. Nurbayeva (✉) · L. B. Aruova · S. S. Kalym · Sh. B. Toleubayeva · Zh. I. Urkinbayeva · Zh. M. Aukazhieva · Zh. N. Ospanova · A. N. Zhakanov
L.N. Gumilyov, Eurasian National University, 2 Satpayev St., Nur-Sultan 010008, Republic of Kazakhstan

A. Budikova
Department of Architecture and Construction Production, Institute of Engineering and Technology, Kyzylorda State University, Kyzylorda, Republic of Kazakhstan

T. Kang and Y. Lee (eds.), *Proceedings of 2021 4th International Conference on Civil Engineering and Architecture*, Lecture Notes in Civil Engineering 201,
https://doi.org/10.1007/978-981-16-6932-3_13

of fine-grained concrete is due to the fact that they do not contain a large aggregate, are indispensable in the manufacture of thin-walled reinforced concrete structures, reinforced cement products, masonry and finishing solutions, etc. The possibilities of modern concrete science allow us to produce various types of concrete.

In recent years, of all the existing technologies for the construction of buildings and structures, the most promising is monolithic construction—a method of constructing structures made of concrete, which allows you to build construction objects of almost any number of floors and shapes in a short time. For monolithic construction, concrete must have a high degree of crack resistance, compressive, tensile and bending strength, low shrinkage, sufficient water resistance and frost resistance. It is necessary to create effective high-strength concretes to ensure such properties.

One of the ways to solve the problems of improving the performance characteristics of fine-grained concrete is the use of active mineral additives, its reinforcement with various types of metallic and non-metallic fibers of mineral or organic origin.

Concrete with fiber is the best solution for construction and production purposes. The saturation of the concrete mass with the filler improves the performance characteristics of the composite: strength properties, resistance to tensile forces, resistance to cracking [1].

Numerous research show that the use of synthetic fibers as a reinforcing material usually leads to a reduction in shrinkage deformations of the brittle cement matrix, an increase in frost resistance, abrasion, weather resistance, and a significant increase in the strength characteristics and fracture toughness of concrete at all structural levels. The increased characteristics of crack resistance and the possibility of its "work" with a significant crack opening are the main features of reinforced fiber concrete. Fiber reinforcement not only increases the characteristics of concrete, but also reduces the labor intensity, the cost of its production, etc. Synthetic fibers are also widely used, which reduce the cost of concrete. The classes of fiber concrete vary and depend both on the ratio of cement and fiber concrete, and on the type of fiber.

Fiber-reinforced concrete in industrial and civil construction is used mainly for the construction of enclosing and load-bearing structures, designed to increase their service life. When working in compression conditions, concrete shows high performance, and when working in tension, its use becomes not so effective and requires additional reinforcement to perceive tensile stresses. In this case, reinforcement is usually performed with the help of rod reinforcement, fiber, or using a combined method that combines both components simultaneously.

The fiberglass added to the concrete mix is distributed in it during mixing, binds the internal structure of the cement stone, as a result of which the concrete becomes stronger, more durable, with high performance properties.

These indicators of the properties of fiber concrete are mainly influenced by the quality characteristics of its constituent raw materials, including fiber.

At the same time, in order to obtain fiber concrete with high performance properties and durability, it is necessary to achieve technological compatibility of fiber and cement stone:

- the fiber must be evenly distributed in the optimal amount in the mortar part of the concrete;
- the fiber-concrete mixture must have sufficient workability in accordance with the production technology of products and structures.

Moreover, the correctly selected fiber content in concrete allows you to get high physical and mechanical performance indicators. An increase in the consumption of polypropylene fiber can increase the porosity of products. According to the researchers [2], there are 2 binding mechanisms in concrete reinforced with polypropylene fiber. First, the interfacial adhesion, which is the result of the formation of a contact zone between the fiber and the concrete matrix, and, secondly, the mechanism of adhesion, which occurs due to the shrinkage of concrete as a result of drying and compression of the reinforcing elements.

Mechanical characteristics, such as brittleness reduction, can also be improved, though not significantly in terms of strength [3–7], except for tensile capacity.

In order to obtain high-quality fiber-reinforced concrete, the effect of introducing various types of reinforcing fibers into the concrete matrix.

Polypropylene fiber fibers are made by a continuous method from pure polypropylene granules by means of extrusion, as well as extraction by heating. When the fibers reach a certain temperature, an oiling compound is applied to their surface. It is this composition that contributes to the adhesion of the surface and the dispersion of polypropylene fiber in the concrete solution. The fiber has the appearance of thin white polypropylene fibers of various sizes, which is an inert substance that is resistant to alkalis and various chemicals.

Polypropylene fiber is a multifilament monofilament that has a circular shape in cross-section. Fibers are also produced in a square or rectangular cross-section, which are less common. The type of fibers can be straight or wavy, having a large anchor capacity.

Glass fiber is an artificial fiber that is made from molten inorganic glass. In this form, the glass does not break, but easily bends without destruction. The diameter of the fibers is 3–100 μm. The physical, mechanical and chemical properties of the fiber depend on its manufacturing technology and the choice of raw materials. There are many varieties of fiberglass, distinguished by specific characteristics: E-glass (electrical), S-glass (strength), AR-glass (alkali resistant), etc. For effective application, the different fibers must meet the following conditions:

- the elastic modulus of the fibers must be higher than the elastic modulus of the composite matrix;
- the fibers must be chemically resistant and not break down in the alkaline environment of the concrete;
- the volume of fiber output should ensure the volume of production of products made of fiber reinforced concrete.

Basalt fiber consists of even thinner fibers. On their surface in the places of defects formed from mechanical influences there is a process of crystallization, there appears a network of thin hexagonal plates and needle crystals, which fuse with the spherical grains of the cement system, further strengthening the effect of the fiber as dispersed reinforcement. The fiber has a hollow structure in the end part of which hydration products penetrate with the formation of crystal aggregates. This leads to an increase in the strength of the cement stone.

Basalt-fiber concrete is characterized by increased water resistance and frost resistance, can withstand the effects of high temperatures longer.

Carbon fibers are carbon filaments cut to a specified length, consisting of the finest carbon fibers. In turn, carbon fibers are nanostructured inorganic material containing 92–99.9% carbon.

Carbon fiber fibers, when added to concrete, increase its impact strength, crack resistance, tensile strength and other properties of concrete. Carbon fiber prevents the appearance of shrinkage cracks both in the early and late stage of hardening and retains its reinforcing properties throughout the life of concrete.

2 Mainpart

Fine-grained concrete is effective in its applications due to a number of features: homogeneous composition, good mobility, chloride permeability, etc. However, for the strength of concrete it is necessary to choose the optimal composition [8].

The addition of fiber in concrete and the evenly distributed fibers in the concrete matrix improves the quality of the concrete mortar, its strength and durability. Also, the percentage of cracking is reduced, which helps to maintain a good appearance [9].

The properties of fiber concrete directly depend on the material used as fiber. The undoubted advantages of fiber-reinforced concrete include its high performance characteristics. Fine-grained concrete with fiber in its composition is much superior to conventional in quality, strength and durability. Products made of it acquire resistance to abrasion and chemical attack, do not deform during operation and have increased tensile and tensile strength. Concrete, with a fiber–fiber filling, has a significantly lower weight than conventional with a metal mesh reinforcement. It can be give any shape, which greatly simplifies the process of building concrete structures. The elimination of the metal mesh reinforcement stage also reduces the thickness of concrete slabs and reduces the consumption of concrete mortar.

High technical characteristics of fibro concrete provide the structures made of it, strength and durability. The service life of such structures exceeds products made of ordinary concrete 15–20 times.

To date, the most widely distributed steel fibers are in the form of needles, cut from thin steel wire with a profiled surface for better adhesion to concrete. Similar fibers can be cuts from a thin steel sheet. The production of fibers from steel waste can significantly reduce the consumption of scarce reinforcing steel in construction.

The steel fiber are protect from corrosion, which can occur during storage, transportation and direct operation inside the concrete. It increases the load-bearing capacity of the structure, reduces its thickness, and increases the ductility and frost resistance of the material.

Due to its geometric features, the steel fiber is evenly distributed over the entire volume of the concrete mix.

The properties of the concrete matrix during fiber reinforcement depend on the structure of the composite. In this regard, in order to analyze these composites and predict their behavior under various loads, it is necessary to study the composition of the concrete matrix and consider the main leaving factors:

- composition of the cement fiber;
- fiber shape and distribution;
- contact zone "cement stone-fiber".

For fibro-reinforced fine-grained concretes, the cement fiber is not different from the fiber in other fine-grained concretes, and usually consists of cement, sand, filler, and water. In fiber concrete, the fiber content varies from one to fifteen percent by weight of cement.

The production of fiber-reinforced concrete is carry out by conventional methods. The use of more fibers is possible due to the use of simple mixing technologies, but with the use of improved matrix compositions based on complex control of the rheology and microstructure of the mixture. The dense microstructure in these composites, as well as the improved rheology, make it possible to combine and unify the dispersion for 2–6% due to the volume of short fibers that can provide effective hardening.

The use of fiber in the composition of concrete has a positive effect on the work of the concrete structure. With its proper distribution in concrete, it is possible significantly improve the physical and mechanical properties, including improving the performance of concrete structures under the influence of bending loads.

To date, many tests of fiber concrete samples with different concentrations and types of fiber, the class of matrix concrete, have been carry out. In this section, we present various compositions of dispersed-reinforced composites based on cement and the results of the study of the influence of fiber on the strength characteristics of fiber concrete.

Many researchers conduct experiments with steel fibers than synthetic fibers, although many fibers have different features [10]. Due to their high strength, low modulus of elasticity, durability and affordability, synthetic fibers are used in fine-grained concrete to improve mechanical properties [11, 12].

Arkadiusz Denisiewicz, Tomasz Socha, Krzysztof Kula, Marcin *Pasula* (University of Zielona Gora, Poland) conducted laboratory tests on the effect of steel and polypropylene fibers on the mechanical and physical properties of fine-grained fiber concrete.

The tests were carried out on samples with different degrees of reinforcement based on steel and polypropylene fibers. For laboratory tests, cubes in size 15 × 15x15 cm were made of concrete mixtures with different steel content (0, 5, 10, 20%) and polypropylene (1.5, 2.5, 3,48%) fibers. After the samples were made, they were kept for 28 days under temperature and humidity conditions. During the first 48 h, the samples were stored under constant temperature and humidity conditions.

The samples were test to determine the compressive strength according to the standard procedure. The results of the compressive strength test calculated from the formula:

$$f_{cm} = \frac{P_{cm}}{A} \tag{1.1}$$

A—cross-sectional area of the sample 225 cm^2, P_{cm}—average destructive force of the sample, fcm, 28—average compressive strength of concrete after 28 days.

As a result of the research and analysis of the data obtained, the authors found that the samples with steel fiber have the best physical and mechanical characteristicsIt can be concluded that the addition of 20% steel fiber can significantly increase the strength characteristics of concrete to an average of 55.65 MPa. The best strength indicators were given by samples containing steel fiber in the amount of 20% [9].

Polypropylene fibers belong to the class of fibers of organic origin. This fiber, despite the lower values of elastic characteristics (compared to steel and glass), is of interest for its research and application as a reinforcing additive for concrete.

Theoretical analysis of the results of experiments on the influence of different dosages of polypropylene fibers showed that concrete samples have different indicators of compressive strength. The highest strength values are achieved at dosages of polypropylene fiber of 2.5% of the concrete volume (see Fig. 1). This is a kind of "efficiency threshold" of the selected fiber, at which the best indicators for the strength of concrete are achieved. At lower dosages of fiber and exceeding these dosages leads to a decrease in strength.

Qin Zhang and his colleagues (Hohai University, China) conducted laboratory tests with the addition of AR glass fibers in varying amounts on the fine-grained concrete. To determine the mechanical properties of concrete, AR-glass fibers with fiber sizes of 6, 12, and 18 mm were selected (Table 1).

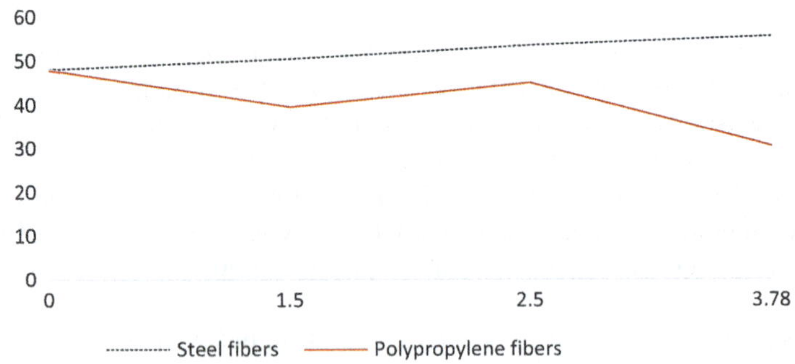

Fig. 1 The results of the compressive strength test

Table 1 Characteristics of AR-glass fibers

Name of the characteristic	Unit	Value	
Relative density	g/m^3	2.7	
Fiber diameter	mm	15	
Ultimate tensile strength	MPa	2500–3500	
Modulus of elasticity	N/mm^2	80.4	

The aggregate, cement, and fiber were mixed until a homogeneous mixture was obtained, closed with the required amount of water, and mixed until a homogeneous mass was formed. The results of the research show that the best strength characteristics are obtained in samples of fine-grained concrete reinforced with AR-glass fiber in the amount of 5%. With the increase in the fiber length, the performance of concrete samples for strength improved. The best results in compressive and flexural strength are shown by fiber concrete samples with a fiber size of 18 mm, 7.79, 61.29 and 80.0% compared to non-reinforced concrete samples, which were generally higher than those with 6 and 12 mm AR-glass fiber.

The compressive strength, flexural strength, and tensile strength of the concrete samples depend on the amount and length (the ratio of length to diameter) of the glass fiber additive. The analysis of the conducted studies shows that the greatest increase in strength characteristics is observed when reinforcing with AR-glass fiber with a length of 18 mm, than with 6 and 12 mm.

Experimental research show that with the addition of reinforcing fibers on fine-grained concrete, the mechanical characteristics of the material change depending on the size of the fibers. With a content of about 5% AR-glass fibers in fine-grained concrete increase the mechanical properties of the material [13].

Shishkanova V. N., Prokofieva Yu. A. (Togliatti, Samara region, Russia) analyzed the results of studies of the dependence of fiber concrete with metal and polypropylene fiber for monolithic structures on the composition of the concrete mixture, the amount of fiber (Table 2).

To determine the physical and mechanical properties of concrete, samples-cubes measuring 70 × 70 × 70 mm and samples-prisms measuring 40 × 40 × 160 mm were made.

Table 2 Characteristics of metal and polypropylene fibers

Name of the characteristic	Unit	Value
Metal fibers		
Fiber diameter	mm	0.3
Length	mm	15
Corrugation step width	PC	4–5
Polypropylene fibers		
Fiber diameter	μm	20–25
Tensile strength	MPa	≥450
Elastic modulus	MPa	≥5100

Batch samples differed from each other by the degree of fiber reinforcement [14].

Before the test, at the age of 7, 14 and 28 days, the samples were kept under normal hardening conditions.

Fiber-reinforced concrete mixtures with the ratio were studied C:N = 1:0.67; C:N = 1:1.43; C:N = 1:2.0 with metal fiber (34, 51, 84, 104, 133, 166 kg/m^3) and polypropylene (14, 8, 10, 11 kg/m^3) fibers.

The changes in the compressive strength from the degree of reinforcement with metal fiber are show in Fig. 2.

The results of the studies (see Fig. 2) show that the best strength characteristics (age of exposure—28 days) are samples of fine-grained concrete reinforced with metal fiber in the amount of 34 kg/m^3. With an increase in the amount of fiber to 84 kg/m^3, there is a gradual decrease in strength by 15% compared to the strength of samples with a metal fiber content of 34 kg/m^3. When the reinforcement is increased to 104 kg/m^3 of fiber or more, the strength of the fiber concrete samples begins to increase, and when the fiber content in the concrete is 166 kg/m^3, the strength of the fiber concrete samples reaches 39.4 kg/m^3.

The changes in the compressive strength from the degree of reinforcement with polypropylene fiber are shown in Fig. 3.

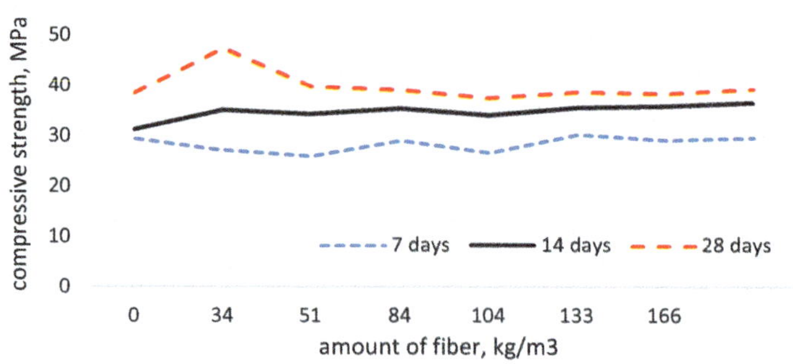

Fig. 2 Changes in the ultimate strength of fiber concrete under compression from the degree of reinforcement with metal fiber

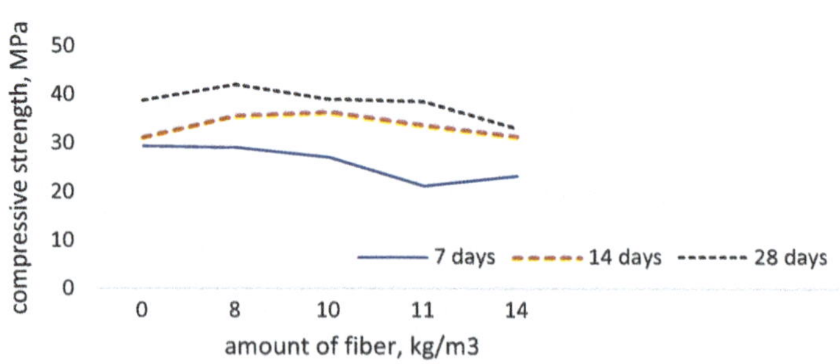

Fig. 3 Changes in the ultimate strength of fiber concrete under compression from the degree of reinforcement with polypropylene fiber

The best results in compressive strength are shown by fiber concrete samples with the amount of polypropylene fiber of 8 kg/m^3. Further increase in the reinforcement of concrete with polypropylene fiber leads to a decrease in their strength. When polypropylene fiber is dosed in an amount of 14 kg/m^3, the strength of fiber concrete samples is reduced by 20% compared to the strength of samples with a fiber content of 8 kg/m^3.

Klyuev A. V., candidate of technical sciences, Klyuev S. V., candidate of technical sciences, assoc., Netrebenko A. V., postgraduate, Durachenko A. V., engineer (Belgorod State Technological University named after V. G. Shukhov) used propylene fiber for the dispersed reinforcement of fine-grained concrete. They conducted experimental research of fiber-reinforced concrete samples.

Sand, binder, polypropylene fiber, and water were mixed until smooth. After forming and compacting, the samples were kept at room temperature for 24 h. The molds were then removed and the concrete samples were moved to a dry place.

To determine the compressive strength, 100 × 100 × 100 mm cubes were prepared. Performance characteristics and durability depend on the diameter of the polypropylene fiber. The most widely used polypropylene fibers are circular in shape with a diameter of 20–30 μm and a length of 3, 6, 12 and 18 mm. The use of polypropylene fiber with a smaller diameter provides high flexural strength and impact strength. The conducted research have shown that the larger the diameter of the fiber, the longer it retains its reinforcing ability.

The analysis of the conducted research shows that the greatest increase in strength characteristics is observed when reinforced with polypropylene fiber with a length of 6 mm. The compressive strength with a length of polypropylene fiber of 3 mm increased to 43.5 MPa, with a length of polypropylene fiber of 6 mm showed 45.4 MPa, respectively, with a length of 12–40.3 MPa.

Fine-grained concrete mixtures can be used instead of coarse aggregate [15].

The fiber used should contribute to the improvement of the defining physical and mechanical properties of concrete, not reduce the quality of products, be safe both in the process of preparing concrete, and in the manufacture and operation of products. The choice of the type of fiber, in addition to its adaptability when mixing concrete, resistance in the alkaline environment of cement stone, availability and price acceptability, depends on the characteristics of the operation of the finished product.

3 Conclusion

We have analyzed the tests conducted by various authors of various types of fiber for fine-grained concrete. We hope that the information published in this article will be in demand and useful for specialists. We conducted a thorough study of the experimental studies conducted and came to the following conclusions:

- One of the promising ways to improve the performance characteristics of concrete and mortars is the introduction of various fibers in their formulation. The use of this material increases the tensile strength of cement stone, which provides higher tensile strength and crack resistance, as well as increases other characteristics of cement-based building materials;
- Numerous studies conducted by scientists from different countries do not give a clear answer about the absolute advantages of a particular type of fiber, but most researchers note the predominant use of steel, glass and polypropylene fibers in concrete;
- the addition of steel fiber can significantly increase the strength characteristics of concrete to an average of 55.65 MPa, and polyropylene fibers reduce the compressive strength of concrete to 30.57 MPa. This opens up wide possibilities for the use of these fibers in fine-grained concrete for certain structures.
- the use of AR-glass fiber with a large diameter ensures high compressive and flexural strength. The conducted research that the larger the diameter of the fiber, the longer it retains its reinforcing ability. The greatest increase in strength characteristics is observed when reinforced with AR-glass fiber with a length of 18 mm, than 6 and 12 mm.
- the content of 5% AR-glass fiber in the concrete mix is the optimal amount for fiber concrete. This is because a further increase in percentage of fiber reinforcement gives a slight increase in performance, and in some cases, and its reduction by reducing the thickness of the concrete layer.
- The effectiveness of the use of fiber for concrete reinforcement is determined, first of all, by the optimally selected dosage for each type of fiber, while the optimal dosages of different types of fiber may differ significantly for the same technological compositions of concrete.
- Comparing different types of fiber, it is impossible to draw an unambiguous conclusion about the advantages of one type of fiber over others. Depending on the type of fiber used, fiber-reinforced concrete has different physical and mechanical characteristics. This material is used more and more often all over the world, it is profitable not only from an economic point of view, but also allows you to ensure the high quality of the structure. In the future, we plan to develop technologies for effective fine-grained fiber concrete through the use of local raw materials. We will design the compositions and investigate the nature of the influence of different types of fibers on the physical and mechanical properties of the material.

References

1. Buratti N, Mazzotti C, Savoia M (2011) Post-cracking behaviour of steel and macro-synthetic fibre-reinforced concretes. Constr Build Mater 25:2713–2722
2. Bentur A, Mindess S, Vondran G (1989) Bonding in polypropylene fibre reinforced concretes. Int J Cement Compos Light-weight Concr 11(3):153–158

3. Kurihara N, Kunieda M, Kamada T, Uchida Y, Rokugo K (2000) Tension softening diagrams and evaluation of properties of steel fiber reinforced concrete. Eng Fract Mech 65:235–245
4. Holschemacher K, Mueller K, Ribakov Y (2010) Effect of steel fibers on mechanical properties of high-strength concrete. Mater Des 31:2604–2615
5. Yang JM, Min KH, Shin HO, Yoon YS (2012) Effect of steel and synthetic fibers on flexural behavior of high strength concrete beams reinforced with FRP bars. Compos Part B 43:1077–1086
6. Meda A, Minelli F, Plizzari GA (2012) Flexural behavior of RC beams in fiber reinforced concrete. Compos Part B 43:2930–2937
7. Soutsos MN, Lampropoulos AP (2012) Flexural performance of fiber reinforced concrete made with steel and synthetic fibers. Constr Build Mater 36:704–710
8. Lam NN (2020) Fine-grained concrete mix design using statistical methods for ultra-thin whitetopping overlay application. Civil Eng J 6, Special Issue "Emerging Materials in Civil Engineering", 33–41
9. Denisiewicz A, Socha T, Kula K (2018) Marcin Pasula (Poland): influence of steel and polypropylene fibers addition on selected properties of fine-grained concrete: civil and environmental engineering reports. E-ISSN 2450–8594:138–148
10. Kosior-Kazberuka M, Berkowski P (2016) Fracture mechanics parameters of fine grained concrete with polypropylene fibres. Procedia Eng 161:157–162
11. Wang Z, Gao D, Zhu H, Zhang Q (2007) Fracture properties of polypropylene fiber reinforced high strength concrete, Kuei Suan Jen Hsueh Pao. J Chin Ceram Soc 35(10):1347–1352
12. Zhang P, Liu C, Li Q, Zhang T (2013) Effect of polypropylene fiber on fracture properties of cement treated crushed rock. Compos Part B 55:48–54
13. Zhang Q, Li S, Gong S, Zhang G, Xi G, Wu Y (2020) Study on flexural properties of basalt fiber textile reinforced concrete (BTRC) sheets including short AR-glass fibers. Front Mater 7, Article 277, pp 1-13
14. Shishkanova VN, Prokofieva YA (2019) Self-sealing fiber concrete for monolithic constructions. Sci Educ New Time. No. 2
15. Aruova LB, Ospanova ZN, Gordienko BS, Alibekova NT, Kalieva ZE, Urkinbaeva ZI, Nurbaeva MN (2018) Properties of fine-grained concrete mixtures during the construction of monolithic structures. In: 2020 IOP Conference Series: Material Science and Engineering 829, 012018, 6

Experimental Study on Concrete Spalling of Post-tensioned Concrete One-Way Slab

S. Y. Park and T. H.-K. Kang

Abstract In this study, concrete spalling of post-tensioned (PT) concrete one-way slab structure was evaluated through fire experiments. Four fire experiments were conducted in simply-supported condition (unrestrained), heated only in one slab surface (bottom). Input fire scenario followed the procedure in ASTM E119-20, targeting 2 h (2) and 3 h (2) of fire-resistance performance. Heating temperature ended up in 1010 °C and 1052 °C, respectively. Concrete strength, duration of concrete curing time, existence of longitudinal reinforcement, and water content of concrete appeared to affect the degree of concrete spalling. Other parameters of the number of longitudinal reinforcing bars and minimum concrete cover were also influential to the slab damage. After the experiment was conducted, the depth and size of the concrete spalling was examined.

Keywords Fire experiment · Fire-resistance performance · One-way slab · Post-tensioned concrete · Spalling

1 Introduction

PT concrete structural systems in buildings are widely used in underground structures, long-span structures, nuclear power plant structures and so on. There are several advantages of PT systems such as reduction of slab thickness and control of crack propagation; however, when fire occurs, the surface where heat is exposed can explode, resulting in section area loss or direct exposure of reinforcing bars and tendons [1]. Change in steel and concrete properties in elevated temperatures is known to be detrimental in the sense of yield strength, ultimate strength and Young's modulus [2–4]. Therefore, prevention and prediction of concrete spalling is important.

S. Y. Park · T. H.-K. Kang (✉)
Department of Architecture and Architectural Engineering, Seoul National University, Seoul 151-744, Republic of Korea
e-mail: tkang@snu.ac.kr

T. Kang and Y. Lee (eds.), *Proceedings of 2021 4th International Conference on Civil Engineering and Architecture*, Lecture Notes in Civil Engineering 201,
https://doi.org/10.1007/978-981-16-6932-3_14

Factors influencing concrete spalling have two categories which are intrinsic factors and external factors [5]. The intrinsic factors include permeability, moisture content, strength of concrete, presence of reinforcement, and water to cement ratio. The external factors include temperature rise rate and external loads. Also, there are three mechanisms of predicting spalling in concrete: (i) hydro-thermal mechanism, also known as pore pressure mechanism, (ii) thermo-mechanical mechanism and (iii) hydro-thermo-mechanical mechanism [6–8]. The hydro-thermal mechanism assumes that spalling occurs when accumulated vapor pressure exceeds the degraded tensile strength of concrete at elevated temperatures. The thermo-mechanical mechanism assumes that thermal stresses develop inside the concrete due to temperature gradients by restrained thermal dilation. This induces tri-axial tension–compression-tension stresses at the exterior surface in concrete and eventually spalling. Lastly, the hydro-thermo-mechanical mechanism combines both pore pressure and stresses induced by thermal gradients and structural loading as a cause for spalling in concrete.

PT concrete structures have two tendon types: bonded and unbonded. Those two types differ in flexural strength, friction coefficient and composition [9, 10]. Due to the difference in flexural strength, required additional reinforcing bars to satisfy target member strength may be different and this may affect the possibility of concrete spalling occurrence. Also, in the bonded PT concrete structures, tendons are confined in grouts, whereas unbonded PT tendons are not confined.

Therefore, in this paper, an experimental study is conducted to examine the spalling damage and extent in PT concrete slab structures. Two separate experiments were done on bonded PT (PF-B series) and unbonded PT (PF-G series), respectively, for 2 and 3 h. The unbonded PT specimens were grouted but not bonded, because high-density polyethylene (HDPE) sheath is coated on each strand (Table 1). Spalling area and depth are measured for each system, and temperature change is also measured using k-type thermocouples.

Table 1 Composition of post-tensioning methods (Shin et al., 2020)

	Bonded (grouted) tendon	Greased sheathed-strand tendon
Composition	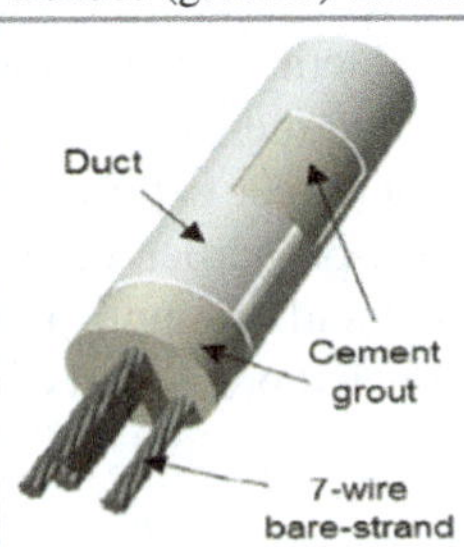	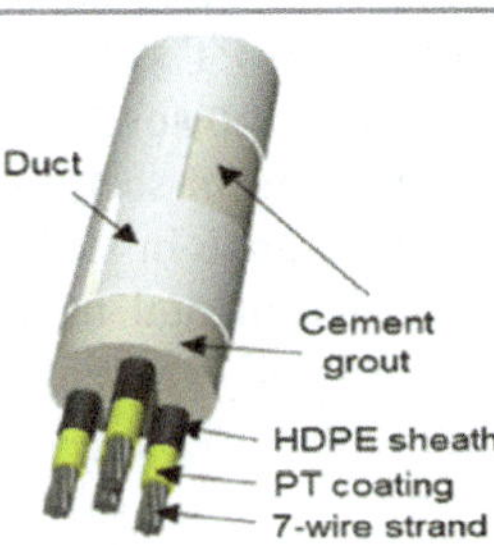

Table 2 Concrete mixture proportions and properties

Property	PF-B2	PF-B3	PF-G2	PF-G3
Cement content Type 1 (kg/m^3)	468	468	468	468
Fine aggregate (kg/m^3)	783	783	783	783
Coarse aggregate (kg/m^3)	953	953	953	953
High range water reduction agent (kg/m^3)	4.78	4.78	4.78	4.78
Water (kg/m^3)	165	165	165	165
Water-cement ratio (%)	35.26	35.26	35.26	35.26
Slump (mm)	150	150	150	150
Concrete compressive strength (MPa)	42.03	46.29	42.72	47.61

Table 3 Steel properties

Property	Tendon	HDPE tendon	Reinforcing bar
Yield strength (MPa)	–	–	400
Ultimate strength (MPa)	1860	1860	–
Diameter (mm)	15.2	18.98	10/13

2 Experiment Outline and Method

2.1 Materials

Table 2 represents concrete mixture proportions and properties. Cement type used in specimen is Portland cement type 1. Compressive strengths of concrete at each test date were 42.03 MPa, 46.29 MPa, 42.72 MPa, 47.61 MPa, respectively, which targeted 40 MPa. Table 3 represents steel properties used in the specimens.

2.2 Specimens

PF-B type and PF-G type had the same size for the same target heating time. Only the number of lateral longitudinal reinforcing bars was different. For the experiment, specimens were placed in simply-supported condition (unrestrained) with heating length of 6000 mm. Details are summarized in Table 4.

The tendon profile was 2nd order parabola shape, where the end height was 125 mm or 140 mm and the height of the tendon at the middle was 55 mm or 70 mm. Each specimen had 6 tendons.

Table 4 Specimen details

Details	PF-B2	PF-B3	PF-G2	PF-G3
Length × width × height (mm)	6700 × 1500 × 250	6700 × 1500 × 280	6700 × 1500 × 250	6700 × 1500 × 280
Number of lateral longitudinal reinforcing bars at bottom	2	2	8	8
Heating length (mm)	6000	6000	6000	6000
Target heating time (h)	2	3	2	3
Concrete cover (mm)	45	60	45	60

2.3 Experiment Method

Experiment was designed to follow the input fire scenario in ASTM E119-20 [11] (Fig. 1). PF-B2 and PF-G2 were designed to be heated for 2 h, while PF-B3 and PF-G3 were designed for 3 h. Before heating started, four point loads of 12.1 tons were applied using an actuator (Fig. 2).

Fig. 1 Input fire scenario

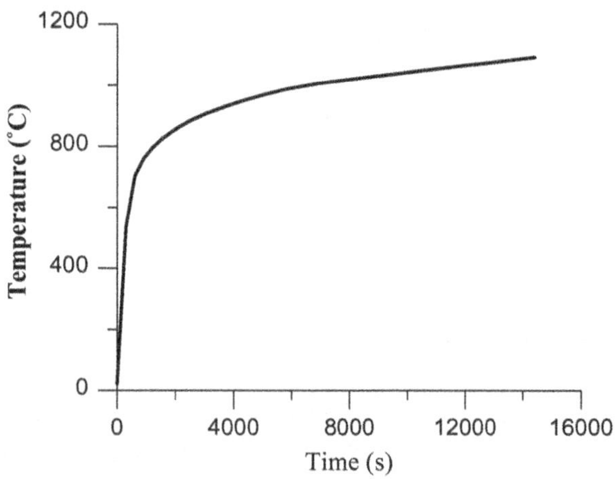

Fig. 2 Actuator setting

3 Experiment Result

3.1 Concrete Surface Temperature

Time–temperature curves of concrete bottom surface are expressed in Fig. 3. PF-B2 and PF-B3 showed similar trends to actual temperature curve till the end of thermocouple measurement. PF-G2 and PF-G3 exhibited some fluctuations at about 20 min after the experiment started. Then, after about 50–60 min, some thermocouples dropped out.

3.2 Spalling Damage and Extent

Spalling damage and extent are illustrated in Fig. 4. PF-B2 and PF-B3 had only a few regional spallings. PF-B3 had spalling concentrated in the anchorage zone (end zone) with a depth up to 185 mm. PF-G2 had spalling in all regions except the support region. The spalling depth at the mid-span was 65 mm, which is just above the reinforcement. In the end regions, spalling occurred with a depth of 100–120 mm, where tendon-sheath existed. PF-G3 experienced extreme spalling in one anchor zone with a depth of 250–260 mm. Spalling did not occur in the center part of the mid-span region, whereas small in the side edge of the mid-span region.

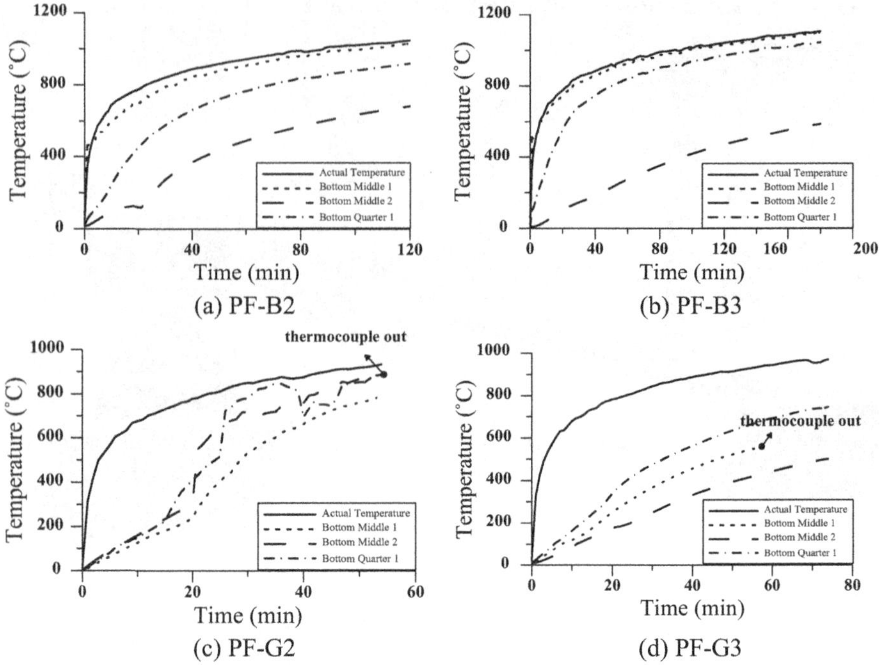

Fig. 3 Concrete bottom surface time–temperature curve

4 Conclusions

Concrete spalling of PT concrete one-way slab structures was evaluated through fire experiments with different variables. Time–temperature curves were obtained using k-type thermocouples in 3 spots. The damage and extent of spalling were quantitatively and qualitatively assessed at the end of the experiment. From the experiments, the following conclusions can be drawn.

- PF-B2 and PF-B3 satisfied target experiment time, whereas PF-G2 and PF-G3 ended at about 50 and 75 min, respectively, much earlier than 120 and 180 min. This was resulted from the difference of damage and extent of the spalling. PF-B2 and PF-B3 had regional spallings with less damage, compared to PF-G2 and PF-G3. PF-B2 had almost no spalling. PF-B3 had some spalling in the anchorage zone, where more reinforcing bars were installed than PF-B2. On the other hand, PF-G2 experienced spalling in all regions. It should be noted that greater amounts of reinforcing bars were provided for PF-G series than PF-B series and it was to make the nominal flexural strengths of the slabs similar.
- By comparing PF-B series and PF-G series, the result shows that formation of concrete spalling is affected by many different variables. The following variables were the same in both systems: concrete compressive strength, water to cement

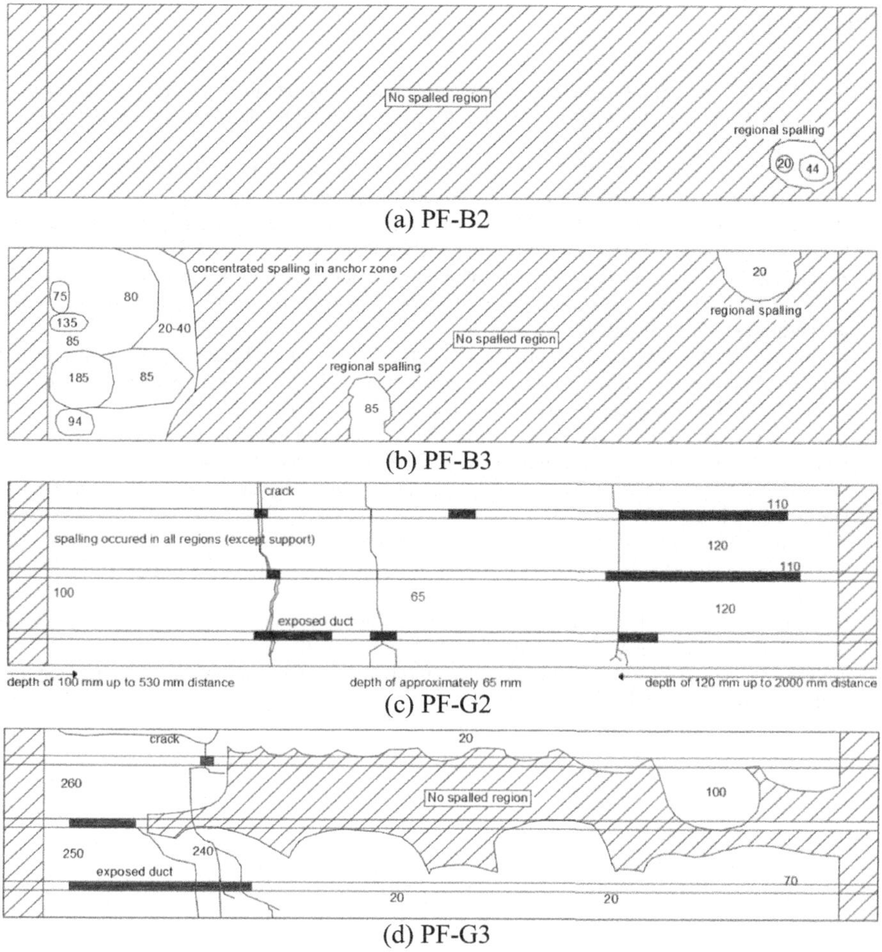

Fig. 4 Concrete spalling damage and extent (number indicates the spalling depth)

ratio, specimen size, and input fire scenario. Whereas, the variables of the number of longitudinal reinforcing bars and tendon type were different.

- Most of the concrete spallings occurred in the region where reinforcing bars existed. There might be more pores near the reinforcing bars, resulting in explosion of concrete cover.
- In the end regions, although deeper spalling occurred, the depth remained about 100–120 mm. which is similar to the height of tendons. The upper part of concrete above the tendon did not experience spalling.
- Further research would be needed to draw a preliminary relationship between concrete spalling and longitudinal reinforcing ratio and/or tendon force for each tendon type.

References

1. Kodur VKR (2000) Spalling in high strength concrete exposed to fire-concerns, causes, critical parameters and cures. In: Advanced Technology in Structural Engineering, Philadelphia, USA ASCE, pp 1–9
2. EN 1993-1-2 (2005) Eurocode 3: Design of steel structures—Part 1–2: General rules—Structural fire design. European Committee for Standardization
3. EN 1992-1-2 (2004) Eurocode 2: Design of concrete structures—Part 1–2: General rules—Structural fire design. European Committee for Standardization
4. Kodur VKR (2014) Properties of concrete at elevated temperatures. ISRN Civil Eng
5. Fu Y, Li L (2011) Study on mechanism of thermal spalling in concrete exposed to elevated temperatures. Mater Struct 44:361–376
6. Bazant ZP (1997) Analysis of pore pressure, thermal stress and fracture in rapidly heated concrete. In: International workshop on fire performance of high-strength concrete proceedings. Appendix B: Workshop papers B10 NIST, Gaithersburg, MD, pp 155–164
7. Kodur VKR, Banerji S (2021) Modeling the fire-induced spalling in concrete structures incorporating hydro-thermo-mechanical stresses. Cement Concr Compos 117:1–20
8. Liu J-C, Tan KH, Yao Y (2018) A new perspective on nature of fire-induced spalling in concrete. Constr Build Mater 184:581–590
9. ACI Committee 318 (2019) Building code requirements for structural concrete (ACI 318-19) and commentary (ACI 318R-19). American Concrete Institute Farmington Hills, MI, pp 378–379
10. Shin H, Kang TH-K, Park JH (2020) Grouted extruded-strand tendons: friction coefficients and differential individual strand forces. ACI Struct J 117:223–233
11. ASTM International (2020) Standard test methods for fire tests of building construction and materials. West Conshohocken, PA

Strength and Workability of Concrete Incorporating Silica Fume and Egg Shell Powder as Cement Replacement

Muhammad Tahir Lakhiar and Sih Ying Kong

Abstract The purpose of this study was to investigate the potential utilisation of silica fume and eggshell powder as cement replacement for concrete production. Cement was replaced using two percentages of silica fume (15 and 20%) combined with three percentages of eggshell powder (5, 10 and 15%) by weight of cement. The workability, compressive, and splitting tensile strengths of concrete incorporating silica fume and eggshell powder were analysed. The results demonstrated that the workability of concrete reduced as silica fume and egg shell powder content increased. Furthermore, compressive and splitting tensile strength increased with the total combined content of silica fume and eggshell powder up to 25%. It was also observed that 15% silica fume combined with 10% eggshell powder cement replacement achieved the maximum compressive and splitting tensile strengths. The replacement of cement with a combination of silica fume and eggshell powder of greater than 25% led to a decrease in compressive and splitting tensile strength.

Keywords Cement replacement · Workability and mechanical properties

1 Introduction

Conventional concrete, which comprises cement as a binder, aggregates and water, has been the most popular construction material because of its strength, durability and feasibility [1]. The environmental problems associated with the construction industry have mostly revolved around two issues: carbon footprint and the greenhouse effect, caused mainly by cement production [2]. Developing and adopting more sustainable concrete is essential to support the ongoing population growth and social and economic development.

Previous research has demonstrated that sustainable construction materials could be developed by incorporating solid wastes into concrete mixes. These solid wastes include construction wastes [3] and industrial wastes [1]. Researches have been

M. T. Lakhiar · S. Y. Kong (✉)
Civil Engineering Discipline, School of Engineering, Monash University Malaysia, Subang Jaya, Selangor, Malaysia
e-mail: kong.sih.ying@monash.edu

T. Kang and Y. Lee (eds.), *Proceedings of 2021 4th International Conference on Civil Engineering and Architecture*, Lecture Notes in Civil Engineering 201,
https://doi.org/10.1007/978-981-16-6932-3_15

performed on cement replacement by different types of pozzolanic materials [1] and industrial by-products [2] in various kinds of concrete. These pozzolanic materials and industrial by-product materials include coal bottom ash [1], sugar cane bagasse ash [4], rice husk ash [5] and slag [1] that could contribute to a decrease in cement usage, and hence moving towards sustainability.

Silica fume demonstrates pozzolanic properties that help to enhance the mechanical properties of concrete [6]. Furthermore, silica fume is typically adopted in the range of 5–20% [7] by mass of the total cementitious material. It is utilised either as a cement replacement or as an additive to improve the mechanical properties of concrete [7]. However, it has been discovered that up to 10% of cement amount can be replaced by silica fume successfully without reducing the strength [6]. A further increase in replacement content could cause a progressive loss in mechanical properties. This might be due to the lack of calcium oxide (CaO) present in the concrete mix that is necessitated during the pozzolanic reaction.

Malaysian Agriculture and Agro-Based Industry ministries [8] reported that Malaysia utilises 2.8 million poultry eggs daily, and this number has been rising over the years. The higher utilisation of eggs results in an increase in eggshell waste, contributing to increased solid waste.

Meanwhile, the eggshell powder is non-pozzolan material but rich in calcium oxide (CaO). The presence of calcium carbonate ($CaCO_3$) might be a suitable cement replacement since Jhatial et al. [9] demonstrated that the eggshell powder has a unique application as an environment-friendly catalyst. Kamaruddin et al. [10] reported that the new trend on CaO-SiO_2 triggered the catalyst reaction. This gives a strong justification that the mixture between silica fume and eggshell powder is feasible as partial cement replacement to initiate the dual purpose in concrete, contributing to the increase in the mechanical characteristics of concrete. Therefore, the purpose of this research is to examine the potential utilisation of silica fume and eggshell powder as cement replacement materials. This study reported experimental results of workability, compressive and splitting tensile strengths of concrete incorporating the silica fume and eggshell powder.

2 Materials and Methods

2.1 Materials

The materials utilised to develop this new innovative concrete are cement, coarse and fine aggregate, silica fume, water and discarded eggshells.

Table 1 Physical properties of aggregates

Physical properties	Fine aggregate	Coarse aggregate
Water absorption (%)	1.15	1.48
Specific gravity	2.63	2.67

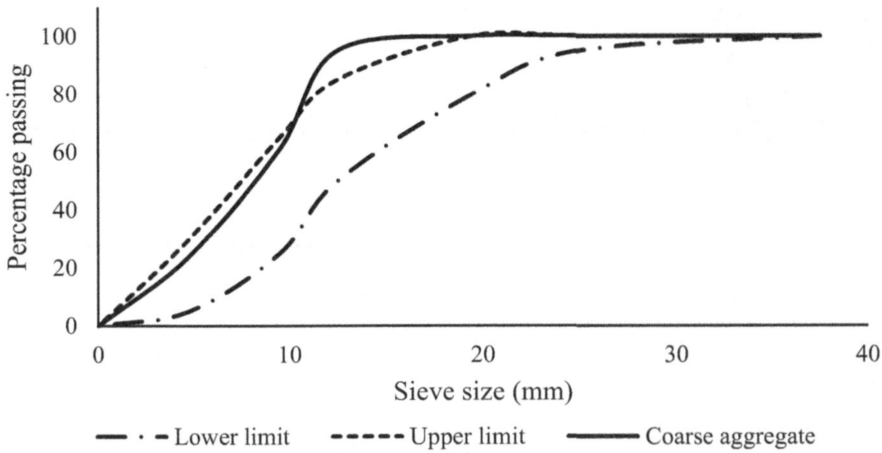

Fig. 1 Sieve analysis of coarse aggregates

2.2 Cement and Aggregates

The cement adopted was Ordinary Portland cement following MS 522 [11]. Crushed stone and sand were utilised as coarse and fine aggregates, respectively. The physical characteristics of coarse and fine aggregates are presented in Table 1. The maximum coarse aggregate size was 10 mm. The sieve analysis results of coarse aggregates and fine aggregates are shown in Figs. 1 and 2. Tap water was utilised for concrete production work.

2.3 Silica Fume

In this research, silica fume showed a specific gravity of 2.32 and a surface area of 341 cm^2/g. It was obtained from a local dealer in Malaysia, as shown in Fig. 3.

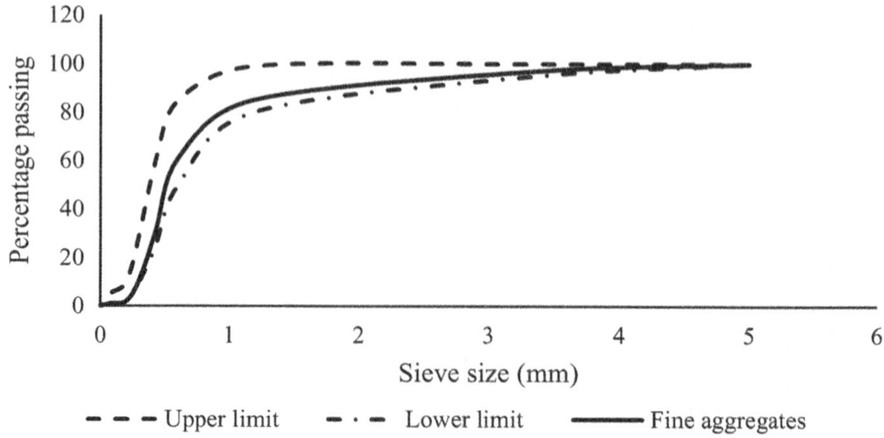

Fig. 2 Sieve analysis of fine aggregates

Fig. 3 Silica fume

2.4 Egg Shell Powder

The eggshells were gathered from various kinds of food courts and cafeterias near Monash University Malaysia. The collected eggshells comprised yellowish fluid detached by washing the raw eggshells with tap water. The washed eggshells were then oven-dried to remove water for 24 h at a temperature of 105 °C ± 5 °C [2]. The oven-dried eggshells were ground utilising Los Angeles Abrasion Machine for 60 min, as shown in Fig. 4. After that, the eggshell powder was ground using a high capacity food grinder and sieved through 63 μm to obtain an eggshell powder of a similar size as cement. Figure 5 shows the final product of the eggshell in powder form.

Fig. 4 Grinding of eggshells

Fig. 5 Eggshell powder

2.5 Mix Design of Concrete

In this research, the DOE mix design method was utilised to develop the concrete mix design. The targeted compressive strength was 30 MPa. The binder-sand-coarse aggregate ratio utilised for this research was 1:1.6:2, while the water-cement (w/c) ratio of 0.5 was adopted. The concrete control specimens were cast for comparison. The quantities of 15% and 20% silica fume (SF) were utilised to replace cement. At the same time, eggshell powder (ESP) content was varied from 5, 10 and 15% by weight, as presented in Table 2.

2.6 Methodology

Coarse and fine aggregates were first to dry mixed using a drum mixer for three minutes. Around half of the mixing water was added to the matrix and mixed for two

Table 2 Quantities needed to develop 1 m^3 concrete according to DOE mix design

Sample	Replacement (%)		Cement	SF	ESP	Sand	CA
	SF	ESP					
M0	0	0	466	0	0	765	936
M1	15	5	372.8	69.9	23.3	765	936
M2	15	10	349.5	69.9	46.6	765	936
M3	15	15	326.2	69.9	69.9	765	936
M4	20	5	349.5	93.2	23.3	765	936
M5	20	10	326.2	93.2	46.6	765	936
M6	20	15	302.9	93.2	69.9	765	936

Note Cement, sand, silica fume (SF), eggshell powder (ESP) and coarse aggregates quantities are presented in kg/m^3

Table 3 Summary of the test methods for each mix design

Test	Standard	Size of samples
Compressive strength	[13]	Cube of 100 × 100 × 100 mm
Splitting tensile strength	[14]	Cylinder of 100 mm diameter and 200 mm height

minutes. The cement, silica fume and eggshell powder were added gradually to the matrix and mixed for two minutes. After that, the remaining half of the water was added to the concrete matrix and mixed for another five minutes. The workability test was conducted using the slump flow test in accordance with ASTM C43 [12].

Meanwhile, cubes and cylinders were cast to determine the compressive and splitting tensile strengths of concrete produced. The samples were de-moulded after twenty-four hours of casting and placed in a water tank for 28 days curing. Experimental testing of the concrete compressive and splitting tensile strength properties were conducted according to ASTM test standards as listed in Table 3.

3 Results and Discussions

3.1 Workability

The workability results are presented in Fig. 6. The results showed that the M0 (control) sample of concrete achieved the maximum workability of 94 mm compared to all samples. As the cement replacement increased, the slump value decreased slightly. According to Faraj et al. [7], the workability of concrete decreased when silica fume was added to concrete. Jhatial et al. [9] reported that concrete workability reduced when the eggshell powder was used. The high surface area of eggshell powder and silica fume particles could be the key factor behind this reduction. M6

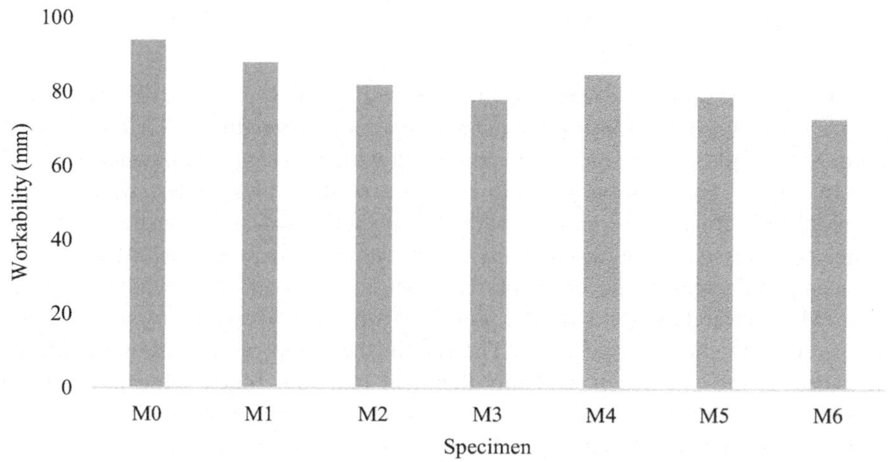

Fig. 6 Workability of concrete

samples (35% replacement) showed the maximum decrease in workability as a large amount of silica fume and eggshell powder absorbed more water.

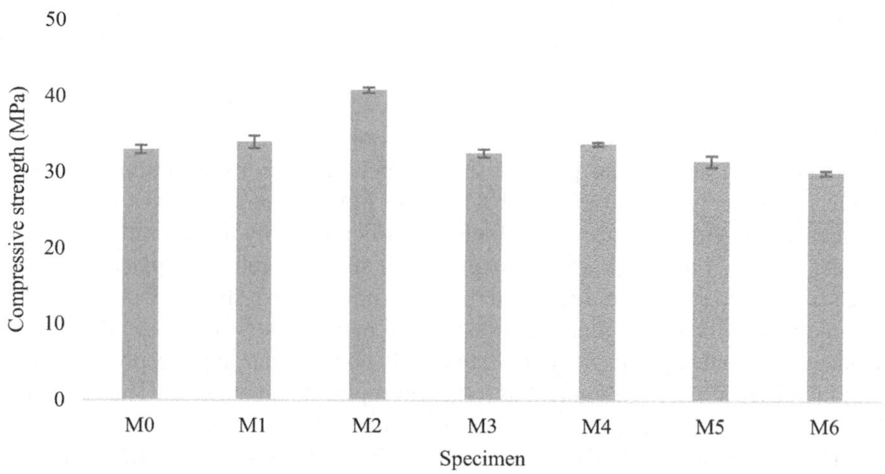

Fig. 7 Average compressive strength of concrete

3.2 Compressive Strength

The average compressive strength results are presented in Fig. 7. The results showed that the control samples (M0) achieved a compressive strength of 33 MPa. Furthermore, M2 samples (15% silica fume combined with 10% eggshell powder) showed 23.63% higher compressive strength compare to the M0 samples, whereas M4 samples (20% silica fume combined with 5% eggshell powder) demonstrated slightly higher compressive strength compare to M0 samples. However, a further rise in replacement decreased the compressive strength. Beyond 25% cement replacement (M3, M5 and M6), the results presented slightly lower compressive strength compared to the control specimen. The rise in the compressive strength could be attributed to the positive pozzolanic chemical reaction. Silica fume is the most popular pozzolan with Silica Oxide (SiO_2), whereas eggshell powder possesses an ample amount of CaO. According to Jhatial et al. [9] and Kamaruddin [10], the CaO, when in contact with water, turns into Calcium Hydroxide, $Ca(OH)_2$, as shown in Eq. 1. When pozzolan appears in contact with water, a pozzolanic reaction happens, as shown in Eq. 2. Lime is consumed to produce extra calcium- silicate-hydride (C–S–H) gels are responsible for getting the strength.

$$CaO + H_2O \rightarrow Ca(OH)_2 \tag{1}$$

$$Ca(OH)_2 + SiO_4 + H_2O \rightarrow CaH_2SiO_4 \cdot 2H_2O \tag{2}$$

Silica fume has been utilised in concrete as supplementary cementing material. During the hydration of concrete, cement can only deliver a limited free lime that silica fume can absorb and produce C–S–H gels. Hence, its adoption as cement replacement has been restricted to 10% due to limited free lime given by cement. Supplementary free lime provided along with silica fume by incorporating eggshell powder during concrete production can further increase cement replacement with silica fume. The combined utilisation of silica fume and eggshell powder increased the amount of silica fume that might be utilised in concrete. It increased the cement replacement content from 5 to 10%; eggshell powder can also be utilised along with increased silica fume content as a cement replacement material. Thus a total of 25% of cement replacement is possible without losing the compressive strength of concrete. The trend observed in this study is similar to previous studies Jhatial et al. [9] and Kamaruddin et al. [10] investigated the material and mechanical characteristics of concrete containing palm oil fuel ash and eggshell powder as supplementary cementing material. These results demonstrated that the addition of eggshell powder increased the positive pozzolanic reaction in a concrete matrix and enhanced the mechanical properties of concrete.

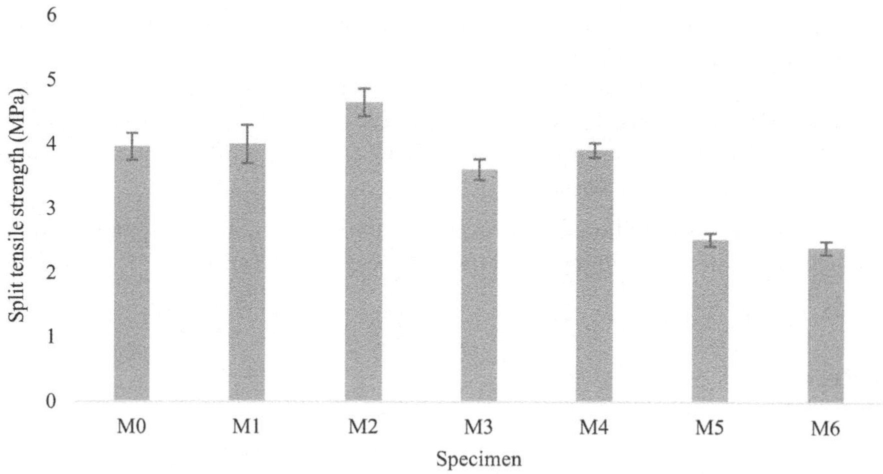

Fig. 8 Splitting tensile strength of concrete

3.3 Splitting Tensile Strength

Figure 8 presents the splitting tensile strength of the concrete specimens incorporating various proportions of eggshell powder and silica fume as partial cement replacement for 28 days. From the results, it could be noticed that the splitting tensile strength increased with the partial cement replacement. This was an indication of the enhancement of splitting tensile strength as a result of the pozzolanic reaction. The highest splitting tensile strength found was 4.65 MPa (M2). The splitting tensile strength results found were in agreement with the compression strength test results. These results are supported by Ghaffar et al. [15] analysed the compressive and splitting tensile strength of concrete containing the palm oil fuel ash and eggshell powder as cement replacement. The outcomes demonstrate that the compressive and splitting tensile strength increased as the cement replacement increased.

4 Conclusion

The purpose of this study was to analyse the potential utilisation of silica fume and eggshell powder as cement replacement materials in concrete. Based on the experimental results, the following conclusions could be drawn:

- The workability of concrete decreased with the addition of silica fume and eggshell powder in concrete due to their high surface area of these powder particles.
- The incorporation of silica fume and eggshell powder significantly impacts the compressive and splitting tensile strength of concrete. Concrete with silica fume and eggshell powder has higher compressive and splitting tensile strength

compared to control samples. The rise in the compressive and splitting tensile strength could be attributed to the positive pozzolanic reaction.

- Furthermore, it was found that the total 25% combined utilisation of silica fume and eggshell powder increased the compressive and splitting tensile strength. The further rise in cement replacement content decreased the compressive and splitting tensile strength of concrete.

It is recommended for future study to analyse the modulus of elasticity and flexural strength of concrete containing silica fume and eggshell powder.

References

1. Lakhiar MT et al (2021) Effect of banana skin powder and coir fibre on properties and flexural behaviour of precast SCC beam. Int J Sustain Eng pp 1–14
2. Kamaruddin S et al (2018) Chemical and fresh state properties of foamed concrete incorporating palm oil fuel ash and eggshell ash as cement replacement. Int J Eng Technol 7(4.30):350–354
3. Sandhu AR et al (2019) Effect of River Indus Sand and REcycled concrete aggregates as fine and coarse replacement on properties of concrete. Eng Technol Appl Sci Res 9(1):3832–3835
4. Gooi S, Mousa AA, Kong D (2019) Efficacy of treatments on coal bottom ash as a cement replacement. In: International Congress and Exhibition "Sustainable Civil Infrastructures". Springer, Berlin
5. Iqtidar A et al (2021) Prediction of compressive strength of rice husk ash concrete through different machine learning processes. Curr Comput-Aided Drug Des 11(4):352
6. Adil G, Kevern JT, Mann D (2020) Influence of silica fume on mechanical and durability of pervious concrete. Constr Build Mater 247:118453
7. Faraj RH et al (2021) Rheological behavior and fresh properties of self-compacting high strength concrete containing recycled PP particles with fly ash and silica fume blended. J Build Eng 34:101667
8. Malaysia Dateline (2019) Hala tuju Kementerian Per Tanian & Industri Asas Tani, P.S. 2019–2020. Editor. Putrajaya
9. Jhatial AA et al (2020) Thermo-mechanical properties of various densities of foamed concrete incorporating polypropylene fibres. Arab J Sci Eng 45:8171–8186
10. Kamaruddin S et al (2021) Development of self-compacting concrete incorporating palm oil fuel ash and eggshell powder as partial cement replacement. In: Proceedings of the sustainable concrete materials and structures in construction 2020. Springer, Berlin, pp 1–12
11. Standard M (2003) Portland cement (ordinary and rapid-hardening): Part 1. Specification (Second revision), Malaysia, MS, p 522
12. ASTM, C (2012) Standard test method for slump of hydraulic-cement concrete. ASTM International West Conshohocken, PA
13. C39, A (2012) Standard test method for compressive strength of cylindrical concrete specimens
14. C496, A (2002) Standard test method for splitting tensile strength of cylindrical concrete specimens. Philadelphia
15. Ghaffar FSA et al (2021) Mechanical properties of concrete containing untreated palm oil fuel ash and egg shell powder. In: Proceedings of the sustainable concrete materials and structures in construction 2020. Springer, Berlin, pp 259–270

Performance Evaluation of Glass Fiber Reinforced Concrete

Prabhakara Rao V. N. Vedula, Yashwant Tatineni, and A. U. Charpe

Abstract Over the period of time, mechanical properties of natural as well as synthetic fiber have been used to reinforce concrete albeit extensively, to enhance the mechanical properties of concrete. Many investigations were carried out to evaluate the engineering characteristics and explore the applications of a variety fibers leaving a few. The present study thus involves the determination of optimum % of glass fiber reinforcement (wt./wt. ratio) with the best compressive strength, split tensile strength, impact strength. With the addition of glass fiber, (1) compressive strength was found to increase by 14% for 0.8% glass fiber, (2) flexural strength was found to increase by 7.84% for 0.8% glass fiber and (3) Schmidt rebound hammer number was found to increase by 14% for 0.8% glass fiber (4), split tensile strength value is found to increase by 37% for 0.8% glass fiber, (5) ultra sonic pulse velocity was found to increase by 24% for 0.8% glass fiber. Glass fiber in the concrete medium thus enhanced the properties of concrete in general but split tensile strength in particular.

Keywords Natural fiber · Synthetic fiber · Concrete · Glass fiber · Strength

1 Introduction

Throughout the last 30 years there were many a study on reinforcing the concrete with synthetic fibers or fibers derived from metals. Most of the synthetic fibers suffer from the drawback of extended life period. Production of synthetic fibers, apart from expensive, leaves lot of carbon footprint. Ready availability over major part of the world renders the natural fibers an obvious choice to researchers even-though they suffer from the drawback of low modulus. Developing countries are blessed with natural fibers which can be used for reinforcing modern day cementing material, i.e.,

P. R. V. N. Vedula (✉) · Y. Tatineni · A. U. Charpe
Velagapudi Ramakrishna Siddhartha Engineering College, Vijayawada, Andhra Pradesh 520007, India

Y. Tatineni
Sri Vasavi Engineering College, Tadepalligudem, Andhra Pradesh 534101, India

T. Kang and Y. Lee (eds.), *Proceedings of 2021 4th International Conference on Civil Engineering and Architecture*, Lecture Notes in Civil Engineering 201,
https://doi.org/10.1007/978-981-16-6932-3_16

concrete. Since ages, fibers derived from vegetable have been expended for reinforcing the dwelling made-up of clay or mud as they are available in abundance locally. Natural fiber can be extracted from plants or animals example straw from shrubs, plants or hair derived from domesticated animals like horse were used to manufacture building materials like bricks. Performance evaluation of natural fiber reinforced polymer and carbon fiber reinforced polymer in the flexural strengthening of reinforced concrete beams was made and it was reported that the epoxy resin increased the environmental impact, flax fiber reinforced plastic at large yielded lower environmental impact than jute fiber reinforced plastic [1]. Hemp fiber of different lengths (6, 12, and 18 mm) was used as a reinforcing medium for a cement mortars in varied proportions (0, 1, 2, 3%) and it was found that cement mortars strengthened with 2–3% amount of hemp fiber and hemp fiber with 12 mm length best characteristics [2]. A variety of structural applications were investigated with fiber derived from kenaf plant and it was used as a reinforcing material in the preparation of concrete and its potential use as a green building material was explored [3]. The postscript of jute and coconut fiber reduces the compressive strength and causes very minimal augmentation of tensile strength [4]. Foam board was used to decrease the self-weight in precast concrete double-tee members [5]. Dispensation of Oil Palm Fibers for amalgamation in concrete is ecologically viable at low cost and high employment potential [6]. Coconut fiber reinforced concrete slabs exhibit nearly the same strength as sheets made up of asbestos cement [7]. The option of exchanging synthetic fibers with natural fibers derived from coconut, sugar cane bagasse, and banana fibers was examined and (i) outstanding performance of coconut fiber in enhancing the fracture and flexural results for polymer concrete, (ii) good endurance in fracture of sugar cane bagasse polymer concrete, (iii) unsuitability of banana pseudo-stem fiber for polymer concrete reinforcement were reported [8]. An investigational appraisal on growth of crack in composites made-up of cementitious media like blast furnace slag cement buttressed with pulped fibers of sisal, banana buttressed with the help of biodegradable natural fibers derived from agricultural by-products was made and stable fatigue crack growth was reported [9]. Structural elements reinforced with natural fiber are bounded by the mechanism similar to structural elements reinforced with metallic fiber was investigated [10]. The effect of impact and the response on cement composite reinforced with unidirectional sisal fiber substantial increase in absorbed energy was studied [11]. Resistance to cyclic loading enhances upon reinforcing with natural fibers jute and coconut fiber was explored [12]. Some of the researchers have experimented even with hybrid fibers. Carbon–steel combination offered concrete, the peak strength and flexural robustness due to the nearly equal modulus and the interface amongst the fibers was synergistic [13]. Nevertheless, the experimental procedures engaged and the extents of routine characteristics documented by numerous investigators are whimsical. So, it there is a great scope to experiment with a new variety of natural fiber within the standards of relevant code of practice in a given country to evolve future application in cementitious composites. There is a lot of scope for alternative natural fibers with low bio-persistence. Glass fiber (Gf) fits into this space.

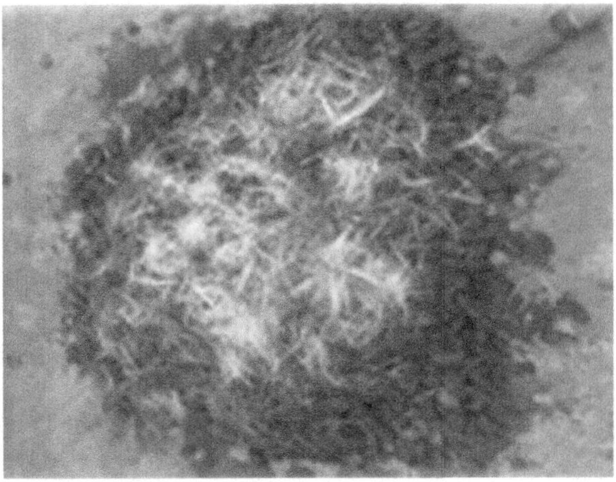

Fig. 1 Glass fiber (Gf) placed on and concrete mix

Table 1 Composition of glass fiber

Parameter	Denomination
Length	12 mm
Filament diameter	14 μ
Specific gravity	2.68
Modulus of elasticity	72 GPa

2 Materials and Methods

2.1 Glass Fiber (Gf)

Glass fiber from natural origin, Fig. 1 was procured from Alminrock, Bangalore, Karnataka, India. The properties of glass fiber were as shown in Table 1.

2.2 Cement, Sand, and Aggregates

In the present investigation, ordinary Portland cement was compatible to IS 12269: 2013 [14], Godavari River sand corroborates zone II off IS 383 was consumed and coarse aggregates of size 12.5 mm (14.8%), 16 mm (27.8%) 20 mm (57.4%) were consumed [15] and setting time of cement was checked [16]. Normal municipal water compatible to IS 456: 2000 was consumed for preparing control test samples [17].

Table 2 Mix proportion—one cum of concrete

Materials	Quantity (kg/m^3)
Cement	377.67
Water	169.95
Sand	478.4
Coarse aggregates	1414.53

2.3 Preparation of Concrete Specimens

Standard size concrete cubes of 150 mm were moulded for determining the compressive strength of designed concrete mix [17]. Concrete mix with water to cement ratio 0.45 was designed for 58.25 MPa target strength at curing age of 28 days [18]. The niceties of concrete mix proportion are mentioned in Table 2.

Fine aggregate was derived from Godavari River. Coarse aggregates of size 12.5 mm (14.8%), 16 mm (27.8%) 20 mm (57.4%) were consumed. Natural glass fiber was obtained from Alminrock, Bangalore, Karnataka, India. Control test sample of concrete (i.e., pure concrete with no additives) and concrete with glass fiber (glass fiber at 0.2, 0.4, 0.6, 0.8, 1.0, 1.2% by weight of cement was incorporated) were used in preparation of test samples meant to assess compressive strength (150 mm cubes), flexural strength (150 mm × 150 mm × 700 mm), Schmidt rebound hammer (150 mm cubes), split tensile strength (150 mm diameter, 300 mm long) and ultrasonic pulse velocity (150 mm cubes). The test samples were cast as per established provisions in the IS code [17]. All test samples were removed from moulds after the lapse of 24 h. The outcome of duration of curing on compressive strength was also observed on 7 day, 21 day and 28 day).

2.4 Curing Conditions

All specimens were prepared using tap water. Control test samples (pure concrete with no additives), test samples with glass fiber additives were kept in a huge tub containing water. During the current investigation, curing periods for 28 days was espoused. Temperature of curing was sustained as (27 ± 2 °C). The curing water was replaced weekly. Ample quantity of water was made available for the test samples to offset heat of hydration during the entire period of curing. For each factor examined, three samples were tested.

2.5 Parameter Examined

For each factor examined, investigations were carried out in triplicate.

2.5.1 Compressive Strength

Compressive strength (CS) assessment exhibits the resistance offered by cementitious ingredients to compressive load. IS 516:1959 (Reaffirmed 2004) was followed for assessing the compressive strength of 150 × 150 × 150 mm concrete cubes at the end of their corresponding curing periods. In order to assess the characteristic compressive strength, concrete cubes were to be tested at the end of 28 day curing [19]. In order to showcase the attainment of strength with respect time upon curing, cubes will be tested upon 7 days of casting, 21 days of casting along with 28 days of casting.

2.5.2 Flexural Strength

Flexural strength (FS) is the materials capability to endure deformation under the applied load. Flexural strength at failure flexural strength is determined in the laboratory either by 3-point. Specimen intended for testing is placed in water tub for curing till the moment of testing. The specimen is placed in the machine. Load is applied. The specimen obviously fails at a certain application of load and which is reported as flexural strength. IS 516: 1959 (Reaffirmed 2004) was followed for assessing the flexural strength of 15 × 15 × 70 cm beams at the end of their corresponding curing periods [19].

2.5.3 Schmidt Rebound Hammer Number

Schmidt rebound hammer test is a technique of indirect assessment of concrete. It entails a spring-controlled mass that slides on a plunger inside a tubular housing. 150-mm concrete cubes were prepared exclusively for this purpose in the same design mix as above. During the process, the nozzle of rebound hammer is constrained against the external vertical face of concrete to rebound back. The magnitude of Schmidt rebound hammer, which is a gauge of surface hardness, was counted on a graduated scale. This counted value was specified as Schmidt rebound number (SRH) [20].

2.5.4 Split Tensile Strength Test

Split tensile strength (STS) test is a destructive process of assessment of tensile strength of concrete using a cylindrical test sample across its diameter, albeit indirectly [21]. Cylindrical specimens are prepared cured and placed on the loading machine along their longitudinal axis perpendicular to the load. Two bearing plates of hardened hardboard of nominal thickness 4 mm compliant to IS 1658 were used. Bearing plates are kept between test samples and both upper, lower bearing blocks of loading machine. Test samples were kept in water for 24 h ahead of testing. The load was applied smoothly and was amplified incessantly at rate of 1.2 N/mm^2/min till fracture, consequent to failure of the test sample. Failure load is noted. The failure load upon computation provides split tensile strength.

2.5.5 Ultrasonic Pulse Velocity Test

Ultrasonic pulse velocity (UPV) test is based on the pulse velocity technique to render knowledge on the homogeneity of concrete and resultant quality of concrete as a consequence. The apparatus to accomplish this assessment entails two transducers, typically metal cylindrical heads. The transducer is secured to vertical surface of the concrete cube to vibrate the concrete. The pulses travel through the concrete and finally get to the receiver. UPV test was performed as per procedure laid down in IS 13311: 1992 [22].

3 Results

3.1 Compressive Strength Test

Compressive strength (CS) assessment of all varieties of test samples, i.e., control test samples and test samples with glass fiber additives, was conducted as per IS 516: 1959 (Reaffirmed 2004) upon the age of the test sample became 7 and 28 days, respectively [19]. On the 7th and 28th day of casting, three concrete cubes were tested for compressive strength for each variety of mix and the average of these three readings were recorded and reported. The average compressive strength of the control test samples (pure concrete with no additives) was 48.44 kN/m^2. Compressive strength results for all proportions of glass fiber, i.e., 0, 0.2, 0.4, 0.6, 0.8, 1.0, 1.2% by the weight of cement on the 7 days and 28 days of curing were depicted in Fig. 2. The strength attainment of all varieties of test sample with age was depicted in Fig. 3.

3.2 Flexural Strength

Flexural strength (FS) assessment of all varieties of test samples, i.e., control test samples and test samples with glass fiber additive, was performed as per IS 516: 1959 (Reaffirmed 2004) [19]. Three concrete beams were tested for measuring flexural strength for each proportion and the average of these three readings were recorded and reported. The average flexural strength of the control test samples (pure concrete with no additives) was 7.77 kN/m^2. Flexural strength results for all proportions of glass fiber, i.e., 0, 0.2, 0.4, 0.6, 0.8, 1.0, 1.2% by the weight of cement on 28 days of curing were depicted in Fig. 4.

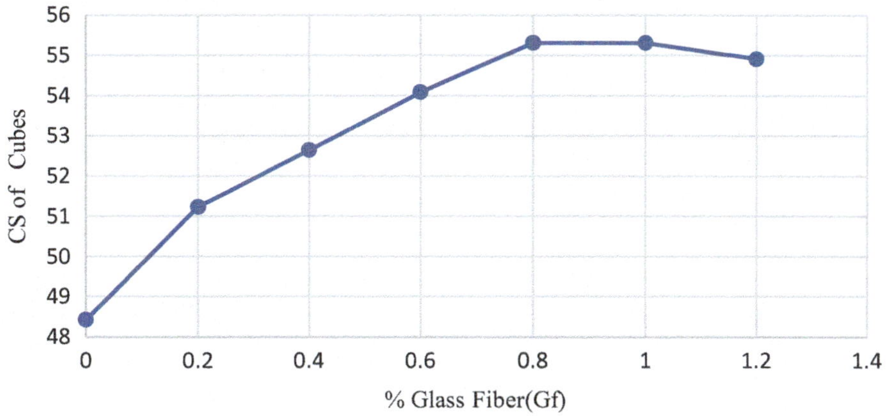

Fig. 2 Variation of CS with % Gf

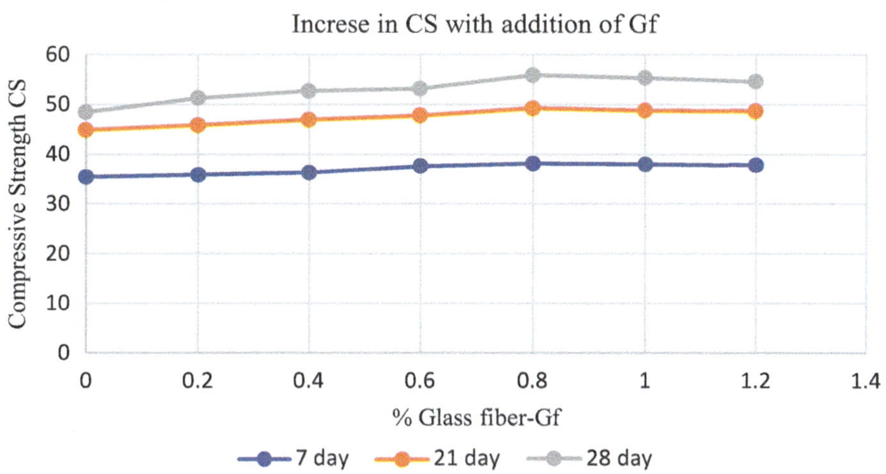

Fig. 3 Increase in CS with addition of Gf

3.3 *Schmidt Rebound Hammer Test*

Schmidt rebound hammer number assessment was conducted on all varieties of test samples, i.e., control test samples and test samples with glass fiber additive, as per IS 13311(2): 1992 [20]. The average Schmidt rebound hammer number of the control specimens (pure concrete with no additives) was 58. SRH for all proportions of glass fiber, i.e., 0, 0.2, 0.4, 0.6, 0.8, 1.0, 1.2% by the weight of cement on 28 days of curing were depicted in Fig. 5.

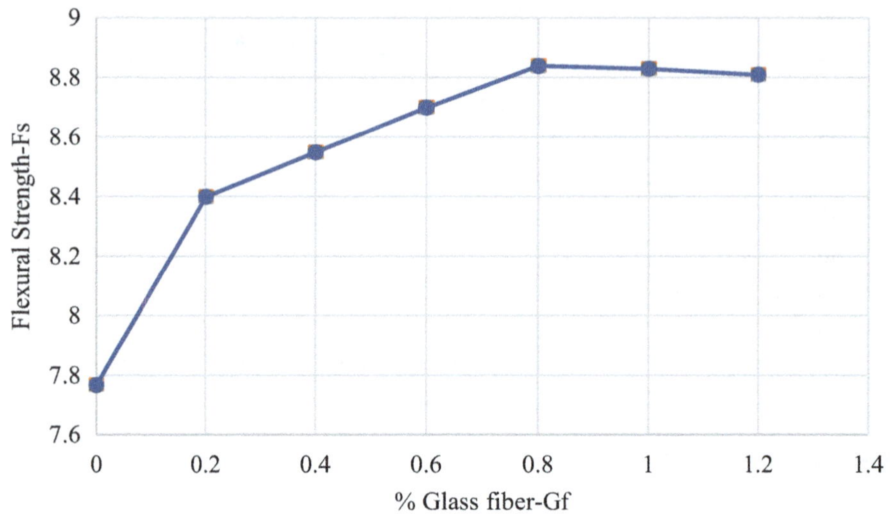

Fig. 4 Variation of FS with % Gf

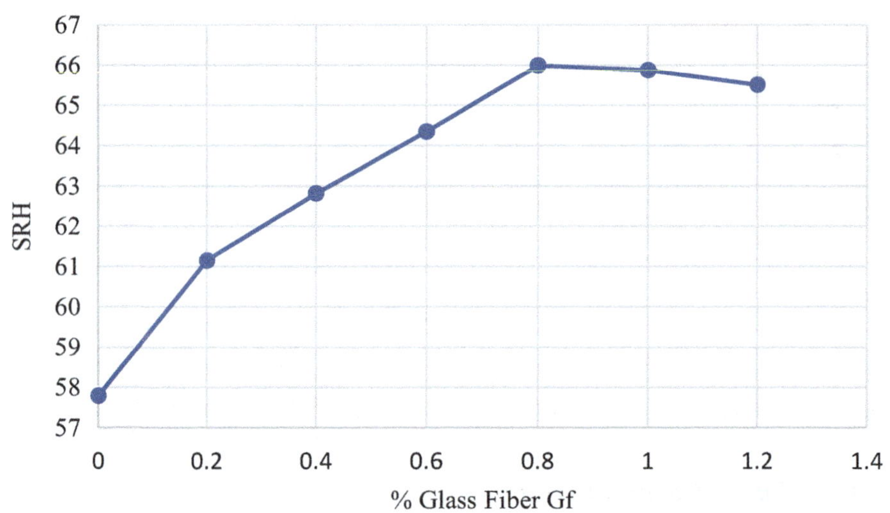

Fig. 5 Variation of SRH with % Gf

3.4 Split Tensile Strength Test

Split tensile strength test of all types of specimens, i.e., control specimens and specimens with glass fiber additive, was performed as per IS 5816: 1999 [21]. The average split tensile strength of the control specimens (pure concrete with no additives) was 5.39 kN/m^2. Split tensile strength results for all proportions of glass fiber, i.e., 0, 0.2,

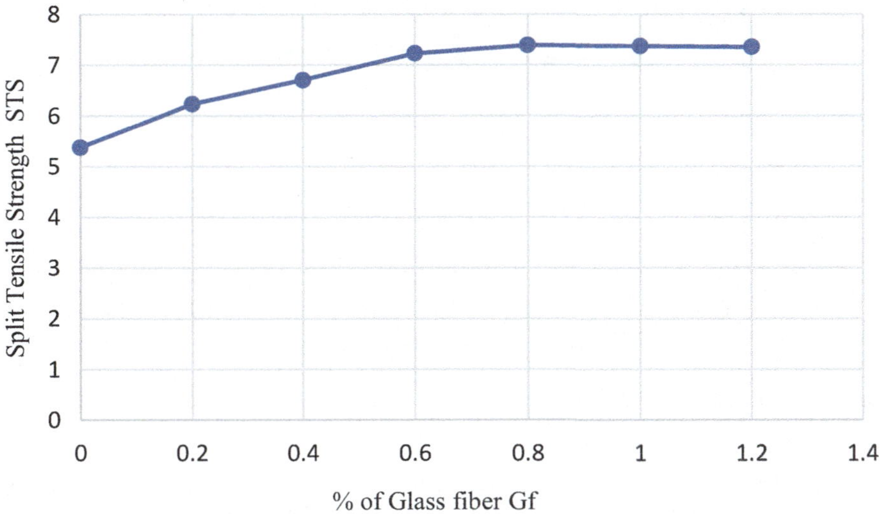

Fig. 6 Variation of STS with % Gf

0.4, 0.6, 0.8, 1.0, 1.2% by the weight of cement on 28 days of curing were depicted in Fig. 6.

3.5 *Ultra-sonic Pulse Velocity*

Test was conducted as per the provisions of IS 13311 (Part 1): 1992 [22]. Ultra-sonic pulse velocity (UPV) test was conducted on 150 mm size test samples as the maximum size of the aggregate used was 20 mm. The surface of the concrete cube was smoothened, levelled and greased on area of the surface where the transducer was placed. The average pulse velocity of the control specimens (pure concrete with no additives) was 3530 m/s. Pulse velocity results for other proportions of glass fiber additive at 0.3, 0.6, 0.9,1, 1.3, 1.6, 1.9, 2, 2.3% by the weight of cement were depicted in Fig. 7.

4 Discussion

Present investigation is focused at observing the effect of glass fiber on the strength of concrete. All varieties of cubes attained strength with the passage of time. For control test samples, gain in compressive strength from 7 to 21 days was 27%, gain in strength from 21 to 28 days was 8% while gain in strength from 7 to 28 days was 36%. Thus it can be observed gain of strength is more in first 21 days and later

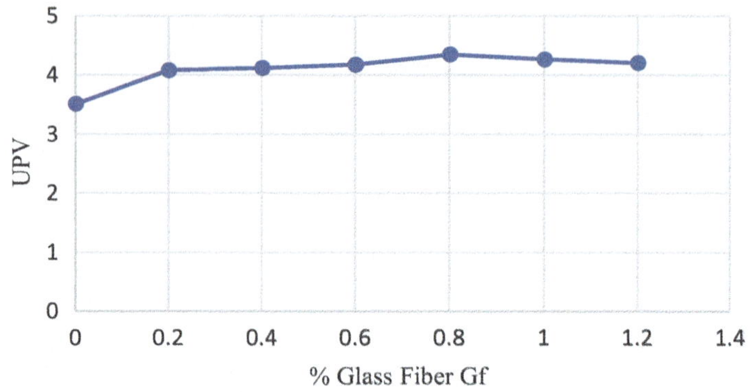

Fig. 7 Variation of UPV with % Gf

on it recedes. For the concrete prepared with 0.8 glass fiber additive, the % gain in compressive strength for 0.8% from 7 to 21 days was 29%, gain in strength from 21 to 28 days was 12% while gain in strength from 7 to 28 days was 44%.

In addition to the compressive strength, flexural strength, split tensile strength, Schmidt rebound hammer test, ultrasonic pulse velocity were performed.

In the present investigation, the average compressive strength of the concrete with glass fiber has shown increment by 14% for 0.8% addition of glass fiber than control test samples (pure concrete with no additives).

Upon 28-day curing, the flexural strength of concrete with glass fiber additive was observed to increase by 7.84% for 0.8% addition of glass fiber.

In this study, Schmidt rebound hammer value of the concrete modified with glass fiber additives increased by 14% for 0.8% glass fiber.

In the present study, split tensile strength of the concrete with glass fiber additives increased by 37% for 0.8% glass fiber.

In the present investigation, pulse velocity for concrete with glass fiber additive wherein the pulse velocity increased by 23% for 0.8% glass fiber.

5 Conclusion

It can be inferred from the tests performed that, in general, all characteristic peaked upon addition of glass fiber to ingredients of concrete at 0.8% by weight of cement. Best merit of addition of glass fiber was observed in enhancement of split tensile strength of concrete and ultrasonic pulse velocity of concrete. There was an improvement in compressive strength and Schmidt rebound hammer number. While, improvement in flexural strength was not substantial.

References

1. Chen C, YangY, Zhou Y, Xue C, Chen X, Wu H, Sui L, Li X (2020) Comparative analysis of natural fiber reinforced polymer and carbon fiber reinforced polymer in strengthening of reinforced concrete beams. J Clean Prod 263, 121572
2. Çomak B, Bideci A, Bideci OS (2018) Effects of hemp fibers on characteristics of cement based mortar. Constr Build Mater 169:794–799
3. Elsaid A, Dawood M, Seracino R, Bobko C (2011) Mechanical properties of kenaf fiber reinforced concrete. Constr Build Mater 25:1991–2001
4. Islam R, Alam AKMK (1987) Study of fiber reinforced concrete with natural fibres. In: Proceedings of ISFRC, Vol. II, Madras, India (1987)
5. Jonnalagadda S, Sreedhara S, Soltani M, Ross BE (2021) Foam-void precast concrete double-tee members. PCI J 66(1)
6. Momoh EO, Osofero AI (2020) Recent developments in the application of oil palm fibers in cement composites. Front Struct Civ Eng 14(1):94–108. https://doi.org/10.1007/s11709-019-0576-9
7. Paramasivam P, Nathan GK, Das Gupta NC (1984) Coconut fibre reinforced corrugated slab. Int J Cem Compos Light Weight Concr 6(1)
8. Reis JML (2006) Fracture and flexural characterization of natural fiber-reinforced polymer concrete. Constr Build Mater 20:673–678
9. Savastano H Jr, Santos SF, Radonjic M, Soboyejo WO (2009) Fracture and fatigue of natural fiber-reinforced cementitious composites. Cement Concr Compos 31:232–243
10. Sethunarayanan R, Chockalingam S, Ramanathan R (xxxx) Natural fiber reinforced concrete. Transport. Res Rec 1226 57–60
11. Silva FA, Zhu D, Mobasher B, Filho RDT (2011) Impact behavior of sisal fiber cement composites under flexural load. ACI Mater J 108(2):168–177
12. Singh RN (1987) Flexural behaviour of notched coir reinforced concrete beam under cyclic loading. In: Proceedings of ISFRC, Vol. II, Madras, India
13. Yao Wu, Li J, Keru Wu (2003) Mechanical properties of hybrid fiber-reinforced concrete at low fiber volume fraction. Cem Concr Res 33:27–30
14. Bureau of Indian Standard (BIS) (2013) Ordinary Portland Cement 53 Grade – Specifications IS 12269 2013
15. Bureau of Indian Standard (BIS) (1970) Specification for coarse and fine aggregate from natural sources for concrete IS 383 1970 (Second revision)
16. Bureau of Indian Standard (BIS) (1988) Methods of physical tests for hydraulic cement IS 4031 1988 (Part 6)
17. Bureau of Indian Standard (BIS) (2000) Plain and reinforced concrete- code of practice IS 456: 2000
18. Bureau of Indian Standard (BIS) (2009) Concrete mix proportioning—guidelines IS 10262: 2009
19. Bureau of Indian Standards (BIS) (1959) Indian standard methods of tests for strength of concrete IS 516: 1959
20. Bureau of Indian Standards (BIS) (1992) Non-destructive testing of concrete—methods of test rebound hammer IS 13311 (Part 2): 1992
21. Bureau of Indian Standards (BIS) (1999) Splitting tensile strength of concrete IS 5816: 1999
22. Bureau of Indian Standards (BIS) (1992) Non-destructive testing of concrete—ultra sonic pulse velocity IS 13311 (Part 1): 1992

Influence of the Use of Recycled Coarse Aggregates on the Compressive Strength for the Production of Concrete of 350 kg/cm^2

Jeffersson Calixto, Doris Esenarro, Willie Rojas, and Ciro Rodriguez

Abstract The objective of this research is the reuse of construction and demolition waste as recycled material, which is presented as one of the leading environmental pollution problems; that is why the use of recycled coarse aggregate has been studied to be able to be used in the preparation of concrete and to be able to generate an option to minimize environmental contamination and seek that the material can be reused so that in turn we obtain concrete of adequate quality. From the results obtained for concrete of 350 km/cm^2, it can be seen that the higher the percentage of recycled aggregate used, its resistance decreases; however, it is still a concrete that achieves an adequate compressive strength due to the safety factor in your design.

Keywords Recycled coarse aggregate (RCA) · Construction and demolition waste (CDW) · Recycled concrete · Recycled aggregate concrete (RAC) · Natural coarse aggregates (NCA)

1 Introduction

During the present century, concrete has become the most widely used construction material in the world due to its extraordinary versatility in terms of form (it can be molded), function (structural and non-structural), and economy, since that the technology developed around it and makes it possible to compete not only with stone and wood constructions but also with steel. Concrete offers, like natural stones, very high resistance to compression forces and minimal traction; therefore, it is unsuitable for forming parts that have to resist pulling or bending. But by supplementing with

J. Calixto (✉) · D. Esenarro · W. Rojas · C. Rodriguez
Federico Villarreal University EUPG, UNFV, Lima, Peru

D. Esenarro
e-mail: desenarro@unfv.edu.pe

C. Rodriguez
e-mail: crodriguez@unfv.edu.pe

D. Esenarro
Federico Villarreal University, FIGAE, UNFV, Lima, Peru

T. Kang and Y. Lee (eds.), *Proceedings of 2021 4th International Conference on Civil Engineering and Architecture*, Lecture Notes in Civil Engineering 201,
https://doi.org/10.1007/978-981-16-6932-3_17

steel rods in the traction zones, this deficiency is made up, thus having a piece resistant to warping. Therefore, reinforced concrete is an artificial stone that can resist compression, traction, and bending forces, a circumstance that does not occur in natural stones. That is why concrete's properties and characteristics are studied to determine the appropriate mix design (proportion of its components) for specific conditions of a given project [1].

Generally, aggregates occupy between 60 and 75% of the total volume of concrete. The global consumption of natural aggregate is approximately 10 billion tons per year, of which 8 billion tons are used in the manufacture of concrete. Therefore, it is estimated that there will be a shortage of natural aggregates in the future since this is a non-renewable natural resource. Conventional concrete, which is the most widely used construction material globally, is not friendly to the environment. Concerns include the depletion of natural resources, high energy consumption for their production, and the elimination of construction and demolition waste (RCD). Therefore, the recycled aggregate concrete (CAR) would help preserve natural resources, mitigate the problem of eliminating the RCD and its environmental impact, and the production of an environmentally sustainable concrete material [2]. This is why an alternative source for aggregates is RCD to produce CAR [3].

Around 25 billion tons of concrete is used globally, which means that more than 3.8 tons per person are generated annually worldwide. Today, there is a trend to reuse demolished concrete as an aggregate to obtain sustainable concrete. Today, there is a more significant trend towards sustainable concrete making, and the reuse of demolished concrete as an aggregate is the way to achieve this sustainability. Although recycled concrete aggregate (ACR) does not lead to a significant reduction in CO_2 emissions, it nevertheless helps reduce the use of natural resources and reduce the disposal of RCD in landfills. Although a good number of investigations have been carried out on the properties of concrete with recycled aggregates (CAR) in a fresh and hardened state, analyses of its structural behavior are limited [4].

From an environmental point of view, rubble recycling is quite attractive because it increases the useful life of sanitary landfills and prevents the degradation of non-renewable natural resources; but, from a purely economic point of view, recycled concrete is attractive when the product is competitive with other materials concerning cost and quality. Recycled materials are generally competitive where there is difficulty in obtaining raw materials and suitable storage locations. With recycled materials, significant savings can be made in transporting construction waste and raw materials. This is especially noticeable in urban areas or on construction projects where demolition and new construction can be brought together. It is also feasible to recycle a large amount of debris on the same job site or in the vicinity [5].

In developed countries, various studies have been carried out in order to know the different physical and mechanical properties of concrete made with recycled aggregates (CAR). However, many of these countries have stringent policies for construction and waste disposal. Therefore, Construction and Demolition Waste (CDW) is assumed to obtain different physical and mechanical properties in other parts of the world, depending on the national reality. That is why this research focuses on identifying those physical and mechanical properties of recycled aggregates in our environment to create concrete with suitable properties for use in the construction industry, taking solid waste in the surroundings as a reference.

2 Methodology

2.1 Materials, Instruments, and Methods

The objective population of this research work is concrete made with recycled aggregates (RCA) that can be made with the construction and demolition waste (CDW) from the Urb surroundings. Pro, in Los Olivos district—Lima, specifically in the Infantas Oval vicinity, where this waste has been continuously used.

For sample, 180 cylindrical car witnesses with C 350 kg/cm^2 will be worked for which the content of natural coarse aggregates (NCA) with recycled coarse aggregates (RCA) will be gradually replaced. It should be noted that the designs will contain from 0, 25, 50, 75, and 100% RCA, where for each design 36 cylindrical witnesses will be made. Tests of two witnesses of each design will be carried out when they reach the age of 3, 7, 14 days, in addition to rehearsing 30 witnesses when you go 28 days of curing [6].

2.2 Procedure

The technique applied in this thesis is correlative, by which concrete specimens with a resistance of 350 kg/cm^2 will be performed using different RCA dosages to study and compare them, thus having to respect the quality standards based on the NTP and ASTM standards.

This technique mentioned will serve in the exploitation of results that each sample we have will give us, therefore the importance of each piece (probe) that is made because these will be subject mainly to the compression effort.

Then, having obtained the required data will be processed to give results and conclusions of the investigation [7].

To test recycled aggregates (CDW):

- Construction and demolition waste (CDW) will be selected from the extraction site, taking care of its composition: concrete, bricks, ceramic materials
- Processing of the material through the selection and transport, manual crushing of the aggregates (CDW), sifting through the 3/4″ and 3/8″ sieves, proportion: 60% pass the 3/4″ sieve and 40% pass the 3/8″ sieve
- Physical properties of recycled aggregates (CDW) will be obtained by performing the following tests:

- Standard test method for total evaporable moisture content of aggregate by drying—(NTP 339.185:2018)—(ASTM C 566)
- Standard test method for materials finer than 75 μm (N° 200) sieve in mineral aggregates by washing—(NTP 400.018: 2013)—ASTM C 117
- Standard test method for sieve analysis of fine and coarse aggregates—(NTP400.012:2018)—ASTM C 136
- Standard test method for bulk density (unit weight) and voids in aggregate—(NTP 400.017:2016)—ASTM C 29
- Standard test method for relative density (specific gravity) and absorption of coarse aggregate—(NTP 400.021:2018)—ASTM C 127

To test concrete (RAC):

- Physical properties of natural and recycled coarse aggregate
- Physical properties of fine aggregate
- Physical properties of hydraulic cement by standard test method for density of hydraulic cement—ASTM C 188
- ACI method for mix design replacing 0, 25, 50, 75, and 100% of natural coarse aggregate by recycled coarse aggregate
- Fresh concrete properties
    - Standard test method for air content of freshly mixed concrete by the pressure method—(NTP 339.080:2017)—ASTM C 231
    - Standard test method for the slump of hydraulic cement concrete—(NTP 3399.035—ASTM C 143
- Mechanical properties of hardened concrete
    - Standard test method for compressive strength of cylindrical concrete specimens—(NTP 339.034)—ASTM C 39

3 Results

3.1 Sampling

Construction and Demolition Waste (CDW) samples were collected from the Infantas Oval vicinity in the Los Olivos district, selected in proportion to their field layout.

Figure 1 shows (a) disposition of the material found around the Infantas Oval and the figure (b) show the processed CDW.

A sample was obtained where there is the presence of non-structural concrete and bricks. These were manually crushed to obtain particles of approximately 1 in. in diameter [8].

Fig. 1 Disposition of the material found around the Infantas Oval and processed CDW

Table 1 Sieve analysis for recycled coarse aggregates. *Source* Author

Sieve (in.)	Sieve (mm)	Retained (%)	Accumulative retained (%)	Passing (%)	% Passing ASTM C 33 HUSO 4
2″	50.00	0.0	0.0	100.0	100
1 1/2″	37.50	9.6	9.6	90.4	90–100
1″	25.00	40.7	50.3	49.7	20–55
3/4″	19.00	23.9	74.2	25.8	0–15
1/2″	12.50	13.6	87.8	12.2	–
3/8″	9.50	3.9	91.7	8.3	0–5
N° 4	4.75	4.7	96.4	3.6	
Bottom		3.6	100.0	0.0	
Fineness module: 7.54					

3.1.1 Test Results for the Recycled Coarse Aggregate (CDW)

In Table 1, we can visualize the percentages of the retained weights of the sieve meshes that determine the coarse aggregate for the recycled material used under the Peruvian technical standard (NTP400.012: 2018).

In Fig. 2, we can visualize the test of the granulometric analysis of the recycled coarse aggregate, which shows the percentage of material retained in each mesh; however, the portion of the owned material does not comply with what is indicated in the standard; it can be used looking for the result end of the research (Table 1).

3.1.2 Test Results for the Natural Fine Aggregate

In Table 4, we can visualize the percentages of the retained weights of the sieve

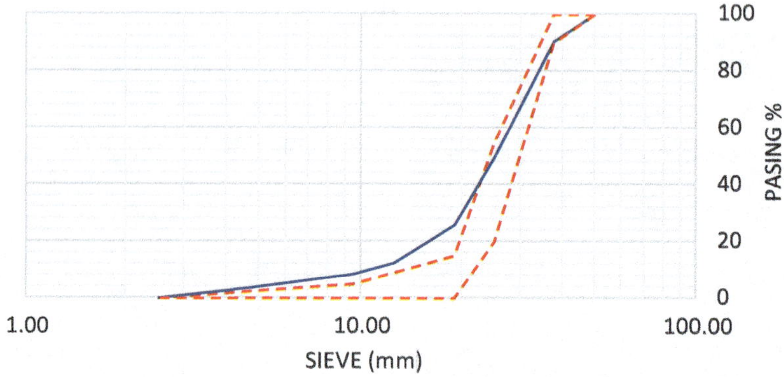

Fig. 2 Sieve analysis curve for recycled coarse aggregate

Table 2 Loose unit weight, compacted unit weight, and moisture content for recycled aggregate. *Source* Author

Sample	Loose unit weight (kg/m^3)	Compacted unit weight (kg/m^3)	Moisture content (%)
RCD	1170	1377	2.37

Table 3 Relative density, SSD relative density, and absorption for recycled aggregate. *Source* Author

Sample	Relative density	Saturated surface dry relative density	Absorption (%)
RCD	2.27	2.42	6.61

Table 4 Sieve analysis for fine aggregate. *Source* Author

Sieve		Retained (%)	Accumulative retained (%)	Passing (%)	% Passing ASTM C 33 Huso fine aggregate
(in.)	(mm)				
3/8″	9.50	0.0	0.0	100.0	100
N° 4	4.75	2.3	2.3	97.7	95–100
N° 8	2.36	13.3	15.6	84.4	80–100
N° 16	1.18	24.9	40.5	59.5	50–85
N° 30	0.60	25.3	65.8	34.2	20–60
N° 50	0.30	17.2	83.0	17.0	5–30
N° 100	0.15	8.5	91.5	8.5	0–10
Bottom		8.5	100.0	0.0	
Fineness module: 2.99					

meshes that determine the fine aggregate under the Peruvian technical standard (NTP400.012: 2018).

In Fig. 3, we can visualize the test of the granulometric analysis of the fine aggre-

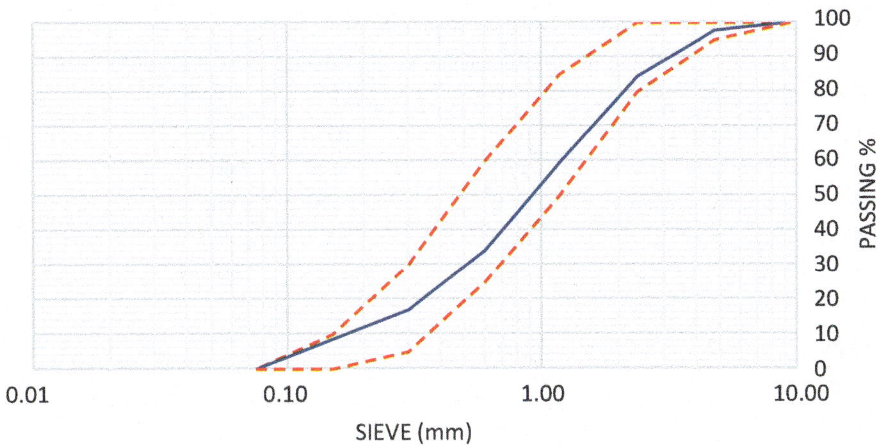

Fig. 3 Sieve analysis curve for fine aggregate

Table 5 Loose unit weight, compacted unit weight, and moisture content for fine aggregate. *Source* Author

Sample	Loose unit weight (kg/m^3)	Compacted unit weight (kg/m^3)	Moisture content (%)
Fine Ag	1658	1803	1.79

Table 6 Relative density, SSD relative density, and absorption for fine aggregate. *Source* Author

Sample	Relative density	Saturated surface dry relative density	Absorption (%)
Fine Ag	2.60	2.64	1.52

Table 7 N° 200 sieve passing material percentage for fine aggregate. *Source* Author

Sample	Passing sieve 75 μm (%)
Fine Ag	8.60

gate, which shows the percentage of material retained in each mesh; these results comply with what is mentioned by the standard (NTP400.012: 2018) (Tables 4).

3.1.3 Test Results for Natural Coarse Aggregate (CDW)

In Table 8, we can visualize the percentages of the retained weights of the sieve meshes that determine the coarse aggregate under the Peruvian technical standard (NTP400.012: 2018).

In Fig. 4, we can visualize the test of the granulometric analysis of the coarse aggregate, which shows the percentage of material retained in each mesh; these results comply with what is mentioned by the standard (NTP400.012: 2018) (Tables 9, 10,

Table 8 Sieve analysis for coarse aggregates. *Source* Author

Sieve		Retained (%)	Accumulative retained (%)	Passing (%)	% Passing ASTM C 33 HUSO 5		
(in.)	(mm)						
1 1/2″	37.50	0.0	0.0	100.0	100		
1″	25.00	5.8	5.8	94.2	90	–	100
3/4″	19.00	70.0	75.8	24.2	20	–	55
1/2″	12.50	22.7	98.5	1.5	0	–	10
3/8″	9.50	0.9	99.4	0.6	0	–	5
N° 4	4.75	0.5	99.9	0.1	–	–	–
Bottom	–	0.2	100.0	0.0	–	–	–
Fineness module: 7.74							

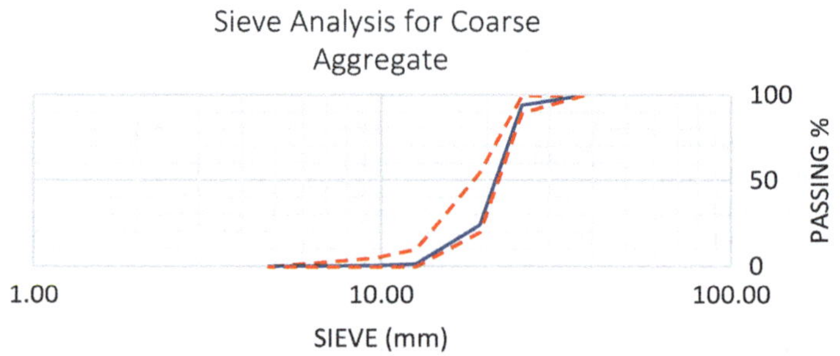

Fig. 4 Sieve analysis curve for coarse aggregate

Table 9 Loose unit weight, compacted unit weight, and moisture content for coarse aggregate. *Source* Author

Sample	Loose unit weight (kg/m^3)	Compacted unit weight (kg/m^3)	Moisture content (%)
Coarse Ag	1430	1605	0.41

Table 10 Relative density, SSD relative density, and absorption for coarse aggregate. *Source* Author

Sample	Relative density	Saturated surface dry relative density	Absorption (%)
Coarse Ag	2.81	2.83	0.73

11).

Table 11 N° 200 sieve passing material percentage for coarse aggregate. *Source* Author

Sample	Passing sieve 75 μm (%)
Coarse Ag	0.17

3.2 Resume

Compressive strength tests were carried out using the Peruvian technical standard procedures, NTP 399.604 and ASTM C140, for 7, 14, and 28 days old (cured) specimens made under the proposed mixture design. Considering that the compressive strength was determined using the equation $\sigma = \frac{\rho m\mathrm{K}x}{A}$.

Table 12 presents the results obtained for the 28-day curing concrete cores, the same procedure being applied for the 3, 7, 14, and 28-day curing pavers.

Table 13 shows the average compression resistance values for the four percentages of recycled coarse aggregate (RCA), for concrete witnesses of 3, 7, 14, and 28 days of curing; resulting in higher compression resistance value for 28 days of healing, with the decrease being, directly proportional to the amount of recycled coarse aggregate (RCA); with this logic, therefore, the lowest compression resistance value was obtained for the replacement of the total to 100% for concrete witnesses of 28 days of curing, with the importance of 351.59 kg/cm^2 being 30% below the allowed design value 350 kg/cm^2 [9, 10].

Figure 5 shows the summary of the compressive strength growth trend curve, having plotted all points shown in Table 12 of the five mix designs at different ages (Table 12). While Fig. 6 shows the development of the compressive strength average growth curve of the five mix designs at different ages (Table 13) [11, 12].

In Table 14, it is shown the statistical data of this research as the average, maximum, minimum compressive strength for every mix design, among others.

4 Discussion

This research suggests determining compression resistance due to the amount of recycled coarse aggregate used for 350 kg/cm^2 concrete processing. The results demonstrated globally that the compression resistance in its different proportions uses the recycled coarse aggregate.

Resistance to conventional concrete compression, the compression resistance values of traditional factual witnesses, prepared for investigation were optimal at 28 days, i.e., met the regulations' requirements. In this sense, a compression resistance value of 424.78 kg/cm^2 was obtained, being more significant than the chosen design value, which is 350 kg/cm^2.

Compression resistance using 25, 50, 75, and 100% recycled coarse aggregate between the four groups of 3.7,14 and 28 days of age; it is shown that the resistance is decreased using more recycled coarse aggregate among which, the percentage of 25% recycled coarse aggregate obtained the highest value within the other portions, obtaining a compression resistance of 387.50 kg/cm^2 that equates to 91.2% of its resistance.

The literature mentions that the greater the use of recycled coarse aggregate, the lower the behavior in its compression resistance [13].

Table 12 Summary of $f'c$ results at different ages. *Source* Author

Item	Age	100% RA ($f'c$)	75% RA ($f'c$)	50% RA ($f'c$)	25% RA ($f'c$)	Standard ($f'c$)
1	3	239.8	257.2	319.9	318.0	355.6
2	3	240.1	274.9	289.7	330.4	363.9
3	7	255.1	270.3	316.2	357.7	389.3
4	7	251.6	286.7	331.9	335.5	407.1
5	14	268.8	296.9	338.7	352.7	422.4
6	14	271.5	295.7	335.8	354.5	414.6
7	28	299.4	297.2	354.1	393.6	435.1
8	28	297.1	300.2	353.7	387.0	415.0
9	28	293.3	308.7	364.6	394.7	424.0
10	28	278.0	277.7	355.3	383.5	429.7
11	28	319.5	306.7	364.3	388.1	417.4
12	28	301.2	309.2	353.0	379.0	413.9
13	28	301.1	314.4	356.2	378.8	413.4
14	28	289.5	308.0	350.5	374.7	434.4
15	28	292.2	312.9	355.0	377.9	431.7
16	28	295.2	315.3	355.9	382.5	407.3
17	28	302.7	306.2	361.8	391.4	434.5
18	28	300.2	319.1	363.9	372.3	407.0
19	28	300.7	318.9	364.9	394.8	432.1
20	28	295.9	308.7	347.7	401.4	415.0
21	28	304.3	305.1	356.0	386.8	424.6
22	28	290.9	311.8	361.0	394.6	421.3
23	28	292.0	299.2	363.1	367.8	460.9
24	28	298.5	308.6	356.8	401.3	420.1
25	28	300.2	296.1	367.0	410.0	421.4
26	28	295.8	315.1	366.3	399.6	420.7
27	28	293.7	302.9	361.7	376.4	421.8
28	28	289.8	308.4	353.4	392.3	427.4
29	28	295.7	315.8	318.1	389.8	411.1
30	28	290.3	302.7	361.1	364.8	434.6
31	28	295.7	300.3	351.4	393.6	430.9
32	28	298.6	315.7	359.3	376.4	425.4
33	28	299.6	304.9	345.6	407.3	423.7
34	28	291.4	308.2	347.2	388.2	429.1
35	28	296.5	311.0	357.1	385.6	425.5
36	28	297.4	295.9	360.2	390.9	434.4

Table 13 Summary of the average *f*'c at different ages. *Source* Author

Age (días)	Standard *f*'c (kg/cm²)	25% RCA *f*'c (kg/cm²)	50% RCA *f*'c (kg/cm²)	75% RCA *f*'c (kg/cm²)	100% RCA *f*'c (kg/cm²)
3	359.77	324.18	304.79	266.00	239.97
7	398.23	346.61	324.07	278.47	253.39
14	418.53	353.60	337.22	296.33	270.16
28	424.78	387.50	356.21	306.83	296.54

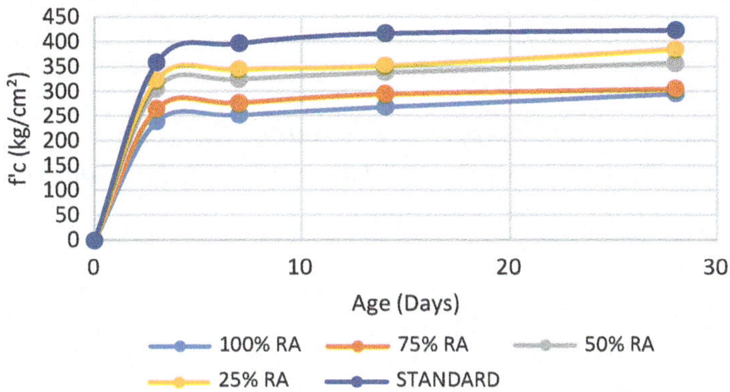

Fig. 5 Average growth curves summary

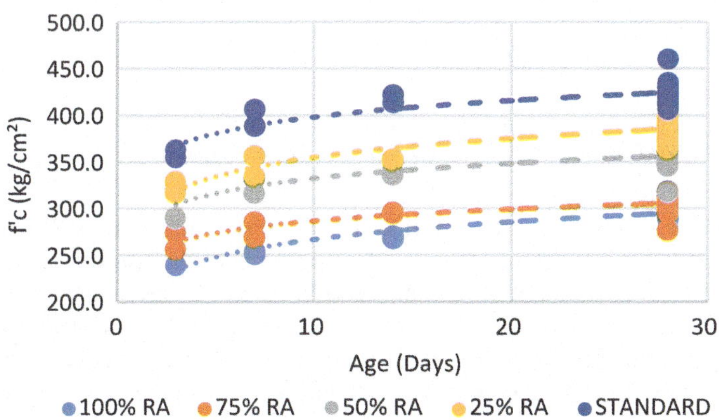

Fig. 6 Growth trend curves summary at different ages

Table 14 Summary table of measures of central tendency and measures of dispersion. *Source* Author

	Standard	25% RCA	50% RCA	75% RCA	100% RCA
Media ($\bar{x}$)	424.78	387.5	356.21	306.83	296.54
Median	424.3	388.14	356.53	308.33	296.21
Max.	460.95	409.99	367.04	319.09	319.54
Min.	407.03	364.77	318.11	277.7	278.01
Range	53.91	45.22	48.93	41.39	41.53
Variance (S^2)	115.8	122.32	85.4	72.89	46.18
Standard deviation (S)	10.76	11.06	9.24	8.54	6.8

4.1 Comparison of Compressive Strength Between Conventional Concrete and Concrete with the Addition of Recycled Coarse Aggregate

The compressive strength of conventional concrete was better than those obtained with the addition of the recycled coarse aggregate of 25, 50, 75, and 100%, having brought a behavior below what was expected. Still, the first two percentages (25 and 50%) complied with the design compression resistance, which was 350 kg/cm^2.

For all cases, especially for the 28-day-old groups, the compressive strength value was higher using the conventional design than concrete with recycled coarse aggregate. Using 25%, RCA was reached 8.8% below compressive strength compared to traditional concrete.

4.2 Contrast the Results with Other Investigations

The present research results are contrasted with those obtained by Bedoya [4] since all analyzed the compressive strength of concrete with the use of recycled coarse aggregate for proportions of 25, 50, and 100%. As in this investigation, he applied the compressive strength test for 3, 7, 14, 28, 56, and 91 days of the age of the concrete, obtaining as results below values compared to conventional concrete; but complying with the design compression resistance, which is complying with those as mentioned earlier. On the other hand, the results of this research show agreement with those obtained [14] since their research also evaluated the effect of the use of recycled coarse aggregate (RCA) on the property of concrete, emphasizing the compressive strength of concrete. Both improved the utilization of the percentage of 25% RCA in the ages of 7, 14, 21, and 28 days. The difference is that the first one applied the study for a 210 kg/cm^2 mixture design and while the second study its mixture design consists of 280 kg/cm^2; in both cases, an improvement was seen in the use of recycled coarse aggregate, causing a slight increase in compressive strength, for the first case, the percentage of 25% RCA reached 1.1%, and for the second case the 40% RCA, achieving a slight increase of 1.9% [15].

5 Conclusions

The compressive strength of conventional concrete was determined, that is, for 0% RCA. For the first, the optimal values were 359.77, 398.23, 418.53 and 424.78 kg/cm^2 for 3, 7, 14 and 28 days of age respectively.

The use of 25 and 50% of RCA are the values that complied with the design compression resistance that exceeded 350 kg/cm^2 that was designed.

When comparing the compressive strength of conventional concrete versus concrete with RCAs use, it is seen that when using said material, the compressive strength decreases, considering that the greater the amount of RCA is used, the more excellent the resistance to compression.

References

1. Almusaed A, Almssad A, Homod RZ, Yitmen I (2020) Environmental profile on building material passports, for hot climates. Sustainability 12:3720. https://doi.org/10.3390/su12093720.(2020)
2. Bhasya V, Bharatkumar BH (2018) Mechanical and durability properties of concrete produced with treated recycled concrete aggregate. ACIMJ [Internet] 115(2):209–217
3. Hamad B, Dawi A (2017) Sustainable normal and high strength recycled aggregates concretes using crushed tested cylinders as coarse aggregates. Case Stud Construct Mater 7:228–239
4. Arezoumandi M, Ghafari E, Sadati S, Fakharifar M (2017) Mechanical properties and structural performance of recycled aggregate concrete: an overview. CERJ [Internet]
5. Malešev M, Radonjanin V, Marinkovic SB (2010) Recycled concrete as aggregate for structural concrete production. Sustainability 2:1204–1225. https://doi.org/10.3390/su2051204
6. Tayeh B, Arafa M, Alqedra MA, Hanoona H (2017) Investigating the effect of sulfate attack on compressive strength of recycled aggregate concrete. J Eng Res Technol 4(4)
7. Sandanayake M, Bouras Y, Haigh R, Vrcelj Z (2020) Current sustainable trends of using waste materials in concrete—a decade review. Sustainability 12:9622. https://doi.org/10.3390/su1222962(2020)
8. Espinoza A (2020) Mechanical properties of an eco-friendly concrete with partial replacement of POC and rubber. IOP Conf Ser Mater Sci Eng 758:012011 (2020)
9. Pani L, Francesconi L, Rombi J, Mistretta F, Sassu M, Stochino F (2020) Effect of parent concrete on the performance of recycled aggregate concrete. Sustainability 12:9399. https://doi.org/10.3390/su12229399(2020)
10. Ramos L, Esenarro D, Rodriguez C, Lagos J (2020) Recovery of public spaces for the conservation of green areasin Tablada Lurin. IOP Conf. Ser Mater Sci Eng 910:012020, doi:https://doi.org/10.1088/1757-899X/910/1/01
11. Hamad BS, Dawi AH (2017) Sustainable normal and high strength recycled aggregates concretes using crushed tested cylinders as coarse aggregates. Case Stud Construct Mater 7:228–239

12. Chandio FA, Oad M, Memon MU, Memon BA (2020) Effect of gap gradation on compressive strength of recycled aggregate concrete. Quaid-E-Awam Univ Res J Eng Sci Technol Nawabshah 18(1):86–92. https://doi.org/10.13140/RG.2.2.20072.29446
13. Nigussie LF, Darwish MM, Ghebrab T (2019) Comparative investigation of the effect of recycled fine aggregate from new and old construction wastes in C-25 concrete in Ethiopia. Sustainability 11:7116. https://doi.org/10.3390/su11247116 (2019)
14. Nigussie LF, Darwish MM, Ghebrab T (2019) Recycling fine aggregate from demolished hollow concrete block for green concrete in Ethiopia. Glob J Eng Sci 3:1–10
15. Park W-J, Ryu H-S, Kim D-M, Shin S-H, Lim S-M (2018) Evaluation on the surface modification of recycled fine aggregates in aqueous H 2 SiF 6 solution. Int J Concr Struct Mater 12:19

Implementation of 3D Concrete Printing Technology in Precast Concrete Mass Production Industry

Astiti Dana Oktaviani and Mohammed Ali Berawi

Abstract Construction industry has been developing rapidly in past years. Adaptation of new technology in this digital era is highly needed to be able to compete within the industry development. Precast concrete implementation used to be one solution for speeding up project performance. But the production of the precast concrete itself also needs to boost to match the speed of construction development. The purpose of this research is to intervene mass production of precast concrete by using 3D printing technology to increase the productivity rate. Model simulation combining the use of 3D printing from COBOD Company and observation on Project X precast production in PT Z factory was conducted using Primavera software based on structured interviews done with related parties from COBOD Company, PT Z, and Project X. Discussion with three experts is also conducted to validate the simulation result. This study results on the productivity improvement by 50% compared to original conventional method. Faster productivity with less environmental impact, the use of 3D printing is able to benefit this industry. With the same area of production yard and the same number of human resources, factory can speed up two times faster only by replacing the casting method to 3D printers.

Keywords Precast concrete · 3D printing · Productivity · Construction industry

1 Introduction

Relationship between construction industry and economy in a country is undeniable. So many countries have grown from the status of developing country to one of industrializing country because of its massive development in construction industry. Singapore was one of the most suitable examples for this case, causing structural changes within industry and direct investment by public sector stated [1]. Political changes occurred as well as physical appearance of the country itself. The most

A. D. Oktaviani (✉) · M. A. Berawi
University of Indonesia, Depok, Indonesia

M. A. Berawi
e-mail: maberawi@eng.ui.ac.id

T. Kang and Y. Lee (eds.), *Proceedings of 2021 4th International Conference on Civil Engineering and Architecture*, Lecture Notes in Civil Engineering 201,
https://doi.org/10.1007/978-981-16-6932-3_18

visible appearance of construction development is the existence of high-rise buildings with advanced and unique features. The city of Singapore now has over 8600 completed high-rises with only 728.3 km^2 area [2]. This kind of industry growth is followed by several countries around Singapore, especially South-East Asia countries like Malaysia and Indonesia. High-rise building construction starts to become very common in big cities of those countries, take Jakarta as examples.

In Jakarta, Indonesia, Government Regulation of Republic of Indonesia No. 36 Year 2005 explains that high-rise buildings are classified as building with eight floors or more, with first floor height is not more than 10 m, and the next floors have less than 5 m floor-to-floor height. The most common structure for high-rise building in Jakarta is reinforced concrete with required calculation of earthquake bearing and wind load bearing. Earthquake bearing structure is strictly considered because the country location itself is in between Australia Plate, Eurasia Plate, and Pacific Plate so that earthquake is possible to take place at any time and any location. This case, the higher the building, the higher the risk. The need of constructing high-rise building as an effort to develop construction industry is assimilating with the urge of facing the higher risk, forcing engineers and related parties of construction industry to innovate. Some innovations have been developed, such as the using of top-down construction method, automatic climbing formwork for shear wall structure, precast concrete elements, and advanced management of heavy equipment like tower crane, passenger hoist, as well as placing boom. The management impacts directly on vertical travelling time, rotation travelling time, and horizontal travelling time, which finally the overall cycle time will be affected. Cycle time is strictly connected to productivity and schedule fulfilment of a construction project [3].

Building a high-rise needs strong considerations since design phase, so that the building is constructable, convenient to build, safe, and convenient to maintain. One common solution that is being developed is the using of precast concrete as one of its structural or architectural features. Precast concrete is often used for structural beam, structural column, structural slab as well as for concrete façade and architectural features. This precast use is increasing project's productivity, make it faster to build while keeping its targeted quality. During the planning process, precast concrete which can be done ex situ can start earlier with not so many predecessors [4].

Since world financial crisis happening in 2012, economic growth has slowly regained its power to reach the original performance before the fall. Economic growth is commonly stated in Gross Domestic Product as well as Total Factor Productivity. Generally, GDP is more enhancing to economic value which express the relationship between economic growth with capital market and other sectors. On the other hand, TFP is more explaining about productivity rate of a country and its relationship with labor management, capital management, and finally express the word management itself. Indonesia is one of the bottom five countries having very low rate of productivity level [5]. Construction sector is one of big contributors on Indonesia's economic growth. This low Total Factor Productivity level represents low productivity in construction industry as well. This case can be explained since more and more high-rises are being constructed leaning on precast construction industry causing a very high demand on precast concrete production. Precast concrete factories are supplying constant products while construction industry demands more, so in the same period of time, produced products are showing declination.

Solution might vary. We can innovate on precast material, labor management, method or its technology. In this 4.0 era, technology is very much used as base of any innovation, name them robots, smart factory, and 3D printing. 3D Printing technology is one technology that is most likely to be suitable method for precast concrete mass production industry. Since 3D printing technology was first developed, there are now several 3D printer suppliers like COBOD from Denmark or Cy-Be from Netherland which start constructing houses using printed concrete in the Europe. 3D printing itself was originally designed and invented for in situ printing, like houses or architectural features. It has limitations on height which only can print up to 8.1 m height. Seeing this opportunity, if we can combine the ability of what 3D printing can bring to the table with precast concrete factory concept, this innovation can surely help overcoming this productivity problem.

1.1 Objectives

Purpose of this study is to develop the method of precast concrete mass production using 3D printing technology and increase the productivity of the industry. The study focuses on how to turn the limit of height in current 3D printing technology in to limitless production of concrete elements so that this technology is not only used for constructing a single house, but also a high-rise building. Implementation of this research makes construction possible to be done with any kind of requirements, including structurally high and strong building, architecturally complicated and curvy building, or large building in vast area. Precast concrete mass production industry can produce more and faster with exactly the same factory area, the same storage yard, the same labor, the same batching plant, just by replacing the technology by 3D printing. Besides, this technology involves more computerized data and automatic system so that human error is minimized.

Finally, the study of this 3D printing innovation will bring a major change in precast concrete mass production industry as well as the construction industry itself.

2 Literature Review

Implementing brand new technology needs carefulness and many considerations. 3D printing technology in construction industry itself is still being developed. Beside its benefits, it also has limitations. Precast concrete mass production industry is struggling as well to overcome its productivity rate to meet demand's expectation. Strong base is needed to merge both technologies to function as one system. Review on some related literatures has been conducted to focus on how precast concrete industry performs within these years and what it needs to improve and overcome.

Basic understanding on how 3D printing works and what concerns it requires in running the technology is also studied as base of this study.

2.1 Performance of Precast Concrete Industry

Precast concrete method has been a solution for slow conventional casting method for years. The labor involved, time used, cost value, and easy process are definitely proven better. In line with technology development, precast concrete production method itself has also implemented several new technologies to keep the productivity increasing. One of the common examples is the using of Building Information Modelling (BIM) which can be combined to other automatic system that can facilitate the precast concrete installation in the future. BIM helps in planning process to produce shop drawing and sequencing, then in executing process to manage construction speed [6].

Despite its major benefits, precast concrete production process unfortunately produces some waste that needs to be controlled due to its massive amount. The waste is produced from motion, waiting time, defect rectification, transportation, overproduction, inventory management, working process, and people. To reduce the risk of productivity decreasing due to the produced waste, lean manufacturing concept has been developed. The process key of this concept is to eliminate waste during production process to optimize production time [7]. Other than the waste, sustainability is also needed during precast production process. Environment has been serious matter nowadays, which precast concrete industry finds hard to achieve. The using of paper works, air pollution due to concrete material and processing, sound pollution due to reinforcement bar fabrication, and other environmental problem due to many workers working in one place need to be considered better. By help of the technology, 3D modelling and digital data transfer is reachable and easy to use for anyone. As for the pollution, 3D printing is one of proposed solutions from many researchers [8].

2.2 3D Printing Technology in Construction Industry

3D printing technology has developed drastically in recent years. Many researchers put their interest on this technology, causing major increase of people's acknowledgement about the technology. Concrete 3D printing itself is developed considering many cases people might face while using the machine. The printer is delivered in several types, two of them are gantry system and robotic arm system. Gantry system is designed to be used for high volume printing, standard construction form, high productivity, repeatable quality and can be used by anyone with no special skills needed. Meanwhile, robotic arm system is designed for low volume printing, single yet complex elements, high agility, complex geometry, and done by skilled contractors.

Business process of 3D printing for concrete generally consists of six sequences; starting stage, order stage, project acceptance stage, construction planning stage, construction execution stage, and finish stage. The business process is as shown at Fig. 1. Most important stages are the planning and executing stage. During planning stage, considerations must take place on how to mobilize the 3D printer machine to the printing location, material availability in the targeted construction area, machine type used, the using of 3D printing software like BIM, and human resources availability. Surveys and studies before implementation are important to occur to prevent waste of time and money. On the other side, during execution stage, it is important to control time as well as construction quality, to maintain the nozzle extruder, to assure the installation of integrated mechanical–electrical–plumbing during 3D printing process, to assure worker's quality and quantity, also to assure the condition of printer machine, controller robot, concrete mixer, and concrete pump machine. Finally, finishing process after printing needs to be done to polish the final look of the product [9].

Material is one of major concern in 3D printing for construction. Researches and tests have been conducted for determining the most suitable mixture for concrete material. The main purpose is to replace the using of reinforcement bar in conventional concrete by fibers such as glass or steel so that the printing will not be disturbed by installation of reinforcement rebar. Other purpose is also to assure the setting and hardening time of the concrete so behavior after printing can be predicted. With these parameters, properties like pump pressure, printer with suitable speed and bead layers are necessary.

By these brief reviews, it is concluded that precast concrete industry has been performing well compared to conventional casting, but construction development occurring nowadays needs more. Wastes and environmental problems need to be fixed as well as its productivity compared to industry's demand. On the other hand,

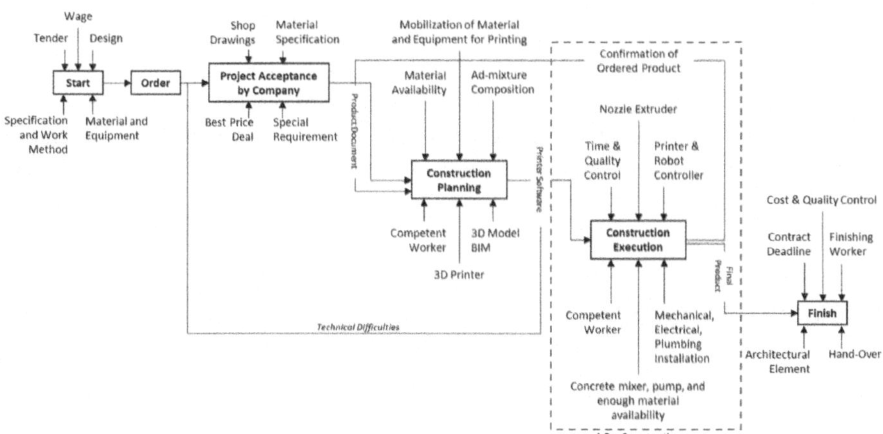

Fig. 1 Business process of 3D concrete printing technology [9]

3D precast printing technology offers solution to overcome the problems, claiming that the using of this machine is more environment-friendly. The business model of 3D concrete printing technology is projected to be faster than the current occurred business model so it can produce more products in the same given period.

3 Methods

This research used a qualitative method to achieve the targeted objectives, by observing precast concrete mass production in PT. Z for specific Project X. Project X is a high-rise building constructed in Jakarta, Indonesia which involved precast concrete use in its structural beam, structural stair, and structural slab. Observation was done by generating typical process of a precast production then the processed is intervened by 3D printing technology. Structured interview was conducted with COBOD Company as a pioneer in 3D concrete printing industry. The process began with a literature review of 3D printing implementation in construction and how it can fill the gap in precast concrete mass production industry to match with today's demands. Basic understanding based on literature review became the base knowledge in keeping the observation on path. There were two objectives: first, to develop the method of precast concrete mass production using 3D printing technology and; second, to increase the productivity of the precast concrete mass production industry.

Data observed from Project X was generated to form typical cycle of precast concrete production process. Typical cycle was then compiled as production simulation of precast concrete used in Project X. Simulation was done using Primavera software towards 100 panels of precast structural beam of Project X. Purpose of the simulation was to explain how the implementation of 3D printing affect to the process of precast concrete mass production industry.

As for second objective, which is to increase the productivity of the precast concrete mass production industry, data from the interviews and observation was generated to: production yard of PT. Z, production rate of PT. Z, required volume for 100 beam panels production of Project X, and required production yard for Project X. Comparison was put between production rate of PT. Z and COBOD 3D Printing speed. Comparing both simulation, productivity increment was concluded. Flow chart of the model simulation process was concluded as shown at Fig. 2.

4 Data Collection

Data was generated from structured interview and direct observation to the precast concrete production factory. From the observation, it is concluded that Project X beam production took 5% of the overall production yard that PT Z had, leaving 250 m^2 area for the production of the beam itself. Assumed that all factory capacity which reaches 5000 m^2 area in total were used for putting the 3D printer machines

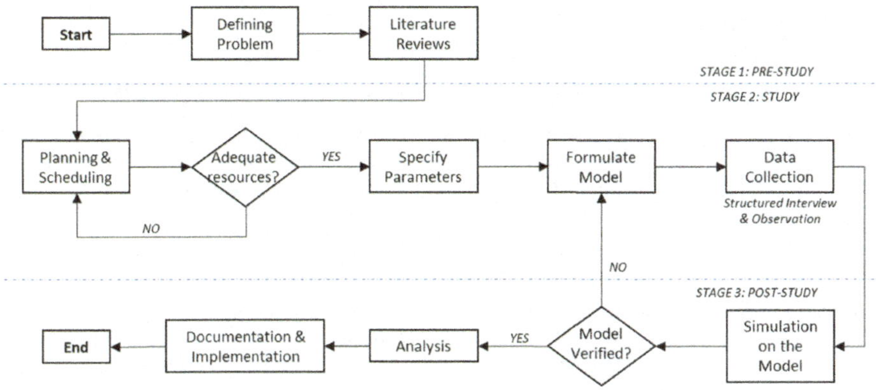

Fig. 2 Flow chart of model simulation process

with maximum gantry type capacity, it would be able to fit up to three machines, with printing dimension equals to 100 m length, 14.5 m width, and 8.1 m height. The height must be added by concrete feet installation for about 1 m height to support the gantry and nozzle load, as shown at Fig. 3.

Total height of the printing area became 9.1 m. Printing speed was stated to be 1.4 m^3 per hour for each machine. Data collections from PT Z, Project X, and COBOD Company were all concluded in Table 1.

Fig. 3 Concrete feet installation

Table 1 Collected data

Items	Unit	Observed parties		
		PT Z	Project X	COBOD
Role	–	Conventional precast concrete mass production factory	Observed project which implements conventional precast concrete	Pioneer company of 3D concrete printer supplier
Source	–	Structured interview direct observation	Structured interview direct observation	Structured interview
Geometry capacity	m	L = 100 m W = 50 m H = 10 m	5% of overall factory capacity	L = 100 m W = 14.5 m H = 8.1 m
Production area	m^2	5000	250	1450
	nos	*3D printer members that fit in the factory*		3
	m^2	*Total production area of 3D printer in the factory*		4350
Concrete production capacity	m^3	–	–	1.4/h/machine
	m^3	1500/22-workday	–	28/20-h/machine
	m^3	68.2/day	–	84/day
Precast beam panel geometry	m	–	L = 12.27 m W = 0.28 m H = 1.2 m	–
Clearance required for precast production	m	2 m at left side 2 m at right side For rebar and molding working platform	–	1 m at left side 1 m at right side For rebar and molding working platform
Concrete setting time	Day	1 day to setting and ready to be stored	–	1 day to setting and ready to be stored

5 Result and Discussion

5.1 Numerical Results

Intervention of 3D printing technology to precast concrete mass production industry affects the productivity of the industry. As collected data mentioned, big difference can be seen in the concrete production capacity, with factory capacity reaches 68.2 m^3/day while 3D printer can afford up to 84 m^3/day. On the other hand, the ability of concrete material to set in conventional case takes place in one whole days, while 3D printing material setting time is only in minutes. Extraordinarily fast setting time is needed in 3D printing technology because 3D printing extrudes concrete layer by layer. When the next layer is about to be extruded, the previous layer has to set first so the strength is enough to bear the layer above. If concrete strength has not

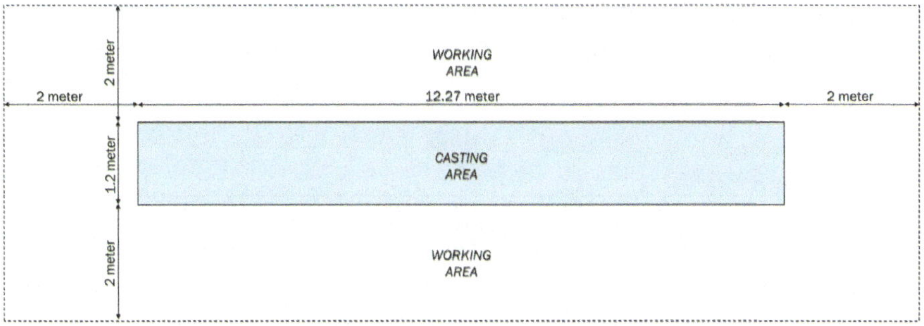

Fig. 4 Clearance for casting area in conventional precast casting method

enough yet, all layers will be melted all over the place and the printing becomes failure.

Another big difference is in the clearance required for precast production. Conventional precast concrete casting needs platform and enough area for worker to measure the precast dimension, do marking, set up molding, install the fabricated rebar, manage concrete bucket when casting and to make sure that the concrete falls inside the molding. For this necessary area, there will be required 2 m width area on each side of precast casting location. For example, the beam casting size is 12.27 m × 1.2 m, it will require 16.27 m × 5.2 m casting area, as shown in Fig. 4.

On the contrary, 3D printing technology does not require such a big area for the workers to proceed the printing. Measurement of the precast dimension as well as the marking process are done by the printer itself. Printing does not need any molding since it prints based on given 3D model data, so setting up molding will not be necessary. Installation of fabricated rebar is needed for structural elements, but the space needed is only for putting the rebar set in to the printing area. Some architectural concrete cases will not need any rebar at all, since the strength of the rebar is replaced by fiber materials. 3D printing material will be extruded directly from nozzle, which is controlled by computer-based machine. So overall, it is concluded that space needed around the printing area is only 1.25 m each side, which is enough for the person in charge to walk around and control the printing process. Clearance area needed for the exact same item using 3D printing technology is 14.77 m × 3.7 m, as shown in Fig. 5.

As the area and production speed is affected, productivity and capacity of the factory is as well affected. Observation was taken place for precast beam production of Project X inside the factory of PT Z. Using conventional process of precast casting, productivity of the beam production is calculated to be:

1. Project X Information

Project X is a high-rise building with 60 stories with reinforced concrete shear wall system. All structures including beam, post, and slab inside the shear wall are constructed using reinforced concrete material. To accelerate shear wall construction as well as to reduce working on height, Project X implements the using of precast

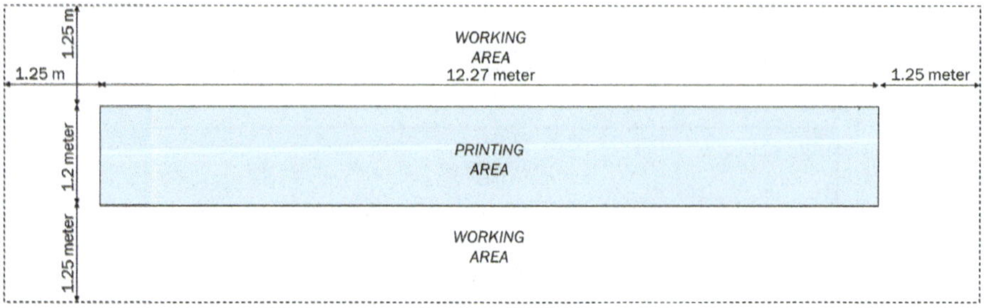

Fig. 5 Clearance for printing area in precast casting using 3D printing technology

structure. There are several types of beams inside the shear wall, but the biggest and critical one is the one that is observed in this research, with dimension of 12.27 m × 0.28 m × 1.2 m. This beam type is used from 6th floor to 58th floor, which in total is 53 floors. Each floor requires six panels of this beam type, which means 318 panels are going to be produced in total. In this observation, simulation is done for 100 beam panels.

2. Beam Panel Data
   Length = 12.27 m.
   Width = 0.28 m.
   Height = 1.2 m.

3. Production Area

$$Casting\,length = 12.27\,\text{m} + 2\,\text{m}\,(left) + 2\,\text{m}\,(right) = 16.27\,\text{m}$$

$$Casting\,width = 1.2\,\text{m} + 2\,\text{m}\,(left) + 2\,\text{m}\,(right) = 5.2\,\text{m}$$

Casting width uses the height data because beam is casted horizontally to prevent overheat at the bottom of molding. This case, casting height is only 0.28 m. Total casting area needed is:

$$Casting\,area = 16.27\,\text{m} \times 5.2\,\text{m} = 85\,\text{m}^2$$

4. Beam Production Rate Using Conventional Method

Data mentions that Project X beam production takes only 5% of overall production yard area, which is:

$$Project \times Production\,area = 5\% \times 5000\,\text{m}^2 = 250\,\text{m}^2$$

Beam that can be produced in the Project × Production area is:

$$Conventional\ production\ rate = \frac{250\,\mathrm{m}^2}{85\,\mathrm{m}^2} \approx 2\ panels/day$$

With calculated production rate for conventional method, assumed 22 working days per month, and 100 panels to be produced, estimated schedule for finishing the beam production is:

$$Estimated\ finish\ time = \frac{100\ panels}{2\frac{panels}{day}} = 50\ working\ days = 2.3\ working\ months$$

It is concluded that factory needs 2.3 months to finish beam production in conventional method.

Intervention of 3D printer machine is simulated by putting three machines in the precast concrete factory of PT Z. Using known data of 3D printing machine, productivity of the beam production is calculated to be:

1. Project X Information

Same project and data are used in the observation, which means 100 beam panels are going to be observed.

2. Beam Panel Data

   Length = 12.27 m.
   Width = 0.28 m.
   Height = 1.2 m.

3. Production Area

$$Casting\ length = 12.27\,\mathrm{m} + 1.25\,\mathrm{m}\,(left) + 1.25\,\mathrm{m}\,(right) = 14.77\,\mathrm{m}$$

$$Casting\ width = 1.2\,\mathrm{m} + 1.25\,\mathrm{m}\,(left) + 1.25\,\mathrm{m}\,(right) = 3.7\,\mathrm{m}$$

Casting width uses the height data because beam is casted horizontally to prevent overheat at the bottom of molding. This case, casting height is only 0.28 m. Total casting area needed is:

$$Casting\ area = 14.77\,\mathrm{m} \times 3.7\,\mathrm{m} = 55\,\mathrm{m}^2$$

4. Beam Production Rate Using 3D Printing Technology

Data mentions that Project X beam production takes only 5% of overall production yard area, which is:

$$Project \times Production\ area = 5\% \times 5000\,\mathrm{m}^2 = 250\,\mathrm{m}^2$$

Beam that can be produced in the Project × Production area is:

$$Conventional\ production\ rate = \frac{250\,\mathrm{m}^2}{5\,\mathrm{m}^2} \approx 4\ panels/day$$

With calculated production rate for conventional method, assumed 22 working days per month, and 100 panels to be produced, estimated schedule for finishing the beam production is:

$$Estimated\ finish\ time = \frac{100\ panels}{4\frac{panels}{day}} = 25\ working\ days = 1.1\ working\ months$$

It is concluded that factory needs 1.1 months to finish beam production in conventional method.

The calculation simulates that by using 3D printer machine, factory is able to produce 25 days faster than the current production method for 100 beam panels, which means this technology boosts up the productivity up to 50%.

5.2 Graphical Results

After productivity increment is calculated to be 50% higher, simulation is done by using Primavera software to implement the case in real time line. Beam production planning process is set to start on 1 June 2021, then simulation continues using detailed activities and work flow. Work flow for 3D printing implementation is much simpler than the conventional ones, as compared in the Figs. 6 and 7.

Based on the activities explained in the workflow, simulation on the schedule is proceeded as shown in Fig. 8.

The simulation shows that since 1 June 2021, through processes mentioned in the work flow, 100 beam panels production is going to finish on 5 August 2021. Production itself takes place from 30 June 2021 to 3 August 2021, which match to the estimated schedule which is 1.1 months.

5.3 Proposed Improvements

Based on the calculation and simulation, the using of 3D printing is projected to be able to overcome productivity problem in precast concrete mass production industry. Calculation proves that replacing the whole production yard by three numbers of 3D printing machine would bring optimum result for the factory productivity increment. Thus, for the next production month, all projects handled by PT Z will increase its speed into 50% faster than originally planned, adding value to the company. On the contrary, 3D printing machine is a new technology that is still seldom used especially in Asian countries. To transport the technology from Europe to Asia surely needs high cost and complicated knowledge transfer. Starting investing on the machine should

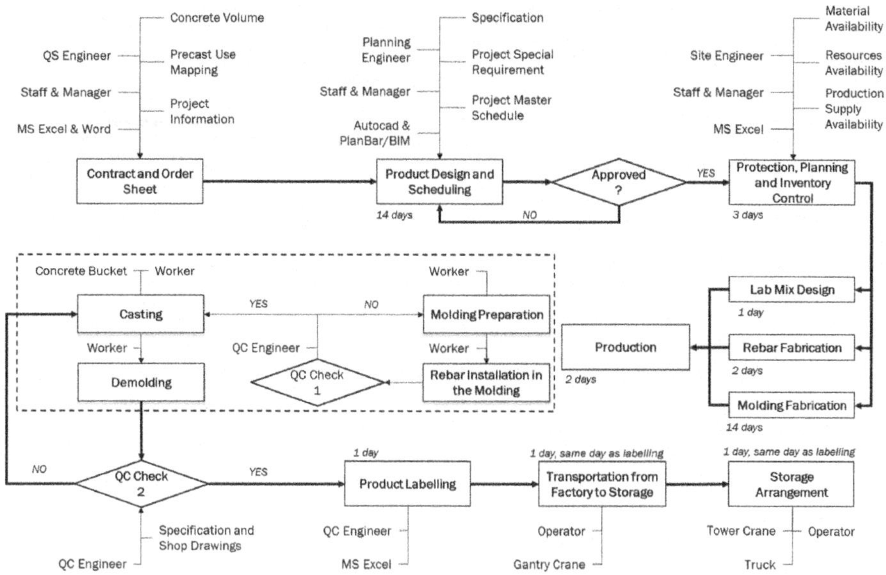

Fig. 6 Work flow for precast concrete production in conventional casting method

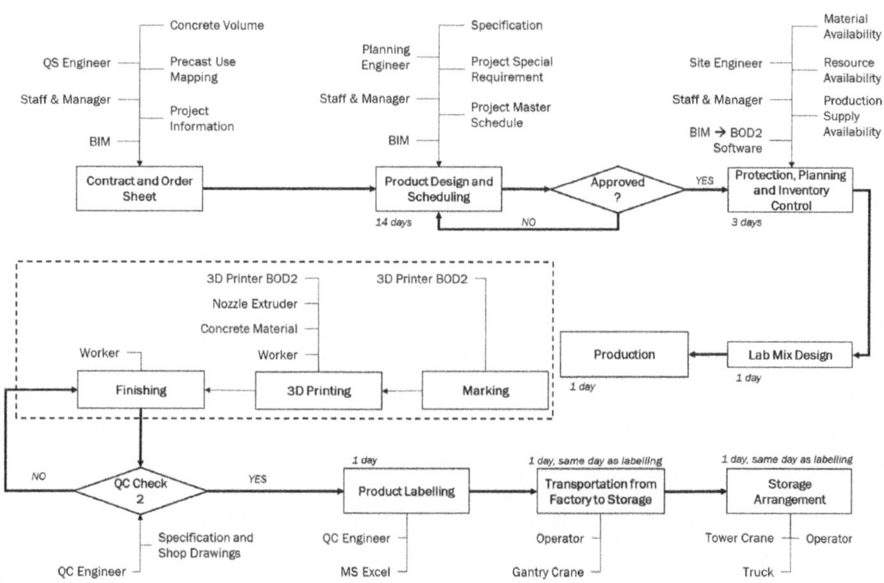

Fig. 7 Work flow for precast concrete production in 3D printing technology

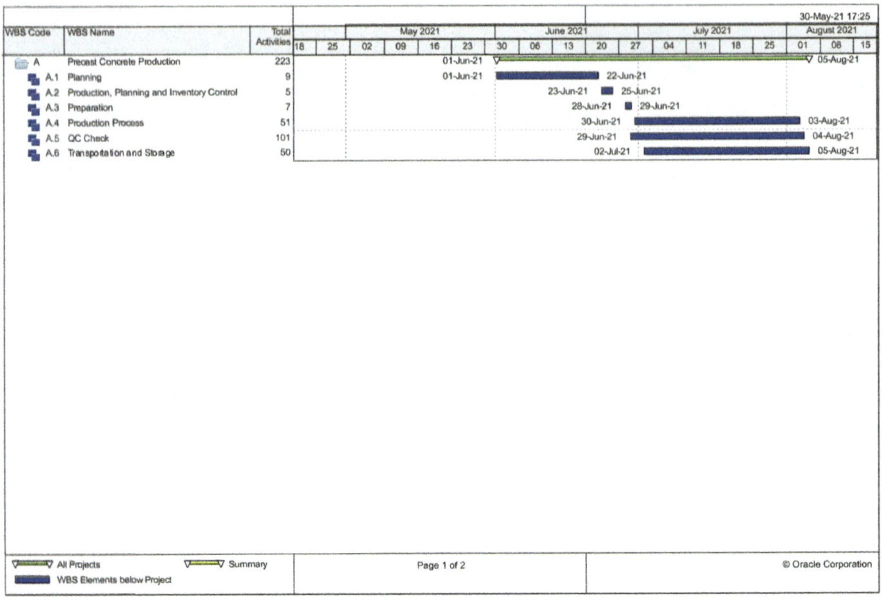

Fig. 8 Simulation done in Primavera software

be well-planned and carefully handled. This research would be surely improved if continued by simulating the investment scheme of 3D printing replacement to conclude if the productivity increment worth the price.

6 Conclusion

Current growth in the construction industry has been steeply developing in many countries compared to ten years ago. High-rise buildings start to appear and flock together in big cities in Asia. Demands for precast concrete production has been increased in line with the industry growth, but on the contrary, the supplies stay constant. Precast concrete mass production industry feels the urge to innovate to increase the productivity and ability in fulfilling the demand. On the other hand, environmental issues are also there to be solved.

3D printing technology offers simple and high-speed solution for concrete production. Originally designed for printing single houses and architectural features, it also has the potential to be used in high-rise projects. Current technology of the printer limits the printing process to maximum 8.1 m height production. Study and simulation are done to turn this limitation into possibilities and result to the using of this technology in precast concrete factory. Production of single elements of high-rise building that is done massively in full-speed is the best solution offered to the precast production industry. It has the potential to increase its productivity rate up to 50%.

Purpose of this study is to develop the method of precast concrete mass production using 3D printing technology and increase the productivity of the industry. Conventional method of precast concrete mass production was studied to be intervened by 3D printing technology by replacing the casting area in PT Z production yard by three 3D precast printer machines while using the same area of yard and the same number of resources. The process of data transfer, marking, preparation, concrete producing, quality and precision control are done by the machine, leaving reduced space for human error. Other than the easiness of the process, productivity is proven to be increased by 50% that it meets the purpose of this study.

Technology revolution has been developing in exponential curve showing that any constant action will not be beneficial for any industry. So is construction industry. This 3D concrete printing implementation will bring so much to the table, and precast concrete industry is strongly benefited by the technology. This precast concrete mass production industry should get ready and take the chance to compete stronger in construction industry.

References

1. Ofori G (2006) Construction industry and economic growth in Singapore. Constr Manage Econ 6:57–70
2. Emporis (2019) Skyline-ranking, 7 Jan. Retrieved from https://www.emporis.com/statistics/skyline-ranking
3. Amalia SD, Purwadi D (2017) Analisis Produktivitas Tower Crane pada Proyek Pembangunan Gedung Tunjungan Plaza 6 Surabaya. Jurnal Rekayasa Teknik Sipil 01(01):144–155
4. Yee AA, Hon P (2001) Structural and economic benefits of precast/prestressed concrete construction. PCI 46:34–42. Precast Concrete Institute
5. APO (2019) APO productivity databook 2019. Asian Productivity Organization, Tokyo
6. Nath T, Attarzadeh M, Tiong RL, Chidambaram C, Yu Z (2015) Productivity improvement of precast shop drawings generation through BIM-based process re-engineering. Autom Constr 54:54–68
7. Ray B, Ripley P, Neal D (2006) Systematic approach to improving productivity in the precast concrete industry. Sustain PCI J 51:62–71. Precast Concrete Institute
8. Hager I, Gollonka A, Putanowicz R (2016) 3D printing of buildings and building components as the future of sustainable construction? Proc Eng 134:292–299
9. Bakri HF (2019) The development of construction business process for applicating the technology of 3 dimensional concrete printing considering investment value and benefits. University of Indonesia, Depok

The Effect of Biochar on Properties of Sulphate Resisting Concrete

Sandra Abbas and Karol S. Sikora

Abstract Concrete is globally the most used material and the cement industry has contributed to 8% of global emissions. To address climatic targets, supplementary cementitious materials (SCM) from waste products have been presented as a partial cement replacement. By doing so, a double benefit is presented, there is a reduction of waste stored and dumped on landfills and the reduction of the amount of cement required; thus, mitigating the environmental impact. One of the products that could potentially satisfy the sustainability criteria is biochar. This study evaluates the integration of biochar and sulphate resisting cement and assesses their interactions. This project utilizes palm tree wood waste biochar as 0.5 and 1% cement replacements compared to a reference mix at ages of 3, 7, and 28 days. Additionally, slump tests were conducted to assess workability. It was noticed that the minimal addition of biochar showed an enhancement to both strength and workability, indicating a potential for practical applications and on-site utilization. However, increasing cement replacement to 1% negatively affected compressive strength. Finally, further studies are essential before any practical recommendations can be made, especially when different combinations of supplementary cementitious materials are used.

Keywords Biochar in concrete · Supplementary cementitious materials · Concrete rheology · Concrete compressive strength

1 Introduction

Nowadays due to urbanization and changes in lifestyle, urban areas are densely populated [1]. To accommodate growing populations, there is a high demand for construction materials to be utilized [1]. Concrete is the most used material in the construction industry primarily for its high mechanical properties in operation; however, its binding composite, cement has shown to contribute to 8% of global emissions in the year 2020 [2, 3]. To address climatic targets, supplementary cementitious materials

---

S. Abbas · K. S. Sikora (✉)
University of Wollongong in Dubai, Dubai, UAE
e-mail: KarolSikora@uowdubai.ac.ae

T. Kang and Y. Lee (eds.), *Proceedings of 2021 4th International Conference on Civil Engineering and Architecture*, Lecture Notes in Civil Engineering 201,
https://doi.org/10.1007/978-981-16-6932-3_19

(SCM) from waste products such as Ground Granulated Blast-furnace Slag (GGBS) and fly ash, have been presented as a partial cement replacement [4–13]. By doing so, a double benefit is presented, there is a reduction of waste stored and dumped on landfills and the reduction of amount of cement required; thus, mitigating the environmental impact [14–16]. Currently, SCM's effect on concrete is well understood and utilization is gaining higher recognition in various construction applications [17–22]. SCM are utilized in sulphate resisting Portland cement (SRC) that are commonly used in coastal areas, where urbanization is developing significantly and the much higher material's ability to withstand penetrations caused by chloride anions in comparison to ordinary Portland cement (OPC) is required [23–26]. To be able to achieve tighter climatic targets, there is a need for other waste materials that can be applied in concrete that does not negatively affect its properties, and desirably even improving them. One of the products that could potentially satisfy the sustainability criteria is biochar. Biochar is the byproduct of biomass gasification, through a technique called pyrolysis [27–30]. As made from organic matter, it acts as a carbon sequester, storing captured carbon dioxide [30]. Nevertheless, biochar is widely known enhancing the soil properties and controlling pollutants from meeting the soil [31–33]. In a recent trial, Qin et al. [34] tested larger amounts of waste plywood biochar and concluded 6.5% replacement of cement was ideal for tested concrete since reducing water runoff. Furthermore, Gupta et al. [35] tested three samples containing 0.5% biochar, 1% biochar, and 2% biochar of cement mass in concrete and concluded that the compressive strength for 28th days for concrete containing 0.5% biochar was higher than for the control mix. However, increasing the content of biochar past 0.5% resulted in a lesser strength [35]. Moreover, the type of cement used impacts the properties of concrete. Even though it is believed that sulphate resisting cement generates a similar compressive strength as using ordinary Portland cement, there is still a lack of data on the effect of various biochar on concrete made with cements containing SCM. To contribute to the current knowledge, this paper investigates the effect of biochar made from palm tree wastes on concrete made with SRC.

2 Materials and Methods

2.1 Mix Preparation

The cement used was Sulphate Resisting Portland-type Cement and the grade is 42.5 N following EN 197 [36] containing Fly Ash. Palm tree wood type biochar was used, see Fig. 1, coarse aggregates, fine aggregates (sand) were used, and tap water.

Coarse and fine aggregates consisted of mixed sizes for coarse aggregates is a mixture between 4 and 10 mm. The fine aggregates are a mixture between 1 and 4 mm. The fine aggregates are a mixture between 1 and 4 mm. There were three samples prepared, see Table 1, where mix composition follows commonly utilized

Fig. 1 Palm tree biochar after sieving used in the study

Table 1 Concrete composition

Materials/quantities for each sample (kg/m^3)	RB	0.5B	1B
SR cement	242.7	242.7	242.7
Sand (fine aggregates)	533.28	533.28	533.28
Coarse aggregates	1066.56	1066.56	1066.56
Water	85.56	87.78	90.22
Biochar	0	1.2135	2.427

by researchers previously [37]. The first sample is the reference mix without biochar denoted RB, the second sample contained 0.5% biochar of cement mass denoted 0.5B, and the third sample contained 1% biochar of cement mass denoted 1B. The composites are mixed in their dry state using a standard concrete mixer with the speed of 12 revolutions per minute followed by slow addition of water during mixing for 30 s to achieve uniformity. Then followed by another 1 min of mixing.

2.2 *Casting and Curing*

To assess workability, a slump test was conducted for each sample just after mixing. For compressive strength testing concrete was placed in 150 mm × 150 mm × 150 mm moulds and covered by a plastic sheet for 24 h. Then, specimens were demoulded and placed in two plastic bags. Moreover, each cube underwent curing until testing at 3, 7, and 28 days after mixing. Similarly, such process was followed since closely mimics curing conditions of a construction site, so outcomes will have practical applicability.

2.3 *Compressive Strength Test*

For every sample, three specimens were tested in compression after 3, 7, and 28 days from mixing to determine the strength development. Compressive testing machine with loading rate of 3 kN/sec. To determine the workability and strength of three samples with 0% content of biochar, 0.5 and 1% content of biochar as an additive in concrete producing 29 specimens in total.

3 Results and Discussion

Three specimens are made for each sample for every age and a mean value was taken for the compressive strength. Table 2 presents the results for strength and slump with the addition of biochar compared to reference mix.

The slump of RB was 10 mm, with the addition of 0.5% biochar it resulted in a 50% increase in workability of the concrete mix. Then, when doubling the content of biochar to 1%, the workability was further increased by 33%. Initially, the results for the average compressive strength on the 3rd day caused by the 0.5% addition of biochar resulted in lowering the strength by 8 and 1% addition of biochar caused the strength to decrease by 35% when compared to the reference sample. The average compressive strength on the 7th day, for 0.5% addition of biochar the same value of strength (MPa) as the reference sample was recorded while for the 1% addition of biochar a decrease in the strength by 24% was recorded. The average compressive strengths on the 28th day were 39 MPa, 30 MPa, and 38 MPa for RB, 0.5B, and 1B, respectively. By adding 0.5% of biochar (0.5B), the 28th day compressive strength was approximately 3% higher in comparison to 0% biochar (RB). When the content of biochar was doubled to 1% (1B), it demonstrated a 21% decrease in the compressive strength in comparison to RB. When comparing 0.5B and 1B by the 28th day, doubling the content of biochar produced a 23% decrease in strength. Figure 2 shows the strength development of samples containing 0.5% biochar and 1% biochar compared to the reference sample. 0.5% biochar addition it has little, but positive impact on the mechanical properties. On the other hand, when doubled biochar amount there is more noticeable impact, however negative.

The pattern of strength development is close to linear. RB and 1B development of strength are at similar rate of growth; however, it is seen in initial testing 1B decreases the strength. Initially, when 0.5B is compared to RB it creates a minor decrease in the strength; however, has a much larger growth (0.56).

Table 2 Slump and strength results

Concrete sample	RB	0.5B	1B
Slump (mm)	10	15	20
Average compressive strength (MPa)/standard deviation			
After 3 days	26/0.5	24/0.7	17/0.4
After 7 days	29/0.9	29/0.0	22/0.5
After 28 days	38/4.5	39/0.5	30/0.5

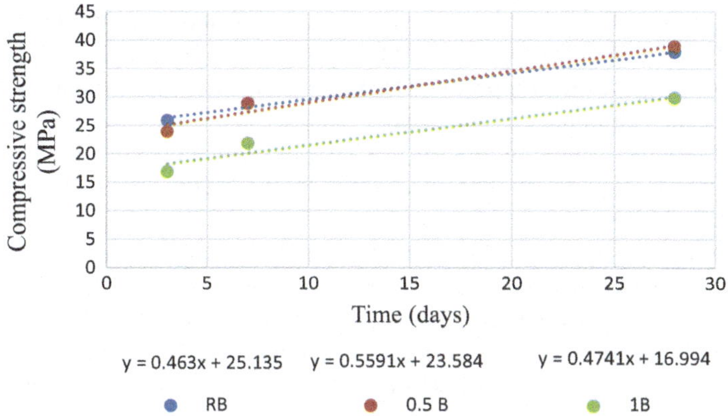

Fig. 2 Compressive strength development

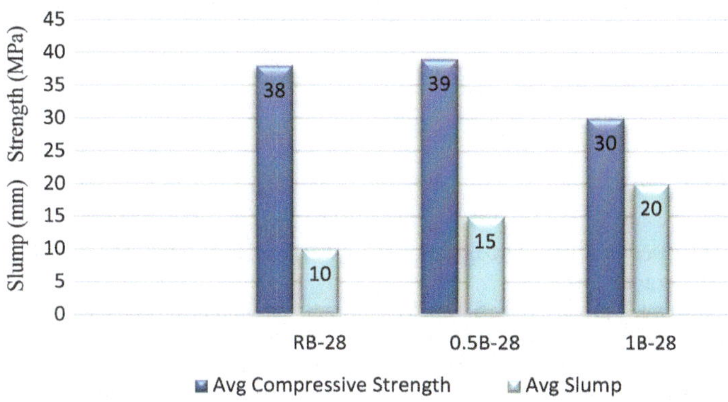

Fig. 3 Average compressive strength at 28 days and workability

Figure 3 presents the relationship between a slump and compressive strength for the samples. Generally, a normally occurring inverse relationship between the workability and the average compressive strength can be noticed, however to different extents. The addition of 0.5% biochar increased both the workability and slight strength too. Doubling the content of biochar to 1% resulted in the highest slump value but the lowest strength in comparison to the two samples.

4 Conclusion

Based on the presented analyses the following conclusion can be formulated:

- The addition of 0.5% biochar provides a slight enhancement to the compressive strength in comparison to the reference mix. By the 28th day, the sample which contained 0.5% of biochar produced a 3% increase in comparison to the reference sample which contained 0% biochar.
- By doubling the content of biochar to 1% by the 28th day of testing, it exhibited a decrease by 23% in the compressive strength and a 21% decrease in comparison to the reference sample.
- Workability was improved with the addition of biochar. Increasing the amount of biochar resulted in higher slump values.
- Even though the results indicating potential in utilization of biochar in concrete, further studies are essential before any practical recommendations can be made, especially when different combinations of supplementary cementitious materials are used.

Acknowledgements The authors would like to express acknowledgements to the International Center for Biosaline Agriculture in Dubai for the provision of the palm tree biochar used in this study.

References

1. Ritchie H, Roser M (2018) Urbanization. Our world in data. [Online]. Available: https://ourworldindata.org/urbanization
2. Ali N, Anwer M, Khan Alwi S (2015) [Online]. Available: https://www.researchgate.net/publication/319292238_The_Greenhouse_Gas_Emissions_Produced_by_Cement_Production_and_Its_Impact_on_Environment_A_Review_of_Global_Cement_Processing
3. Lei Y, Zhang Q, Nielsen C, He K (2011) An inventory of primary air pollutants and CO_2 emissions from cement production in China, 1990–2020. Atmos Environ 45(1):147–154
4. Nath P, Sarker PK (2014) Effect of GGBFS on setting, workability and early strength properties of fly ash geopolymer concrete cured in ambient condition. Constr Build Mater 66:163–171
5. Nguyen KT, Ahn N, Le TA, Lee K (2016) Theoretical and experimental study on mechanical properties and flexural strength of fly ash-geopolymer concrete. Constr Build Mater 106:65–77
6. Sikora KS, Klemm AJ (2014) The effect of superabsorbent polymers on performance of fly ash cementitious mortars exposed to accelerated freezing/thawing conditions. Int J Comput Methods Exp Meas 2(3):255–268
7. Klemm AJ, Baker P, Sikora K (2012) The effect of super absorbent polymers on the performance of immature cementitious mortars. In: Brittle matrix composites 10. Woodhead Publishing, pp 21–31
8. Klemm A, Sikora KS (2012) The effect of cement type on the performance of mortars modified by superabsorbent polymers. In: Concrete repair, rehabilitation and retrofitting III. Taylor & Francis Group, pp 210–216
9. Sikora KS, Klemm AJ (2015) Effect of superabsorbent polymers on workability and hydration process in fly ash cementitious composites. J Mater Civ Eng 27(5):04014170
10. Klemm AJ, Sikora KS (2013) The effect of superabsorbent polymers (SAP) on microstructure and mechanical properties of fly ash cementitious mortars. Constr Build Mater 49:134–143
11. Klemm A, Sikora KS (2012) Freeze/thaw performance of polymer modified cementitious mortars exposed to NaCl solution. Restor Build Monum 18(3/4):1–8

12. Czarnecki L, Klemm AJ, Sikora K (2010) The effects of mineral fillers and superplasticizers on rheology and the heat of hydration of cementitious mortars. In: International RILEM conference on use of superabsorbent polymers and other new additives in concrete
13. Sikora K, Jaafar K, Nical A (2020) The effect of superabsorbent polymers on consistency and strength of mortars with GGBS and FA. In: International conference on civil, architectural and environmental engineering (ICCAEE 2020), AIP (2021)
14. Tommaso MD, Bordonzotti I (2016) NO_x adsorption, fire resistance and CO_2 sequestration of high performance, high durability concrete containing activated carbon. Book of abstracts
15. Xiong X, Yu IKM, Cao L, Tsang DCW, Zhang S, Ok YS (2017) A review of biochar-based catalysts for chemical synthesis, biofuel production, and pollution control. Biores Technol 246:254–270
16. Lima C, Caggiano A, Faella C, Martinelli E, Pepe M, Realfonzo R (2013) Physical properties and mechanical behaviour of concrete made with recycled aggregates and fly ash. Constr Build Mater 47:547–559
17. Martin A, Pastor JY, Palomo A, Fernández Jiménez A (2015) Mechanical behaviour at high temperature of alkali-activated aluminosilicates (geopolymers). Constr Build Mater 93:1188–1196
18. Fang G, Ho WK, Tu W, Zhang M (2018) Workability and mechanical properties of alkali-activated fly ash-slag concrete cured at ambient temperature. Constr Build Mater 172:476–487
19. Klemm AJ, Almeida FCR, Sikora KS (2016) Application of superabsorbent polymers (SAP) in cementitious materials with blended cements. Concr Plant Int
20. Sikora KS, Liu Z (2018) Shear strength of different connection and concrete types for timber concrete composites (TCC). In: World conference on timber engineering, Seoul
21. Richard P, Liu Z, Descamps T, Sikora KS (2020) Effect of concrete modification on shear of connections for timber–concrete composites. Proc Inst Civ Eng Struct Build 173(5):326–339
22. Klemm AJ, Sikora KS, Wiggins DE (2014) Superabsorbent polymers in cementitious composites with fly ash cements: selected aspects. Constr Mater Struct 57–67. IOS Press
23. Akhtar A, Sarmah AK (2018) Novel biochar-concrete composites: manufacturing, characterization and evaluation of the mechanical properties. Sci Total Environ 616–617:408–416
24. Lai S, Loke LHL, Hilton MJ, Bouma TJ, Todd PA (2015) The effects of urbanisation on coastal habitats and the potential for ecological engineering: a Singapore case study. Ocean Coast Manag 103:78–85
25. Yildirim H, Ilica T, Sengul O (2011) Effect of cement type on the resistance of concrete against chloride penetration. Constr Build Mater 25(3):1282–1288
26. Kroviakov S, Zavoloka M, Dudnik L, Kryzhanovskyi V (2019) Comparison of strength and durability of concretes made with sulfate-resistant portland cement and portland cement with pozzolana additive. In: Elektronički časopis građevinskog fakulteta, Osijek, pp 81–86
27. Choi WCY (2012) Mechanical properties of mortar containing bio-char from pyrolysis. J Korea Inst Struct Maint Insp 22
28. Kua HW, Low CY (2018) Use of biochar as carbon sequestering additive in cement mortar. Cem Concr Compos 87:110–129
29. Sirico A et al (2020) Mechanical characterization of cement-based materials containing biochar from gasification. Constr Build Mater 246:118490
30. Akinyemi BA, Adesina A (2020) Recent advancements in the use of biochar for cementitious applications: a review. J Build Eng 101705
31. Varjani S, Kumar G, Rene ER (2019) Developments in biochar application for pesticide remediation: current knowledge and future research directions. J Environ Manage 232:505–513
32. Hunt J, Duponte M, Sato D, Kawabata A (2010) Soil and crop management SCM-30. In: The basics of biochar: a natural soil amendment
33. Kamali M, Jahaninafard D, Mostafaie A, Davarazar M, Gomes APD, Tarelho LAC, Dewil R, Aminabhavi TM (2020) Scientometric analysis and scientific trends on biochar application as soil amendment. Chem Eng J 395:125–128
34. Qin Y, Pang X, Tang K, Bao T (2021) Evaluation of pervious concrete performance with pulverized biochar as cement replacement. Cem Concr Compos 119:104022

35. Gupta S, Kua HW, Pang SD (2020) Effect of biochar on mechanical and permeability properties of concrete exposed to elevated temperature. Constr Build Mater 234:117338. https://doi.org/10.1016/j.conbuildmat.2019.117338
36. Dhir R, Jones MR (1994) Euro-cements: impact of ENV 197 on concrete construction. CRC Press
37. Zhao M, Jia Y, Yuan L, Qiu J, Xie C (2019) Experimental study on the vegetation characteristics of biochar-modified vegetation concrete. Constr Build Mater 206:321–328

Characteristics of Concrete Made with Treated Domestic Wastewater

Abdelrahman Abushanab and Wael Alnahhal

Abstract The dramatic increase in the worldwide production of concrete and its consequences in depleting freshwater resources has become an urgent challenge among researchers. Many countries use the desalination treatment method to secure their needed quantities of fresh water. However, desalination has negative environmental and economic impacts. For that reason, various alternatives to fresh water have recently been proposed by researchers. One of them is the tertiary treated domestic wastewater (TWW). From a sustainability perspective, utilizing TWW in concrete applications decreases the worldwide freshwater depletion and reduces the costs and energy associated with wastewater disposal and seawater desalination. In this study, two concrete mixes made entirely with fresh water and TWW were tested for their mechanical and durability characteristics. The experimental tests include the compressive strength and electrical resistivity of concrete. Test results revealed that TWW concrete had slightly lower compressive strength than freshwater concrete. Moreover, the corrosion susceptibility of TWW concrete was similar to that of freshwater concrete.

Keywords Treated wastewater · Sustainability · Concrete · Durability · Compressive strength · Electrical resistivity

1 Introduction

The worldwide demands for concrete have dramatically increased over the last few years due to the rapid population growth and urbanization of cities. Globally, concrete's consumption rate has exceeded 9 billion tons per year [1]. Such a high consumption has resulted in 10% CO_2 emissions in the environment and rapid depletion of fresh water, raising the global concerns about the sustainability of concrete [2, 3]. In fact, the global consumption of fresh water for concrete mixing and curing is 2 billion tons per year [2]. Given the facts that fresh water accounts for only 2.5%

A. Abushanab · W. Alnahhal (✉)
Department of Civil and Architectural Engineering, Qatar University, Doha, Qatar
e-mail: wael.alnahhal@qu.edu.qa

T. Kang and Y. Lee (eds.), *Proceedings of 2021 4th International Conference on Civil Engineering and Architecture*, Lecture Notes in Civil Engineering 201,
https://doi.org/10.1007/978-981-16-6932-3_20

of total water [4] and approximately half of the world's population will face water shortages as early as 2025 [5], there is a pressing need to investigate the viability of replacing fresh water with tertiary treated domestic wastewater (TWW) in concrete applications. Utilizing TWW in concrete applications will not only save fresh water but also decreases the costs and energy associated with TWW disposal and seawater desalination.

Various studies have already investigated the influence of TWW on the characteristics of concrete and acknowledged the feasibility of TWW in concrete applications. Al-Ghusain and Terro [6] and Noruzman et al. [7] demonstrated that TWW had little-to-no effect on the fresh concrete workability and density. Tay and Yip [8], Lee et al. [9], and Shekarchi et al. [10] studied the compressive strength of TWW and revealed that TWW concrete specimens had 15% higher compressive strength than those made with fresh water. However, Asadollahfardi et al. [11] pointed out that the compressive and flexural strengths of concrete with TWW were comparable with those made with fresh water. Conversely, Arooj et al. [12] reported that TWW decreased the compressive strength of concrete by 15%. Regarding the durability properties of TWW concrete, Shekarchi et al. [10] revealed that TWW concrete had 16% lower electrical resistivity than conventional concrete.

Although previous studies have addressed the influence of TWW on concrete mechanical and durability properties, still further studies are needed to confirm its applicability in national and international standards, as the properties of TWW are influenced to a large degree by TWW origin and pollutant contents. To this end, the aim of this study was to investigate the impacts of using TWW (i.e., produced in the State of Qatar) on the mechanical and durability properties of concrete.

2 Experimental Program

2.1 Materials

Two concrete mixes were prepared in the current study, namely mixes A and B. Mix A was made entirely with fresh water and considered as the reference mix, while mix B was made entirely with TWW. The TWW used was obtained from Doha North Sewage Treatment Plant in Qatar. Chemical characteristics of both fresh water and TWW are shown in Table 1. It could be emerged from the table that TWW has significantly higher chloride, sulfate, and total dissolved solids (TDS) than fresh water. However, both fresh water and TWW showed comparable pH.

Concrete with a 28-day target compressive strength of 50 MPa was used to prepare the test specimens. Following ASTM C192/C192M-19 [13], the concrete ingredient per cubic meter were 156, 350, 1076, and 708 kg for water, cement, coarse aggregates, and fine aggregates, respectively. The chemical characteristics of cement used were obtained using X-ray fluorescence (XRF) analysis [14] and listed in Table 2.

Table 1 Chemical characterization of fresh water and TWW

Characteristic	Unit	Fresh water	TWW
Chloride (Cl^-)	mg/l	14.1	511
Sulfate (SO_4^{-2})	mg/l	6	490
TDS[a]	mg/l	90	1690
pH	–	8.1	7.8

[a]TDS = total dissolved solids

Table 2 Chemical compositions of cement

Characteristic	Weight (%)
MgO	2.5
Al_2O_3	3.48
SiO_2	15.84
SO_3	2.5
Cl	0.05
K_2O	0.47
CaO	68.91
TiO_2	0.28
V_2O_5	0.05
Cr_2O_3	0.05
MnO	0.09
Fe_2O_3	4.8
SrO	0.05

2.2 Methods

Concrete compressive strength was measured based on procedures provided in ASTM C39/C39M-20 provisions [15]. Three cylinders with dimensions of 100 × 200 mm were prepared from each mix and tested at 28 days. Concrete electrical resistivity was tested according to AASHTO TP 95 standards [16]. After curing for 28 days, three specimens with dimensions of 100 × 200 mm were tested for each mix.

3 Results and Discussion

The compressive strength and electrical resistivity results for mixes A and B are listed in Table 3. The presented results showed that completely replacing fresh water with TWW for concrete mixing slightly decreased the 28-day compressive strength by 6%. The decrease in the compressive strength of mix B was due to the higher amounts of suspended solids and dissolved oxygen in TWW compared to that in

Table 3 Compressive strength, flexural tensile strength, and electrical resistivity results at 28 days

Mix	Compressive strength (MPa)	Electrical resistivity (kΩ cm)
A	50.9	18.5
B	47.8	17.5

fresh water. The suspended solids and dissolved oxygen, in turn, have increased the porosity of concrete specimens and decreased the bond strength between aggregates and cement. The results of Asadollahfardi et al. [11] also confirmed that mixing concrete with TWW slightly decreased the compressive strength by 4–7%.

Regarding the effect of TWW on concrete durability properties, it could be seen that mix B had 5.4% lower electrical resistivity than mix B. According to AASHTO TP 95 standards [16], mixes A and B have moderate resistance to corrosion, indicating an insignificant impact of using TWW on concrete corrosion. The slight decrease in the electrical resistivity results in mix B is due to the suspended solids in TWW, which increased the concrete's pores and allowed for more electrons to pass. The obtained results agreed with Shekarchi et al. [10] and Asadollahfardi et al. [11].

4 Conclusions

This study investigated the impacts of replacing fresh water with TWW for concrete applications in Qatar. Utilizing TWW in concrete applications reduces the depletion of fresh water and costs associated with wastewater disposal. Two mixes were prepared to investigate the effect of TWW on concrete compressive strength and electrical resistivity. Based on the obtained results, the following conclusions could be drawn:

- Replacing fresh water with TWW for concrete mixing has slightly decreased the compressive strength of concrete.
- The electrical resistivity corresponding to the mix made with TWW was slightly lower than its counterpart made with fresh water.
- The mechanical and durability properties of TWW concrete are mainly affected by the suspended solids in TWW.

Acknowledgements This publication was made possible by GSRA grant GSRA6-1-0509-19022 from the Qatar National Research Fund (a member of Qatar Foundation). Also, the financial support from Qatar University through grant no. QUST-1-CENG-2021-20 is acknowledged.

References

1. Aprianti SE (2017) A huge number of artificial waste material can be supplementary cementitious material (SCM) for concrete production—a review part II. J Clean Prod 142:4178–4194
2. Younis A, Ebead U, Suraneni P, Nanni A (2020) Performance of seawater-mixed recycled-aggregate concrete. J Mater Civ Eng 32:4019331
3. Miller SA, Horvath A, Monteiro PJM (2018) Impacts of booming concrete production on water resources worldwide. Nat Sustain 1:69–76
4. World Water Assessment Programme (United Nations). Water: a shared responsibility, vol 2. United Nations Educational Scientific and Cultural (UNESCO)
5. Farhadkhani M, Nikaeen M, Yadegarfar G, Hatamzadeh M, Pourmohammadbagher H, Sahbaei Z, Rahmani HR (2018) Effects of irrigation with secondary treated wastewater on physico-chemical and microbial properties of soil and produce safety in a semi-arid area. Water Res 144:356–364
6. Al-Ghusain I, Terro M (2003) Use of treated wastewater for concrete mixing in Kuwait. Kuwait J Sci Eng 30
7. Noruzman AH, Muhammad B, Ismail M, Abdul-Majid Z (2012) Characteristics of treated effluents and their potential applications for producing concrete. J Environ Manage 110:27–32
8. Tay J, Yip W (1987) Use of reclaimed wastewater for concrete mixing. J Environ Eng 113:1156–1161
9. Lee OS, Salim MR, Ismail M, Ali MI (2001) Reusing treated effluent in concrete technology. J Teknol 34
10. Shekarchi M, Yazdian M, Mehrdadi N (2012) Use of biologically treated domestic waste water in concrete Kuwait. J Sci Eng 39:97–111
11. Asadollahfardi G, Delnavaz M, Rashnoiee V, Ghonabadi N (2016) Use of treated domestic wastewater before chlorination to produce and cure concrete. Constr Build Mater 105:253–261
12. Arooj MF, Haseeb F, Butt AI, Irfan-Ul-Hassan DM, Batool H, Kibriya S, Javed Z, Nawaz H, Asif S (2021) A sustainable approach to reuse of treated domestic wastewater in construction incorporating admixtures J Build Eng 33:101616
13. ASTM C192/C192M—19 (2019) Standard practice for making and curing concrete test specimens in the laboratory
14. Wheeler BD (1983) Chemical analysis of portland cement by energy dispersive X-ray fluorescence. Cem Concr Aggr 5:123–127
15. ASTM C39/C39M—20 (2020) Standard test method for compressive strength of cylindrical concrete specimens
16. AASHTO TP 95 (2011) Standard method of test for surface resistivity indication of concrete's ability to resist chloride ion penetration. American Association of State Highway and Transportation Officials, Washington, DC

Building Materials and Structural Engineering

Corrosion Behavior of Hot-Dip Al Coating Contact with Concrete During Hardening Process of Concrete

Y. Gao, S. Kainuma, M. Yang, S. Ishihara, M. Hashimoto, and D. Inoue

Abstract Hot-dip Al coating is an effective and economical protection method for steel structures. Hot-dip Al coated steel members have also been used in the case that contact with concrete. Al is an amphoteric metal that can react with acid and alkaline. Concrete is an alkaline material. The corrosion behavior of hot-dip Al coating contact with concrete is studied by measurement OCP and monitoring the change of pH during the hardening process of concrete. And the atmospheric test that simulates the actual structures. The corrosion products are analyzed by the diffraction of X-rays (XRD). The hot-dip Al coating can be dissolved during the hardening process. With the decreasing pH of concrete, the Al coating also has a relatively stable due to the passivation effect. In the dry and no stagnant water environment, the hot-dip Al coating has a better anti-corrosion behavior due to the passivation and shielding effect of complex calcium-aluminum compounds.

Keywords Hot-dip Al coating · Concrete · Corrosion behavior · Atmospheric environment · OCP · XRD

1 Introduction

Hot-dip Al coating is an effective method to protect the steel structures from a corrosive environment. It is one of the most economical protection methods for coating large tonnages of steel. Especially in humid, marine, and industry atmosphere. Many ancillary facilities of concrete structures, for example, guardrails, platforms, these

Y. Gao (✉) · S. Kainuma · M. Yang
Department of Civil Engineering, Kyushu University, Fukuoka 819-0395, Japan
e-mail: gao@doc.kyushu-u.ac.jp

S. Ishihara
Research and Development Center, Mitsui E&S Machinery Co., Ltd., Tamano 706-8651, Japan

M. Hashimoto
Shinko Alumer Industries Co., Ltd., Kitakyushu 808-0074, Japan

D. Inoue
Sumitomo Mitsui Construction Steel Structures Engineering Co., Ltd., Osaka 550-0004, Japan

T. Kang and Y. Lee (eds.), *Proceedings of 2021 4th International Conference on Civil Engineering and Architecture*, Lecture Notes in Civil Engineering 201,
https://doi.org/10.1007/978-981-16-6932-3_21

steel structures were coated by hot-dip Al coating. Al is an amphoteric metal, can be dissolved in both acid and alkaline environments. The pH of the pore solution of concrete is also an important parameter for the durability of concrete structures. The pH of concrete made based on Portland cement range from 12.5 to 13.8. The pH depends on the alkali metal content of the cement. The corrosion behavior of hot-dip Al coating contact with the concrete is also unclear. This study evaluated the effect of an alkaline environment of concrete on the hot-dip Al coated steel plate during the hardening process and after the hardening process [1].

2 Experimental

2.1 *Measurement of Open Circuit Potential (OCP)*

The hot-dip Al coating specimens were made by the round bar (Φ12 mm, Length: 30 mm JIS G 3101 SS400). The specimens were exposed to the $Ca(OH)_2$ aqueous solution. The temperature of the solution was about 20 °C, dissolved oxygen by direct contact with the laboratory atmosphere. The pH of the solution was initially about 12.6, change the pH of the solution by adding pure water. The measurement range of pH from 10.25 to 12.6. The specimens were used as the working electrode, and an Ag/AgCl electrode was used as the reference electrode.

2.2 *The Corrosion Behavior of Hot-Dip Al Coating During the Hardening Process of Concrete*

The concrete mixes of the hardening process were shown in Table 1 (Fresh concrete). The concrete was injected into the stainless steel thimble (Φ100 mm, Height: 200 mm JIS A 1132), the round bar (Φ19 mm, Length: 100 mm JIS G 3101 SS400) coated by hot-dip Al coating was inserted into the concrete. The part of round bar that was immersed into the concrete is about 70 mm. The specimen was fixed by the device, the temperature of the ambient is about 20 °C. After 24 h, the specimen was placed in the constant temperature and humidity cabinet (20 °C, 60%*RH*) about 28 days, a polythene sheet was cover on the surface of the flash concrete in order to keep

Table 1 Concretes compositions (kg/m^3)

Test type	Water	Cement	Sand	Gravel	*w*/*c* ratio
Fresh concrete	290	579	1304	0	0.50
pH measurement	160	242	872	1007	0.66
Exposure test	154	285	783	1087	0.54

moisture. The concrete mixes were shown in Table 1 (pH measurement). pH is an important parameter to evaluate the alkalinity of concrete. The pH was measured at a certain time interval.

2.3 *Atmospheric Exposure Experiment*

Hot-dip Al coated steel members are usually used in the ancillary facilities of bridge substructure. In order to simulate the actual situation of long-term contact between Al coated steel members and concrete. The atmospheric exposure test was carried out for one year at Univ. of the Ryukyus and Kyoda (Lat. 26° 32′ N, Long. 127° 57′ E). The environment data of each exposure site is shown in Table 2. The schematic diagram of the exposure test was shown in Fig. 1. The specimen consists of three parts, concrete, bolt, and hot-dip Al coated plate. The concrete mixes of atmospheric exposure specimens were shown in Table 1 (Exposure test). The concrete base (Φ150 mm, Height: 150 mm) and Al coated steel plate (JIS G 3106 SM490A 60 × 60 × 12 mm) were fixed by a bolt (M12 × 125 mm). The bolt consists of a screw, washer, and nut. All the tiny members of bolt were coated by hot-dip 55Al-Zn coating. In this test, in order to study the influence of hot-dip Al coated specimen contact with the concrete. In particular, a comparative test was set up, the Al coated plate and concrete were isolated by poly tetra fluoroethylene (PTFE).

Table 2 Atmospheric environment of University of the Ryukyus and Kyoda sites

Exposure site	Temperature T (°C)	Humidity RH (%)	Airborne salt w (mdd)	Offshore distance (m)	Average precipitation (mm/day)	Rain washing effect
University of the Ryukyus	23	82	0.3	2300	5.2	Yes
Kyoda	23	83	0.5	30	5.3	No

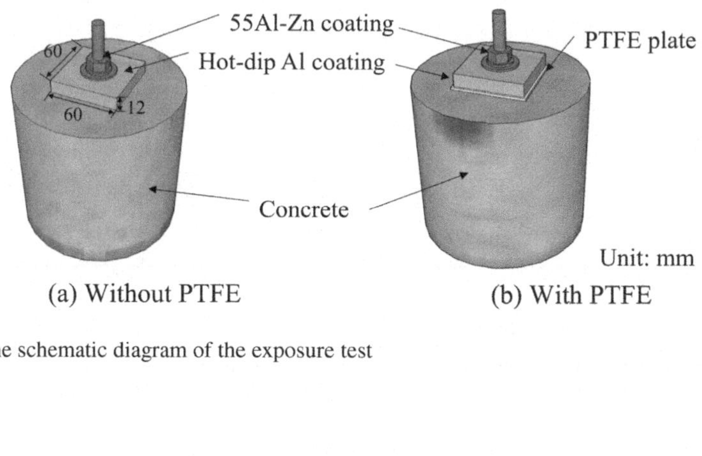

Fig. 1 The schematic diagram of the exposure test

3 Results and Discussion

The OCP measurement of the specimen at different pH was shown in Fig. 2. The OCP decreased (compare with pH from 10 to 12) suddenly when the pH exceeded 12.1. At pH 11, the dissolve behavior of Al is similar to neutral pH solutions. The formation of the surface film has a high ionic conduction resistance. When the pH near 12, the surface film resistance is significantly reduced, the dissolution current was increasing by the influence of ohmic resistance and mess transport [2]. In this situation, Al dissolution and weight loss were also significantly [3]. The change of pH measured on the surface of concrete with days during the hardening process was shown in Fig. 3. The pH drop at the surface of the concrete during the first days was due to diffusion of hydrogen and hydroxide ions, which is attributed to the difference of their concentrations [4]. In the early stage of concrete hardening, the pH of fresh concrete is nobler than that of 12. Therefore, the dissolution behavior of

Fig. 2 OCP in different pH

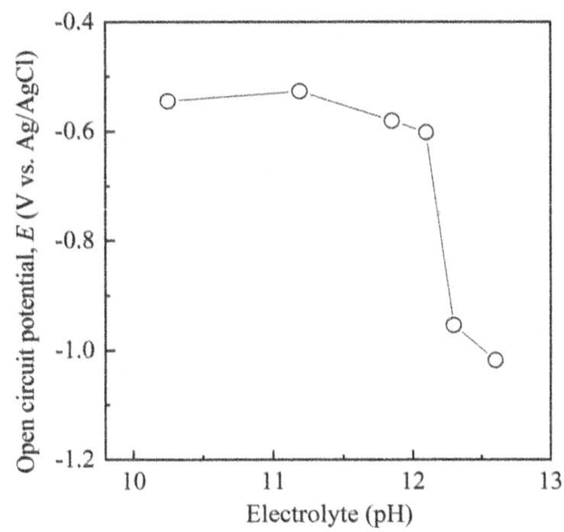

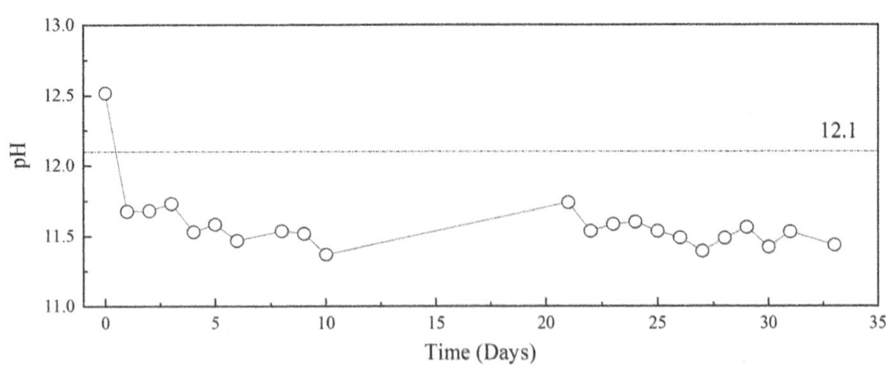

Fig. 3 The change of pH on concrete surface during hardening process

Fig. 4 Cross-section observation

Al has occurred in fresh concrete. In order to verify the dissolution behavior during the hardening process, the Al coated round bar was immersed into fresh concrete and contact directly. After the curing of the concrete specimen was finished, cut the specimen into half, the observation of cross-section as shown in Fig. 4. There are many voids in the interface surface of Al coated round bar and concrete. Moreover, the hot-dip Al coated steel members can't be used in the sites which contact with the fresh concrete due to the dissolution behavior of Al in alkaline media. In the ancillary facilities of bridge substructure, hot-dip Al coated steel members still in contact with the hardened concrete. The hot-dip Al coated steel plates were shown in Fig. 5. The skyward of Al coated plate was exposure to the atmosphere, the groundward was contacted with the concrete specimens or PTFE sheet (prevent contact with concrete tightly). The corrosion behavior of specimens indicated that the corrosion behavior of skyward, specimen of Kyoda is more serious than that of Univ. of the Ryukyus due to no rain-washing environment has more salt accumulation. But the contact part of the specimen, the specimen of Univ. of the Ryukyus has a certain corrosion due to the rainy accumulation in the gap of steel plate and concrete. On the contrary, the specimen of Kyoda did not show any corrosion behavior. The reason for this behavior is in the weakly alkaline and absence of stagnant water environment, the passive film and compound formed in the alkaline media, will firmly be attached to the interface of plate and concrete. In the comparative test carried out with and without PTFE, the effect of PTFE sheet in this test isolated the Al coating and concrete. In this situation can avoid the corrosion behavior caused by concrete. The results shown that, in Kyoda, the corrosion behavior of PTFE was more serious than that without PTFE. Hence, Al coating has a certain protection provide by concrete in no stagnant water environment. On the contrary, the PTFE specimen of Univ. of the Ryukyus, corrosion behavior of without PTFE was more serious than that of PTFE.

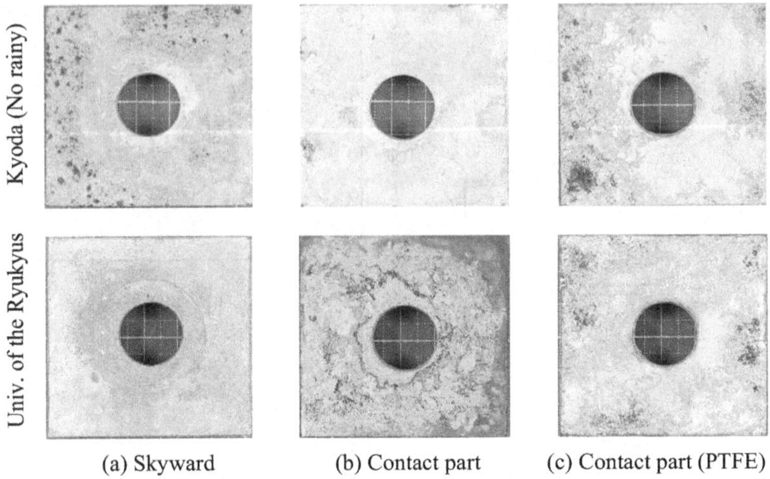

Fig. 5 The corrosion behavior of exposure specimens coated by hot-dip Al coating

The XRD analysis of the contact part was shown in Fig. 6. Through the composition analysis of corrosion products indicated that the corrosion products of contact parts that with PTFE were similar, consist of Al, Al_2O_3. This indicated that the Al coating was isolated with concrete, the corrosion behavior is not obviously. Conversely, the specimens of without PTFE, there were many differences of corrosion products' composition. The contact part of Univ. of the Ryukyus, in addition oxide, $Al(OH)_3$ was also found due to the stagnant. The contact part of Kyoda, because of no rain-washing, only can found Al. However, $Ca_2Al_2O_3 \cdot 8H_2O$ and $CaAl(OH)_2$ also be found. Those complex calcium and aluminum compounds were the most reactive

Fig. 6 The XRD analysis of contact part with and without PTFE

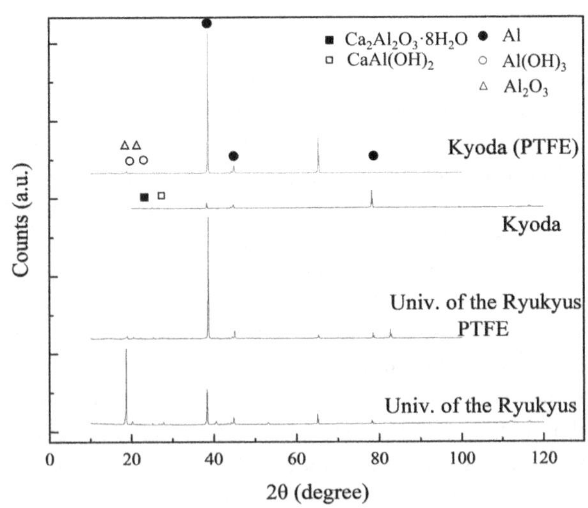

aluminates in the hydration of Portland cements, indicated that Al coating contact with concrete, the components of the cement will remain in the surface of Al and have a certain protect function to Al coating [5].

4 Conclusions

(1) In this study, the hot-dip Al coating can be dissolved in the fresh concrete environment. So the cover of rebar can't be made by hot-dip Al coating.
(2) With the hardening of concrete, the pH on concrete surface is gradually decreasing. The hot-dip Al coating can't be dissolved and can be used for the structures that contact with the hardened concrete.
(3) Compare with the stagnant environment, the Al coated steel have a better anti-corrosion behavior in dry environment.

Acknowledgements This work was supported by JSPS KAKENHI under Grant Number JP19K15074 and JP19H02227. Any opinions, findings, and conclusions expressed in this study are those of the authors and don't necessarily reflect the view of the JSPS KAKENHI.

References

1. Plusquellec G, Geiker MR, Lindgård J, Duchesne J, Fournier B, De Weerdt K (2017) Determination of the pH and the free alkali metal content in the pore solution of concrete: review and experimental comparison. Cem Concr Res 96:13–26
2. Adhikari S, Hebert KR (2008) Participation of aluminum hydride in the anodic dissolution of aluminum in alkaline solutions. J Electrochem Soc 155:C189
3. Zaid B, Saidi D, Benzaid A, Hadji S (2008) Effects of pH and chloride concentration on pitting corrosion of AA6061 aluminum alloy. Corros Sci 50:1841–1847
4. Behnood A, Van Tittelboom K, De Belie N (2016) Methods for measuring pH in concrete: a review. Constr Build Mater 105:176–188
5. Jensen TR, Christensen AN, Hanson JC (2005) Hydrothermal transformation of the calcium aluminum oxide hydrates $CaAl_2O_4 \cdot 10H_2O$ and $Ca_2Al_2O_5 \cdot 8H_2O$ to $Ca_3Al_2(OH)_{12}$ investigated by in situ synchrotron X-ray powder diffraction. Cem Concr Res 35:2300–2309

Experimental Study on the Use of Plastic Waste as Natural Aggregate Replacement in Pervious Cement Concrete

Steve W. M. Supit, Priyono, and Harry Sumajow

Abstract Pervious concrete is a type of concrete that is normally created for pavement that allows rainwater to percolate into the layer below the concrete. This paper reports the experimental results on the effects of partial replacement of coarse aggregate by plastic waste type Polyethylene Terephtalate (PET) in pervious concrete. The coarse aggregate from plastic waste was prepared with size of 10–20 mm after heating the plastic at the temperature of 150–180 °C. Three different percentage of plastic which are 5, 10 and 15% by weight of coarse aggregate were used in pervious concrete mixtures with 0.3 of water to cement ratio and coarse aggregate to cement ratio was 3:1. The results show that the pervious concrete containing 5% of plastic waste as coarse aggregate increases the 7th day compressive strength reaching the value of 16.4 MPa, which is 115% higher when compared to the control sample. In addition, the void ratio results also show that pervious concrete with plastic waste addition increases the continuous voids of samples at 7 and 28 days. Furthermore, it can be concluded that plastic waste at suitable proportion under well-prepared method can potentially be used as a new alternative construction material especially when environment, energy-conserving and low-cost benefits are considered.

Keywords Pervious concrete · Plastic waste · Compressive strength · Void ratio · Infiltration

1 Introduction

As the increase of sustainable global growth in construction field, the use of sustainable materials like by-products or waste materials and natural resources to be used

S. W. M. Supit (✉)
Civil Engineering Department, Manado State Polytechnic, Manado, Indonesia

Priyono
Mechanical Engineering, Manado State Polytechnic, Manado, Indonesia

H. Sumajow
Graduate Student of Civil Engineering Department, Manado State Polytechnic, Manado, Indonesia

T. Kang and Y. Lee (eds.), *Proceedings of 2021 4th International Conference on Civil Engineering and Architecture*, Lecture Notes in Civil Engineering 201,
https://doi.org/10.1007/978-981-16-6932-3_22

in mortar and concrete for structural and non-structural application has been extensively investigated. One of the innovations that can be developed is the potential use of plastic waste as material in concrete mixtures for building construction. Plastics are biodegradable material made by monomers like ethylene and propylene, derived from fossil hydrocarbons. The use of plastics have become an integral part in human lives mostly as packaging material, artificial implant, automotive application and many other uses. With abundant of applications, the volume of solid waste stream has been rapidly increasing and unfortunately contaminating the marine environment, freshwater system and terrestrial habitats [1]. Therefore, the optimization in using plastic waste as construction material can be one solution upon the environmental problem impacted from disposing plastic into the landfills or natural environments. The use of pervious concrete has been growing in popularity as permeable pavements that benefits stormwater runoff reduction, groundwater rechargement and application in transportation facilities such as pedestrian walkways, parking lots, roadways, rigid drainage layers, structural wall application, bridge embankments, green houses and others [2, 3]. Among other things, pervious concrete is also suitable for natural aggregates resources preservation and energy savings due to low compaction needed during installation into a pavement construction [4]. ACI 522R-10 defines pervious concrete as concrete with near zero-slump with little or no sand and consisting of cement, coarse aggregate, admixtures and water. These ingredients will form hardened concrete with typical connected pores of 2–8 mm that allow water to pass through. Compared to the typical concrete with voids of 3–5%, the void content of pervious concrete is about 15–25% with a flow rate of 0.34 cm/s [5]. Some factors that generally affect the characteristics of pervious concrete are the optimum value of water/cement ratio, the use of admixture, aggregate sizes, and aggregate/cement ratio [6]. Additionally, the incorporation between cement and pozzolan materials like fly ash [6], metakaolin [7], carbon fiber [8], nano-silica [9] has been investigated to improve the strength properties of pervious concrete.

In order to optimize the use of plastic waste, this research aims to investigate how the plastic wastes effects the pervious concrete properties, when replacing partial content of natural coarse aggregate in pervious concrete mixtures. Some studies have been conducted and reported on the evaluation of plastic waste as coarse or sand replacement in normal concrete. Based on these studies some findings have been reported as follows:

1. The use of plastic waste as a coarse aggregate for structural concrete was studied in Ref. [10]. It was found that the compressive strength and split tensile strength of concrete with 20% plastic aggregate are comparable with the controlled concrete. However, increasing the plastic waste content up to more than 20% lead to reduce the strength development. On the percentage replacement of 15% plastic waste aggregate, Ref. [11] reported positive results on compression test with the value of 18.34 MPa at 28 days after curing and concludes the application of this concrete for structural application where as Ref. [12] recommended the percentage of using plastic aggregate as partial replacement of natural coarse aggregate to be 5–10%. This study also reported the increase of workability

values on concrete mixtures containing plastic aggregate since plastic does not absorb any water during mixing.

2. As a replacement of fine aggregate, the study conducted in Ref. [13] reported that 10% plastic waste showed satisfactory performance on the improvement of mechanical properties of concrete and potential to save millions of sand every year from being used in concrete mixture [14].
3. Using plastic waste as replacement of coarse aggregate in concrete increases the ductility, therefore, potential to be used in severe environment like in harsh weather [15]. In addition, concrete mixture with plastic waste can be useful for applications that requiring non-bearing lightweight concrete such as concrete panels used in facades. However, bonding effect and the adhesive between the surface of plastic and cement paste should be taken into account since it effects the reduction of concrete strength properties due to the hydrophobic characteristics of plastic that restricts the hydration of cement [16].
4. In term of the type of plastic waste in concrete, a study in Ref. [17] found that aggregate made from the waste of plastic bottles had more influence on supporting strength rather than aggregate made from the waste of plastic bags. It was then recommended the use of waste plastic bottles and bags for non-structural concrete elements. Moreover, Ref. [18] reported the effect of using recycled HDPE plastic as a partial replacement of coarse aggregate in pervious concrete. It was concluded that HDPE aggregates increased the permeability and porosity of a pervious concrete as the percentage replacement was increased up to 30%. On the other hand, Ref. [19] commented on his study on strength and permeability of pervious composite using post-consumer plastic waste bottles. This study concluded the influence of soil/aggregate ratios in binding the aggregates that contributed the strength development. In this research, they concluded the potential application of pervious plastic composite for low-strength construction such as driveways, sidewalks and parking lots.

Among the results mentioned above, most of the research was conducted on the use of plastic waste in normal concrete while very less reports are available on the use of plastic waste in pervious concrete. Therefore, this research aims to investigate more on the properties of pervious concrete containing plastic waste aggregate as partial substitution of the natural aggregate. In this research, the plastic waste sourced from plastic bottle type Polyethylene Terephtalate (PET) is considered. The method on the preparation of plastic waste aggregate was also discussed. The effectiveness including the adhesive between plastic surface and cement paste are evaluated based on the compressive strength, void ratio and infiltration rate test. The experimental results would benefit the recommendation for the application of using pervious concrete containing plastic waste as construction material.

(a) (b)

Fig. 1 Image of **a** plastic wastes before heating and **b** plastic coarse aggregate

2 Materials

Cement type Portland Composite Cement (PCC), coarse aggregate with maximum size of 10–20 mm and water are the main materials used in this experiment. Plastic waste type Polyethylene Terephtalate (PET) sourced from plastic bottle was prepared as seen in Fig. 1.

The cut plastics were melted traditionally under temperature of 150–180 °C to form hardened plastic block that was pressed under maximum load pressure of 10 ton. The plastic paving block was then placed in the bucket containing water for 10–20 min until it gets hardened. The hardened plastic in block form was then crushed manually using the steel hammer to form plastic aggregate with the same size of the natural coarse aggregate. Figure 1a shows the image of plastic wastes before being heated while Fig. 1b is the image of plastic coarse aggregate after being crushed and sieved to gain the desired size with maximum size of 20 mm. Superplasticizer type F from SIKA Ltd. was also used as additive in pervious concrete mixtures with percentage of 0.5% by weight of cement. The use of superplasticizer is important as a water reducer that can also contributes the bonding performance between aggregate and cement paste.

3 Experimental Methods

Three parts were conducted in this experimental work. Part 1 investigated the maximum compressive strength at 7 and 28 days of pervious concrete with different type of mixes containing 5, 10 and 15% of plastic aggregate as a replacement of natural coarse aggregate. Part 2 evaluated the void of ratio of pervious concrete containing plastic aggregate that achieved highest compressive strength to be compared with normal pervious concrete after water curing at 7 and 28 days. Part 3 evaluated the infiltration rate of pervious concrete containing plastic aggregate that achieved highest compressive strength on Part 1, to be compared with normal

Table 1 Mixture proportions of different type of pervious concrete (kg/m^3)

Type of concrete	Cement	Natural coarse aggregate	Plastic aggregate	Water	Superplasticizer
NC	400	1200	0	120	2
PC-5%	400	1140	60	120	2
PC-10%	400	1080	120	120	2
PC-15%	400	1020	180	120	2

Notes NC = Natural aggregate Concrete, PC-5% = 5% Plastic in Concrete, PC-10% = 10% Plastic in Concrete, PC-15% = 15% Plastic in Concrete

pervious concrete without plastic aggregate at 28 days of testing period. In the mixing process, Portland Cement Composite, natural coarse aggregate and plastic coarse aggregate were firstly prepared and mixed all together in concrete mixer with 0.3 of water to cement ratio and the ratio of coarse aggregate: cement was 3:1. Water mixed with superplasticizer with dosage of 0.5% by weight of cement was then poured into the mixtures. After around 15 min of mixing and reached the homogeneous level, concrete mixtures were then cast in 100 mm/200 mm concrete cylinder and demoulded after 24 h for water curing until the days of testing. The mixture proportions of different type of pervious concrete are shown in Table 1.

3.1 Compressive Strength

The testing of compression strength procedure was followed the ASTM C39/C39M-18 Standard test method for testing compressive strength of cylindrical concrete specimens [20]. The concrete sample was placed on the compression machine and the load was applied until fracture. The ultimate load of each pervious concrete samples were then recorded and the average of three pervious concrete samples on each mixture was used as the final result. Figure 2 presents the pervious concrete sample when tested using a compression machine.

Fig. 2 Pervious concrete sample when tested using a compression machine

3.2 Void Ratio

This test was proposed by Japan Concrete Institute (JCI) standard test method (Report on Eco-concrete committee on design, construction and recent applications of porous concrete) and conducted in order to calculate the total void ratio that is defined as the percentage of the total volume of voids to the total volume of the specimen [15]. In this test, the pervious concrete cylinder specimens with size of 100 × 200 mm were prepared after curing at 7 and 28 days. The volume of specimens was measured as V_1. After water curing, the mass of samples in water was recorded (W_1). The samples were then left for 24 h at room temperature and then weighted as W_2. On the next day, the measurement in water was taken again (W_3) to calculate the total void ratio (A_t) based on the formula in Eq. (1):

$$A_t\,(\%) = 1 - \frac{(W_2 - W_1)/\rho_w}{V_1} \times 100 \tag{1}$$

Additionally, the continuous void ratio (A_c) can be also calculated using the formula in Eq. (2):

$$A_c\,(\%) = A_t - \frac{(W_1 - W_3)/\rho_w}{V_1} \times 100 \tag{2}$$

3.3 Infiltration Rate

The infiltration rate was conducted using infiltration ring according to ASTM C1701M-17a Standard test method for infiltration rate of in place pervious concrete [21]. Total water of around 18.0 ± 0.05 kg was poured into the ring and the elapsed time was recorded to the nearest of 0.1, started from the time when water impacted the pervious concrete surface until the free water disappeared from the surface. The infiltration rate test conducted on pervious concrete slab can be seen in Fig. 3. The formula used to calculate the infiltration rate is shown in Eq. (3):

$$I = \frac{KM}{(D^2 t)} \tag{3}$$

where

- I Infiltration rate (mm/h)
- M Mass of infiltrated water (kg)
- D Inside diameter of infiltration ring (mm)
- t Time required for measured amount of water to infiltrate the concrete (s)
- K 4,583,666,000 in SI units.

Fig. 3 Infiltration rate test

4 Results and Discussions

4.1 Compressive Strength

The compressive strength results of different type of pervious concrete at 7 and 28 days are shown in Fig. 4. It can be observed that the compressive strength of pervious concrete with 5% of plastic aggregate (PC-5%) has higher strength when compared to PC-10% and PC-15% concrete samples. This is an indication that the maximum of replacing natural coarse aggregate with plastic aggregate could not be higher than 5% by weight of total coarse aggregate used in the mixture. Interestingly, the compressive strength of PC-5% is higher than compressive strength of normal pervious concrete without plastic inclusion (NC), particularly at 7 days. The strength

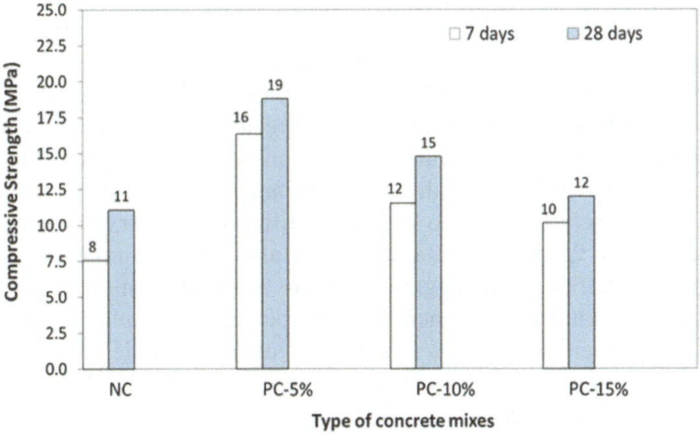

Fig. 4 Compressive strength of pervious concrete at 7 and 28 days

of PC-5% is 16.4 MPa, more than two times of the strength of NC which is 7.6 MPa. In addition, the 28th day compressive strength results show that the resistance of pervious concrete containing plastic aggregate on compression load is also higher than the strength of NC concrete. The strengths of pervious concrete with plastic aggregate at 5, 10, and 15% replacement of natural aggregate after curing at 28 days are in the range of 12–19 MPa. However, it can be seen that increasing the plastic aggregate content up to 15% led to decrease the compressive strength development at 7 and 28 days. The compressive strength is only around 10 and 12 MPa at 7 and 28 days, respectively, lower than PC-5% and PC-10% pervious concrete samples, but still higher than normal concrete (NC) sample.

On the other experiment conducted by Ref. [22], the partial replacement of coarse aggregate up to 10% with plastic waste type PET in concrete resulted in increasing the compressive strength at 7 and 28 days. The compressive strength was found higher than the minimum compressive strength required for structural concrete. In this experiment, the percentage of replacement plastic aggregate up to 30% did not give significant improvement at all days of testing periods due to the flat, and shiny surface of plastic material that could not be easy in forming good bonding with concrete ingredients. Furthermore, this study confirms the potential use of artificial coarse aggregate from plastic waste as construction material in the production of composite cement and concrete including road, highways, and pavements applications.

4.2 Void Ratio

The void ratio of previous concrete samples was evaluated based on the procedure described in Sect. 3.2. Two samples of each concrete mixture were taken after 7 and 28 days of water curing to evaluate the percentage of voids including total voids (At), continuous voids (Ac) and discontinuous voids (Ad) of all different types of concrete samples. In this case, the continuous voids become the most important factor in determining the permeability performance of variation types of pervious concrete. Figure 5 shows the percentage of voids of NC, PC-5%, PC-10%, and PC-15% pervious concrete sample. All values are then tabulated in Table 2. Based on the results, it can be seen that using plastic aggregate lead to increase the total voids of the concrete. As seen in the figure, PC-5% concrete sample reached the value of 18.4% of total voids at 7 days with continuous and discontinuous voids are 17.5% and 0.7%, respectively. There is no significant different is observed when the curing was extended up to 28 days. Similar trends are also observed on concrete samples containing 10 and 15% plastic aggregate. There is a reduction of total voids when 10% plastic was included while increasing the percentage of plastic up to 15% led to increase the percentage of voids in pervious concrete sample. Interestingly to be noted that even there are some reductions on total voids in PC-10% and PC-15% when compared to the total voids of PC-5%, the continuous voids of PC-5% are still higher indicating the adhesive between cement paste and the aggregate is more effective in PC-5% concrete. In comparison to the total voids of normal pervious

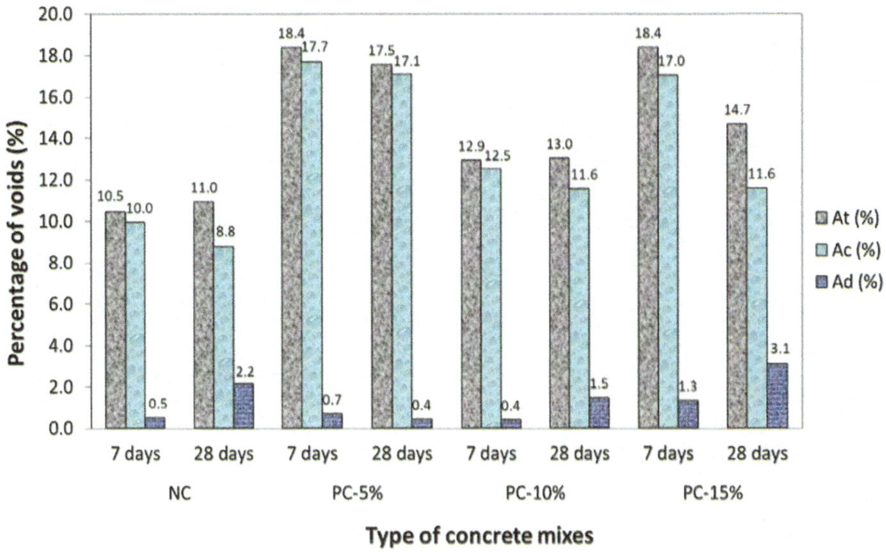

Fig. 5 Percentage of voids of pervious concretes at 7 and 28 days

Table 2 Percentage of voids in different types of concrete

Type of mixes	7 days			28 days		
	Total voids—At (%)	Continuous voids—Ac (%)	Discontinuous voids—Ad (%)	Total voids—At (%)	Continuous voids—Ac (%)	Discontinuous voids—Ad (%)
NC	10.5	10.0	0.5	11.0	8.8	2.2
PC-5%	18.4	17.7	0.7	17.5	17.1	0.4
PC-10%	12.9	12.5	0.4	13.0	11.6	1.5
PC-15%	18.4	17.0	1.3	14.7	11.6	3.1

concrete, concrete with plastic aggregate have more total voids but the continuous voids are more obvious and shows almost linear relationship between compressive strength and total continuous voids ratio. The improvement of bonding between cement matrix and plastic aggregate was also found on the study conducted by [22] through Scanning Electron Microscopy (SEM) analysis. This is an indication that plastic waste under treatment can be very promising material to be used as partial replacement of natural coarse aggregate that could improve the bonding by reducing the gaps between aggregate and cement paste thus contributes the enhancement of strength properties.

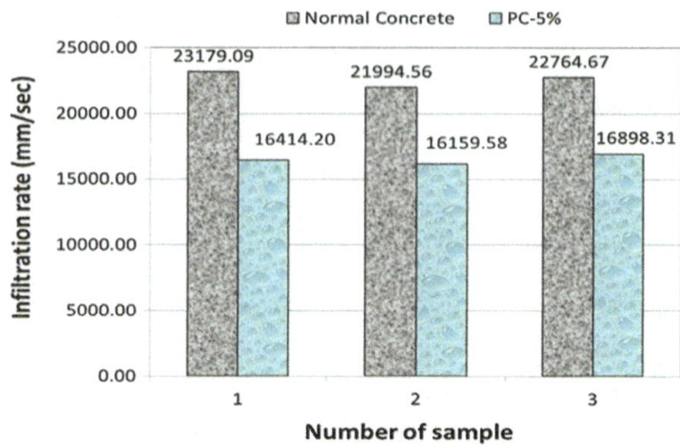

Fig. 6 Infiltration rate of NC and PC-5% pervious concretes

4.3 Infiltration Rate

The comparison of normal pervious concrete and 5% plastic aggregate (PC-5%) pervious concrete on infiltration rate measured for three times can be seen in Fig. 6.

The results of infiltration rate determine the permeability of different type of pervious concretes in correlation with the aggregate size and density of the mixture. In this case, the permeability of pervious concrete containing 5% plastic aggregate can be evaluated when compared with normal pervious concrete. Based on Fig. 6, the infiltration rate shows that pervious concrete containing plastic aggregate has lower infiltration rate compared to the normal concrete. The average of infiltration rate of PC-5% is 16,490 mm/s while normal concrete is 22,646 mm/s. From this result, the infiltration rate of PC-5% is around 27% lower than the infiltration rate of NC. This is an indication that the density of pervious concrete is more when plastic aggregate was included to replace 5% natural aggregate in pervious concrete mixture. The results are also in line with continuous void results as shown in Table 2 showing that the continuous voids of PC-5% is more than the continuous voids of normal pervious concrete that influence the infiltration rate in mm/s.

5 Conclusions

Based on the limited results from this experimental works, some results could be drawn as follows:

1. The pervious concrete specimens containing various amount of plastic waste sourced from plastic bottle type Polyethylene Terephtalate (PET) exhibited higher compressive strength after 7 and 28 days of water curing, compared to normal pervious concrete. The maximum strength was achieved by pervious concrete containing 5% plastic aggregate with the strength values are 16 and 19 MPa at 7 and 28 days, respectively.

2. The optimum percentage of replacement using plastic waste as coarse aggregate in pervious concrete is to be 5% by total coarse aggregate. Increasing the plastic aggregate content up to 15% led to reduce the compressive strength development.
3. Percentage of total voids of pervious concretes with the inclusion of plastic aggregates is higher than normal pervious concrete. However, the continuous voids of pervious concrete with plastic waste are more and indicating the connection between plastic aggregate and concrete ingredients can be formed due to the shape of plastic aggregate that increases the density of pervious concrete specimens.
4. The infiltration rate of pervious concrete is lower than normal concrete. The average of infiltration rate of PC-5% is 16,490 mm/s while normal concrete is 22,646 mm/s. The results confirm the formation of more continuous void in pervious concrete containing plastic waste type PET.
5. The use of artificial plastic waste aggregate type PET in pervious concrete could produce acceptable strength properties and permeability for application on drainage structure and road construction. However, there is a concern such as the controlled temperature when heating the plastic could not be more than 180 °C and the challenge on shaping the plastic block to form into the coarse aggregate size. Further analysis on the bonding characteristics between the plastic aggregate with the mortar in concrete is still needed.

Acknowledgements The authors would like to thank *Direktorat Jenderal Pendidikan Tinggi Direktorat Sumber Daya, Kementerian Pendidikan, Kebudayaan, Riset, dan Teknologi Republik Indonesia* for the research funding under scheme *Penelitian Dasar Unggulan Perguruan Tinggi* Year 2021 No. 095/E4.1/AK.04.PT/2021—12 Juli 2021.

References

1. Geyer R, Jambeck JR, Law KL (2017) Production, use, and fate of all plastics ever made. Sci Adv 3(7):25–29
2. Malegaonkar N (2019) Permeable pavement by using waste plastic bottles. Int J Res Appl Sci Eng Technol 7(5):590–596
3. Toghroli A et al (2018) A review on pavement porous concrete using recycled waste materials. Smart Struct Syst 22(4):433–440
4. Aoki Y (2009) Development of pervious concrete, pp 1–138
5. Hemalatha L, Keerthana S, Jagadesh P, Karthik Prabhu T (2018) Study on pervious concrete by using cost effective materials. Int Res J Eng Technol 671–675
6. Uma Maguesvari M, Sundararajan T (2017) Influence of fly ash and fine aggregates on the characteristics of pervious concrete. Int J Appl Eng Res 12(8):1598–1609
7. Supit SWM, Pandei RW (2019) Effects of metakaolin on compressive strength and permeability properties of pervious cement concrete. J Teknol 81(5):33–39
8. Rangelov M, Nassiri S, Haselbach L, Englund K (2016) Using carbon fiber composites for reinforcing pervious concrete. Constr Build Mater 126:875–885
9. Mohammed BS, Liew MS, Alaloul WS, Khed VC, Hoong CY, Adamu M (2018) Properties of nano-silica modified pervious concrete. Case Stud Constr Mater 8:409–422

10. Subramani T, Pugal VK (2015) Experimental study on plastic waste as a coarse aggregate for structural concrete. Int J Appl Innov Eng Manag 4(5):144–152
11. S.E.E. Profile, S.E.E. Profile (2018) Concrete production using recycled waste plastic as aggregate, June, pp 10–17
12. Patil B, Agrawal PSD (2018) An experimental study on partial replacement of coarse aggregate by using recycled plastic aggregate on strength of concrete. Int J Eng Techn Res 0869(5):97–99
13. Kore SD (2018) Sustainable utilization of plastic waste in concrete mixes—a review. J Build Mater Struct 212–217
14. Thorneycroft J, Orr J, Savoikar P, Ball RJ (2018) Performance of structural concrete with recycled plastic waste as a partial replacement for sand. Constr Build Mater 161:63–69
15. Ashwini Manjunath BT (2016) Partial replacement of E-plastic waste as coarse-aggregate in concrete. Procedia Environ Sci 35:731–739
16. Jaivignesh B, Sofi A (2017) Study on mechanical properties of concrete using plastic waste as an aggregate. IOP Conf Ser Earth Environ Sci
17. Jibrael MA, Peter F (2016) Strength and behavior of concrete contains waste plastic. J Ecosyst Ecogr 6(2):4–7
18. Lopez N, Collado E, Diacos LA, Morente HD (2019) Evaluation of pervious concrete utilizing recycled HDPE as partial replacement of coarse aggregate with acrylic as additive. MATEC Web Conf 258:01018
19. Solanki P, Bhattarai S (2018) Strength and permeability of pervious composite prepared by using post-consumer plastic waste bottles. MATEC Web Conf 174
20. ASTM International (2012) Standard specification for compressive strength of cylindrical concrete specimens. ASTM c39-12
21. ASTM International (2009) Standard test method for infiltration rate of in place pervious concrete. Annu B ASTM Stand
22. Lee ZH, Paul SC, Kong SY, Susilawati S, Yang X (2019) Modification of waste aggregate PET for improving the concrete properties. Adv Civ Eng 2019

Research on Optimal Design Strategy of Sport Building Facade in Cold Region Based on Intelligent Interaction

Y. Quan and H. Mei

Abstract The participation of the crowd has gradually become an important indicator of building facade, and the building facade design has changed from the aesthetic stage to the interactive stage. In this study, applications of intelligent interaction in architectural facade design were investigated using qualitative and quantitative methods and building facade design was optimized based on intelligent interaction. The methods of intelligent facade intervention in architecture were reviewed. Based on intelligent facade design methods, we propose aesthetic optimization strategies for facades, optimization strategies for facade lighting and ventilation, optimization strategies for energy efficiency of facades, and a systematic optimization design strategy of building facades with intelligent interactions. In the era of information and interaction, building facades with intelligent interactions will provide more options to interactions of people, buildings and the environment and become the future direction of building facades.

Keywords Building facade · Sport building · Intelligent interaction

1 Introduction

The comfort of building occupants and building energy consumption are important issues in the field of building design and building technology. The demand for thermal and visual comfort of building occupants has greatly increased, leading to increased energy consumption, CO_2 emissions, and corresponding cost of building operating energy [1]. Applying passive design strategies in the early stages of design benefits the architects and engineers the most by prioritizing the rational use of sunlight [2]. However, optimizing the comfort of building occupants through passive use of

Y. Quan (✉) · H. Mei
Key Laboratory of Cold Region Urban and Rural Human Settlement Environment Science and Technology, School of Architecture, Ministry of Industry and Information Technology, Harbin Institute of Technology, No. 66 Xidazhi Street, Harbin 150001, China

H. Mei
Architectural Design and Research Institute of HIT, No. 73 Huanghe Road, Harbin 150090, China

T. Kang and Y. Lee (eds.), *Proceedings of 2021 4th International Conference on Civil Engineering and Architecture*, Lecture Notes in Civil Engineering 201,
https://doi.org/10.1007/978-981-16-6932-3_23

daylight is a daunting challenge, because the building (interior) and the environment (exterior) are two relatively independent spaces. Due to the conflict between them, it is difficult to optimize visual and thermal comfort. In particular, the sun movement will continuously change the indoor comfort conditions. However, through interactive architectural design methods and related interdisciplinary research, we may solve the above contradictions.

Building facade is a complex medium between the interior of the building and the external environment, which has a protective or regulating effect against the fluctuations of the external climate [3]. The facade is a part of building facing the street or open space, and the pattern is a specific form, shape, or structure [4]. The facade is the most politically and culturally significant element [5]. "The understanding of the role of the facade in architectural history is a conservative element that conveys a sense of stability and persistence." From the fully automatic Girasole villa launched by Invernizzi in the 1930s, the complexity of the exterior walls began to increase. "The buildings were inspired by sunflowers, rotating along mechanical tracks, and the inner glass courtyard ran along the path of the sun". For environmental adaptation, this showed the ability to adapt to changing environmental conditions while maintaining stable internal conditions [6]. This solution was an innovative work of architectural interactive facade, which pushed the architectural facade to become a dynamic facade from a static facade. Sample Heading (Third Level). Only two levels of headings should be numbered. Lower-level headings remain unnumbered; they are formatted as run-in headings.

The concept of intelligent interactive facade has reached a consensus in the construction industry, that is, using digital control technology to create interactive intelligent solutions for solving design problems through multidisciplinary strategies [7]. The intelligent interactive facade is designed with multi-disciplinary factors, which can be used to explore a wide range of parameters across fields to detect multi-objective solutions to multi-objective problems [8]. In general, the intelligent interactive facade may be an important strategy for optimizing the adjustment function of the external facade, which facade should respond to the demand of user. These goals will be studied by identifying and changing important facade factors that affect comfort conditions to control climate fluctuations.

2 Literature Review

The database of related articles was created by searching through Google Scholar and Scopus using keywords and association technology. The related problems of intelligent building interactive buildings were discovered by searching the following keyword combinations: "intelligent interaction", "building facade", and "building skin". After comprehensive analysis, the related issues of the facade were refined, including design issues (e.g., response, adaptation, optimization algorithms, parameterization, and generative design), natural interaction issues (e.g., morphology, form-finding, bionics, morphology, dynamics, simulation, and natural mimicry; microclimate, wind, daylight and other resource utilization issues), zero net energy buildings, multidisciplinary, daylight performance evaluation, visual and thermal comfort, energy efficiency and other issues. These keyword combinations were further used in Google Scholar and Scopus research databases to perform more precise searches.

Architects and architectural physics experts considered the versatility of intelligent interactive facades and their potential for dynamic response to changes in daylight. In this research, we were guided by the following research questions: (1) What are the interdisciplinary themes that correspond to the concept of interactive facades for creating innovative architectural forms? (2) On the basis of multidisciplinary research, what is the most effective factor in the interactive facade that aims to improve the visual and thermal comfort of building occupants at the same time? The interactive facade is a multi-functional and complex building component. Through the study of different fields with parametric design mindset, an adaptive facade can be realized. The development of an adaptable and responsive facade requires a historical cognition beyond appearance. In addition, façade as the only medium between the interior of the building and the surrounding environment, the facade is affected by multiple factors, which makes façade complex and heterogeneous.

3 Significance of Interactive Facades for Sports Buildings in Cold Region

According to statistics, the average annual energy consumption per unit area of ice and snow sports buildings in northern China has reached 316 kWh/m^2, far exceeding the energy consumption level of 192 kWh/m^2 for commercial buildings and 138 kWh/m^2 for office buildings. Therefore, the energy consumption of heating and cooling accounts for a large proportion of the energy consumption of sports buildings in cold regions, and seeking a stable balance in the indoor and outdoor environments is the key to the design of sports buildings in cold regions.

The design of the intelligent interactive ecological facade can effectively resist the unfavorable external climate environment, alleviate the adverse impact of the external environment, and utilize of natural resources effectively. For large public buildings such as sports buildings, the ecological facade can respond to the changes of environmental conditions, which can effectively use the natural environment while shielding harmful factors. In the design of the facade of sports buildings in cold regions, seeking a balance between basic functions such as thermal insulation, ventilation, lighting, and sun shading should be considered. However, how to organize these factors in the design of cold sports buildings, while satisfying the basic functions, and creating a beautiful and comfortable building environment, is the arduous task that architects must face in the ecological and energy-saving design. Therefore, in the face of different climates and different geographical conditions, the design method should be adapted to local conditions and optimized strategy combination, so that the material can give full play to its own characteristics to meet the basic functional requirements.

4 Optimization of Intelligent Interactive Facades of Sports Buildings in Cold Region

4.1 The Improvement of the Aesthetic Taste in the Facade Form

The architectural design concept takes inspiring ideas and morphological methods as important themes, applying geometry and materials to create intelligent interactive facade elements, transferring to static technology and intelligent systems from static technology and intelligent systems [9]. Integrating architectural design concepts and mechanisms through morphological methods to create responsive and adaptive forms for dynamic facades to meet the needs of occupants [10]. The level of interactivity plays an important role in the intelligent interactive facade that affects the configuration of the facade form. Several functions should be evaluated according to the metrics. Therefore, choosing appropriate metrics to evaluate the comfort of the occupants of the building will greatly affect the design of facade form. All in all, combining architectural design concepts, mechanisms and evaluations may lead to the creation of a reliable cycle for exploration and inventing the movable dynamic facade.

As the earliest example of intelligent interactive facade elements, the sliding Japanese-style shoji doors require manpower to move and serve different purposes. Their physical structure is composed of translucent paper and a wooden lattice frame. The intelligent interactive facade feature of the shoji door provides a high degree of flexibility for the building, which can enhance the spaciousness and daylight performance of any room. The evolution of the facade stems from the historical perception of movable elements, such as the sliding shoji doors in traditional Japanese architecture, to responsive and programmable media, such as the Al-Bahar Tower in Aedas, Abu Dhabi, using foldable panel to control the daylight, which can enhance the function of the exterior wall (Fig. 1). The supervisory functions of the smart building facade include control and response. These are achieved through dynamic responses to environmental stimuli based on certain parameters that meet the desired goals.

In this study, the dynamic interactive facade can be transformed according to the dynamic daylight and the location for the occupants of buildings (based on functional scenarios) to meet visual comfort. In addition, the daylight parameter simulation also uses climate-based daylight metrics to study the visual comfort performance provided by the interactive facade form. The simulation results prove the high performance of the dynamic interactive facade, which can improve the visual comfort under

a. Shoji Gate of Japan b. Al-bahar Tower

Fig. 1 Interactive architectural facade derived from Shoji Gate

basic conditions. In particular, the change of three-dimensional shape provides better visual comfort compared to the two-dimensional shape changes of Useful Daylight Illuminance (UDI), Exceed UDI and Daylight Glare Probability (DGP) metrics.

For the Ice Ribbon of the National Speed Skating Hall for the 2020 Winter Olympics (Fig. 2), speed is the theme of the facade design. The athletes galloped past in circles, and the ice skates left sparkling nicks on the ice surface. These lines are the imprints of speed, which rise along the contour of an ellipse, bend, and interlace inside and outside to form an "ice ribbon" facade. From the edge of the hyperboloid roof, the basic form of the façade slopes from top to bottom, forming the largest plane on the second floor to accommodate the needs of the audience lounge. After testing the number of ribbons, density and glass folding method of the façade pass parameters through modeling, it is finally decided to bend 4 times on the datum line

a. 2020 Winter Olympics National Speed Skating Hall b. Free-form surface curtain wall

Fig. 2 Ice ribbon free-form surface curtain wall

of the facade, forming stacked folds, just like the cascading roof of the Temple of Heaven. After that, a total of 22 printed glass sunshade tubes with a diameter of 35 cm diameter are installed on the folds, which distribute densely in the folds and loosely in the recesses. The glass tubes are translucent, reflecting each other with the highly transparent curved glass. The ultra-white glass used in the ribbon integrates small-radius semicircular bending, tempering, laminated SGP film and gradient color glaze printing and other processes, which challenge the application of curtain wall glass. The structure of the curved curtain wall adopts a single-cable supporting net shell system. Trying to reduce the vertical curtain wall keel section to emphasize the horizontal dimension of the ice ribbon through the synergy of the diagonal curtain wall cables. The connection between the cable and the keel adopts a specially designed shuttle node to show the delicate details of the curtain wall structure.

4.2 Enhancement of Lighting and Ventilation of Facades

Since the concept of comfort develops over time, the concept of changing the facade configuration over time is a way to improve thermal and visual comfort. The three-dimensional shape change of the facade elements has the potential to control the microclimate (sun and wind) (Fig. 3). For example, the Kiefer Technic Showroom is interactive due to the functional scenarios, which can reduce glare and thermal performance. Michael Fox, the head of the intelligent interactive facade design team at MIT, proposed a classification of intelligent interactive facade control, which could be converted into three behaviors in the intelligent interactive facade: manual, autonomous and responsive. In the process, the responsive envelope responds to natural forces and is programmed to produce the desired visual effect or move according to certain parameters. Through the application of sensors and actuators, the responsive intelligent interactive facade interacts with the surrounding environment morphologically, thereby improving interesting functions according to certain parameters, such as thermal comfort and visual comfort. Compared with the fixed facade, the solar heat energy in the smart interactive facade of Al Bahar Tower and

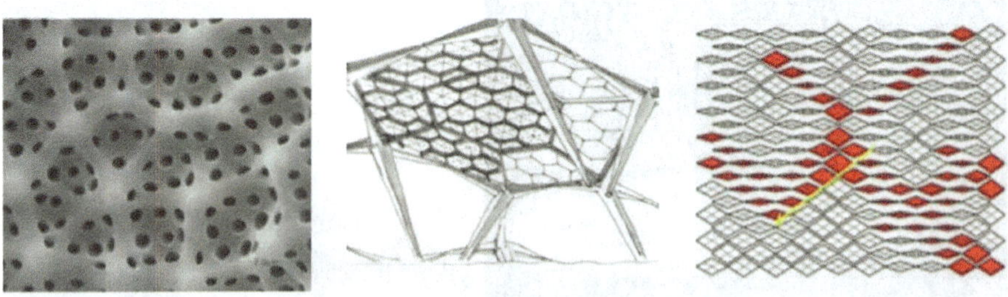

a. Organic grid b. Hexagonal grid c. Parametric grid

Fig. 3 Various types of facade elements

the Helio Trace Building Center are reduced by 50% and 81%, respectively. Similarly, the Media ICT Building shrinks and expands according to the use of sunlight, resulting in a heat and ultraviolet filter rate of 85%. In addition, dynamic shading provides great indoor comfort for the cooling and heating seasons, ranging from 13.3 to 26.9%.

Interaction with functional scenes can reduce glare and provide visual comfort for building occupants. Indeed, the three-dimensional movement of the intelligent interactive facade is the outstanding response to the functional scene. In different situations, the dynamic facade operation has fully met the requirements of thermal comfort and daylight. Regarding Mahmoud and Elghazi [11], compared to the base case, dynamic shadows in summer and spring increase daylight by about 50% and autumn and winter by about 30%, respectively. Moreover, they claim that rotating motion can improve illumination more than translational motion.

It is necessary to study the human visual and thermal comfort metrics for parameter and algorithm design. Therefore, exploring the comfort status literature helps to find suitable measurements to evaluate the comfort status of building occupants. In fact, the use of measurable metrics for comfort evaluation can accelerate the regeneration process of parameter facade configuration. Natural light has a positive effect on health and energy efficiency. However, in order to obtain visual comfort, some issues should be considered, such as increased heat and visual discomfort (sunlight glare). "It is always a challenge to strike a balance between visual comfort and daylight", which needs to study the relationship between human needs and ambient light based on the following specific factors: "The amount of light, the uniformity of light, the light quality of rendering colors, the prediction of glare risk for buildings occupants". Therefore, by using these factors to carry out parametrical simulations, it is possible to prepare data for the fast and accurate evaluation for daylight performance.

Thermal comfort is a complex topic due to human and physical reasons. Human factors affect the physical and psychological parameters that need to be investigated face-to-face, such as clothing insulation, metabolic rate, age, and gender. However, physical factors point to indoor operating temperature, relative humidity, and air velocity. Optimizing thermal comfort with passive technology means natural ventilation and sunlight control systems. Regarding wind, the form of the facade and its surface affect the wind speed and airflow pattern, thereby improving the temperature regulation ability of the building. The performance of shading devices and windows play an important role in reducing energy consumption and preventing overheating. It is possible to achieve a trade-off solution to solve the contradiction between visual and thermal comfort using the complex window opening system promoted by kinetic shading equipment. Finally, a balance between geometry, orientation, window to wall ratio, glass material, color, reflectivity, and emissivity will improve visual and thermal comfort.

The Aviva Stadium on Lansdowne Road in Dublin, Ireland was designed by POPULOUS (Fig. 4). The architects first constructed and determined the basic shape and surface of the stadium based on the Rhino modeling platform. Subsequently, the parametric platform was adopted to optimize the geometric form, standardize the outer skin component, and optimize the structural form of the curtain wall support

a. External interface of Aviva Stadium

b. Digital adjustable curtain wall unit

Fig. 4 Framework of execution strategy by CNN

component. In addition, the designer chose a translucent skin material to provide a smaller canopy on the auditorium and its surroundings. In order to ensure a good ventilation environment of the stadium, the curtain wall panel was designed as a rotatable unit, which achieved good visual effects and high spatial quality.

4.3 Enhancement of Energy Efficiency of Facades

The operation of buildings accounts for about one-third of global energy use, and they account for a similar proportion of greenhouse gas emissions. The energy consumption of cold large-space buildings is particularly prominent. From the perspective of full-life environmental load, large-scale public buildings such as cold sports buildings generate construction waste proportional to their scale during the construction process. During operation after completion, their environmental air-conditioning and site cooling devices generate a large amount of waste gas and wastewater. At present, the traditional HVAC system is generally used to adjust the indoor environment of the building, thereby increasing energy consumption. In the net zero energy building (NZEB), insulation and heat recovery (winter), shading (summer) and natural ventilation can reduce energy through the appropriate envelope structure design [12]. NZEB refers to a building that the energy required for the operation is generated by renewable energy on or near the building. In addition, the European Parliament [13] and the Massachusetts have defined the NZEB as the following principles: (A) application of on-site renewable energy, (B) annual energy consumption equal to or less than on-site energy production, (C) annual functional capital cost equal to or less than energy grid, (D) reduction of carbon emissions and dependence on fossil fuels. However, in the early stages of conceptual design, passive strategies are sustainable ways to balance energy consumption through smart and renewable resources. A prominent concern for architects and engineers is providing residents with a comfortable indoor environment while keeping optimal energy consumption. However, using HVAC systems to provide comfortable conditions will lead to a large amount of operating energy consumption. Therefore, the capital cost of greenhouse gas emissions and energy consumption will increase.

Using technology in an appropriate way in the early stages of design. Passive continuous design leads to the combination of passive strategies and active technologies to balance energy consumption and provide comfort for building occupants. Generating energy from on-site renewable resources (Solar radiation) is a smart function that can be applied to different locations and climates. Therefore, designing architectural elements, such as movable shading devices or smart interactive exterior walls, can optimize the comfort of the occupants and the building. For example, Manitoba Hydro has a dual facade with biological interaction, which can increase energy efficiency by up to 70%. In addition, dynamic shadows have great potential to save 20–30% of the energy use of public buildings, and dynamic photovoltaic cells produced by tracking the sun can generate 30–40% more energy than similar stationary systems.

The intelligent interactive facade has a variety of unfolding methods, and its basic prototype is often based on the traditional rotational mechanism (Fig. 5). Passive strategies refer to precise and accurate solar design with a purpose of reducing heat gain, which requires artificial cooling, heating, and visual glare. In addition, there are some passive design parameters that affect thermal and visual comfort, including orientation, plane size, window-to-wall ratio of several faces, materials, colors and

a. Arabe b. SDU Campus c. Sharifi-ha House

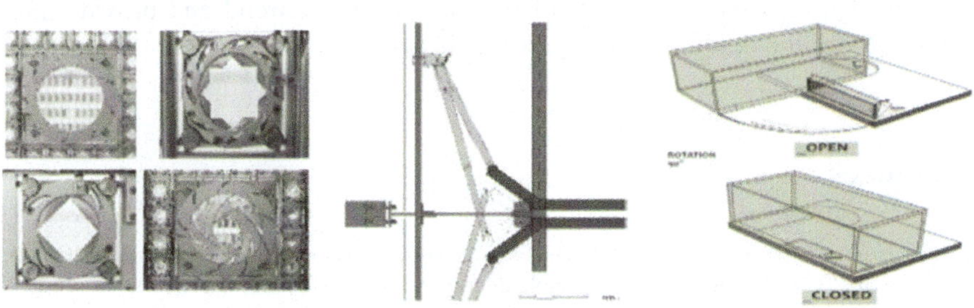

d. Principle of contraction and expansion e. Folding and unfolding principle f. Rotational unfolding principle

Fig. 5 The dynamic expansion method of window

shadow types. The benefit of using active technology is its intelligence and responsiveness, such as using smart electrochromic glass can improve the energy efficiency of the building, the thermal and visual comfort of the occupants in the buildings. This is achieved by dynamic movement, which changes the structure of the building regarding the real-time position of solar energy, to regulate daylight and radiant heat.

For passive strategies, applying a responsive surface with the ability to change direction, form, color and material is an innovative idea to achieve better performance in buildings. In addition, the optical properties of the material (including reflectance and emissivity) will greatly affect the thermal balance and visual comfort of the facade. Therefore, it is necessary to study the changing form, direction, and material properties to improve the energy efficiency of the building. In terms of active strategies, buildings benefit from reconfigurable geometric shapes, multifunctional mechanisms, dynamic and responsive skins to meet the needs of users. Therefore, the use of sensors, actuators, and automatic folding surfaces as facades can provide opportunities to sense and respond to the surrounding environment in real time in sports buildings.

5 Conclusions

The intelligent building facade design process based on multidisciplinary research was supported by the concept of building form as a microclimate modifier, and this process helps design ideas evolve from static to dynamic. The intelligent interactive facade used biologically inspired geometric shapes and undergoes topological transformation to control the microclimate of surrounding environment. In this paper, the development process of the intelligent interactive facade was systematically described, the core related issues and its application in cold ice and snow sports buildings were analysed. The design strategies to enhance the aesthetic of the facade form, promote the lighting, ventilation and energy efficiency of the facade were proposed. The development potential of intelligent interactive architectural facades is presented in this paper, which will become the future trend and provide more possibilities for the design of cold sports buildings.

References

1. International Energy Agency (2013) Transition to sustainable buildings—strategies and opportunities to 2050. Directorate of Sustainable Energy Policy and Technology (SPT), Paris. https://www.iea.org/publications/freepublications/publication/Building2013_free.pdf
2. Wang L, Gwilliam J, Jones P (2009) Case study of zero energy house design in UK. Energy Build 41(11):1215–1222. https://doi.org/10.1016/j.enbuild.2009.07.001
3. Herzog T, Krippner R, Lang W (2004) Facade construction manual. Birkhauser, Publishers for Architectures, Basel, Boston, Berlin. 3-7643-7031-9

4. Oxford Dictionary Press. Façade, morphology. https://en.oxforddictionaries.com/definition. Accessed 21 Nov 2018
5. Simulation modelling and analysis (2012) Int J Prod Res 50(7):2058–2078
6. Leng H, Yuan Q (2014) International experience and enlightenment on urban microclimate environment control and optimization. Urban Plan Int 29:114–119
7. Ding W, Hu Y, Dou P (2012) Study on the correlation of urban morphology and urban microclimate. Archit J 2012(7):16–21
8. Badarnah L (1970) Form follows environment: biomimetic approaches to building envelope design for environmental adaptation. Buildings 7(2):40. https://doi.org/10.3390/buildings7020040
9. Zuk W, Clark RH (1970) Kinetic architecture. Van Nostrand Reinhold Company, New York, pp 14–27
10. Hong C, Baofeng L, Weining Z (2015) Study on the correlation between urban microclimate regulation and form factors of blocks. Urban Archit 2015:31
11. Mahmoud AHA, Elghazi Y (2016) Parametric-based designs for kinetic facades to optimize daylight performance: comparing rotation and translation kinetic motion for hexagonal facade patterns. Sol Energy 126:111–127
12. Hongyuan M, Fei W, Yuliang Z (2013) Study on the design for architectural form adapted to cold regions for low energy consumption. Archit J 11:88–93
13. Aldawoud A (2008) Thermal performance of courtyard buildings. Energy Build 40(5):906–910. https://doi.org/10.1016/j.enbuild.2007.07.007

Bond Load-Slip Behaviors of Strand Assemblies-in-Cement

Shih-Tsung Hsu, Chiao-Hsin Shih, Ke-Ting Chen, Yung-Chen Yen, Wei-Cheng Chang, and Chung-Pin Li

Abstract A series of laboratory pulled-out tests is performed to elucidate the effect on ultimate bond load/stress for strands-in-cement under different conditions. The bond load-slip curves are drawn and discussed. In addition, this study establishes and explores the typical curve of bond stress-slip with various parameters. The seven-wire strands are commonly used for constructing anchors in Taiwan, this study adopts the strand assemblies and cement paste to produce specimens. The pulled out tests are carried out via 500 kN Material Test System. Parametric studies include a variety of strand quantity, cement compressive strength (water/cement ratio, specimen age) and bond length. This study applied the Hyperbolic Model to simulate the compressive strength of cement cubic with different age, the estimated results are in good agreement with those tests results. The pulled-out tests of strand assemblies-in-cement reveal that the ultimate pulled load increase with increases in cement compressive strength, bond length and quantity of strand. Ultimate bond stress increases with an increase in cement compressive strength; however, a longer bond length and more strands obtain a smaller ultimate bond stress. The test results depict that the bond stress-slip curve can be divided into five stages: micro-slip section, slip section, crack generation section, lowering section and residual section.

Keywords Strand assemblies-in-cement · Bond behavior · Laboratory tests

1 Introduction

To overcome problems of settlement and bearing capacity, the construction of high-rise buildings in urban districts frequently employs deep excavations. Deep excavations are currently supported by diaphragm walls with internal braces or tieback

S.-T. Hsu (✉) · C.-H. Shih · K.-T. Chen · Y.-C. Yen · C.-P. Li
Department of Civil and Construction Engineering, Chaoyang University of Technology, Taichung City, Taiwan
e-mail: sthsu@cyut.edu.tw

W.-C. Chang
Department of Aeronautical Engineering, Chaoyang University of Technology, Taichung City, Taiwan

T. Kang and Y. Lee (eds.), *Proceedings of 2021 4th International Conference on Civil Engineering and Architecture*, Lecture Notes in Civil Engineering 201,
https://doi.org/10.1007/978-981-16-6932-3_24

anchor support. A system of tieback anchors offers substantial economic advantages in excavations of a large area. Difficulty working in an available space and increases in expenditure are commonly associated with a complex internal brace system in excavations with an irregular area or an unsymmetrical cross section. In such situations, the use of a tieback anchor system is most effective.

According to the Standards and Comments for Design and Construction of Ground Anchors [1], it defines the safety factor about the ultimate bond stress of strands-in-cement for temporary and permanent anchors. Additionally, JSSMFE Regulation [2] provides the allowable bond stress for varies kind of strand and cement strength. However, the regulation o JSSMFE has not considered the quantity and bond length of strands, the ultimate bond stress is not to be considered as well. As a result, this study adopts the strand with 12.7 mm in diameter to study the bond behavior of interface between the strand assemblies and cement pasta. The bond load-slip curves are drawn and discussed. Moreover, the typical bond stress-slip curve is established and explored in this study.

2 Background

According the definition of ACI Committee 318 [3], the bond stress is a kind of shear stress of the interface between the strand and cement pasta. ACI 1954 [4] indicates the bond stress is come from chemical adhesion between the strand and cement pasta, Hoyer's effect (relaxing the pre-stress of strand induces dilatant effect) and mechanical interlock. The failure modes of pulled strand-cement can be categorized into two types, the pulled out of strand and several cracks of the cleavage fracture in cement pasta. Lee et al. [5] adopts strand with 12.7 mm in diameter and cement pasta to make specimens for pulled-out test, bonded stress-slip curve are established.

3 Test Program

Samples are made in term of bonded length and number of strand, cement strength and the age to figure out the bonded behavior of strands-in-cement, the bonded stress-slip curves are established. This study employs cement pasta with water/cement ratio (W/C) of 0.45 and 0.55 to produce cubic specimens, the compressive strengths of cubic specimens are established in term of W/C and age of cement cubic with 7, 14 and 28 days. The diameter of cement cylinder is 15 cm; lengths of the cylinders are 15 cm, 30 cm and 45 cm, respectively. The quantities of strand are 1, 3, 5 respectively which are putted into cement cylinder. After curing, strand assemblies-in-cement cylinder samples are tested by 500 kN-MTS universal testing machine.

The sample and test device are demonstrated in Fig. 1. The data logger is KYOWA UCAM-60A containing 10 channels. Additionally, test devices include 10 mm LVDT and load cell. As displayed in Fig. 1, the test procedures are descripted as follows:

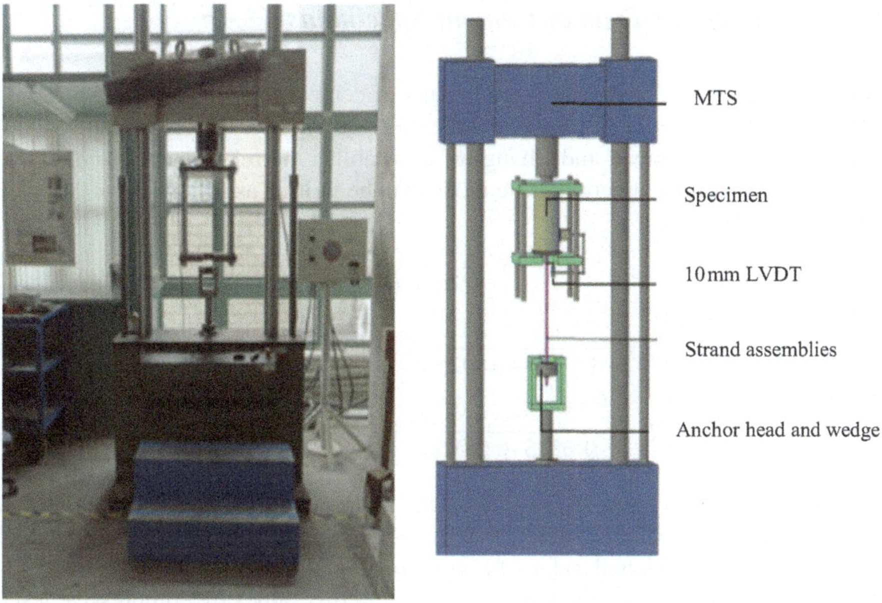

Fig. 1 500 kN MTS and test equipment

(1) Cubic specimens for compressive test

According to the Standards and Comments for Design and Construction of Ground Anchors [1], the W/C of cement mortar should be between 0.45 and 0.55, this study produces the dimension of 50 * 50 * 50 cubic specimens of age 7, 14 and 28 days for compressive test, five specimens are prepared for each mixture ratio and age.

(2) Strand assemblies-in-cement specimens for pulled-out test

This study produces the strand assemblies-in-cement specimens with a series of various bonded length, quantity of strand, compressive strength and age. The diameter of cement cylinder is 15 cm, diameter of strand is 12.7 mm, lengths of cylinder are 15 cm, 30 cm and 45 cm, quantities of strand are 1, 3 and 5, and the W/C are 0.45 and 0.55, respectively.

(3) Other test devices include cement mixer, compressive test machine, 500 kN MTS, Load Cell, LVDT, KYOWA UCAM-60 data logger, etc.

4 Analysis and Discussion of Results

To study the bond mechanism of strand-in-cement, the bonding load-slip curves are drawn and discussed in terms of different strand quantity, cement grout compressive strength (specimen W/C and age) and bond length in this paper.

4.1 Compressive Tests of Cement Specimens

Table 1 displays the compressive test results of cement specimens (5 * 5 * 5 cm) with different water/cement ratio and age. This study adopts the Hyperbolic Model that suggested by Duncan and Chang [6] to establish the relations of compressive strength and specimen age (day), the model can be written as follows:

$$f_c' = \frac{t}{a + bt} \tag{1}$$

$$f_c' = \frac{t}{0.0317 + 0.0256t} \quad \text{for W/C} = 0.45 \tag{2}$$

$$f_c' = \frac{t}{0.0375 + 0.029t} \quad \text{for W/C} = 0.55 \tag{3}$$

Equations (2) and (3) are the estimated equations about compressive strength and age relationships of cement cubic for W/C = 0.45 and 0.55, respectively. Figure 2 demonstrated the predicted values for compressive strength and age relationships by Eqs. (2) and (3) are in good agreement with those values by compressive tests. Figure 2 also display that the less W/C can obtain more compressive strength at same specimen's age.

Table 1 Compressive test results of cement cube

Age, t (day)	Compressive strength (MPa)	
	W/C = 0.45	W/C = 0.55
3	28.42	25.04
7	32.83	28.36
14	35.80	31.54
28	37.50	32.94

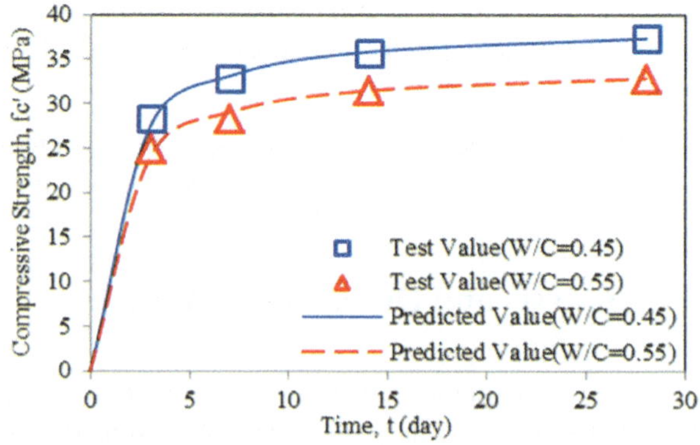

Fig. 2 Compressive strength and age relationships of cement cube

4.2 Bond Load/Stress and Slip Relationships

In laboratory tests, the displacement of strand measured by LVDT contents elastic elongation and slip movement, therefore the total displacement should be subtracted the elastic elongation, and could lead slip movement of strand in this study. However, the slip movement dominates most the total displacement, Fig. 3 reveals that the elastic elongation accounts 2.7% of the total displacement, and can be neglected.

Figure 4 shows that when the bond length increases, the ultimate bond load also increases. Figures 4 and 5 depict that the more compressive strength of cement the more ultimate bond load. Figure 5 illustrates that bond load increase with an increase in quantity of strand. These phenomena can be found for W/C = 0.55 as well, Chen [7] present a more detail about the behavior of ultimate bond load.

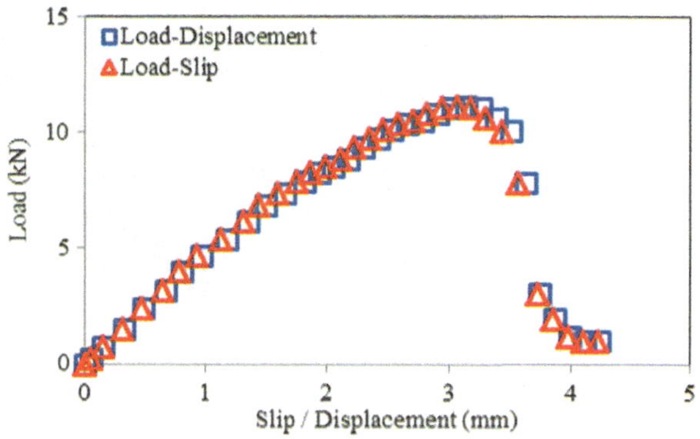

Fig. 3 Bond load-slip/total displacement relationships for W/C = 0.45, bond length = 15 cm, fc′ = 32.83 MPa, one strand

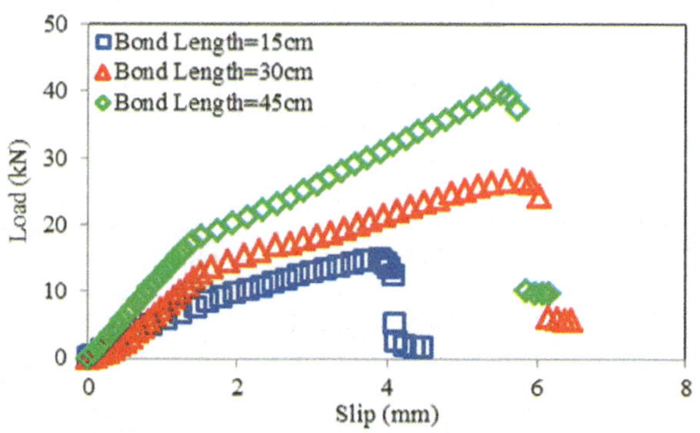

Fig. 4 Bond load-slip relationships for W/C = 0.45, fc′ = 32.83 MPa, one strand, various bond lengths

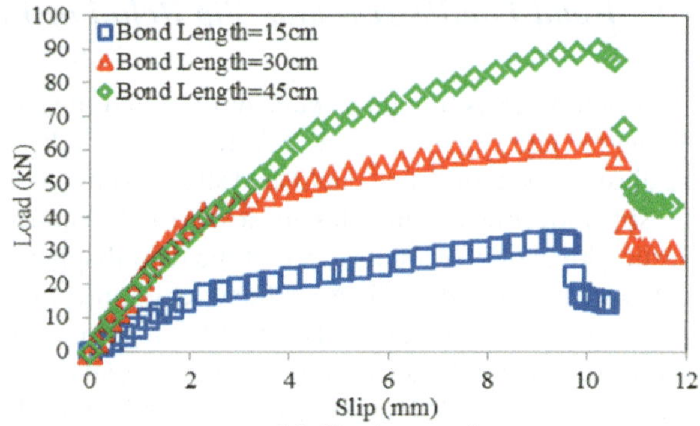

(a) Single strand

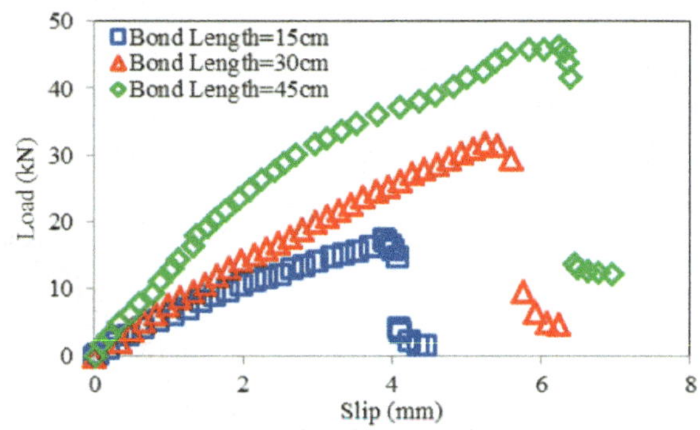

(b) Five strands

Fig. 5 Bond load-slip relationships for W/C = 0.45, fc′ = 37.5 MPa, various bond lengths

4.3 A Model of Bond Stress-Slip Curves

The bonding stress can be obtained by following equation, therefore bonding stress-slip curve could be established as shown in Figs. 6, 7 and 8.

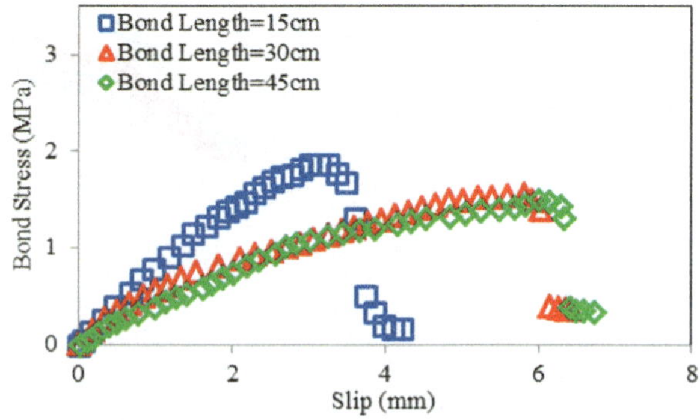

Fig. 6 Bond stress-slip displacement relationships for W/C = 0.45, fc′ = 32.83 MPa, one strand

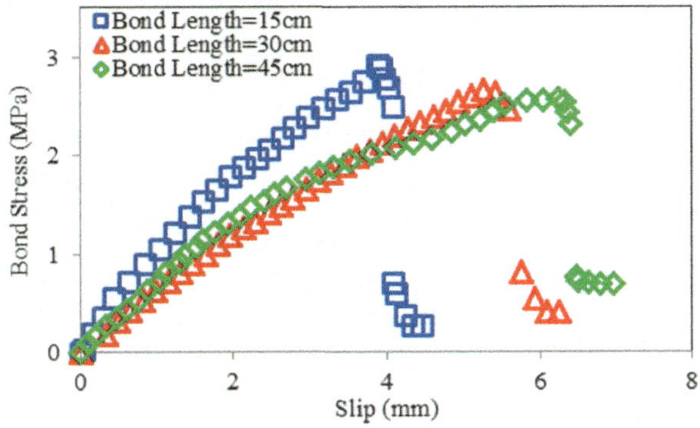

Fig. 7 Bond stress-slip displacement relationships for W/C = 0.45, fc′ = 37.5 MPa, one strand

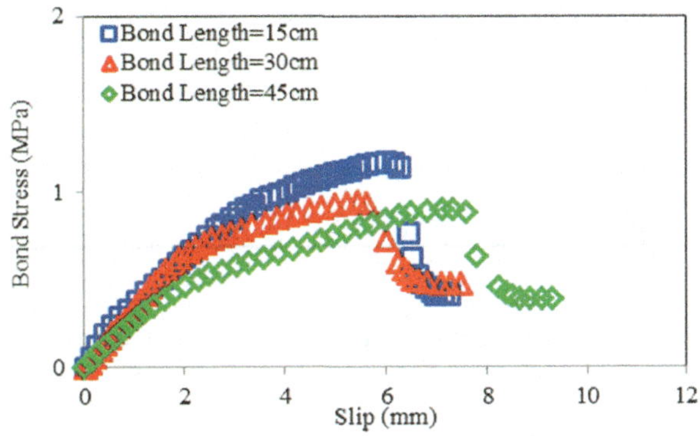

Fig. 8 Bond stress-slip displacement relationships for W/C = 0.45, fc′ = 32.83 MPa, three strand

$$\tau = \frac{Q}{\pi d_s l_b} \tag{4}$$

where τ denotes the bond stress, d_s is the nominal diameter of a strand, l_b represents the bond length of strand-in-cement.

Figures 6 and 7 display that the higher compressive strength could obtain higher bond strength (Ultimate bond stress), the bond strength indicates a little bit decease with an increase in bond length. From Figs. 6 and 8, this study finds that the more the strand, the less the bond strength.

Form the pulled-out test processes, this study divides the bond stress-slip curve into five stages, the phenomena are similar to the bond stress-slip model which established by Lee et al. [5]. Figure 9 demonstrates the five sections of bond stress-slip model, and are explained as follows:

(1) Micro-slip section: The specimen is stressed to 0.4–0.5 times the ultimate load. In this section, the chemical adhesion between strand and cement is failure, the micro-slip occurs.
(2) Slip-section: Surpassing the micro-slip section, the relative slip increase until the mechanical interlocking mechanism losses between strand and cement.

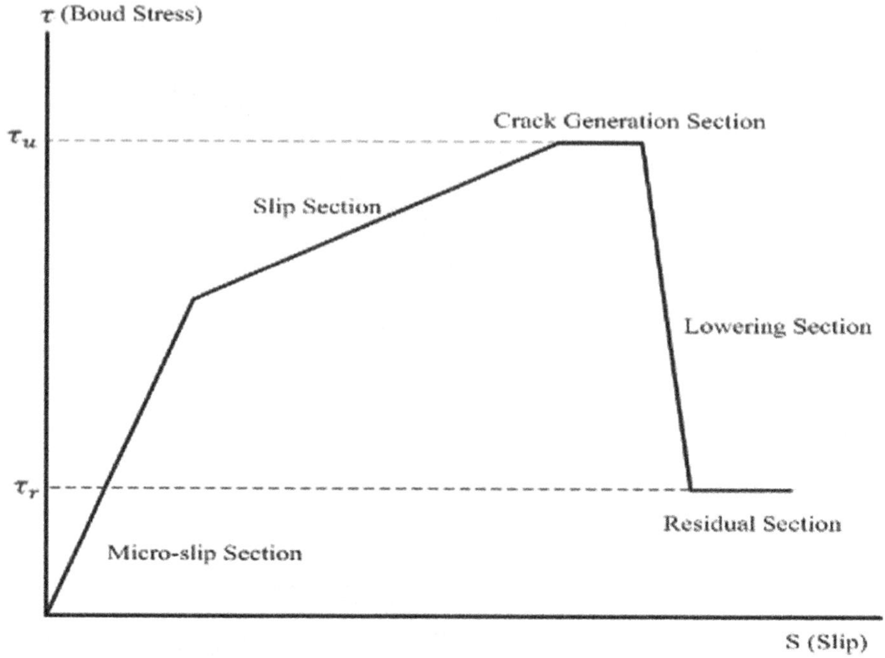

Fig. 9 Typical bond stress-slip relationships for strand assemblies-in cement system

(3) Crack generation section: The specimen stressed to peak load, the bond stress-slip curve appears horizontally. In this stage, the crack occurs nearly by strand initially, and extends to the surrounding of cement cylinder.
(4) Lower section: When the stress of specimen surpasses the ultimate bond stress τ_u, the cement which located in thread of strand is crushed, the mechanical interlocking mechanism almost losses, the bond stress-slip curve decrease rapidly in this stage. This phenomenon differs from the research results by Lee et al. [5], whose bond stress-slip curve decrease more slowly.
(5) Residual section: After the lower stage, the mechanical interlocking mechanism losses completely, only the friction between strand and cement offers the residual stress τ_r.

5 Conclusion

This study conducted a series of laboratory pulled out tests to elucidate the effect on bond strength between strands and cement grout under different conditions. The results lead to the following conclusions.

(1) The Hyperbolic Model could simulate the compressive strength of cement cubic with different age and water-cement ratio, the estimated results were

in good agreement with those tests. As W/C decreases and age increases, the compressive strength of cement cube increased.

(2) The pulled-out tests of strand-in-cement revealed that the ultimate pulled load increase with increases in cement compressive strength, bond length and quantity of strand.
(3) Ultimate bond stress increased with cement compressive strength; however, a longer bond length and more strands obtained a smaller ultimate bond stress.
(4) The test results depicted that the bond stress-slip curve can be divided into five stages: micro-slip section, slip section, crack generation section, lowering section and residual section.

The authors suggest that the bond length should be increased to fit the field scale, the predicted equation on bond strength in terms of compressive strength of cement cube, bond length and quantity of strand should be developed in the future.

References

1. Chinese Institute of Civil and Hydraulic Engineering (CICHE) (2001) Standards and comments for design and construction of ground anchors. Taiwan
2. Japan Society for Soil Mechanics and Foundation Engineering (JSSFME) (1990) Ground anchor design, construction standards and solution. Japan
3. ACI Committee (2005) Building code requirements for structural concrete (ACI 318-05) and commentary (318R-05). ACI, Michigan
4. ACI (1954) Nature of bond in pre-tensioned prestressed concrete. ACI
5. Lee C, Shin S, Lee S (2017) Modeling of bond stress–slip relationships of a strand in concrete during steam curing. Int J Concr Struct Mater 11(3):477–487
6. Duncan JM, Chang CY (1970) Nonlinear analysis of stress and strain in soils. J Soil Mech Found Div ASCE 96(SM5):1629–1653
7. Chen KT (2018) A preliminary study on the bonding mechanism of strand assemblies. Master thesis, Chaoyang University of Technology, Taichung

Dynamic Analysis of Bridge Pier Considering Hysteresis Characteristics of High Damping Rubber Bearings

K. H. Park, Y. Fujiwara, T. Mazda, and Y. Kajita

Abstract The aim of this research is to develop an advanced hysteresis model of High Damping Rubber (HDR) bearing. Based on the experimental research so far, various dependences and nonlinearities can be confirmed in the experimental hysteresis curves of HDR. In particular, the gap in the stiffness between the first cycle and the cycle after the second of the hysteresis curve is confirmed due to the dependence of the maximum shear strain. The advanced hysteresis model of HDR, tentatively named the bilinear Double Target model, was proposed by conducting the dynamic loading test in this research. To assess the technical feasibility of this hysteresis model, three phases were performed in this research. The preliminary proportion was carried out to confirm the mechanical behaviors of HDR by performing a dynamic loading test. Subsequently, based on the experimental results, a modeling procedure of the bilinear Double Target model was proposed and the hysteresis curve of HDR was modeled. Finally, dynamic analyses by design seismic motion are performed on the structural model of the actual bridge pier. The validity of the proposed model was confirmed by investigating the behavior of the bearing part and the lower part of the pier. As a result, it was possible to propose a hysteresis model considering the dependence of HDR and to conduct seismic response analysis.

Keywords High damping rubber · Hysteresis model · Dynamic analysis

1 Introduction

In 1995, the Hyogo-ken Nanbu Earthquake with a seismic intensity of 6.9 occurred, causing enormous damage to civil engineering structures such as road bridges. Since then, seismic isolation structures have begun to be actively introduced into the design to minimize the damage to the structure due to earthquakes and improve durability and safety. The seismic isolation system is a structure that used seismic isolation members such as laminated rubber bearings to block the path through which the energy of

K. H. Park (✉) · Y. Fujiwara · T. Mazda · Y. Kajita
Faculty of Civil Engineering, Kyushu University, 744 Motooka, Nishi-ku, Fukuoka, Japan
e-mail: 3TE21920S@s.kyushu-u.ac.jp

T. Kang and Y. Lee (eds.), *Proceedings of 2021 4th International Conference on Civil Engineering and Architecture*, Lecture Notes in Civil Engineering 201,
https://doi.org/10.1007/978-981-16-6932-3_25

seismic motion is transmitted to the structure. Since then, about 90% of new bridges in Japan have been introduced with seismic isolation structures and seismic isolation members (laminated rubber bearings, seismic isolation dampers, etc.). As a result, it was confirmed that no serious damage was caused to the bridge to which the seismic isolation structure was applied in the Great East Japan Earthquake that occurred in 2011, and its effectiveness was highly evaluated. Since the laminated rubber bearing, which is one of the seismic isolation members, has high rigidity against vertical load and large deformation performance in the horizontal direction, it is possible to minimize shaking due to an earthquake while supporting the superstructure. This research was focused on a high damping rubber bearing called HDR. In particularly, the high damping rubber bearings (HDR) have a relatively simple structure but are easier to install than other laminated rubber bearings and have a great advantage in terms of maintenance costs. Furthermore, by using environmentally friendly materials, their usability has been gradually increasing recently. However, there are few cases of research on high damping laminated rubber bearings in the research to date, and there is a limit to efficient application during construction and design due to the characteristics of HDR. Therefore, it is necessary to develop a hysteresis model for more accurate seismic isolation design using HDR. The hysteresis models considering the dependence of the maximum shear strain so far include the modified HD model of Takenaka [1], the extended bilinear model of Yoshida [2, 3], and the Double Target model of Fujisawa [4]. Also, there are improved Double Target models of Ohtori [5, 6]. The restoring force model called the modified HD model by Takenaka et al. was developed aiming at a curved hysteresis model that has practicality and versatility for LRB and HDR. The extended bilinear model of Yoshida et al. is an extension of the bilinear model, and a method is proposed to accurately and easily model the hardening and dependence of maximum shear strain that the seismic isolation laminated rubber bearing shows at the time of large displacement. The Double Target model of Fujisawa et al. is the hysteresis model that considered changes in rigidity due to the influence of mechanical characteristics of seismic isolation rubber bearings. These models can approximate the mechanical characteristics and dependence of HDR within the applicable range, but the parameter setting is complex for practical use. It cannot be easily used in seismic design. Therefore, Matsuda et al. proposed a bilinear Double Target model that eliminated the complexity with reference to the improved Double Target model, conducted seismic response analysis for seismic isolated structures [7]. In this research, dynamic loading test and dynamic analysis were conducted to develop an advanced hysteresis model of HDR [8, 9]. Firstly, dynamic loading test with high damping rubber specimens was conducted. From the test results, the mechanical properties were evaluated and the effects of the dependence of the maximum shear strain were examined. Based on the test results, the modelling procedure of the bilinear Double Target model which is the hysteresis model that considered the properties of HDR was proposed [3, 10, 11]. Finally, to evaluate the technical feasibility of hysteresis model considering the mechanical characteristics of HDR, dynamic analysis was performed on an actual bridge as a structural model [12, 13]. In this dynamic analysis, the bearing part applied the characteristics of HDR, and as the hysteresis model, the bilinear Double Target model which is modeled according to the modeling procedure proposed in the previous step was applied. As a result, it was possible to confirm the mechanical properties and dependence of HDR. In addition, since the hysteresis model considering the hysteresis characteristics of HDR was applied to the dynamic analysis, the behavior of the bearing part and the bridge base part could be more accurately predicted.

2 Dynamic Loading Test

The outline of the experimental test conducted in this research is the dynamic loading test considering the vertical load equivalent to the dead load of the structure. Shear deformation is induced in the specimen according to the determined shear strain, and its historical data is measured. The purpose of this test is to understand the mechanical properties of the HDR and to investigate the effect of maximum shear strain dependence. Furthermore, it aims to obtain historical data to be used for modeling the hysteresis model considering the characteristics of HDR.

2.1 Test Specimen and Apparatus

Figure 1 shows the test specimen for dynamic loading test. In this research, the HDR specimen manufactured by Yokohama Rubber, Tokyo in Japan was used. Table 1 shows dimension and material properties of HDR specimen. The Cross section of the test specimen is 110 mm × 110 mm, the thickness of the rubber on one layer is 4 mm, and the total thickness of the rubber on all 4 layers is 16 mm. Shear modulus of elasticity is G8 (0.8 N/mm^2), and shape factor and aspect ratio are equal to 4.0. The thickness of the coated rubber of the test specimen is 5 mm. This specimen is in a virgin state that has never undergone shear deformation before the dynamic loading test.

Figure 2 shows the 20 kN hybrid actuator used in the dynamic loading test. The

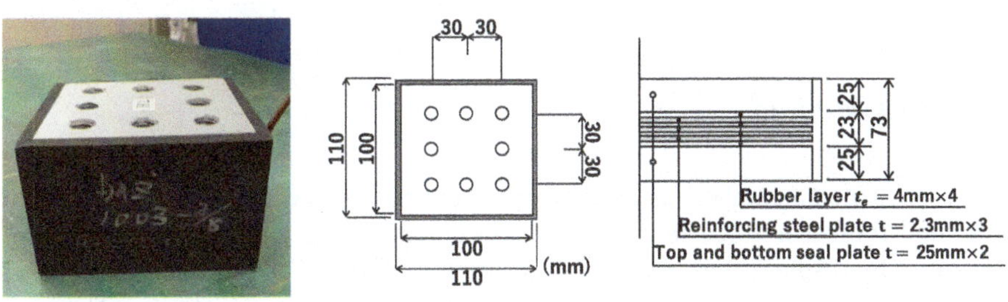

(a) specimen (b) plan view (c) side view

Fig. 1 Test specimen for dynamic loading test

Table 1 Dimension and material properties of HDR specimen

Name of the material	High damping rubber
Cross section (mm^2)	10,000
Number of rubber layers	4
Total rubber thickness h (mm)	16
Thickness of one steel plate (mm)	2.3
Shear modulus (N/mm^2)	0.8
Shape factor, S_1	4.0
Aspect ratio, S_2	4.0

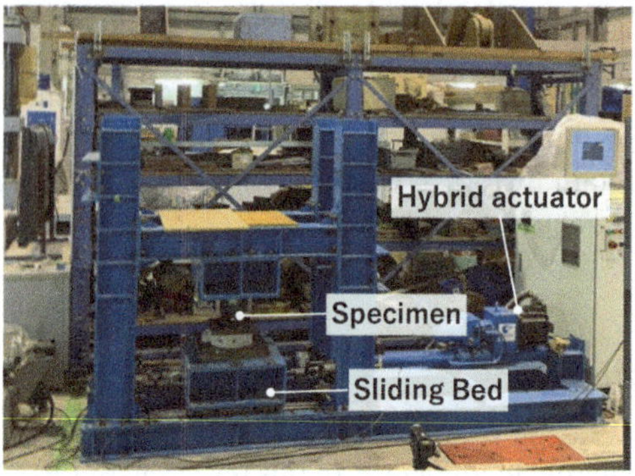

(a) side view

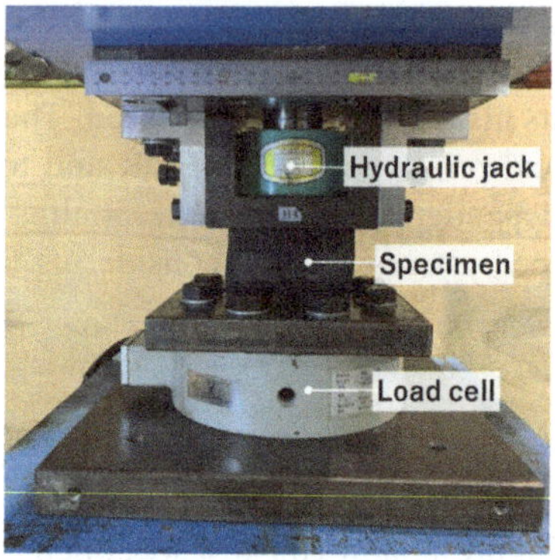

(b) mounting of specimen

Fig. 2 Test setup of dynamic loading test

shear strain is given by this hybrid actuator, and shear deformation of sine wave is controlled by the loading test program. The vertical load was measured by the load cell installed in the actuator. In addition, assuming the dead load of the structure, a vertical load of 60 kN was applied to the test piece by a hydraulic jack. Because of the temperature dependence of the test specimen, the temperature of the atmosphere was kept constant by the heater.

2.2 Test Method

As an experimental method, a vertical load is applied to HDR specimen by a hydraulic jack, and the shear force and shear displacement of the HDR test specimen are measured by displacement measuring instrument. Table 2 shows the conditions of dynamic loading test. In this test, a 0.5 Hz sine wave is set as the input wave. The cases of shear strain are 50% (8 mm), 100% (16 mm), 150% (24 mm), 200% (32 mm), and 250% (40 mm). The loading plan of this experiment aims to confirm a phenomenon

Table 2 Conditions of dynamic loading test

Specimen	HDR
Input displacement type	Sine wave
Amplitude [strain γ (%)]	50, 100, 150, 200, 250
Frequency (Hz)	0.5
Number of cycles	5
Temperature (°C)	21

in which the historical characteristics of HDR differ depending on whether the test specimen has experienced it in the past, even if the shear strain is the same. To consider the dependence on the heat generated by internal friction. After the shear deformation was generated in the test specimen by the actuator, the experiment was conducted with a time interval of 30 min.

2.3 *Test Results*

Figure 3 shows the hysteresis curve of each case of shear strain obtained by the dynamic loading test. Comparing the results of the first cycle and the next cycle in each curve, it was confirmed that the shear force decrease was remarkable even in the hysteretic curve of the same shear deformation. From the experimental test results, it was found that there was the influence of dependence of the maximum shear strain of HDR. Table 3 shows the area of each cycle calculated. From the experimental results, the hysteresis curve for each shear deformation was calculated and compared for each cycle. As a result, it was confirmed that the area of the hysteresis curve gradually decreased after 1 cycle. This is the result of the dependence of the maximum shear strain of HDR. The area of each hysteresis curve is reduced by about 30% on average. In addition, the case of 50% shear strain among each shear strain showed the largest rate of degradation in area.

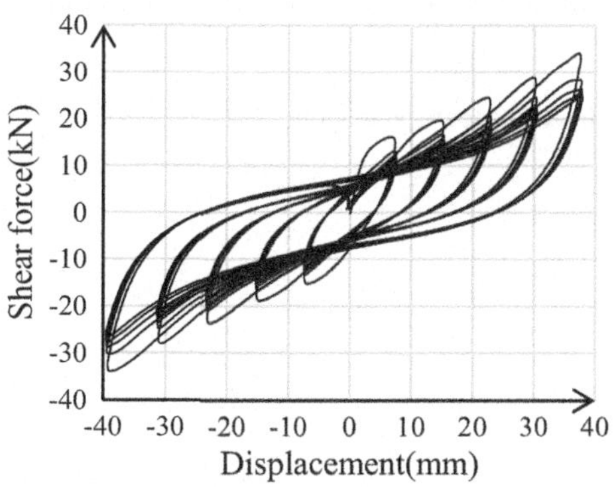

Fig. 3 Hysteresis curve of dynamic loading test (50, 100, 150, 200, 250% shear strain)

Table 3 Comparison of area of hysteresis curve for each cycle

	Shear strain (%)				
	50%	100%	150%	200%	250%
1 cycle	140.45	291.06	496.87	828.74	1281.69
2 cycle	95.33	247.87	428.77	750.67	1184.40
3 cycle	81.36	226.49	399.38	714.07	1110.21
4 cycle	74.15	213.43	380.98	688.44	1051.781
5 cycle	69.56	202.86	365.66	665.25	1000.731
Rate of degradation (%)	50	30	26	20	22

3 Dynamic Analysis

This dynamic analysis begins with modeling a hysteresis curve capable of expressing the characteristics of HDR based on the experimental results, and then performs a dynamic analysis for the modeled actual bridge piers. The purpose of the dynamic analysis is to verify the effectiveness of the proposed hysteresis model and to predict the behavior of the bearings and the lower part of a pier in response to earthquake response with high accuracy. This analysis was a non-linear time history response analysis, and the input ground motions were basically evaluated by response of each of the level 2 seismic motion TYPE I and TYPE II shown in the Specifications for Highway brides in Japan.

3.1 Hysteresis Model

In this research, a hysteresis model of HDR was proposed from the results of dynamic loading test [10]. Figure 4 is a conceptual diagram of the bilinear double target model applied as a hysteresis model of HDR. Unlike general bilinear models, this model is that considers the characteristics of the initial stiffness of HDR and divides the stiffness after the first yield into two [11]. Table 4 shows each parameter calculated

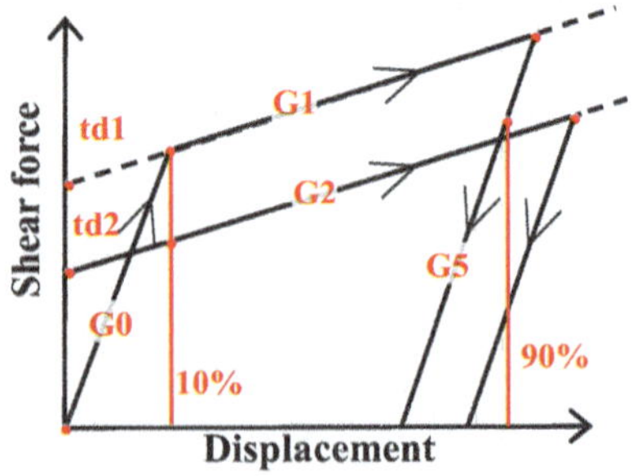

Fig. 4 Conceptual diagram of model

Table 4 Parameters of the bilinear double target model

G0 (N/mm)	G1 (N/mm^2)	G2 (N/mm^2)	G5 (N/mm^2)	td1 (N/mm^2)	td2 (N/mm^2)
1785.47	1297.4	944.1	10,830.7	758.2	736.8

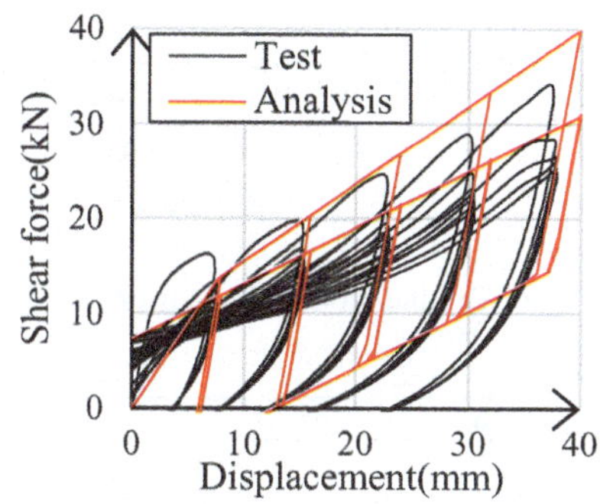

Fig. 5 Analysis results

based on the experimental result of 250% shear strain in the dynamic loading test. For each parameter, a skeletal curve was established by connecting several points of the measured historical data. G0 is the spring constant, and the rest of parameters are calculated in units of stress from the test results. Figure 5 shows the comparison between the test results and the analysis results.

3.2 Analysis Conditions

Figure 7 shows a cross-sectional view of the superstructure of the bridge. This superstructure is supported by five HDR bearings installed on the pier for each span. One pier was taken out from the bridge [12, 13], the cross-sectional configuration of the pier is shown in Fig. 8. The bridge to be analyzed is located on the TYPE II ground described in the seismic design of road bridges in Japan. Figure 6 shows ground motion time history. Table 5 shows the specifications of bearings. The allowable design shear strain of HDR bearings is 250%, and the hysteresis model of HDR is the bilinear double target model. The hysteresis model of the pier is shown in Fig. 9.

3.3 Analysis Results

Figures 10 and 11 show the response history of the bearings to TYPE I and TYPE II design ground motions, and Fig. 12 shows the response history of the lower part of pier. Table 6 shows the maximum and minimum values of HDR displacement and shear force, and the plasticity ratio of the lower part of pier. As a result of the analysis

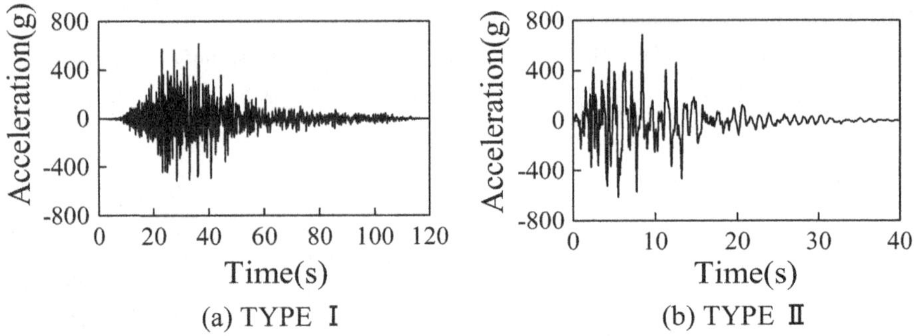

Fig. 6 Ground motion time history

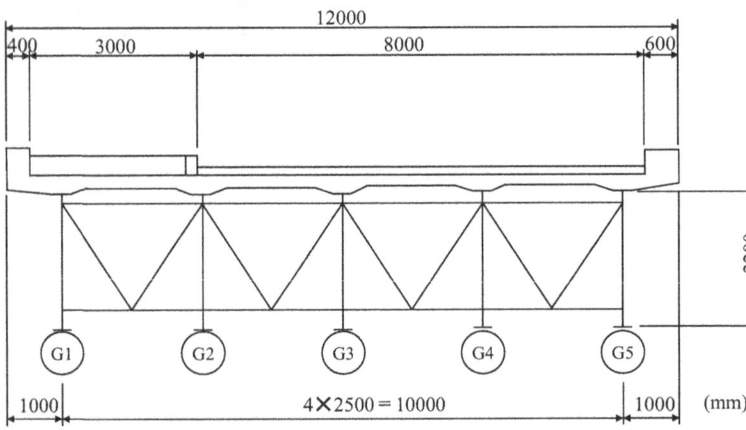

Fig. 7 Cross-sectional view of the superstructure of the bridge

Fig. 8 Analytical model of the bridge pier

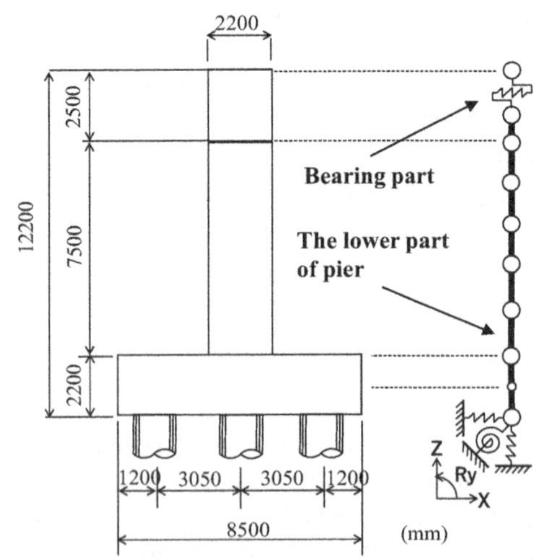

Table 5 Specifications of bearings

Name of the material	HDR
Cross-section (mm^2)	540,225
Total rubber thickness (mm)	117.6
Number of bearings	5
Design displacement u_e (mm)	320
Design shear strain γ_e (%)	250%
Hysteresis model	Bilinear double target model

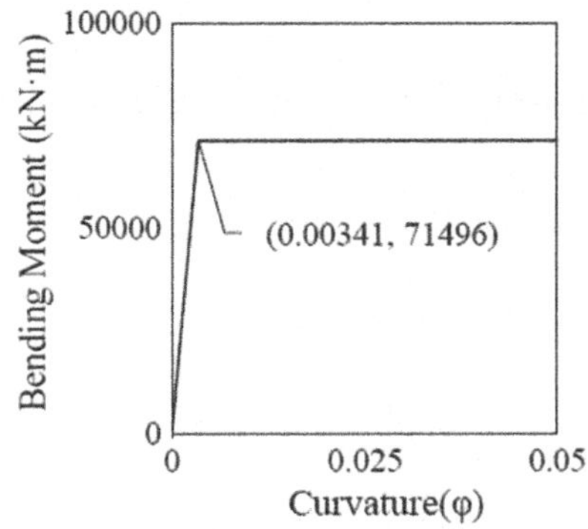

Fig. 9 Hysteresis model of the bridge pier

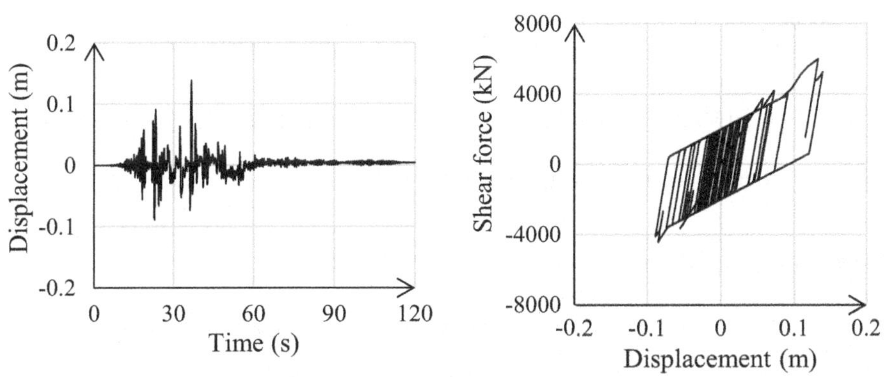

Fig. 10 Analysis results of HDR bearing (TYPE I ground motion)

of the bearing part, it was confirmed that the shear displacement of HDR was within the design range when the bilinear Double Target model was applied. In addition, by setting the bilinear Double Target model that considered the initial history of HDR, it can be confirmed that if the shear deformation of HDR is larger than the shear deformation experienced in the past, the hysteresis curve is drawn by the set initial skeletal curve.

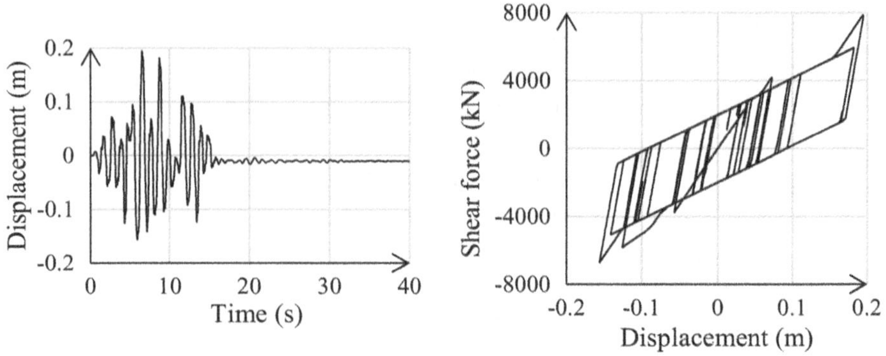

Fig. 11 Analysis results of HDR bearing (TYPE II ground motion)

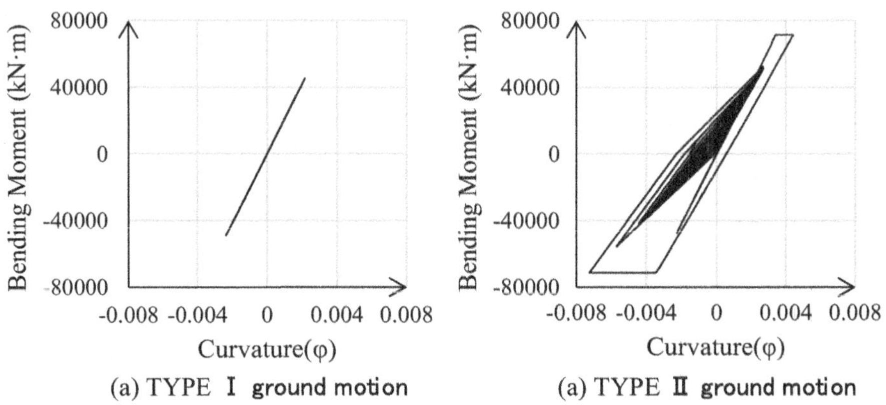

Fig. 12 Analysis results of the lower part of pier

Table 6 The results of dynamic analysis

	TYPE I		TYPE II	
	MAX	MIN	MAX	MIN
Displacement (m)	0.139	−0.089	0.195	−0.156
Shear force (kN)	6012.2	−4415.2	7865.4	−6706.1
Plasticity rate	0.667		1.696	

4 Conclusions

In this research, the mechanical properties of HDR bearings introduced as seismic isolation members for bridges were investigated through dynamic loading experiments. Based on the dynamic loading test results, the validity of proposed hysteresis model that reflected the dependence of HDR was examined by conducting dynamic analysis. The major findings are summarized as follows.

1. As a result of the dynamic loading test, it was experimentally clarified that the mechanical properties such as the area of hysteresis curves and the shear force of HDR change depending on the magnitude of the experienced shear strain in the past. In particularly, if the HDR experienced the same shear deformation after experiencing the shear deformation for the first time, the mechanical properties tended to decrease. It was also confirmed that when the shear strain applied to the HDR exceeded 200%, a hardening phenomenon occurred and the shear force suddenly increased.
2. From the results of the dynamic loading test, the bilinear Double Target model was modeled to consider the characteristics related to the initial history of the HDR. In this research, parameters were calculated using the experimental results with a shear strain of 250% of HDR. As a result, in the dynamic analysis of the sine wave as in the dynamic loading test, the initial stiffness showed high agreement with the experimental results. However, in the region of 50% shear strain, it was confirmed that the difference between the analysis result and the experimental result was large, and correction of the initial stiffness was required to calculate the parameter.
3. Based on the modeling results of the HDR hysteresis model, the pier of the actual bridge was modeled as the analysis target, and dynamic analysis was performed by design ground motion. In the behavior of the bearing, the bilinear Double Target model was applied as the HDR hysteresis model, so it was confirmed that the analysis results were obtained along the initial skeletal curve for the first history. As a result, it was possible to analyze the dynamics considering the mechanical properties of HDR, and the behavior of the lower part of the pier was also confirmed. In particularly, it was found that the displacement of the bearings of HDR and the displacement of the lower part of the pier occurred significantly in the results of the dynamic analysis by the TYPE II ground motion with a relatively shorter time than the TYPE I ground motion.

Acknowledgements This study is part of the results of research conducted in collaboration with the HDR Study Group (Kawakin Core-Tech Co., Ltd., Sumitomo Riko Company Ltd., NIPPON CHUZO K.K., Bridgestone Corporation, THE YOKOHAMA RUBBER CO., LTD.).

References

1. Takenaka Y (2001) Modified H-D model: a new smooth-curve hysteresis model of laminated rubber bearings for base isolation. AIJ J Technol Des 14:87–92
2. Yoshida J (2006) Optimal bilinear models of super high damping rubber bearing for seismic response analyses. J JSCE 62(2):454–466
3. Yoshida J (2014) Extended bilinear hysteretic models for laminated rubber bearings. J JSCE 70(2):238–251

4. Fujisawa K (1993) A study on analytical modeling for seismic isolator. In: Summaries of technical papers of annual meeting Architectural Institute of Japan. B. Structures 1, pp 505–506
5. Ohtori Y (1994) Earthquake response analysis of base isolated building with modified double target model. In: Summaries of technical papers of annual meeting Architectural Institute of Japan. B. Structures 1, pp 793–794
6. Ohtori Y (1995) Effect of experienced shear strain dependency of high damping rubber bearing on earthquake response of isolation structure. J Struct Constr Eng (Trans AIJ) 60(472):75–84
7. Mazda T (2015) Evaluation on seismic response considering characteristics of high damping rubber bearing. J JSCE 71(4):198–209
8. Kato H (2015) Analytical model for elastoplastic and creep-like behavior of high-damping rubber bearings. J Struct Eng 141(9)
9. Iriyama T (2010) Experimental study on dynamic behavior of two-story wooden structure including shear-wall with high damping rubber using shaking table test. J Struct Constr Eng AIJ 75(651):967–976
10. Grant DN (2004) Bidirectional modelling of high-damping rubber bearings. J Earthquake Eng 08(spec01):161–185
11. Paramashanti (2011) Experimental investigation of aging effect on damping ratio of high damping rubber bearing. J Struct Eng 57A:769–779
12. Naito N (2017) Seismic performance evaluation of LRB considering Mullins effect and hardening. J JSCE 73(4):499–510
13. Chaudhary MTA (2001) Performance evaluation of base-isolated Yama-agé bridge with high damping rubber bearings using recorded seismic data. Eng Struct 23(8):902–910

Bending Analysis of Continuous Slab Plates in Two Directions

Filomeno Pereira, Yoshitaka Suetake, and Saliha Shukri Muhammed

Abstract Continuous plate members are often used as structural elements of building such as roofs, parking and deck of bridges. When designing such structures, approximation methods are usually employed, because analytical approaches based on the plate theory are complex and require a lot of mathematical manipulations. In the present investigation, a simplified method obtained by using plate theory is proposed for analyzing continuous plates over rigid supports. This simplified method for continuous plates are obtained by considering the force method technique, introducing hinge at intermediate supports and apply redundant moments at those supports to substitute the removed restraints. Afterwards, a three-moment equation and seven-moment equations for continuous plates are derived by using superposition of the Lévy's solution and the Navier's one as deflections of plates, and consideration of the continuity of the slope at adjoining edges. To confirm the validity of the present method, numerical examples are performed and the results are compared with those by Finite Difference Method and solutions obtained by means of Marcus Table.

Keywords Continuous slab plates · Three-moment equation · Seven-moment equations

1 Introduction

The reinforced concrete slabs are among the most common structural elements, but despite the large number of slabs designed and built, the details of the elastic behavior of slabs are not always appreciated or properly taken into account. This occurs at least partially because of the mathematical complexities of dealing with elastic plate equations, especially for support conditions which realistically approximate those in multi-panels building floor slabs [1]. Therefore, since, the classical analysis based on

F. Pereira (✉) · Y. Suetake
Construction and Environmental Engineering, Ashikaga University, Ashikaga, Japan

S. S. Muhammed
Construction Technology and Managements, Wolaita Sodo University, Sodo, Ethiopia

T. Kang and Y. Lee (eds.), *Proceedings of 2021 4th International Conference on Civil Engineering and Architecture*, Lecture Notes in Civil Engineering 201,
https://doi.org/10.1007/978-981-16-6932-3_26

the plate theory is quite cumbersome, numerical or engineering methods are usually used to obtain usable approximate solutions of these plate problems [2]. However, it is expected that the analytical solution based on plate theory is more accuracy and much closer to the exact solution than those approximate methods [3]. Continuous slab plate has been research object for many researchers either by numerical methods or approximate methods as given in Refs. [4–10].

Continuous plates are externally statically indeterminate members and the well-known methods developed in structural mechanics can be used for their analysis [11]. The classical methods used in the analysis of continuous plates fall into two distinct categories, force and deformation methods. Following either approach, the continuous plate is subdivided along the intermediate supports into single-span panels. The analysis is based on the equilibrium conditions of the individual panels and the compatibility of displacements at the adjoining edges [2]. In this paper, force method technique is considered.

2 Governing Equation of Plates and Its Analytical Solutions

The theory of plate is based in integration of Lagrange differential equation also known as governing equation of plate. The governing equation of thin plates based on Kirchhoff's theory can be written as Eq. (1), where D represents the flexural rigidity, P the lateral load and w the deflection of plate. If the w in the governing equation of plate is found, then, bending moments and other parameters can be calculated.

$$\frac{\partial^4 w}{\partial x^4} + 2\frac{\partial^4 w}{\partial x^2 \partial y^2} + \frac{\partial^4 w}{\partial y^4} = -\frac{P}{D} \tag{1}$$

In this paper, two solutions for the governing equation of plate (Eq. 1), are assumed. One is the double trigonometric series given by Eq. (2). This is well-known Navier's solution which represents deflections of simply supported rectangular plates [2].

$$w(x, y) = \sum_{j=1}^{\infty}\sum_{k=1}^{\infty} W_{jk} \sin\frac{j\pi x}{a} \sin\frac{k\pi y}{b} \tag{2}$$

The other solution is the Lévy's method given as Eq. (3). This represents the homogeneous solutions for the deflections of rectangular plates with two opposite edges parallel to y axis simply supported, and the others two opposite edges parallel to x axis with arbitrary conditions, where the constants are determined by prescribed boundary conditions [11].

$$w_h(x, y) = \sum_{j=1}^{\infty} \left(\left(A_j + B_j y\right) \sinh \frac{j\pi y}{a} + \left(C_j + D_j y\right) \cosh \frac{j\pi y}{a} \right) \sin \frac{j\pi x}{b} \quad (3)$$

3 Continuous Plate Over Rigid Support

3.1 Continuous Plate in One Direction

Let us consider two successive panels of a continuous plate in one direction represented in Fig. 1. To solve the problem, first, fictitious hinges at intermediate supports are introduced, and redundant moments are used to replace the removed restraints. When the lateral loads are symmetric to the center of the individual panels, the distribution of the moments along the edges $i-1$, i and $i+1$ (Fig. 1) can be represented by the simple sinusoidal functions [2], as shown in Eq. (4).

$$M_y^{(i-1)}(x) = \sum_j^{\infty} M_j^{(i-1)} \sin \frac{j\pi x}{a}; \quad M_y^{(i)}(x) = \sum_j^{\infty} M_j^{(i)} \sin \frac{j\pi x}{a};$$
$$M_y^{(i+1)}(x) = \sum_j^{\infty} M_j^{(i+1)} \sin \frac{j\pi x}{a}. \quad (4)$$

The analysis is based on expressing slope continuity at the adjoining edges, as in the classical analysis of continuous beams [2]. According to Fig. 1, the expression

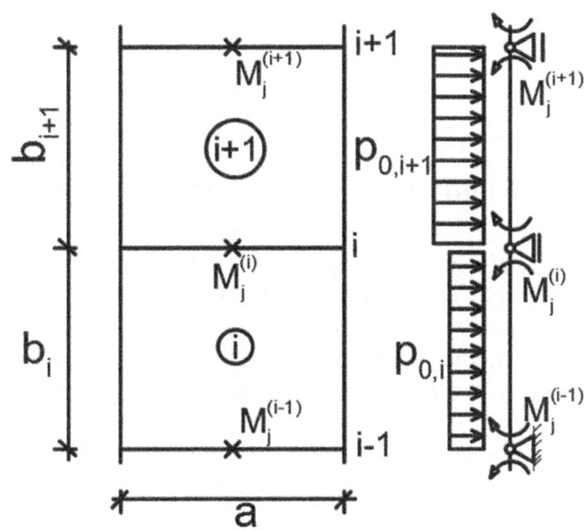

Fig. 1 Internal panels of a continuous plate in one direction

of compatibility between panels i and $i + 1$ at adjoining support can be expressed as Eq. (5).

$$\left(\frac{\partial w_i}{\partial y_i}\right)_{y_i=b_i} = \left(\frac{\partial w_{i+1}}{\partial y_{i+1}}\right)_{y_{i+1}=0} \tag{5}$$

Since, the governing equation of plate is linear, the equation of compatibility are formulated by using the superposition principle [2], where the deflections of individual panel are given as summation of the solution proposed by Navier for simply supported plates under lateral load and Lévy solution for rectangular plates with two opposite edges simply supported and the other two edges subjected to distributed edge moments. In Ref. [3], the author used same approach, for lateral uniform distributed load and found Eq. (6), which is known as three-moment equation for continuous plate in one direction. Applying the three-moment equation between each two successive panels and solve the system of equations, the unknow redundant moments can be found. Consequently, the deflection and stress of plate at any point can be computed.

$$\beta_{j,i} M_j^{(i-1)} + (\alpha_{j,i} + \alpha_{j,i+1}) M_{1,j}^{(i)} + \beta_{j,i+1} M_j^{(i+1)} = -\Theta_{j,i+1} - \Theta_{j,i} \tag{6}$$

where,

$$\alpha_{j,i} = \sum_j \frac{1}{D\lambda_{j,i} sh_{j,i}} \left(ch_j - \frac{\lambda_j b}{sh_{j,i}}\right); \ \beta_{j,i} = \sum_j \frac{1}{D\lambda_{j,i} sh_{j,i}} \left(\frac{\lambda_{j,i} b_i}{th_{j,i}} - 1\right);$$

$$\Theta_{j,i} = \frac{32}{\pi^6} \frac{p_{0,i} a_i^4}{D} \sum_{j=1}^{\infty} \sum_{k=1}^{\infty} \mu_{k,i} \frac{\sin \lambda_{j,i} x_i}{jk\left(j^2 + k^2/\gamma^2\right)^2}; \ sh_{j,i} \equiv \sin(\lambda_{j,i} b_i); \tag{7}$$

$ch_{j,i} \equiv \cos(\lambda_{j,i} b_i)$; $\mu_{k,i} = k\pi/b_i$; $\lambda_{j,i} = j\pi/a$; $j = 2m - 1$; $k = 2n - 1$; $m, n = 1, 2, 3\ldots$

3.2 Continuous Plate in Two Directions

The analogous procedure adopted in continuous plate in one direction, is used to continuous plate in two directions. Hinges and redundant moments are introduced in adjoining edges in all directions. The continuity of slope can be considered only at the middle of each adjoining edge [2]. Considering Fig. 2 the interior panels of continuous plate in two directions, under uniformly distributed load and taking the continuity of slope in the middle of adjoining edges between panels i and $s + 1$, and also between panels i and $i + 1$, it resulted in Eqs. (8) and (9), which is called here "seven-moment equations" for continuous plates.

Fig. 2 Internal panels of a continuous plate in two directions

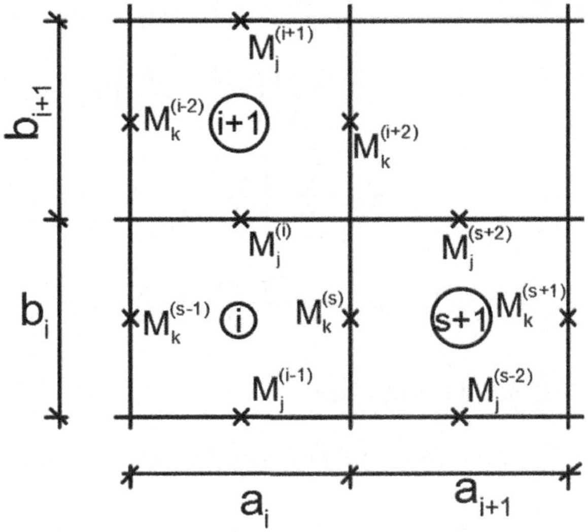

$$\beta_{k,i}M_k^{(S-1)} + \left(\alpha_{k,i} + \alpha_{k,S+1}\right)M_k^{(S)} + \beta_{k,S+1}M_k^{(S+1)} + \phi_{j,i}M_j^{(i-1)} + \psi_{j,i}M_j^{(i)} + \phi_{j,S+1}M_j^{(S-2)} + \psi_{j,S+1}M_j^{(S+2)} = -\Theta_k^{(S+1)} - \Theta_k^{(i)} \tag{8}$$

$$\beta_{j,i}M_j^{(i-1)} + \left(\alpha_{j,i+1} + \alpha_{j,i}\right)M_j^{(i)} + \beta_{j,i+1}M_j^{(i+1)} + \phi_{k,i}M_k^{(S-1)} + \psi_{k,i}M_k^{(S)} + \phi_{k,i+1}M_k^{(i-2)} + \psi_{k,i+1}M_k^{(i+2)} = -\Theta_j^{(i+1)} - \Theta_j^{(i)} \tag{9}$$

where,

$$\alpha_{j,i} = \sum_j \frac{1}{D}\cdot\frac{(-1)^{m-1}}{\lambda_{j,i}sh_{j,i}}\left(ch_{j,i} - \frac{\lambda_{j,i}b_i}{sh_{j,i}}\right);\ \beta_{j,i} = \sum_j \frac{1}{D}\cdot\frac{(-1)^{m-1}}{\lambda_{j,i}sh_{j,i}}\left(\frac{\lambda_{j,i}b_i}{th_{j,i}} - 1\right);$$

$$\phi_{k,i} = \sum_k \frac{1}{D}\left\{\left(\frac{-a_i}{sh_{k,i}^2} - \frac{a_i}{2}\right)sh_{0.5k,i} + \frac{a_i}{2th_{k,i}}ch_{0.5k,i}\right\};$$

$$\psi_{k,i} = \sum_k \frac{1}{D}\left\{\frac{a_i}{sh_{k,i}th_{k,i}}\mathrm{sh}_{0.5k,i} - \frac{a_i}{2sh_{k,i}}\mathrm{ch}_{0.5k,i}\right\};$$

$$\Theta_j = \frac{32}{\pi^6}\frac{p_{0,i}a_i^4}{D}\sum_{j=1}^{\infty}\sum_{k=1}^{\infty}\mu_{k,i}\frac{(-1)^{m-1}}{jk\left(j^2 + k^2/\gamma^2\right)^2}$$

$$\alpha_{k,i} = \sum_k \frac{1}{D}\cdot\frac{(-1)^{n-1}}{\mu_{K,i}sh_{k,i}}\left(ch_{k,i} - \frac{\mu_{K,i}a_i}{sh_{k,i}}\right);\ \beta_{k,i} = \sum_k \frac{1}{D}\cdot\frac{(-1)^{n-1}}{\mu_{K,i}sh_{k,i}}\left(\frac{\mu_{K,i}a_i}{th_{k,i}} - 1\right) \tag{10}$$

And,

$$\phi_{j,i} = \sum_j \frac{1}{D}\left\{\left(\frac{-b_i}{sh_{j,i}^2} - \frac{b_i}{2}\right)\text{sh}_{0.5j,i} + \frac{b_i}{2th_{j,i}}\text{ch}_{0.5j,i}\right\};$$

$$\psi_{j,i} = \sum_j \left\{\frac{b_i}{sh_{j,i}th_{j,i}}\text{sh}_{0.5j,i} - \frac{b_i}{2sh_{j,i}}\text{ch}_{0.5j,i}\right\};$$

$$\Theta_{k,i} = \frac{32}{\pi^6}\frac{p_{0,i}a_i^4}{D}\sum_{j=1}^{\infty}\sum_{k=1}^{\infty}\lambda_{j,i}\frac{(-1)^{n-1}}{jk\left(j^2 + k^2/\gamma^2\right)^2} \tag{11}$$

It should be noted that:

$$sh_j \equiv \sin(\lambda_j b);\ ch_j \equiv \cos(\lambda_j b);\ sh_k \equiv \sin(\mu_k a);\ ch_k \equiv \cos(\mu_k a);$$

$$sh_{0.5j} \equiv \sin(\lambda_j b/2);\ ch_{0.5j} \equiv \cos(\lambda_j b/2);\ sh_{0.5k} \equiv \sin(\mu_k a/2); \tag{12}$$

$$ch_{0.5k} \equiv \cos(\mu_k a/2);\ \lambda_j = j\pi/a;\ \mu_j = k\pi/b;\ \gamma = b/a$$

$j = 2m - 1$; $k = 2n - 1$; $m, n = 1, 2, 3\ldots$

For the panel i (represented in Fig. 2) with the origin of coordinate located at left-down corner, the deflection is represented by Eq. (13), where the stresses can be found using the relation between stresses and the deflections. This expression can be generalized for the other panels.

$$\begin{aligned} w_i(x, y) &= \sum_j \left\{(A_{j,i} + B_{j,i}y)\sinh\lambda_{j,i}y + (C_{j,i} + D_{j,i}y)\cosh\lambda_{j,i}y\right\}\sin\lambda_{j,i}x \\ &+ \sum_j \left\{(E_{K,i} + F_{K,i}x)\sinh\mu_{K,i}x + (G_{K,i} + H_{K,i}x)\cosh\mu_{K,i}x\right\}\sin\mu_K y \\ &+ \frac{16}{\pi^6}\frac{p_{0_i}a_i^4}{D}\sum_{j=1}^{\infty}\sum_{k=1}^{\infty}\frac{\sin\lambda_{j,i}x\cdot\sin\mu_{k,i}y}{jk(j^2 + k^2/\gamma^2)^2} \end{aligned} \tag{13}$$

where

$$A_{j,i} = -\frac{B_i}{2D\lambda_{j,i}sh_{j,i}^2}\left(M_j^{(i-1)} - ch_{j,i}M_j^{(i)}\right);\ B_{j,i} = -\frac{M_j^{(i-1)}}{2D\lambda_{j,i}};$$

$$D_{j,i} = \frac{1}{2D\lambda_{j,i}sh_{j,i}}\left(ch_{j,i}M_j^{(i-1)} - M_j^{(i-1)}\right);\ C_{j,i} = 0;$$

$$E_{k,i} = -\frac{a_i}{2D\mu_{k,i}sh_{k,i}^2}\left(M_k^{(S-1)} - ch_{k,i}M_k^{(S)}\right);\ F_{k,i} = -\frac{M_k^{(S-1)}}{2D\mu_{k,i}}$$

$$H_{k,i} = \frac{1}{2D\mu_{k,i}sh_{k,i}}\left(ch_{k,i}M_k^{(S-1)} - M_k^{(S-1)}\right) \text{ and } G_{k,i} = 0 \tag{14}$$

4 Numerical Application, Results and Discussions

In order to evaluate the validity of the proposed method (PM), four numerical examples are performed, as shown in Fig. 3, and the results are compared with solutions from two others methods. The models (Fig. 3) are reinforced concrete slabs over rigid support, having same flexural rigidity, D, and subjected to distributed uniform load P. The Poisson's ratio (ν) adopted in the calculations are zero, since, according

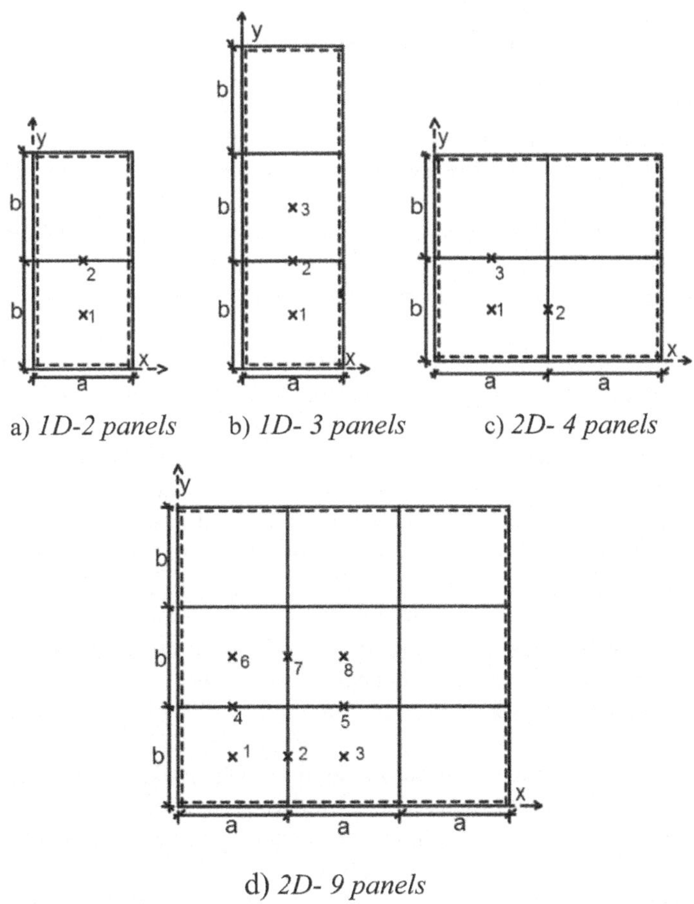

Fig. 3 Models of numerical analysis

Table 1 Deflections and bending moments in terms Pa^4/D and Pa^2 respectively, at middle span, for b/a = 1 and $v = 0$, with increment of m and n (PM)

m = n	w	M_x	M_y
1	0.0029421	0.029037	0.036394
2	0.0027891	0.023679	0.031187
3	0.0027902	0.024619	0.032164
4	0.0027867	0.024221	0.031775
5	0.0027864	0.024391	0.031947
6	0.0027859	0.024290	0.031848
7	0.0027858	0.024347	0.031906
8	0.0027857	0.024308	0.031867
9	0.0027856	0.024334	0.031893
10	0.0027856	0.024315	0.031874
20	0.0027855	0.024321	0.031881
30	0.0027855	0.024322	0.031881
50	0.0027855	0.024322	0.031882

to Ref. [1], when dealing with reinforced concrete slabs, the Poisson's ratio has often taken value of zero, because all slabs become cracked when the load approaches to the collapse load.

To solve the models (Fig. 3) the redundant moments at supports are determined, applying the three-moment and the seven-moment equations. Thus, for each panel, deflections are obtained, consequently, bending moments in terms of Pa^2 could be also computed.

When numerical calculation is performed by using trigonometric series, we have to confirm whether the series solutions are convergent or not. Therefore, convergence of the series should be tested also for the present method. In this paper, the models in Fig. 3a are adopted in order to test numerical convergence of the solution. Table 1 shows the variation of deflections and bending moments at middle span with the increment of the number of terms, for aspect ratio $b/a = 1$ and $v = 0$. We should mention that the displacement w is given in terms of Pa^4/D and bending moments in terms of Pa^2.

If we compare the results obtained by taking $m = n = 1$ and $m = n = 50$, the percentage differences are 5.32%, 16.24% and 12.43% respectively for deflection, bending moments with respect to x axis and bending moments with respect to y axis. Depending on accuracy that each problem requires, the number of terms can be taken less or more. According to Table 1 it can be noted that results by taking $m = n = 1$ are indeed, higher than the exact solution therefore, when taking only one terms when designing slabs, we are in safety side. Comparing the results taking $m = n = 10$ and $m = n = 50$, it can be seen that the results are almost the same presenting percentage difference less than 0.03%. Therefore, for the next calculations, m and n equal to 10 is adopted.

To validate the results of the proposed method and to use the method in the design of reinforced concrete slabs and others plates, two other methods are applied to the problems in Fig. 3, and the results of maximum bending moments are compared between these three methods. One of chosen methods is finite difference method which is based in plate theory where the derivatives in the governing equation is replaced by difference quantities at some selected points of the plates [2]. The other solutions are obtained by using the well-known Marcus Table.

Figures 4, 5 and 6 show graphically comparison of maximum moments obtained from the model in Fig. 3c, between the proposed method (PM) with finite difference method (FDM) and solution by means of Marcus table (MT). We should mention

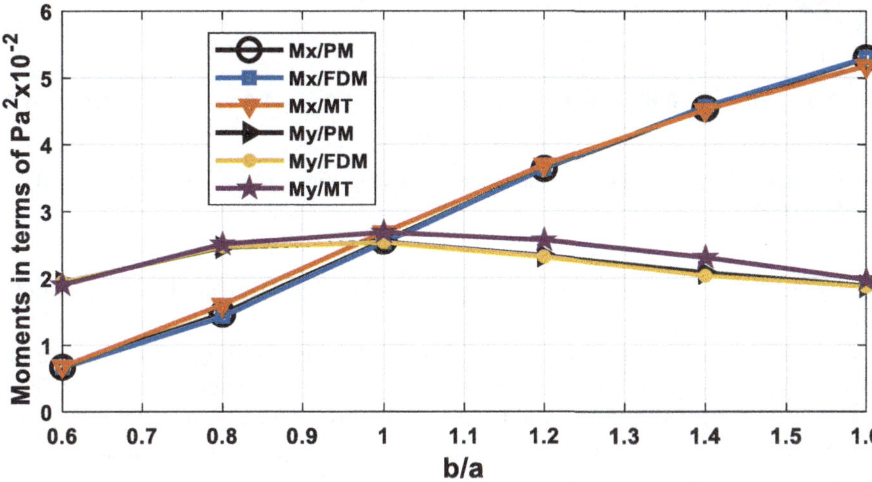

Fig. 4 Graphically representation of maximum positive moments between PM, FDM and MT of Fig. 3c at point 1

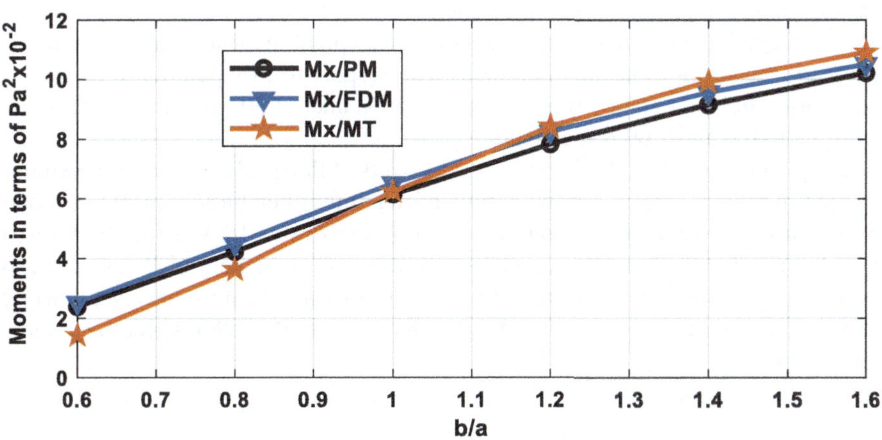

Fig. 5 Graphically representation of maximum negative moments between PM, FDM and MT of Fig. 3c at point 2

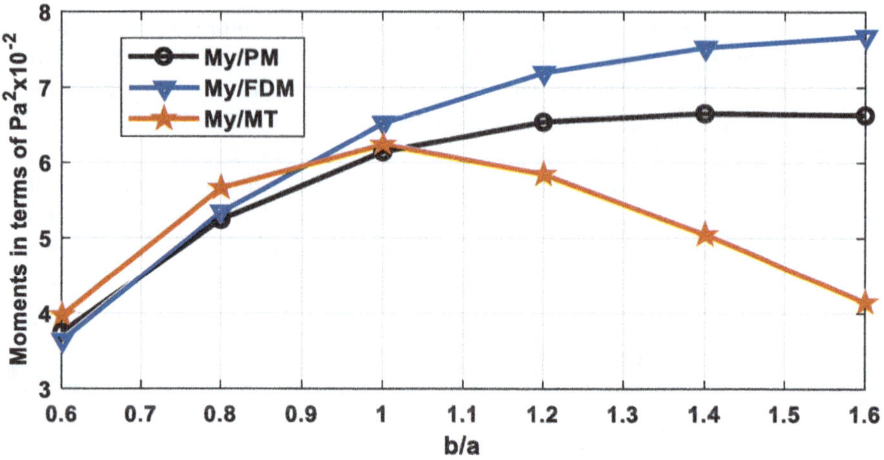

Fig. 6 Graphically representation of maximum negative moments between PM, FDM and MT of Fig. 3c at point 3

that the maximum positive moments (represented in Fig. 4) are not located at middle span.

Figures 4, 5 and 6 show good approximation between PM and FDM, where the maximum percentage difference for positive and negative moments found are 2.72% and 13.71% respectively. We should mention that for implementing FDM, the slabs were divided into square mesh of width of $a/10$. Dividing the mesh in much smaller elements even better approximation to PM can be achieved.

Regarding to the solution by means of Marcus table the results have been found to be less accurate especially, maximum negative moments with respect to the axis parallel to longer dimension for aspect ratios b/a not close to 1 (one). In this case, it was found maximum percentage difference of 9.95% for maximum positive moment, 7.7% for negative moments with respect to the axis parallel to the smaller dimension. However, regarding to negative moment with respect to axis parallel to the longer dimension, the percentage difference are 39.89% and 37.30% for the aspect ratios b/a equal to 0.6 and 1.60 respectively.

It might be of interest also to analyze the effects of Poisson's ratio to the maximum bending moments. Table 2 shows results of maximum bending moments of model in Fig. 3a for $a = b$ and different value of Poisson's ratio, by PM for $j = k = 10$. It is obvious that the bending moments increase with increment of Poisson's ratio, however, at intermediate support the Poisson's ratio has influenced only Mx in which the value can be obtained by multiplying bending moment with respect to y axis by Poisson's ratio.

Table 2 Maximum moments in terms Pa^2 for b/a = 1 with respect to v (PM)

v	Point 1		Point 2	
	M_x	M_y	M_x	M_y
0	0.025018	0.034051	0	−0.083865
0.1	0.028392	0.036522	−0.008387	−0.083865
0.2	0.031767	0.038996	−0.016773	−0.083865
0.3	0.035153	0.041473	−0.025160	−0.083865
0.4	0.038539	0.043958	−0.033546	−0.083865
0.5	0.041926	0.046443	−0.041933	−0.083865

5 Conclusions

In this work simple formulae are derived for analyzing bending problems of continuous slab plates. These equations are derived using force method technique, similar to that used to derive the well-known three-moment theorem for continuous beams. To do that, hinge at intermediate supports are introduced, redundant moments at those supports are placed to substitute the removed restraints, and the deflection of each individual plate is taken as superposition of Lévy and Navier's solutions. Taking the continuity of the slope at adjoining edges, three-moment and seven-moment equations are derived for analyzing continuous slab plates.

For the numerical examples presented in this work, the following conclusion can be drawn:

(a) The proposed method has a good accuracy when comparing it with the results of finite difference method. The percentage difference between the results of those two methods show to be low. Therefore, we can expect that the proposed method can be utilized in designing of continuous slab plates.
(b) When compare the results of proposed method with the Marcus table, significant percentage difference in bending moments at supports with respect to the axis parallel to longer directions when the aspect ratio b/a of the panel is not closed to one. However, for all aspect ratio b/a, comparison between the positive moments and moments at support with respect to axis parallel to the shorter dimension or main direction, good agreements are observed i.e. percentages difference is low. Therefore, for the statistic the results are considered acceptable.
(c) Deflection and stresses for the proposed method are trigonometric series and from the numerical examples it is evident that the convergence of Serie is fast, then, good approximation can be achieved by taking few numbers of terms, for the case of distributed uniformly lateral load.

5.1 *Suggestions for Next Studies*

In this paper, the lateral loads are restricted to only a uniformly distributed load and exterior support of panels are considered simply supported. For next studies, the following suggestions are given by the authors:

(a) Study of continuous slab plates under other types of lateral load as for example triangular load, parabolic or others.
(b) Adaptation of the propose method for analysing continuous plate slabs with exterior edges free.
(c) Development of computer programs based on the theory proposed in this paper, using software as Matlab, capable to simulate continuous slab plates with different panel dimension and several spans.

References

1. Park R, Gamble W (2000) Reinforced concrete slabs, 2nd edn. Wiley, New York
2. Szilard R (2004) Theories and applications of plate analysis. Wiley, Hoboken, NJ
3. Muhammed SS (2017) A study of design bending moments for reinforced concrete continuous slab plates. Master thesis submitted to Construction and Environmental Engineering Department, Ashikaga University, Japan
4. Adah I, Ibearugbulemo, Ubi S, Idagu F (2021) Analysis of a two-way continuous plate based on beam analogy. IOSR J Mech Civ Eng (IOSR-JMCE) 18(1 Ser. I):59–67
5. Ohga M, Shigematsu T (1990) Analysis of continuous plates by a combined boundary element-transfer matrix method. Comput Struct 36(1):81–89
6. Adah I (2016) Development of computer programs for analysis of single panel and continuous rectangular plates. Master thesis submitted to Postgraduate School Federal University of Technology, Owerri
7. Timoshenko S, Woinosky-Krieger S (1959) Theory of plates and shells. McGraw-Hill Book Company Inc., New York
8. Dallam LN (1966) Analysis of continuous rectangular plate systems by the slope deflection method. Master thesis submitted to the Faculty of the Graduate School of the Oklahoma State University
9. Golley B, Petrolito J (1984) Method for analyzing tanks and continuous plates. Comput Struct 18(6):1141–1151
10. Alharazin BS (2018) Comparative study of analysis of two-way solid RC slab using approximate and finite element methods. Master thesis submitted the Islamic University, Gaza
11. Ventsel E, Krauthammer T (2001) Thin plate and shell: theory analysis and application. Dekker Inc., New York

Architecture and Environmental Engineering

Utilisation of Organic Waste Materials as Biocover for Landfills

S. Rahim, M. Jeludin, S. Shams, and M. H. Isa

Abstract Brunei Darussalam is Asia's second-largest generator of municipal solid waste (MSW) (1.46 kg/capita/year). MSW in Brunei, is mainly disposed at the Sungai Paku landfill. The landfill receives about 0.189 million tonnes of waste per year. By 2025, the Sungai Paku landfill is estimated to have exceeded its design capacity. The current landfill uses clay as daily cover. The objective of this study is to look for potential daily cover by varying biocover mixtures as an alternative landfill cover and evaluate its geotechnical properties for landfill applications. Biocover is an effective method to minimise methane (CH_4) emissions from landfills through natural processing of microbial oxidation of CH_4. Experiments were carried out in accordance with BS 1377:1990 to study the organic content, liquid limit, compaction, consolidation and strength properties of biocovers. The biocover mixtures (compost-clay, recycled soil-clay and crushed rice husk-clay mixtures) were investigated at percentage mixtures of biocovers at 25, 50 and 75%. These organic waste materials are abundant and inexpensive in Brunei Darussalam. The results had shown that using 25% mixtures of compost and crushed rice husk, and 50% mixture of recycled soil into the clay had significant effect on the compaction, consolidation and strength properties.

Keywords Landfill · Methane emissions · Solid waste · Biocover

1 Introduction

The amount of municipal solid waste (MSW) received at Sungai Paku landfill in Tutong District of Brunei Darussalam has increased from 0.184 million tonnes in 2015 to 0.245 million tonnes in 2019, with a population of 459,500 people and a growth rate of 3.9% [1, 2]. The total amount of municipal solid waste received from Sungai Paku landfill was 0.184 million tons in 2015, 0.176 million tons in 2016, 0.174 million tons in 2017, 0.205 million tons in 2018 and 0.245 million tons in

S. Rahim (✉) · M. Jeludin · S. Shams · M. H. Isa
Civil Engineering Programme, Universiti Teknologi Brunei, Tungku Highway, Gadong BE1410, Brunei Darussalam
e-mail: P20190001@student.utb.edu.bn

T. Kang and Y. Lee (eds.), *Proceedings of 2021 4th International Conference on Civil Engineering and Architecture*, Lecture Notes in Civil Engineering 201,
https://doi.org/10.1007/978-981-16-6932-3_27

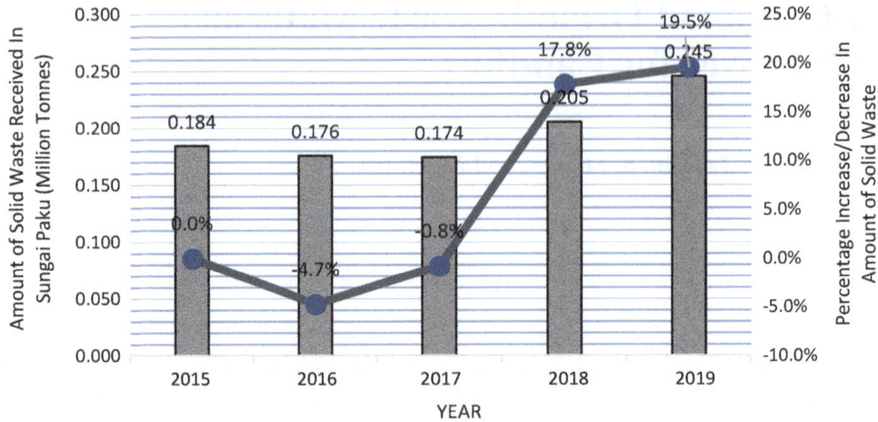

Fig. 1 Total collection of solid waste from the Sungai Paku landfill. *Source* JASTRE [2]

2019 [2], as shown in Fig. 1. This is due to rapid urbanisation, population growth, industrialisation and improved lifestyle [3]. Sungai Paku is expected to exceed its design capacity by 2025 to an estimated amount of 4.5 million tonnes, which is 23 times of the present volume of MSW generated. Currently, Sungai Paku landfill collects municipal solid wastes from 3 out 4 Brunei Darussalam Districts. The total capacity of Sungai Paku landfill is nearly four million tonnes as of June 2019 [2]. The landfill at Sungai Paku uses clay materials as daily cover.

The Intergovernmental Panel on Climate Change (IPCC) (2001) [4] reports that household-generated municipal solid waste is considered as the third largest (19%) anthropogenic source of methane emissions. Globally, landfills can be responsible for 19% of CH_4 emissions into the atmosphere [4]. Without proper waste management, waste collected in the landfill will keep on decomposing and releasing methane. This would create harmful impacts to our surrounding environment because methane has 23 times higher global warming potential (GWP) than carbon dioxide. Mitigation actions using organic waste materials as biocovers are recommended to reduce methane emissions and leachate generation due to rainwater infiltration in landfill.

Biocover helps to control the oxidation rate of methane through gas exchange process, where they rely on a group of bacteria living in cover soil to oxidise methane to carbon dioxide. Daily cover system is considered as one of the important components in landfill as the system separates waste from the surface environment, control infiltration of water into the waste, and limits release of gas from the waste. Biocovers must be physically stable and have sufficient shear strength. Sudden changes in the stress in biocovers lead to consolidation settlement, reduction in porosity and development of cracks in the biocover [5, 6]. Good knowledge of engineering properties of cover materials is required for the appropriate design of biocover.

Many researchers have used organic materials as biocovers for landfills. These include paper mill sludge, rice husk, waste tire chips, and waste sludge [6–10]. It has been shown that the mixture of organic waste materials and soil affect both the soil's moisture content and maximum dry density. The plasticity index of the soil decreased significantly, which indicates high shear strength of the soil with the increase in organic waste mixtures into the soil [6–8]. In terms of compaction properties, the value of optimum moisture content had increased, and the maximum dry density value decreased with increasing organic waste mixture into the soil [6–9]. In terms of the consolidation properties, increasing the percentage of organic waste materials to 50% in clay increased the permeability coefficient by up to 100% organic waste mixture into the clay [6–10].

2 Experimental Programme

2.1 Materials Used

Natural clay: The soil studied in this experiment was obtained from the Sungai Paku Landfill site, Tutong District, Brunei Darussalam. The basic characteristics of clay used as daily cover for landfill in this experiment are documented in Table 1. The soil is categorised as clay with intermediate plasticity (CI) according to the British Soil Classification System (BSCS) [11].

Compost (green waste): Compost used in this experiment was obtained from A.E AgroMart, Kampung Kilanas, Brunei Muara District, Brunei Darussalam. It is one of the organic waste materials studied in this research to be mixed with clay. It is a product of decomposition of organic municipal solid waste fermented from plant waste such as dead leaves, kitchen scraps and skins of fruits which cannot be processed like other wastes. One of the advantages of using compost as a biocover is it can reduce bad odours. Besides, it is environmentally friendly, pollution-free and economical. In Brunei Darussalam, about 38% of the waste comes from kitchen or food waste [2], as illustrated in Fig. 2, therefore, one way to reduce this waste is by recycling this organic material through composting. Furthermore, compost is excellent for use as a daily cover because it can oxidise the CH_4 biologically better than ordinary soil. Compost supports the growth of methanotrophic (methane oxidising) bacteria.

Table 1 Properties of clay obtained from Sungai Paku

Clay properties	Values
Specific gravity, G_s	2.70
Liquid limit, W_{LL} (%)	39.2
Plastic limit, W_{PL} (%)	22.5
Optimum moisture content, W_{opt} (%)	17.2
Maximum dry density, ρd_{max}	17.95

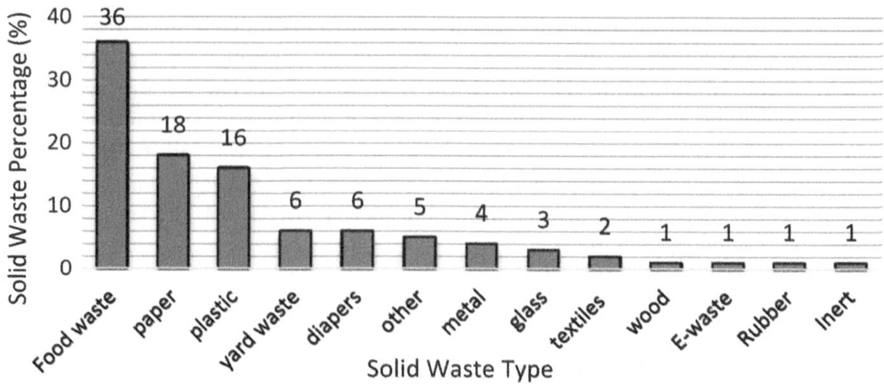

Fig. 2 Brunei's composition of solid waste. *Source* JASTRE [2]

Crushed rice husk: Rice husk, also known as rice hull, is the coating on a seed or rice grain. It is formed from hard materials such as silica and lignin to protect the seed during the growing season. According to Department of Agriculture and Agrifood statistics (2017), Brunei Darussalam produces about 2307.62 metric tonnes of paddy per year [5]. Rice husk contributes about 20% of the total rice crop [6]. Hence, the suitability of rice husk as a biocover is investigated as it is considered a valuable organic waste product. Furthermore, rice husk is environmentally friendly, economical and available in abundance in Brunei Darussalam.

Recycled soil: Recycled soil is a treated soil obtained from a local petroleum storage facility that was contaminated through the seepage of hydrocarbon into the ground. Consequently, a remediation project on the contaminated site was proposed using low-temperature thermal desorption (LTTD) technology. LTTD is an ex-situ remedial technology that uses heat to physically separate petroleum hydrocarbons from excavated soils. It is designed to heat soils to temperatures sufficient (of up to 649 °C) to cause constituents to volatilise and desorb from the soil [10]. After leaving the desorber, treated soils were cooled down, re-moistened to control dust, carbonised, and stabilised to prepare them for disposal or reuse. Hence, instead of disposing the treated soil, the soil can be recycled and used as a biocover for landfill.

2.2 *Sample Preparation*

Natural clay: In this experiment, the collected soil sample, in accordance with BS 1377:1990, was first oven-dried at 105 °C for 24 h and crushed into a powder form using a hammer. After air-drying, it was ground into a fine powder and filtered through a 425 μm sieve.

Table 2 Percentage of sample variant

Test No.	Sample variant	Test No.	Sample variant
1	100% natural clay (control)	6	50% crushed rice husk and 50% clay
2	25% compost and 75% clay	7	75% crushed rice husk and 75% clay
3	50% compost and 50% clay	8	25% recycled soil and 75% clay
4	75% compost and 75% clay	9	50% recycled soil and 50% clay
5	25% crushed rice husk and 75% clay	10	75% recycled soil and 25% clay

Waste materials: Compost and crushed rice husk and recycled soil used for in this experiment were first oven-dried at 105 °C for 24 h [12]. After waste materials were oven-dried, they were crushed into a powdered form and then filtered using a 425 μm sieve size. The soil sample (clay), waste materials, and water were then mixed thoroughly, and left for 24 h inside airtight containers. The mixtures were then prepared for organic content test, Atterberg limit tests, standard Proctor compaction tests, one-dimensional oedometer consolidation tests and unconsolidated undrained triaxial tests.

2.3 Testing Programme

All tests performed were carried out in accordance with BS 1377:1990. The tests were conducted on a varying percentage of 0, 25, 50 and 75% of organic waste materials mixtures to the dry unit weight of clay (shown in Table 2). Percentage mixtures of up to 75% were investigated in this experiment as previous studies indicate that 100% mixture of waste materials are not suitable to use as biocover for landfill [3–6].

3 Results and Discussion

This section elaborates the results of the performed experimental investigations to study the geotechnical properties of organic waste materials mixed with natural clay. The effects on the organic content, Atterberg limit characteristics, compaction characteristics, consolidation characteristics and strength characteristics were evaluated.

3.1 Organic Content Test

Ignition of a dry test sample was used to determine the organic content in the waste materials. It was observed that rice husk has the highest amount of organic content with a value of 80.35%, followed by recycled soil, with a value of 69.42%. Compost

Fig. 3 Effect of organic waste materials as biocovers on organic contents (%)

has an organic content of 64.09% (Fig. 3). Previous researches have shown that waste materials with a high organic content are more likely to encourage the growth and development of methanotrophic bacteria [7, 11]. Organic matter provides the bulk of nutrients in the soil, more dissolved organic carbon may be present in the soil, and help to increase the rate of methane oxidation [13–16].

3.2 *Atterberg Limit Test*

Figure 4 shows the liquid limit (LL) results at different percentage mixtures of compost, recycled soil and crushed rice husk into the clay. It was found that the value of LL of compost-natural clay mixture increased with an increasing compost content from 29.20% at 25% compost content and to 33.00% at 50% compost content. The LL value rose further to 39.00% at 75% compost content. Similarly, the LL values of recycled soil-natural clay mixtures at 25 and 50% were slightly higher with values of 30.60 and 38.00%. The LL value rose further at 75% recycled soil content to a value of 42.00%. When crushed rice husk was used as the organic waste mixture to clay,

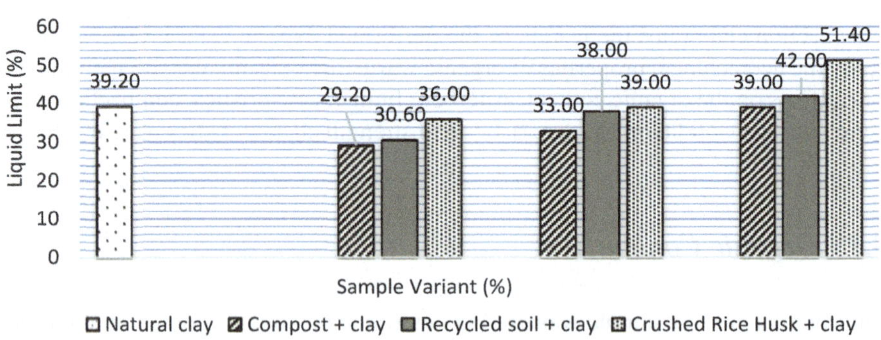

Fig. 4 Effect of organic waste materials as biocovers on percentage of liquid limit (%)

the LL values at all percentage of crushed rice husk content increased drastically with values of 36.00, 39.00 and 51.40% at 25, 50 and 75% crushed rice husk content, respectively. However, as compared to clay with LL values of 39.20%, the introduction of these organic waste materials into the clay lowered the LL values. This suggests that there is a low compressibility that could be caused by stronger bonds between waste material and soil, which increases soil compression resistance [8]. Nevertheless, due to its potential for high compressibility, the use of 75% crushed rice husk and clay mixtures as an alternative landfill cover is not recommended.

3.3 2.5 kg Standard Proctor Compaction Test

As shown in Figs. 5 and 6, the moisture content (w_{opt}) at all percentage of organic waste materials into the clay increased compared to the soil with no organic waste

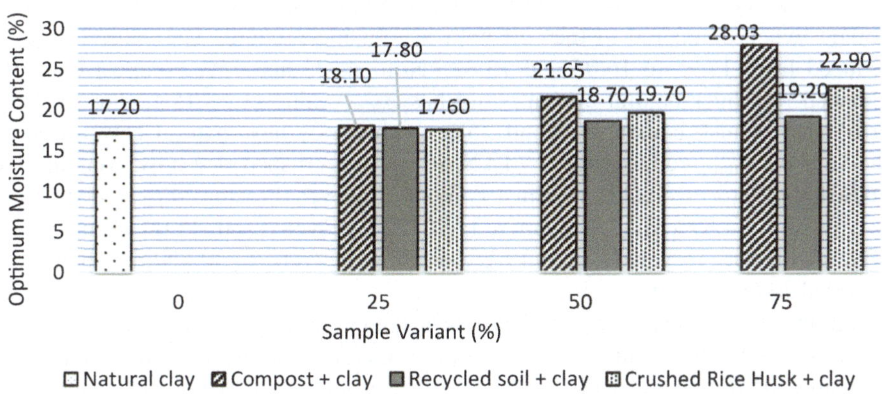

Fig. 5 Effect of organic waste materials as biocovers on optimum moisture content (w_{opt})

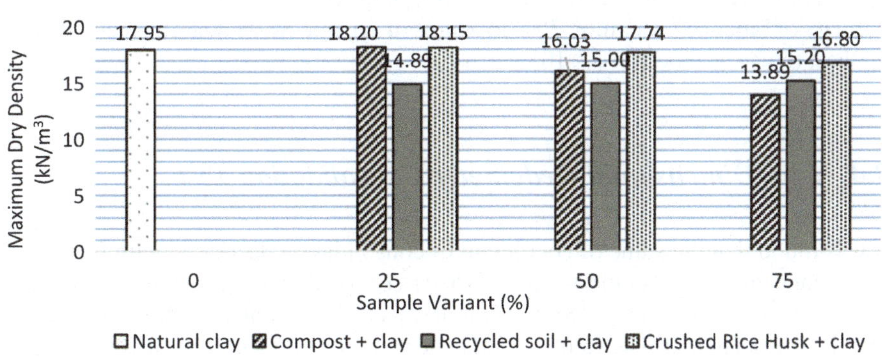

Fig. 6 Effect of organic waste materials as biocovers on maximum dry density (ρ_{dmax})

mixtures. The value of w_{opt} of compost-natural clay mixture increased from 17.20 to 18.10 and 21.65% at 25 and 50% of mixture of compost into the clay. The value of w_{opt} increased to 28.03% at 75% compost content. As shown in Fig. 5, the value of w_{opt} of recycled soil-natural clay mixture slightly increased from 17.20 to 17.80% at 25% recycled soil content and further increased to 18.70% at 50% recycled soil content and 19.20% when 75% recycled soil was used. When crushed rice husk was used as organic mixture into the clay, w_{opt} increased to a value of 17.60% at 25% crushed rice husk content, to 19.70% at 50% crushed rice husk content and 22.90% when 75% crushed rice husk was used.

These results correlate well with the observation made by Khoshand and Fall [8] where they stated that at a very high w_{opt} value, the air-filled voids are no longer interconnected and oxygen must diffuse on the liquid phase, which is about 10,000 times slower than oxygen diffusion in the gas phase. This means that the slowing down of gaseous exchange process will decrease the oxidation of CH_4.

It was discovered that the value of maximum dry density (ρ_{dmax}) increased at 25% compost-natural clay mixture, with a value of 18.20 kN/m^3 (Fig. 6). However, the ρ_{dmax} value reduced at 50% compost content, and further reduced at 75% compost content when compared to the soil with no organic mixture. The value of ρ_{dmax} reduced from 17.95 to 16.03 and 13.89 kN/m^3 at 25 and 75% compost content. Crushed rice husk-natural clay mixture has ρ_{dmax} value of 18.15 kN/m^3 at 25% crushed rice husk content. Beyond 25% crushed rice husk-natural clay mixture, the value of ρ_{dmax} reduced to 17.74 kN/m^3 at 50% crushed rice husk content and 16.80 kN/m^3 at 75% crushed rice husk content. However, the value of ρ_{dmax} decreased to a value of 14.89 kN/m^3 when mixed with 25% recycled soil, 15.00 kN/m^3 when mixed with 50% recycled soil and ρ_{dmax} value reduced to 15.20 kN/m^3 when 75% recycled soil was used.

Varghese and Prakash [6] and Hermawansyah et al. [7] obtained similar results, where they observed that the value of ρ_{dmax} increases when the soil content was increased in the mixtures with organic waste material. The increase in maximum dry density of clay mixed with 25% organic waste content into the clay was attributed to strong bonding of the soil matrix with waste materials. As the oxidation rate of CH_4 in biocovers is controlled by gas exchange process and strongly affected by the porosity of biocovers medium, 25% mixtures of organic waste materials into the clay are best to use as biocover [7–10].

3.4 One-Dimensional Oedometer Consolidation Test

It was found that the value of coefficient of consolidation decreased with increasing percentage mixtures of compost and crushed rice husk into the clay (Fig. 7). The value of c_v increased from only 1.468 m^2/year at 0% compost content to 5.460 m^2/year at 25% compost content. The value started to decrease to 3.703 m^2/year at 50% compost content and to 2.985 m^2/year at 75% compost content. When crushed rice husk was used as organic mixture into the clay, the value of c_v increased to 11.882 m^2/year at

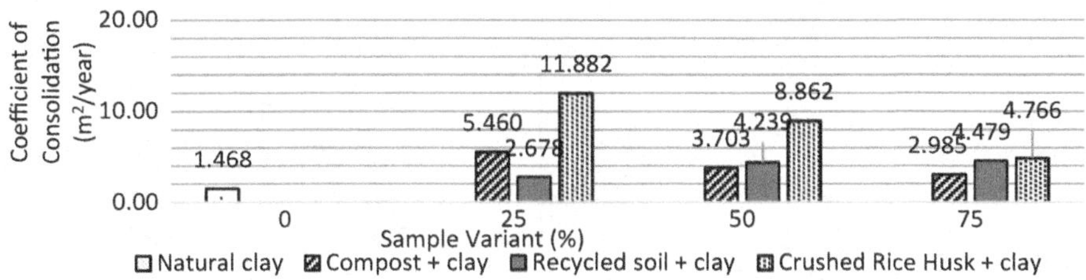

Fig. 7 Effect of organic waste materials as biocovers on coefficient of consolidation (m^2/year)

25% crushed rice husk content, and decreased to 8.862 m^2/year at 50% crushed rice husk content, and to 4.766 m^2/year at 75% crushed rice husk-natural clay mixture. However, when recycled soil was mixed with natural clay at percentage of 25%, 50% and 75%, the value of c_v increased from 2.678 m^2/year to 4.239 m^2/year and to 4.479 m^2/year, respectively.

These results were in agreement with studies by Varghese and Prakash [6] and Khoshand and Fall [8]. The increase in c_v values at 25% compost and crushed rice husk content, and at 50% recycled soil into the clay might be attributed to the better bonding between the waste material-natural clay mixture, which increased the permeability of the mixture and allowed more water to flow along the interface of waste materials and clay. Increasing the percentage of waste materials into the clay increased the coefficient of consolidation, and thus, the time required to reach 90% consolidation decreased significantly (illustrated in Fig. 8). Hence, consolidation settlement could be achieved faster at higher percentages of organic waste mixtures into the clay.

The effect of mixing organic waste mixture into the clay on the compression index values, c_c is illustrated in Fig. 9. It was observed that increasing percentage of compost content at 25% reduced the c_c values from 3.178 (soil with no organic waste mixtures) to 2.043 at 25% and slightly increased to 3.163 at 50% compost content. Similarly, using crushed rice husk-natural clay mixture decreased the value of c_c from 8.450 at 0% crushed rice husk content to 6.329 at 25% crushed rice husk content, to 3.339 at 50% and 75% crushed rice husk content. When recycled soil was used as organic waste mixture into the clay, the value of c_c increased to 2.133 at

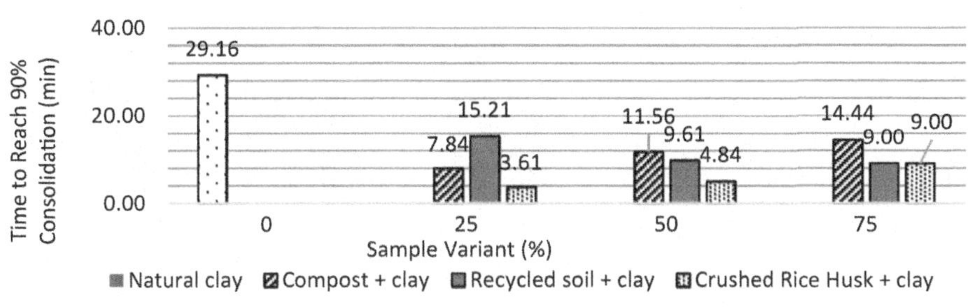

Fig. 8 Effect of organic waste materials as biocovers on time to reach 90% consolidation (min)

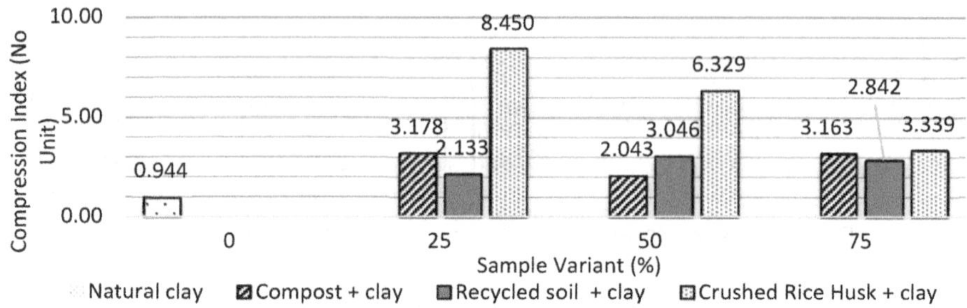

Fig. 9 Effect of organic waste materials as biocovers on compression index (no unit)

25%. However, the value started to reduce at 50 and 75% recycled soil content, with a value of 3.046 and 2.842. The result in this investigation agreed with Varghese and Prakash [6] where they stated that the value of c_c reflects the compressibility of the mixture. The increase in c_c values may be caused by stronger bonds between waste material and soil that increase the soil's compression resistance. This will further increase soil strength and reduce settlement.

It was discovered that increasing the percentage of compost and recycled soil up to 75% into the soil increased the value of permeability coefficient of the soil. According to Fig. 10, the permeability coefficient increased from 0.944×10^{-10} m/s at 0% compost content to 2.178×10^{-10} m/s (at 25% compost content), and slightly reduced to 2.043×10^{-10} m/s at 50% compost content. When 75% compost content was used, the value of k increased to a value of 2.842×10^{-10} m/s. The value of k for recycled soil-natural clay mixture increased to 2.133×10^{-10} m/s when mixed at 25% recycled soil content. Subsequently, the value increased further to 3.046×10^{-10} and 2.842×10^{-10} m/s at 50 and 75% recycled soil content. Similarly, increasing crushed rice husk content into the clay increased the value of k. The value of k increased to 8.450×10^{-10} m/s at 25 rice husk content and to 6.329×10^{-10} m/s at 50% rice husk content, and to 3.339×10^{-10} m/s at 75% rice husk content. Soil covers with low hydraulic conductivity or permeability are more suitable to use as biocover as it leads to a low amount of leachate.

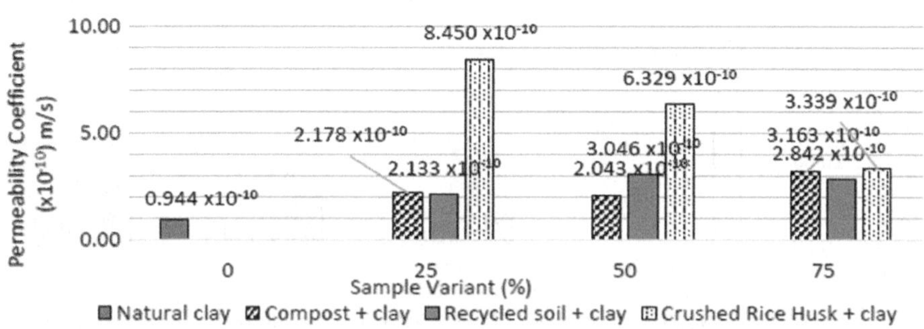

Fig. 10 Effect of organic waste materials as biocovers on permeability coefficient (m/s)

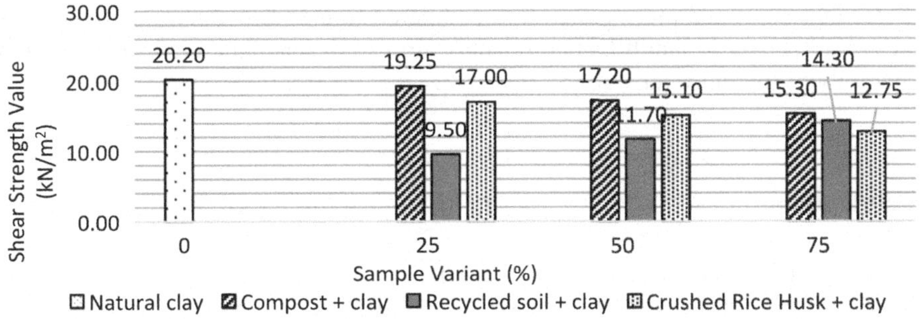

Fig. 11 Effect of organic waste materials as biocovers on shear strength value (kN/m^3)

3.5 Unconsolidated Undrained Triaxial Test

Figure 11 shows the effect of mixing the studied organic waste materials at different percentage mixtures into the clay in terms of undrained shear strength, c_u. It was found that the values of c_u were 20.20 kN/m^2 at 0% compost content and 19.25 kN/m^2 at 25% compost content. However, the c_u value slightly decreased at 50% and 75% compost content with values of 17.20 kN/m^2 and 15.30 kN/m^2, respectively. Similarly, when crushed rice husk was used as organic waste mixture into the clay, the c_u values have slightly reduced to 17.00 kN/m^2 at 25% crushed rice husk content, and to 15.10 kN/m^2 at 50% crushed rice husk content, and further reduced to 12.75 kN/m^2 at 75% crushed rice husk content. However, when 25% recycled soil-natural clay was used, the value of c_u decreased from 20.20 kN/m^2 when natural clay (with no organic waste mixture) was used to 9.50 kN/m^2 at 25% recycled soil content, to 11.70 kN/m^2 at 50% recycled soil content and 14.30 kN/m^2 at 75% mixture of recycled soil into the clay. This finding correlates with Ng and Lo [10] when using compost and crushed rice husk as organic waste mixture into the soil. The increase in strength values at 25% compost and crushed rice husk mixture into the clay might be due to all the void spaces in the waste material were sufficiently filled by the waste material-natural clay mixture. However, increasing the percentage of compost and crushed rice husk content beyond 50% into the mixture reduced the bulk density and strength of the mixture as there was not enough sludge to support and fill the void spaces in the individual waste material aggregates.

4 Conclusion

Based on the results obtained, 25% compost-clay and crushed rice husk-clay, and 50% recycled soil-clay had significant effect on the organic content, liquid limit, compaction, consolidation and strength characteristics. From the standpoint of geotechnical properties, it can be concluded that the biocover materials studied are lighter in weight and more compacted compared to clay and can be used as alternative daily cover in landfills. Future study aims to evaluate the methane oxidation capability of the biocover.

References

1. DEPD (2018) Population estimates of Brunei Darussalam for 2018, Oct
2. JASTRE (2020) Solid waste management in Negara Brunei Darussalam
3. Guo Z, Shams S, Juani RHM (2014) Integrated and sustainable solid waste management for Brunei Darussalam. In: 5th Brunei international conference on engineering and technology (BICET 2014), Nov 2014, p 1.12 (6.). https://doi.org/10.1049/Cp.2014.1066
4. Intergovernmental Panel on Climate Change (IPCC) (2007) Fourth assessment report: climate change 2007 (AR4). IPCC, Geneva
5. MPRT (2020) Brunei Darussalam agriculture and agrifood statistics in brief 2019, Feb 2020
6. Varghese J, Prakash VM (2017) Study of geotechnical properties of organic waste materials as landfill cover. Int Res J Eng Technol. Available at: https://www.irjet.net/archives/V4/IRJET-V414508.pdf
7. Hermawansyah Y et al (2018) Study of compost use as an alternative daily covering Sukawinatan landfill Palembang. Int J Geomate 15(51):47–52. https://doi.org/10.21660/2018.51.50249
8. Khoshand A, Fall M (2014) Geotechnical characterization of compost based biocover materials. Geotech Geol Eng 32(2):489–503. https://doi.org/10.1007/S10706-014-9728-9
9. Chen P, Zhan L, Wilson W (2014) Experimental investigation on shear strength and permeability of a deeply dewatered sewage sludge for use in landfill covers. Environ Earth Sci 71(10):4593–4602. https://doi.org/10.1007/S12665-013-2851-8
10. Ng KTW, Lo IMC (2007) Engineering properties of MSW landfill daily covers using waste tire chips and paper sludge. In: Proceedings Sardinia Margherita di Pula, 1(5)
11. Puma S et al (2013) Reuse of MSW bottom ash mixed with natural sodium bentonite as landfill cover material. Waste Manag Res 31(6):577–584. https://doi.org/10.1177/0734242x13477722
12. BS 1377:1990 (1990) Methods of test for soils for civil engineering
13. Kareem MSA (2018) Effects of industrial wastes on geotechnical properties of effects of industrial wastes on geoenvironmental engineering is the application of newly developed. J Eng Des Technol
14. EPA (2016) How to evaluate alternative cleanup technologies for underground storage tank sites: a guide for corrective action plan reviewers, Nov 2016
15. Abichou T et al (2015) Use of evapotranspiration (ET) landfill covers to reduce methane emissions from municipal solid waste landfills. J Water Resour Prot. https://doi.org/10.4236/Jwarp.2015.713089
16. Ammar AM (2000) Designing sustainable landfill for Jeddah city, Saudi Arabia. Bull Eng Geol Environ 58

Daylighting Comfort of Verandah in a Traditional House, Thailand

Kornkamon Tantiwanit and Pimonmart Wankanapon

Abstract Traditional architecture usually had its characteristic related to its users and climate. For Thai traditional houses, daylighting sequences between spaces were considered as their dominant characteristics to provide visual comfort during the daytime. The highlight is the daylighting characteristic of the verandah, a blurred space between brightness and darkness. The average luminance in the field of view, vertical illuminance at the eye level, and several daylighting glare evaluation indices were analysed to understand daylighting conditions of the verandah. The methods included a field study, an image processing technique, and a discomfort glare evaluation. The results found that DGP and DGI scores reported satisfying condition of discomfort glare. This semi-outdoor verandah showed to have a low level of luminance, which help reduced discomfort glare sensation along with its building form and wooden construction. The results also confirmed that the setting of Thai traditional houses among tropical gardens is vital to ensure visual comfort to lesser the glare sources. The understanding of daylighting ambiance would help designers and architects to recreate local identity and aesthetics.

Keywords Daylighting · Visual comfort · HDR image · Tropical region · Traditional house

1 Introduction

Traditional architecture is usually constructed based on its local climate and its users. They accumulate problem-solving solutions and suitable architectural design strategies that fit their inhabitants [1]. In Thailand, where daylight availability is plenty, the daylighting level between interior and exterior space is extremely contrast. It is necessary to gradually decrease the brightness level for human eye adaptation

K. Tantiwanit (✉) · P. Wankanapon
Faculty of Architecture and Planning, Thammasat University, Bangkok, Thailand
e-mail: kornkamon@ap.tu.ac.th

P. Wankanapon
e-mail: pimonmart@ap.tu.ac.th

T. Kang and Y. Lee (eds.), *Proceedings of 2021 4th International Conference on Civil Engineering and Architecture*, Lecture Notes in Civil Engineering 201,
https://doi.org/10.1007/978-981-16-6932-3_28

when approaching a room from the outdoor area. For traditional Thai house, it used daylighting sequences between spaces to achieve visual comfort. The interior luminance levels can be categorized into three zones related to its building form and wooden structure. The first zone is the highest luminance of an uncovered terrace, while the second zone provided a dimmed luminance level of the verandah in front of a room. In the third zone, the private bedroom showed to have the lowest luminance. The veranda, a transition space between exterior and interior, is considered a dominant characteristic of Thai house identities. To understand its visual comfort, a blurred space between brightness and darkness condition was chosen in this study. Nevertheless, this study is a part of the research on daylight quality as an aesthetic valuation of traditional Thai houses.

2 Background

Thailand is located at latitude 15 N and longitude 100 E. The country is geographically divided into four regions; Central, Northern, North-eastern, and Southern area. These factors make traditional houses vary in each region. Considering a traditional house in the central region, it has a unique and certain layout, building form, and materials. It is typically a group of wooden structure buildings raised about 1.5 m above the ground on a single platform. The ground floor is a clear space without any walls or partitions for ventilation. On the second floor, individual buildings are joined to one single house by a large terrace. Each of the buildings is assigned for a specific function, for example, common, living, sleeping, and cooking space with its toilet facility normally built outside. All buildings have a similarity in their form of a high gable roof, responding to a high ambient temperature and amount of rainfall in the tropic. In addition, all building construction materials are made of wood from the tropical forest such as Teak wood for each column, beam, floor, wall, and roof structure.

Ruen Norn, or a bedroom, is usually constructed separately as one building with its roof and used by an individual or a couple. For a wealthy family, the number of bedrooms can be added up to the number of family members. This type of building is normally divided into two areas by wall and split-step floor. The first area is an enclosed sleeping space, and the second area is a verandah as a private living space. The length and width of the building are generally fixed to 3-bays with a total length of 9 m and 1-span of 4 m in width, due to the limitation of the used wood. For the verandah, stretching full length along the bedroom, the width is limited to 2–2.5 m. The verandah is a semi-outdoor space covered with a roof and extended eaves. The area is wrapped by 3-wooden walls. The first wall separates this space from the bedroom without visual connection and used double solid wooden doors to link both parts together. The perpendicular wall on both sides normally has one window on each. However, one side of the verandah is left open without any walls (Figs. 1, 2, 3 and 4).

Fig. 1 Selected traditional houses no. 1

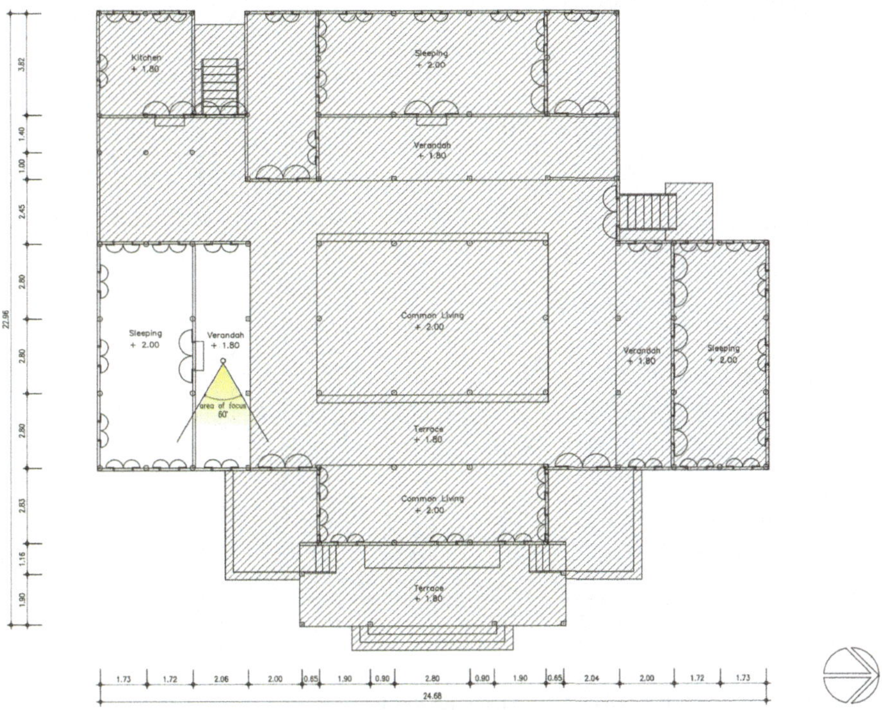

Fig. 2 Plan and research setup position facing East of house no. 1

Fig. 3 Selected traditional houses no. 2

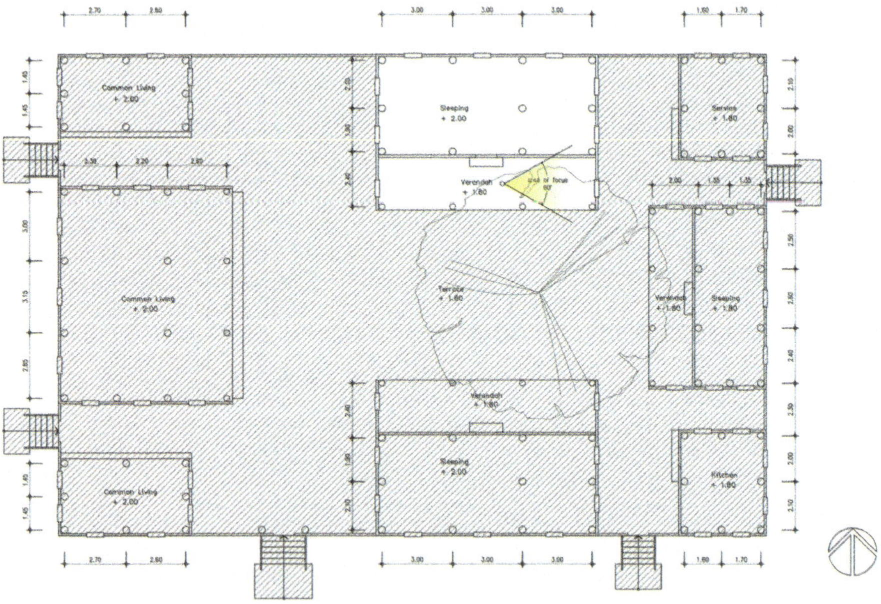

Fig. 4 Plan and research setup position facing East of house no. 2

3 Visual Comfort Evaluation

Visual comfort has contributed to human sensation relating to the satisfaction of lighting conditions. For Thai traditional houses, a condition of luminous environment developed overtimes to perfectly suitable for the local climate, building functions, and

Table 1 DGI range and DGP range for glare sensation rating categories

Glare sensation rating	DGI range	DGP range
Imperceptible glare	< 18	< 0.35
Perceptible glare	18–24	0.35–0.40
Disturbing glare	24–31	0.40–0.45
Intolerable glare	> 31	> 0.45

user habits. This would imply that lighting characteristics in the traditional house met occupants' overall environmental satisfaction. To understand the visual environment of a house, visual comfort should be quantitatively determined. Nowadays, such daylighting metrics were available to evaluate certain visual comfort conditions. The two most common metrics were Daylighting Glare Index (DGI) and Daylight Glare Probability (DGP) [2]. While DGI was developed by Hopkinson in 1972 [3] and calculated according to Eq. (1), DGP was developed by Wienold in 2006 [4] and expressed in Eq. (2). Both metrics reported the glare sensation categories with the different results showing either in percentage or numeric number as seen in Table 1.

$$\mathrm{DGI} = 10 \times \log 0.48 \sum_{i=1}^{n} \frac{L_{s,i}^{1.6}\omega_{s,i}^{0.8}}{L_b + (0.07\omega_{s,i}^{0.5}L_{s,i})} \quad (1)$$

$$\mathrm{DGP} = 5.87 \times 10^{-5}E_v + 9.18 \times 10^{-2}\log\left(1 + \sum_{i=l}^{n} \frac{L_{s,i}^2\omega_{s,i}}{E_v^{1.87}P_i^2}\right) + 0.16 \quad (2)$$

4 Methods

The method of this study was a combination of a field study, an image processing technique, and a discomfort glare evaluation. [5] The field study took place in two selected traditional houses in the central region of Thailand. The one-day data was collected from each location on 11st and 27th September 2020, respectively. On both days, there was a clear sky condition in the morning and a partly cloudy sky in the afternoon. The series of luminous environment images, vertical illuminance, and vertical luminance were recorded every two hours from 10:00 to 16:00. All research equipment was set up at the center of the verandah facing East. The data were collected at 0.8 m above the verandah floor representing an eye-level of a person's sitting position viewing the window. Series of images were obtained from a Fuji GFX50R camera. For each scene, the shutter speed was adjusted automatically with a remote control ranging from 1/4000 to 1 s recording twenty-four images per set. The vertical illuminance and luminance data were manually collected from a digital LUX meter (INS DX-200) and a luminance meter (Minolta LS-100) at the same position of images taken. Then, images were merged into a single High Dynamic

Range (HDR) image for each set using Photosphere. All HDR images were processed and calculated using Evalglare with the collected vertical illuminance and luminance [6]. In addition, the annual glare analysis results were derived from DIVA [7].

5 Results and Discussion

The results were shown as HDR images, false-color representation, and discomfort glare evaluation of the two traditional houses. The first house was shown in Fig. 5, and the second house is shown in Fig. 6. Moreover, the average luminance, vertical illuminance, and glare scores including DGI and DGP for both houses were shown in Tables 2 and 3 respectively.

For house number one, the average luminance in the field of view derived from software, showed a minimum of 20 cd/m^2 at 16:00 and a maximum of 155 cd/m^2 at 10:00. The vertical illuminance at the eye-level, derived from field measurements, showed a minimum of 55 lx at 16:00 and a maximum of 378 lx at 10:00. While the DGI showed as a perceptible glare for sometimes, the DGP values on were considered an imperceptible glare at all the times. Therefore, it shows to have different DGP and DGI scores of discomfort glare sensation.

For house number two, the average luminance level showed a minimum of 5 cd/m^2 at 10:00 and a maximum of 17 cd/m^2 at 16:00. The vertical illuminance measured showed a minimum of 55 lx at 10:00 and a maximum of 181 lx at 14:00. The glare scores, both the DGI and DGP, were reported as an imperceptible glare.

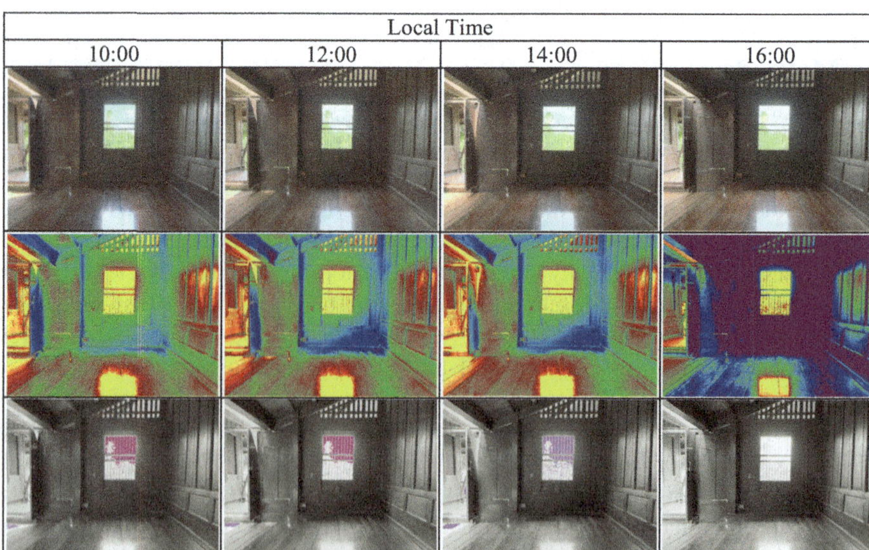

Fig. 5 Images and glare evaluation of house no. 1

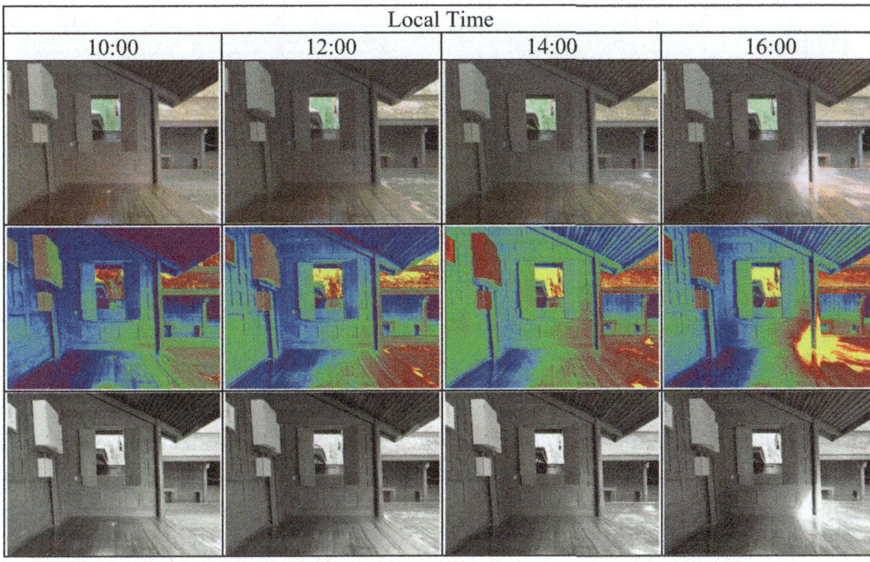

Fig. 6 Images and glare evaluation of house no. 2

Table 2 Data and results from house no. 1

Local time	Average luminance (cd/m^2)	Vertical illuminance (lux)	DGI	DGP
10:00	155	378	22.27	0.25
12:00	105	316	21.08	0.22
14:00	142	242	20.06	0.19
16:00	20	55	0.00	0.01

Table 3 Data and results from house no. 2

Local time	Average luminance (cd/m^2)	Vertical illuminance (lux)	DGI	DGP
10:00	5	55	0.00	0.01
12:00	13	152	14.88	0.07
14:00	12	181	4.19	0.10
16:00	17	123	0.00	0.04

To be noted that, the DGP calculation showed to underestimate the glare source because the vertical illuminance was under 380 lx and the DGP was under 0.2 [8]. According to the DGP calculation, the discomfort glare evaluation tends to report an imperceptible glare when the vertical illuminance is low [9]. The low light level of the verandah was a consequence of the building form and construction materials. Since the construction of traditional houses was made of wood which was a low reflectance material. For both houses, the reflectance values of both a wooden wall

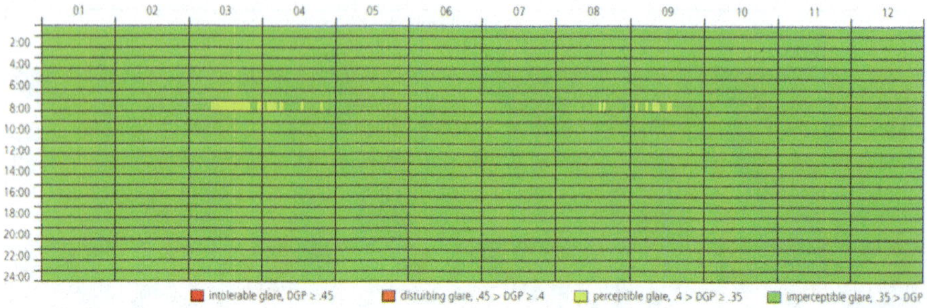

Fig. 7 Annual glare analysis of house no. 1

and ceiling is at 4%, while typical reflectance values of a wall and ceiling of modern construction is at 50 and 80% respectively. Nevertheless, the reflectance value of a wooden floor of the house no. 1 is 20% and the house no. 2 is 17%, which is equal to a floor's reflectance value as 20% of a modern construction. As a result, it can be said that the building form and construction materials played an important role in minimizing indoor luminance and reducing the chance of visual discomfort in a person's field of view of the verandah.

While house no. 2 was always reported the discomfort glare evaluations for both the DGI & DGP as an imperceptible glare, house no. 1 was reported the DGP as a perceptible glare from 10:00 to 14:00. In this case, the glare source is a window with an unobstructed sky view. To be noted that there was no sky view from house no. 2. Hence, the least sky view from window is crucial to provide daylight without glare. Therefore, the location of the house among the lush tropical garden should be considered.

Apart from point-in-time glare evaluation, the annual glare analysis, DGP, was analysis through DIVA software. The results found that both houses were experience imperceptible glare most of the times. Moreover, they were never encounter a disturbing glare as well as intolerable glare. However, for house no. 1, the perceptible glare occurred only a short period in a very specific time of the year. In this case, it existed in the morning from 8:00 to 9:00 for some days in March, April, September, and October as shown in Fig. 7. Considering house no. 2, the perceptible glare rarely took place. It happened solely in the morning from 8:00 to 9:00 for some specific days on March, April, and September as shown in Fig. 8. The analysis confirmed the daylighting design strategies and visual comfort of the verandah in a traditional house.

6 Conclusion

Traditional houses had their unique identities owing to the climate and users. Thai traditional houses had unique characteristics due to its daylight sequencing. It is

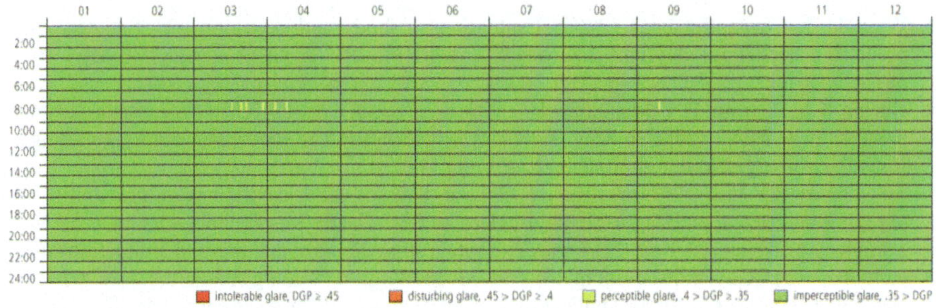

Fig. 8 Annual glare analysis of house no. 2

crucial to quantify and analyse the uniqueness of the semi-outdoor verandah in terms of its visual comfort and the factors affecting it. The study approached daylighting comfort using field measurement data from two traditional Thai houses located in the central region of Thailand incorporated with HDR images technique as well as supporting analysis from Evalglare and DIVA software. In contrast to the outdoor environment, the verandah showed to have a low average luminance in the occupant's field of view. These helped reduce the discomfort glare significantly as seen on the DGP results of an imperceptible glare while the DGI results showed to have a perceptible glare due to the unobstructed sky view from the East facing window in the morning. In this case, the building form, the wooden structure, and construction materials showed to be the contributing factors on a low light level environment and a glare reduction due to its low reflectance. Furthermore, the setting of the houses among the tropical garden is contributed to visual comfort. This is due to lesser luminance coming to the occupant from the glare source, in this case, sky, from the view out of the window.

The results from this study would help designers and architects to recreate local identity and aesthetics in modern architecture. The distinctiveness of daylighting would be valuable for a concept design of hospitality buildings such as hotels and resorts. And with the quantifying results and contributing factors, this could be a key ingredient in maintaining the Thai identity in contemporary architecture.

Acknowledgements This study was supported by Thammasat University Research Fund, Contract No. FT006/2563.

References

1. Maiyah SA, Elkadi H (2015) Turkish D-light: accentuating heritage values with daylight. Proc Instit Civil Eng Eng Hist Heritage 168(4):139–149
2. Suk JY (2019) Luminance and vertical eye illuminance thresholds for occupants' visual comfort in daylit office environments. Build Environ 148:107–115
3. Hopkinson RG (1972) Glare from daylighting in buildings. Appl Ergon 3(4):206–215

4. Wienold J, Christoffersen J (2006) Evaluation methods and development of a new glare prediction model for daylight environments with the use of CCD cameras. Energy Build 38:743–757
5. Kurnia KA, Azizaha DN, Mangkuto RA, Atmodipoero RT (2017) Visual comfort assessment using high dynamic range images under daylight condition in the Main Library Building of Institut Teknologi Bandung. Procedia Eng 170:234–239
6. Suk JY, Schiler M (2013) Investigation of Evalglare software, daylight glare probability and high dynamic range imaging for daylight glare analysis. Light Res Technol 45(4):450–463
7. Jakubiec JA, Reinhart CF (2011) DIVA 2.0: integrating daylight and thermal simulations using Rhinoceros 3D, Daysim, and EnergyPlus. In: Proceedings of building simulation, pp 2202–2209
8. McNeil A, Burrell G (2016) Applicability of DGP and DGI for evaluating glare in a brightly daylit space. In: Proceedings of ASHRAE/IBPSA-USA building simulation, pp 57–64
9. Pierson C, Wienold J, Bodart M (2018) Review of factors influencing discomfort glare perception from daylight. LEUKOS 14(3):111–148

Vernacular Passive Strategies for Sustainable Houses in the Andean Ecuadorian Context

Marcelo Villacis Ormaza, Susana Moya Vicuña, and Emilia Monar

Abstract Sustainability and thermal comfort in developing countries could be hard to achieve due to the high cost of standardized construction methods, such as reinforced concrete and steel framed structures. As a result, most dwellings ignore the basic principles of livability and sustainability promoting a high demand of external construction resources and using elevated quantities of energy during its operative life. The purpose of this paper is to analyze passive and vernacular strategies for typical active thermal comfort as alternatives for low rise housing in the Andean region of Ecuador by using rammed earth walls as thermal mass elements. This research aims to reach the minimum comfort levels stablished by INER and ASHRAE 90.1 envelope requirements for climate zone 4C (temperate continental).

Keywords Architecture · Andean · Raw earth wall · Thermal performance · Vernacular · Comfort

1 Introduction

Worldwide, construction practices represent the 40% of energy production, 40% of raw materials and extraction of natural resources and 35–40% of CO_2 emissions. These are all cogent reasons for prompting sustainable architecture by improving construction practices and achieving comfort conditions for its inhabitants with a low to zero environmental impact. However, it must be recognized that the socioeconomic conditions of a country or a region are decisive when planning and building sustainable housing projects [1]. The conditions of a developing country like Ecuador and its Andean zone are no exception.

Most of the structural systems used in the country for singular housing are reinforced concrete with lightened concrete block for walls and steel framing with lightened concrete block for walls as well. These two systems can be found in the construction of formal and informal housing. However, in both scenarios, thermal comfort

---

M. V. Ormaza (✉) · S. M. Vicuña · E. Monar
Faculty of Architecture, Arts and Design (FAAD), Research Centre DITES, Universidad Tecnológica Indoamérica, Quito, Ecuador
e-mail: marcelovillacis@uti.edu.ec

T. Kang and Y. Lee (eds.), *Proceedings of 2021 4th International Conference on Civil Engineering and Architecture*, Lecture Notes in Civil Engineering 201,
https://doi.org/10.1007/978-981-16-6932-3_29

standards are ignored, resulting in houses that do not adapt to bioclimatic conditions of the region.

This paper focuses on analyzing passive vernacular systems which can help to produce houses that meet local comfort standards.

2 Methodology

Research methodology has been developed into two phases. The first phase consisted on a literature review which provided background information about climate and thermal comfort in the Andean Ecuadorian context, as well as an overview about current sustainability issues. Findings allowed to consider vernacular techniques as alternatives to solve thermal comfort problems in the area.

The second phase provides a quantitative method for analyzing the performance of the first phase findings, in order to provide proper data which allows to establish alternative construction systems for projects within the study area.

3 Local Conditions Analysis

The dialogue between architecture and environment has been a constant from the most primitive architecture vestiges to the most contemporary constructions through the application of sustainability principles, which are simple and effective adaptative responses to the natural conditions of the environment by defining materials, systems, and strategies of design. One of the mistakes but at the same time one of the most common practices in architecture has been importing these elements or 'construction recipes' to different contexts regardless of the site characteristics. For this reason, the architecture-environment integration requires a redirection of architectural thinking, based on ecological, social, and economic criteria of each region [2].

3.1 Climatology of the Andean Region

According to INAMHI (Ecuador's National Institute of Meteorology and Hydrology), the minimum average air temperature in the inhabited Andean region of Ecuador is 7.6 °C, the maximum average is 21 °C and the monthly average is approximately 14.6 °C. Relative air humidity is 79%, which fluctuates from 86% to a minimum of 70%. The study area is located at an altitude of between 2500 and 3200 m [3]. In this region, climate is characterized by being highly variable throughout the day and night, but relatively constant all over the year with fluctuations on its rainfall, having as a result a wet season and a dry season [4]. This is due

Table 1 Climate zones of Ecuador according to INER and ASHRAE 90.1

Climate zone (Ecuador)	Climatic zone (ASHRAE 90.1)	Definition (name)	Thermal criteria
1	1A	Very hot and humid	5000 < CDD10 °C
2	2A	Hot and humid	3500 < CDD10 °C ≤ 5000
3	3C	Rainy continental	CDD10°C ≤ 2500 y HDD18 °C ≤ 2000
4	4C	Temperate continental	2000 < HDD18 °C ≤ 3000
5	5C	Cold	CDD10 °C ≤ 2500 y HDD18 °C ≤ 2000 2000 < HDD18 °C ≤ 3000 3000 m < Altitude (m) ≤ 5000 m
6	6B	Very cold	CDD10 °C ≤ 2500 y HDD18 °C ≤ 2000 2000 < HDD18 °C ≤ 3000 5000 m < Altitude (m)

Source Inamhi meteorological records

to its proximity to the Andes mountain range and its association with the equatorial line.

According to the INER (Institute for Geological and Energy Research), the inhabited Andean zone presents the following types of climate: (1) Hot Humid, (2) Rainy Continental, (3) Temperate Continental and (4) Cold. For this research, the temperate continental climatic zone (4) was used as a case study [5] (Table 1).

A very important meteorological variable in the thermal and energy analysis of materials and systems to be used in the construction of buildings in the Andean zone is solar radiation. In 2008, the Ministry of Electricity and Renewable Energies published the Solar Atlas of Ecuador where the solar resource is estimated through a climatological solar radiation model. This document stipulates the distribution of the solar resource over the Ecuadorian territory, showing an average global insolation of 4575 Wh/m^2/day [6, 7]. Commonly, this value is used only for calculations of renewable energy generation, leaving aside the efficiency that these levels of insolation can reach in elements with high thermal mass.

3.2 *Comfort in the Andes—Principles for the Design of Bioclimatic Architecture*

When we talk about thermal comfort, we refer to a combination of environmental, physiological, psychological, and cultural parameters that generate a feeling of well-being for the inhabitant of a building. This feeling of comfort must achieve a good long-term performance to assess how sustainable the system used for this purpose was.

To have greater clarity about the comfort strategies, the climatic annual data were inserted in givoni abacus diagram, which allow to find that thermal oscillations during the day are not too high. In general, the thermal needs are for heating during night hours. Zone 4 or 4C has a Specific Space Heating Demand of 14.2 kWh/m^2/yr (4.5 kBtu/ft^2/yr) [8], this demand is characterized by being low and can be achieved without the need to use active strategies, improving building's thermal comfort conditions and optimizing energy demand at the same time.

At this point, it can be defined that according to the natural, economic, and cultural conditions of the place of study, it is recommended to rely on passive strategies to achieve minimum standards of thermal comfort [9]. The passive strategy must allow solar gain during the day by a high thermal inertia element.

On the other hand, the building envelope can be generated from local vernacular construction techniques. This will reduce carbon footprint of imported materials and revalue ancestral construction techniques in the Andean zone.

4 Vernacular Thermal Mass Performance

4.1 Properties of Non-stabilized Raw Earth Wall in Low Rise Construction

Some of the advantages of rammed earth walls as partitions and structural elements compared to concrete, brick and cement blocks rely on their low embodied energy or zero carbon [10]. The whole materials are recyclable and locally available. These walls offer many textures, colors or finishes, and construction costs could be considerably lower [11] (Table 2).

In addition, this type of wall has an excellent behavior when stabilizing the internal humidity of the building's air, provides excellent acoustic insulation qualities, and

Table 2 Materials characteristics

Natural materials	Density ρ [kg/m^3]	Thermal conductivity λ [W/(mK)]	Embodied energy [4] [MJ/(kg)]
Wood	400–1000	0.075–0.25	1.19–9.19
Stone	1500–2000	1.16–1.4	0.5
Rammed earth	1000–2200	0.20–0.95	0.5–1.2
Hollow concrete masonry	< 1680	0.49–0.76	1.5

Source Salihbegovic and Salihbegovic [2]

is able to store lots of energy due to its heat capacity (1830 J/m^3 °C) and thermal diffusivity.

4.2 *Thermal Performance Comparison of Rammed Earth Wall and Concrete Masonry*

Nowadays the most common construction practice in the Ecuadorian context for partitions and perimeter walls of low-rise housing is the concrete masonry system (0.20 m thick). Each concrete masonry block of the system contains an internal cavity which increases the thermal insulation.

For research purposes, stabilized rammed earth walls (thickness 0.40 m) were used inside and around the perimeter of the house. Walls are exposed to direct solar radiation between 5 and 6 h every day. The proposed stabilizers are lime, straw, alabaster, and pneumatic fibers [12]. This composition allowed to achieve better levels of insulation and thermal inertia. Consequently, thermal comfort conditions were improved during the day and night.

For the correct functioning of these elements, it is necessary to avoid thermal bridges with the outside and to avoid damp problems by capillarity from the base of the wall. On days with low radiation, thermal mass can be heated through a low-scale fireplace.

Regarding thermal performance comparison, a computational analyzes was performed in IDA Indoor Climate and Energy software (IDA-ICE) [13, 14] as shown in Figs. 1 and 2. A simulation was carried out with each construction system (rammed earth walls and concrete masonry walls), the layout and surface in both was the same (140 m^2). The software allowed the use of a weather file containing annual information on wind speed and direction, diffuse radiation, direct radiation, dry air temperature and relative humidity [13].

For indoors climate studies the heat emission of selected areas of the house was 125 W per inhabitant, the maximum load for equipment was 3.00 W/m^2, the envelope air leaks were 32% for doors, windows, and roof joints [15].

5 Results

Figure 1 shows the performance of both systems in wet and cold months, where rammed earth walls allow us to stay within the comfort range, it means 66% of the total time while concrete masonry wall only 25% of the time. Figure 2 shows the simulation results of a dry and temperate month, it is evidenced that rammed earth walls keep the house 87.5% of the time within the comfort range while the concrete wall only allows 25% of hours of comfort again.

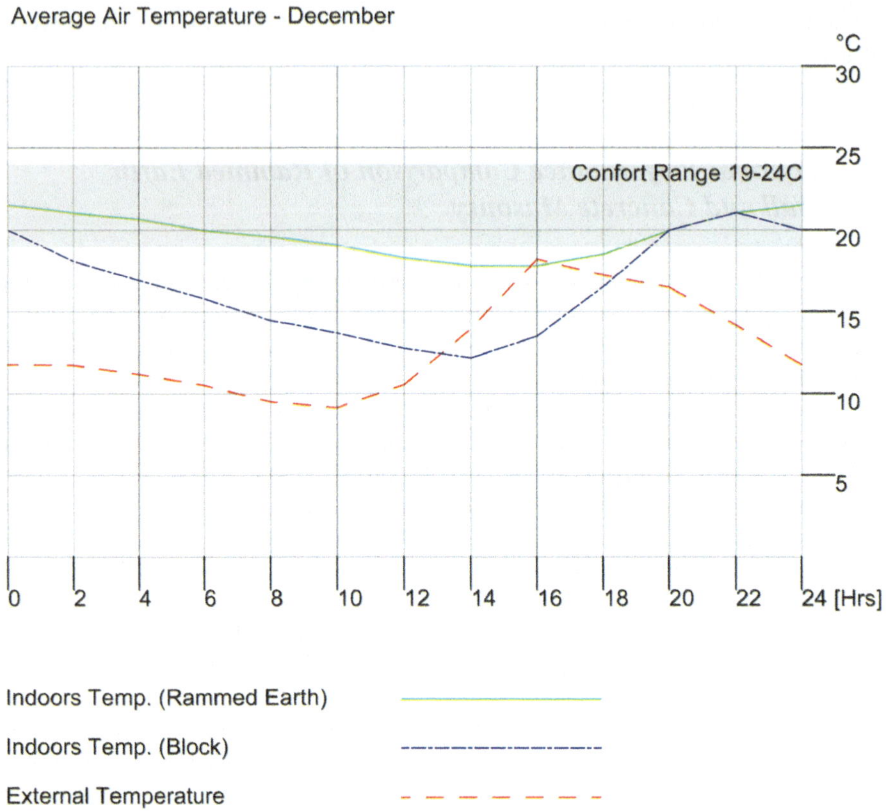

Fig. 1 Average indoors temperature December

This study led to suggest that rammed earth walls can be part of an effective sustainable system to reach the optimal thermal comfort for low-rise housing located in the Andean region (4C). Subsequently, it is possible to recover a neglected vernacular technique. Furthermore, within the Ecuadorian economic context, the cost of a rammed earth wall is considerably lower than standardized construction methods. The environmental impact is also lesser considering that the suggested system is biodegradable and free of embodied carbon.

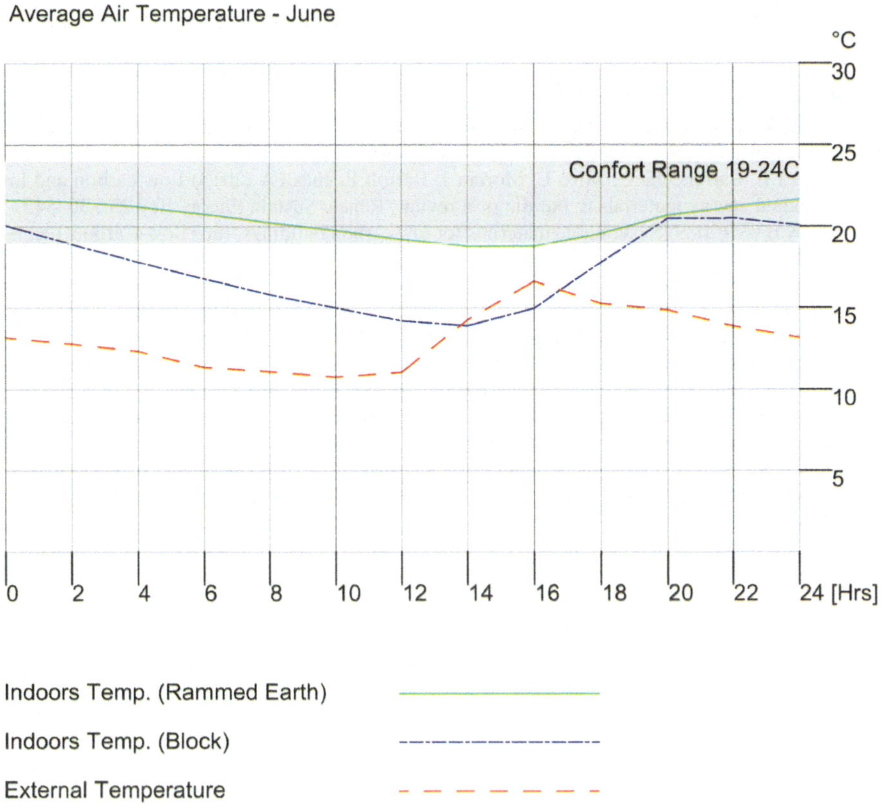

Fig. 2 Average indoors temperature June

References

1. Serpell A, Kort J, Vera S (2013) Awareness, actions, drivers and barriers of sustainable construction in Chile. Technol Econ Dev Econ 19(2):272–288
2. Salihbegovic A, Salihbegovic A (2020) Natural materials in contemporary low-tech architecture. IOP Conf Ser: Mater Sci Eng 960:042012
3. Instituto Nacional de Meteorología e Hidrología (2013) Anuario Hidrológico. [Hydrological Yearbook] N51 edn. INAMHI, Quito, Pichincha
4. Morán-Tejeda E, Bazo J, López-Moreno J, Aguilar E, Azorín-Molina C, Sanchez-Lorenzo A et al (2016) Climate trends and variability in Ecuador (1966–2011). Int J Climatol 36(11):3839–3855
5. Godoy F, Palme M, Villacreses G, Beltrán RD, Gallardo A, Almaguer Ochoa M, Lobato-Cordero A, Kastillo J (2017) Estrategias para mejorar las condiciones de habitabilidad y el consumo de energía en viviendas (Actualización) [Strategies to improve living conditions and energy consumption in homes (Update)]
6. Vaca Revelo D, Ordóñez F (2020) Mapa Solar del Ecuador 2019 [Solar Map of Ecuador 2019]
7. OLADE (2008) Atlas solar del Ecuador con fines de generación eléctrica. Organización Latinoamericana de Energía. [2008 Solar Atlas of Ecuador for electricity generation purposes.

Latin American Energy Organization] Online. Available: https://www.researchgate.net/deref/https%3A%2F%2Fgoo.gl%2FxyQQTt
8. Pless S, Torcellini P (2005) Energy performance evaluation of a low-energy academic building. Preprint
9. Hasper W, Kirtschig T, Siddall M, Johnston D, Vallentin G, Harvie-Clark J (2020) Long-term performance of passive house buildings. Energy Efficiency 14(1)
10. Cabeza L, Barreneche C, Miró L, Morera J, Bartolí E, Inés FA (2013) Low carbon and low embodied energy materials in buildings: a review. Renew Sustain Energy Rev 23:536–542
11. Reddy BVV (2009) Sustainable materials for low carbon buildings. Int J Low-Carbon Technol 4:175–181
12. Serrano S, Barreneche C, Rincón L, Boer D, Cabeza LF (2013) Optimization of three new compositions of stabilized rammed earth incorporating PCM: thermal properties characterization and LCA. Constr Build Mater 47:872–878
13. Hilliaho K, Lahdensivu J, Vinha J (2015) Glazed space thermal simulation with IDA-ICE 4.61 software—Suitability analysis with case study. Energy Build 89:132–141
14. Crawley DB, Hand JW, Kummert M, Griffith BT (2008) Contrasting the capabilities of building energy performance simulation programs. Build Environ 43:661–673
15. Simson R, Rebane T, Kiil M, Thalfeldt M, Kurnitski J (2020) The impact of infiltration on heating systems dimensioning in Estonian climate. In: E3S Web of conferences, vol 172, p 05004

Building Technology Sustainablism—New Vision

Abeer S. Y. Mohamed

Abstract The concern of the world at present to preserve the environment and the life of human societies on earth represents the most important global trends, and the subsequent trend towards sustainability that aims at the development that enables the enjoyment of the environmental resources and natural values that we use now. Therefore, development must be dealt with broad insight in terms of the temporal dimension, as well as advanced methods and creative ideas for dealing with natural resources require the concerted efforts of both architects, constructors, planners, and researchers in cooperation with decision-makers to focus on environmentally friendly technologies, especially in the field of building and construction technology. According to this vision, the paper analyzes the technical revolution in building technology from a sustainable perspective, then monitored its implications on architecture and associated urbanism, considering the promotion of sustainable architecture and the rationalization of building methods and energy consumption, to propose new pillars Building Technology Sustainablism Framework.

Keywords Building technology · Sustainable architecture · Construction technology · Sustainablism · Framework

1 Introduction

The world was drifting after the industrial revolution in the last century and the development of the machine that followed it to the total dependence on the industry in the field of building science and materials and the subsequent scientific and technological development accelerated, especially in the second half of the last century, all of this, in turn, led to dependence on oil as a source of energy. Makes humans deplete this resource significantly and depend on it mainly for the conduct of life.

A. S. Y. Mohamed (✉)
Architecture Department, College of Architecture and Design, Effat University, PO Box 34689, Jeddah 21478, Saudi Arabia
e-mail: asmohameddawod@effatuniversity.edu.sa

Architecture Engineering Department, Faculty of Engineering, Tanta University, Tanta, Egypt

T. Kang and Y. Lee (eds.), *Proceedings of 2021 4th International Conference on Civil Engineering and Architecture*, Lecture Notes in Civil Engineering 201,
https://doi.org/10.1007/978-981-16-6932-3_30

When the energy crisis arose in 73 AD of the last century, it was a turning point in the construction and design process, as trends began to appear in the field of energy in buildings, the field of preserving natural resources, the field of the environment and combating pollution. As a result of this crisis, several intellectual trends have resulted, including the birth of what is termed the post-modernism trend. Trends have also emerged that rejects industry and call for a return to nature. Other trends support modern technologies in architecture (Hi-Tech) and sustainable architecture. They have become schools of thought that graduate pioneers in these fields [1]. At the end of last century, it became common to use the term sustainable development, considering sustainability a prerequisite for the feasibility of development efforts in the long run. Although the term is still without an agreed precise procedural definition, the historical fact clarifies that caring for the environment is at the heart of sustainable development, whether it is within the framework of social, economic, political, or technical considerations necessary to sustain development. The conditions of construction in the context of sustainable development are the mirror that reflects what each era of architecture and urbanization has achieved within the framework of the reciprocal relationship between it and culture and society, and the accompanying use of technological capabilities and building materials (Fig. 1).

World's development scientists have concluded that the rapid growth rate accelerates the decline. Then he started thinking of new concepts that achieve continuity of development and prevent deterioration or maintain the level of what has already been achieved. Thus, stemmed from the concept of sustainable development, which achieves optimal utilization of available resources in development projects, without the need to exhaust other resources, which is a condition for the continuity of these projects [2]. The study is to develop a knowledge framework through which construction technology deals with the issues and potentials of the era from a sustainable perspective aiming at reducing the depletion of the natural environment.

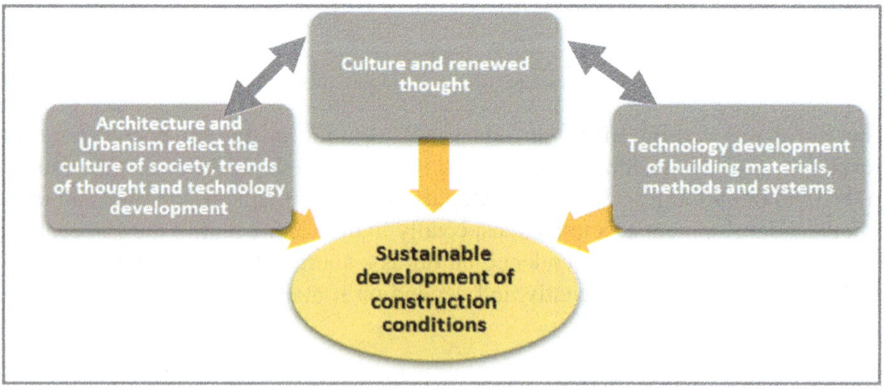

Fig. 1 Shows the reciprocal relationship between sustainable development of community conditions that accompany renewed thinking, and the development of building materials technology and systems that are reflected in the architecture and urbanization of society

1.1 Research Problem

This paper study the consequences and effects of the developing construction technology on the architecture field that could use the benefit of this technology and the architectural outputs to evaluate the impact of sustainable development on individual and societal institutions.

1.2 Research Objectives

Registration and analysis of the guidelines for development and architectural reframing of the emerging architectural technology.

This research is aimed at supporting the role of sustainable technology in developing an ideal design system which takes into account the implications of sustainable technology on future construction technology development, in particular in the Arabic world.

1.3 Research Methodology

This research follows the descriptive technique depending on the inclusion of sustainable features of building technology and architecture ideas, and on the connection of architecture-intensive technologies. This is accomplished by a field survey which shows the current viewpoints on sustainable development and the improvement of the Arab community. Cultural, economic, technical, and natural.

2 Sustainable Development

At the of the last century, the use of the term sustainable development became popular. Sustainable development is a development that meets the requirements of the present without diminishing the ability of future generations to conform to meeting their requirements [3]. Development according to this definition includes two basic contents:

- It is not limited to several sciences and regions, but rather to denote the whole world now and in the future.
- There is no specific concept of sustainable development, but the purpose is to continue that development.

That is the concept of sustainable development.

It is the development that is based on a set of policies and measures taken to achieve a balance between the interaction of the three ecosystems:

- The Natural Environment
- The Manufactured Environment
- The Social Environment: Social Environment.

Consequently, sustainable development has three main axes which are considered the main pillars of it according to the difference of one of them being affected by the main goals of development or sustainability in general (Fig. 2) and these axes are:

- Environment
- Economy
- Society.

For the success of the sustainable development process, these axes must be linked and integrated due to the close connection between the environment, the economy, and social security, making economic improvements and raising the level of social life in proportion to the preservation of the natural basic components of life, which is considered a long-term process. A good condition for future generations is better than it was. If a person maintains activity and performs it without depleting the natural materials or wasting the natural environment, then this activity is naturally sustainable [4], and this is achieved by (Fig. 3).

- Low consumption of natural materials.
- Using materials that are completely recyclable after consumption and are renewable and are collected without damaging the environment or depleting its resources.

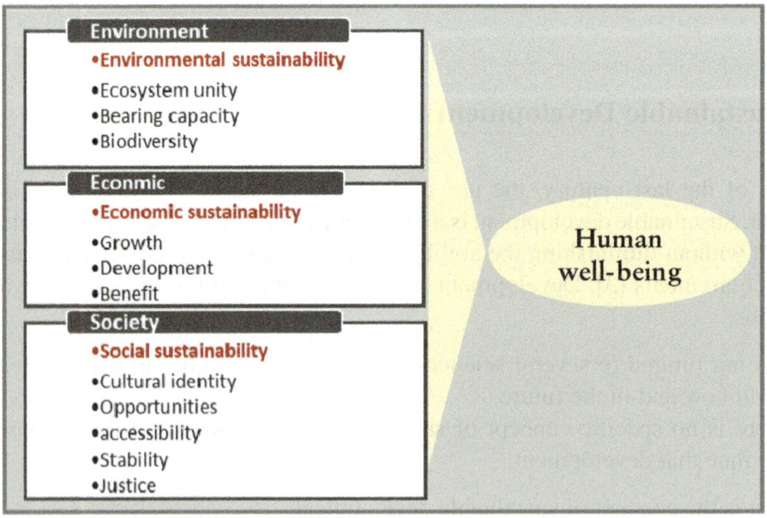

Fig. 2 Key indicators of sustainability

- 100% waste recycling rate.
- Energy conservation, the ability of its reserves to be renewable, and the preservation of the environment.

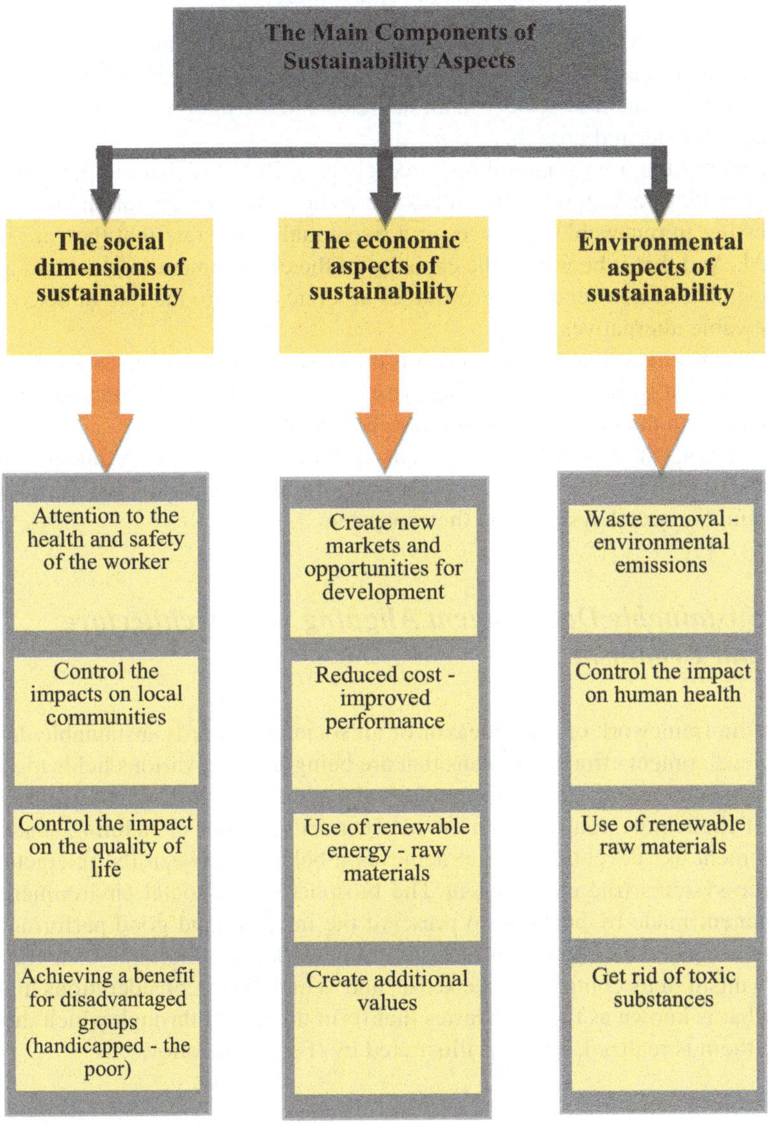

Fig. 3 Sustainability pillars and means of achieving them

2.1 Characteristics of Sustainable Development Processes

Sustainable development has a set of features and characteristics that distinguish it from development in its traditional concept, namely:

- Continuity: What is meant by the process of sustainability and communication in development because it is a criterion for the success of the development process in the development of society in all its fields and the integration of all its goals to achieve the desired growth.
- Regulating the use of natural resources: especially those that are permeable and renewable, in a way that guarantees the right of future generations to them, by investing in renewable resources at a rate equal to the rate that they are replenished. And that it be within the capacity of the environment to absorb it, and the investment of non-renewable resources at a rate equal to the rate of discovery of renewable alternatives.
- Achieving environmental balance: it is the guiding criterion for sustainable development, that is, preserving the integrity of natural life and producing renewable wealth, with the fair use of non-renewable wealth.
- Integration: Sustainable development is based on achieving integration and exchange between the different objectives of three basic systems: the economic system, the social system, and the ecosystem.

2.2 Sustainable Development Aligning with Architecture and Urbanism

Within the framework of the endeavor of all societies towards sustainable development, development efforts and plans that are being made in various fields to support them have emerged, to achieve the goal of raising the standards of living and productivity of their individuals, to achieve the general concept of defining sustainable development as "development that achieves a balance between the interactions of the three systems (the environment The biosphere—the social environment—the environment made by humans (to preserve the integrity and good performance of ecosystems [5], which intertwines in its framework the conditions of architecture and urbanism in its comprehensive sense, and Which these considerations highlight to be what is known as the boundaries that form them, and through which the basic idea of them is realized, which is illustrated by (Fig. 4) as follows:

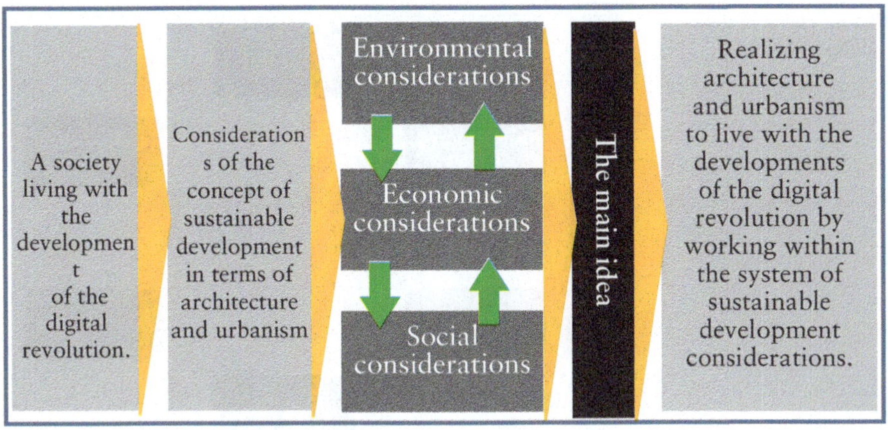

Fig. 4 Explains the concept of sustainable development in the context of the architecture and urban living with the developments of the digital revolution

3 Sustainable Architecture

The concepts of sustainability and their synonyms in the conceptual language of sustainable cities, green buildings, and environmentally friendly buildings constitute an important reference in dealing with various architectural styles of buildings. Sustainability is concerned with the natural environment and the built environment (buildings, squares, residential compounds, and spaces that contain areas for human use) in a way that serves future generations as the city is a spiritual element and a sustainable cultural expression from one generation to the next. At the level of architecture, sustainable architecture meets the functional, aesthetic, and symbolic needs. It is an architecture that goes beyond the functional concept of making the physical and spatial content of the building, and it does not limit its art to the building capable of continuity with its components without relying on alternative systems, which confirms that every architecture meets the utilitarian need. It is a sustainable architecture [6], because architecture is unique in the field of sustainability, as architectural projects consume large quantities of materials and generate larger quantities of waste and waste Earth pledge.

3.1 Principles of Sustainable Architecture:

The principles of sustainable architecture are as follows:

- Reducing the consumption of non-renewable resources.
- Beautify the natural environment.
- Eliminate or reduce the use of toxic substances.

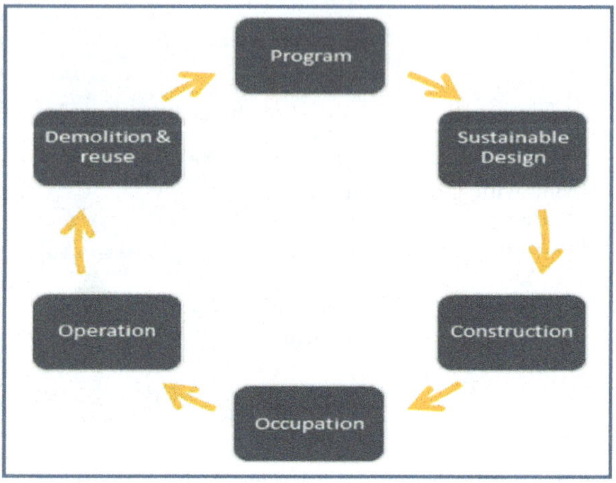

Fig. 5 The complete building life cycle

The application of these principles reduces the negative impact on the natural and constructed environment in terms of buildings and their immediate, regional, and global surroundings, as a sustainable building based on these principles is defined as building practices that seek integrated quality (economic–social–environmental) clearly, the logical use of natural resources and the appropriate management of buildings contribute In saving scarce resources [7, 8], reducing energy consumption, and improving the environment, taking into account the entire life cycle of the building (Fig. 5), and in light of the change in the global trend from concern for quantity to quality, the main objectives of sustainable architecture have evolved as follows:

- Resource efficiency
- Energy efficiency
- Pollution prevention
- Compatibility with the environment.

According to these new goals, which in turn affect the environmental, functional, and aesthetic quality, and the future values of the building and architectural products [9], a set of perceptual standards must be available within any architectural product to be considered a sustainable architectural product. These standards are as follows:

- Site parameters.
- Linking the site to the design curriculum.
- The spatial quality of the buildings.
- External spaces and squares.
- Landscaping elements: green areas and water elements.
- Climate solutions and orientation of the building or buildings.
- Building materials and traditional building techniques.

4 Modern Trends of Building Technology in Architecture from Sustainable VISION

Sustainable construction has been defined as the innovation and management responsible for building a healthy environment based on effective resources and environmental principles (Resource Efficient & Ecological Principles) and the goal of this type of architecture is the limit. From the negative impact on the environment through energy and resource efficiency [10, 11]. The implications of the development and application of construction and building technology have varied among societies of developed and developing countries, and then their impact on architecture and urbanism, as the modern trends of building technology in sustainable design architecture are closely related to the social, economic, and building conditions of the existing environment Where building technology is represented in the framework of sustainability (Fig. 6).

In detail, from the social point of view, these considerations are represented in serving the needs of the people within their homes and respond to their way of life, the social systems surrounding them, urbanization, culture, and education, and from the economic point of view, it is represented in the actual contribution to the construction industry process, moving the economic wheel and getting acquainted with economic policies, the standard of living and investment plans. Building conditions are represented in the environmental aspect concerned with treatments of drinking water, sanitation, heat, sound, and wastes resulting from the use of natural resources and have different impacts on the environment, as represented in urban conditions concerned with building formation and its connection to the site and the environment, and the use of roads and means of transportation as shown (Fig. 7), proposed Building Technology Sustainablism framework.

Thus, the role of technology increases, in being the main axis of the development of architecture and urbanization, both material and moral. Within the framework of the concept of comprehensive sustainable development, those in whose scope the construction conditions are intertwined and inseparable from it so that building

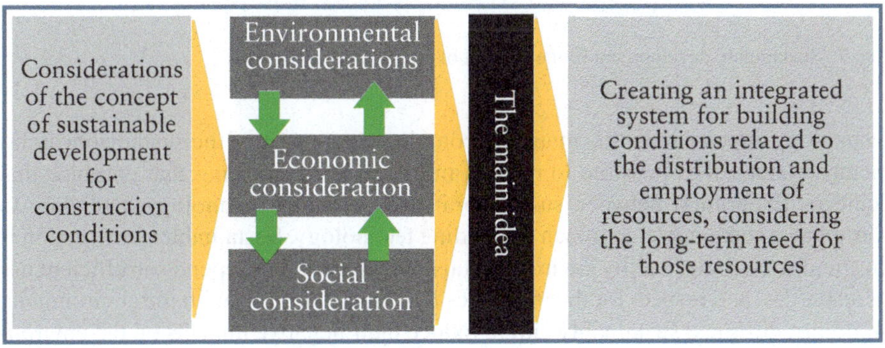

Fig. 6 Sustainable development concept considerations for building conditions

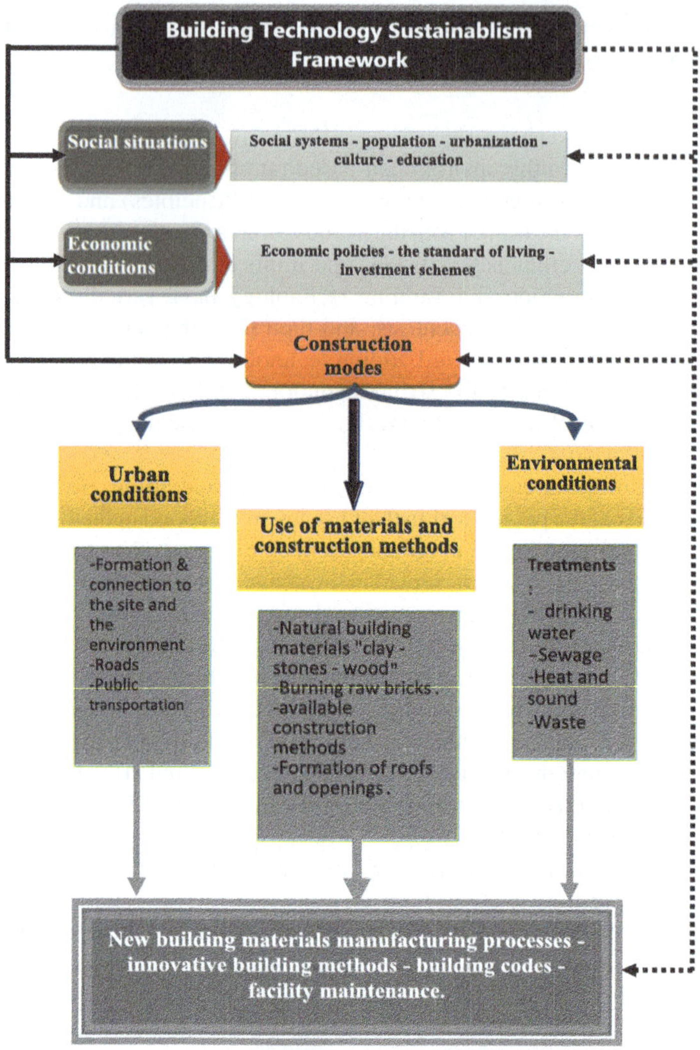

Fig. 7 Sustainable development in the context of construction processes

technology appears to be the main and complementary axis of the sustainable architecture system, it is possible to reach a mechanism for working that entrance that achieves Integration between sustainability and building technology, intending to formulate a sustainable approach to building technology. Sustainable construction is an attempt to create healthy environmental management that depends on efficient use of resources, and respect for the principles leading to harmony with the environment. Buildings designed sustainably aim to reduce their negative impacts on the environment. Sustainable building is a building method that seeks to integrate in terms of quality with the economic, social, and environmental performance of the building. In this way, the rational use of natural resources and the appropriate management of the building lead to the preservation of limited resources, reduce and conserve energy consumption, and improve the quality of the surrounding environment considering the use of building technology for the discarded era to serve that and complete the dimensions of sustainability. That is, it includes considering the life span of the building, the quality of the environment it contains, the quality of the performance of the building systems, and the future values expected in the community.

One of the architectural trends that apply modern technologies in the field of building technology in line with the idea of sustainability in architecture is what is known as Eco-Tech: a term given to architecture that is concerned with the fields of environment and modern technologies in architecture, and it is now one of the leading fields in architecture in the modern era. This term denotes the abbreviation of two words:

- Eco: Ecology, which stands for Environmental Sciences.
- Tech: Technology, which means technology.

The interests of this field are related to Sustainable Architecture and High Technology (Fig. 8).

5 Construction Technology Sustainablism—New Vision

Through the previous analysis, it is clear that the sustainable view of building technology is crystallized in the establishment of a built environment with a holistic approach to the design process as it takes into account all the principles constituting sustainable architecture in an integrative pattern and the technological thought directed to the construction process in all its dimensions (materials, systems and construction methods), so the building technology must be derived and expressive of the environment to which it belongs, add to it and not harm it in light of the optimal and efficient exploitation of the scientific potential of the third millennium. Accordingly, this paper is trying to introduce a new theory to achieve sustainability in the environment, its architecture must consider all the resources contained in the building, whether it is energy materials or a contribution to providing the needs of building users under the holistic approach to sustainable architecture from the perspective of building technology Sustainablism.

From the above, a set of principles on which the design process was based can be drawn to direct the architects' view of the sustainable design process within the framework of building technology and based on interest in the following elements:

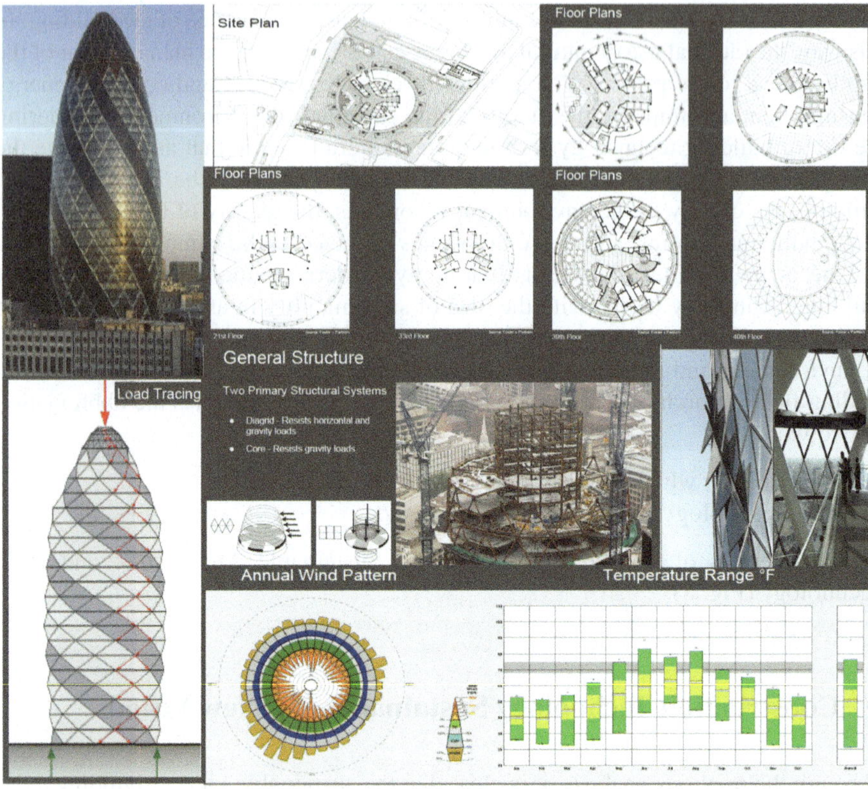

Fig. 8 This 30 St Mary Ax Office Tower—London England 2004 AD Designer: Norman Foster the use of modern technical methods in creating the building with a distinct structural system that allowed the implementation of that distinctive mass and adapted to the surrounding environment. The building is naturally ventilated through an inner courtyard through which natural air passes through the entire height of the building emerges from openings in the facade. This reduces the dependence of the tower on the air conditioning system [12]

5.1 Site Consideration

- The initial study of the site. Any sustainable design begins with studying the spatial content of the building to preserve it without destroying it by studying the site's determinants and linking the site to the design approach.
- That is the authentication of nature. Contact with nature, whether it is a natural or built environment, gives life to the building, and the structural content is integrated with a coexistent and user environment.
- Valuation and appraisal of the site's wealth and its optimum utilization with the use and reuse of existing buildings.

5.2 Design Consideration

- The design of the building should be continuous in its positive interaction with the environment and play its role all the time and be resistant to natural disasters, in addition to the design relying on sun, daylight, and natural cooling as natural sources of supply and creating a suitable atmosphere for the user.
- Planning integrity, design integration (self-operation) and early planning decisions have a strong influence on energy efficiency, and designers thus contribute to appropriate design activities such as guidance and preservation of the natural environment and its conformance with the design and access to integration between the building and its built environment.
- Integration of the design as each piece is part of the entire structure and is required for this design to succeed.
- The collaboration of all design participating disciplines in the early design decision-making phases with inclusion of sustainable buildings, and the involvement of users, local community, and adjacent areas in decision-making.
- Sustainable design interests in the study of user naturalities and the characteristics of the built environments and fulfilling the needs of the people and of society as a sustainable architecture require the integration of the esthetic, environmental, social, political, and ethical values as well as the use of user expectations and technology.
- Considering sustainable design as a building philosophy rather than a proposed building model, as the buildings that are built with this idea are not defined in shape or character.
- The cost of sustainable buildings should not be greater than traditional buildings, nor do they differ from them in the simplicity and uncomplicated design in integration and construction technology.
- Considering the rationalization of energy consumption and the improvement of the user's health as one of the basic elements in the design, followed by the other elements. Modern design trends should be directed to the forms of energy conservation and effectiveness and the integration of technology with its positive dimension to preserve the human being and the environment.

5.3 Construction Consideration and "Building Technology Used": (Fig. 9)

- The building achieves the maximum utilization rates of energy, water, and materials and the ability of the building to be self-sufficient in energy, with the building accepting future modifications and expansion, considering avoiding the design of the building for health damage.

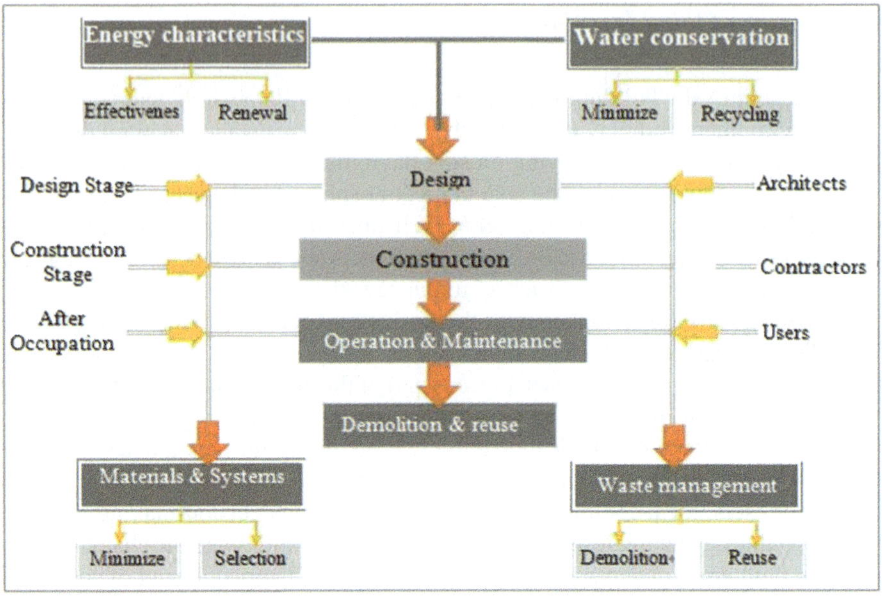

Fig. 9 Building technology uses mapping with a sustainable framework

- Optimum utilization of materials, with the employment of renewable building materials, the use of durable materials and products, and the choice of energy-saving materials.
- Sustainable design seeks to realize the environmental impact of the design by evaluating, understanding, and trying to achieve negative factors through use of sustainable materials and the use of recyclable onsite materials and instruments by evaluating locations, energy, materials, energy efficiency and construction techniques.
- Sustainable structures are expected to cost more during the construction phase but are not more difficult than traditional buildings and economical throughout the operating phase.
- An emphasis on sustainable design is added to the importance of understanding the environmental impact on design through site assessment, studying the energy used in all activities related to construction operations, whether it is for the manufacture of construction materials or equipment or the extraction of metals used in the manufacture of this equipment, as well as assessing the toxicity of the materials used in Construction to mitigate the negative impacts resulting from the users of these buildings

6 Results

Through this research, a set of results were reached, including the following points:

- Sustainable architecture emerged as realistic solutions with environmental, social, and developmental dimensions that simulate society and achieve the requirements of its groups. This was evident in the structural models that the research reviewed. Sustainable architecture is an integrated scientific approach that looks at information and analyzes it, keeps pace with developments, and develops them, as the community reads with its variables and aspirations of its groups Different.
- Sustainability from a technological perspective does not mean confining and imitating ancient and historical technologies and enriching a strict commitment to heritage, but rather the harmonization between originality and contemporary, between beauty and reason, between economy and luxury in a way that serves the categories of users, their aspirations and the plans set.
- The development of architecture and urbanism from a sustainable perspective is due to the technological development of building technology, and in this way, it represents the continuity of the relationship between the development of materials technology and construction methods and the contemporary architectural ideas of it, and the building technology from a sustainable perspective is the mirror that reflects the reciprocal relationship between the intellectual and cultural development of society and what It is accompanied by the use of technological capabilities and building materials.
- Hence, recent developments in technology and the possibility of transferring energy sources and advanced building materials have helped to find building solutions differently, but many of these solutions are depleting the planet despite modern technologies.
- The great challenge facing architects and planners now is choosing and modifying the advanced technology in a way that does not negatively affect the environment, and at the same time developing the performance and comfort requirements of the environmentally compatible building.

7 Recommendations

The research paper recommends a set of recommendations as follows:

- The necessity of continuing to find architectural solutions that keep pace with the development of technology in all fields, as it is the expression of the development of architecture and urbanism in both material and moral parts, to avoid the dangers of environmental pollution and the importance of linking the requirements of the human with his environment within the scope of building construction with ecological systems and building materials that can be reused in a way that means less depletion Natural Resources.

- The building must be balanced and fulfill three basic principles to improve environmental sustainability:
- Continuous design, resource-saving, and design life cycle (design, construction, operation, and maintenance,
- Recycling and reuse of architectural resources).
- The necessity to encourage sustainable architecture through a new vision of building technology, considering the rationalization of building methods and energy consumption, is considered one of the pillars on which the success of sustainable development in any society depends.
- Considering the role of information technology in achieving sustainable development in this era in which technologies determine competitive capabilities, and information technology can play an important role in sustainable development, as the endless potentials it provides can be harnessed to bring about sustainable economic, social and environmental development. By promoting building technology for sustainable development in architecture.
- The field of sustainable architecture and advanced technology is one of the vital areas important to every individual in society and all environments. Social awareness of the concept of sustainability must be increased and education about the features of modern technology in the field of construction.

References

1. Ahmed MASH, Waleed FO (2006) Toward buildings more sustainable. In: Conference: Saudi Urban Science Society Conference, At Jeddah, K.S.A., vol 1 January
2. Vezzani R et al (2015) A general-purpose sensing floor architecture for human-environment interaction. ACM Transactions on Interactive Intelligent Systems. Softech-ICT, University of Modena and Reggio Emilia
3. Menachem D 2019 Smart Home Systems Based on Internet of Things February 28th, www.intechopen.com/books
4. Thiago VC (2018) Virtual reality in the architecture, engineering, and construction industry proposal of an interactive collaboration application. Porto University, Portugal
5. Rasooldeen M (2013) Green buildings concept growing (Arab News). Retrieved, from http://www.arabnews.com/news/487676
6. Miranda L, Marulanda L (2001) Sustainable construction in developing countries. A Peruvian Perspective. Retrieved 21 Sept 2010, From http://www.sheltercentre.org/sites/default/files/CIB_Agenda21ForSustainableConstructionInDevelopingCountries.pdf
7. Silverman N, Mydin A (2014) Green technologies for sustainable building. Acta Technica Corviniensis–Bull Eng Tome 7(3):87–94
8. Olawumi TO, Chan DWM (2018) A scientometric review of global research on sustainability and sustainable development. J Cleaner Prod 183(2018):231–250. https://doi.org/10.1016/J.JCLEPRO.2018.02.162
9. Baggs S (1996) The healthy house. Thames & Hudson, London
10. Baxter A (2011) Natural landscape areas of London and their natural signatures. London's Natural Signatures: The London Landscape Framework
11. Smith F (2001) Architecture in a climate of change a guide to sustainable design. Architectural Press, Oxford, pp 205–206

12. https://faculty.arch.tamu.edu/media/cms_page_media/4433/30%20St%20Mary%20Axe_barker_lawrence_lopez_teal_wendlandtPresent.pdf

Embodied Energy and Embodied GHG of Architectural Works on Low-Cost Apartment Construction in Indonesia

Dewi Larasati, Noveryna Dwika Reztrie, Try Ramadhan, Yulita Hanifah, and Irma Handayani Lubis

Abstract The development of lower to middle-class apartments in big cities in Indonesia continues to increase due to urbanization. The increase in the growth of apartments in this class will of course directly affect the rise in building material needs for construction. This is also elevating energy consumption and carbon emissions. This research will explain how the use of materials in apartment building construction can affect embodied energy and carbon emissions. This study uses the IO table method as method for calculating embodied energy and carbon emissions. This research selects three case studies of low-cost and middle-class apartment construction in several big cities in Indonesia. The results indicate that the EE value and GHG emissions in low-cost apartments have a similar pattern. Works that use cement and concrete materials are the ones that have the biggest contribution to architectural work, such as screeding work and precast work. The volume reduction and the selection of the right construction technique such as high-intensity cement materials can be an option in reducing the value of embodied energy and carbon.

Keywords Embodied energy · Embodied GHG · IO Table · Architectural works · Low-cost apartment

1 Introduction

The construction of lower to middle-class apartments in several big cities in Indonesia continues to increase as the government has established a million houses program since 2015 [1]. Apartment development in urban area of Indonesia is also expected to continue to grow due to high urbanization. By the end of 2020, the percentage

D. Larasati (✉) · N. D. Reztrie · I. H. Lubis
Department of Architecture, SAPPD, Institut Teknologi Bandung, Bandung, Indonesia
e-mail: dewizr@ar.itb.ac.id

T. Ramadhan
Department of Architecture, FPTK, Indonesia University of Education, Bandung, Indonesia

Y. Hanifah
Department of Architecture, FTUP, Pancasila University Indonesia, South Jakarta, Indonesia

T. Kang and Y. Lee (eds.), *Proceedings of 2021 4th International Conference on Civil Engineering and Architecture*, Lecture Notes in Civil Engineering 201,
https://doi.org/10.1007/978-981-16-6932-3_31

of the urban population will be up to 60%, or 217.47 million [2]. This increase in population will encourage vertical housing development in urban areas in Indonesia.

The increase in the growth of low-middle class apartments undeniably will affect the raise in building material needs for construction. According to the data from Statistics Indonesia (BPS), there was an increase in the value of building material expenditures from 2010 to 2016 [3]. As long as the building material needs is increasing, the energy consumption and carbon emissions certainly escalating. This is proved by 50% of energy consumption and 33% of carbon emissions in the world are derived from the buildings. Mohammed, et al. also shared the same presumption, which states that buildings in the world consume 30–40% of primary energy for the construction, operation, and maintenance phases, and also take responsible for 40% of carbon emissions in the air [4].

The energy use of the uncontrolled and massive construction practices can give negative impacts on the environment. One of the environmental impacts of energy use is the emergence of carbon (CO_2) emissions. Carbon emissions are a type of greenhouse gas that is considered to be one of the factors causing the increase in world temperature and climate change [5]. Indonesia was once considered the third-largest contributor to greenhouse gases (GHG) in the world after China and the United States and, or fourth if the European Union was included in the list [6]. This study will explain how the use of materials in apartment building construction can affect embodied energy and carbon emissions.

2 Research Methodology

This study uses the LCA method as a method for calculating embodied energy and carbon emissions. In the LCA calculation, there is a scope of research that can be studied, including the product phase, construction, maintenance to demolition [7]. As for this study, the research was limited to the product and construction phases only, namely the A1–B2 phase. In calculating embodied energy and carbon emissions, there are three types of methods that can be used, including input output, process, and hybrid [8]. One method that is considered comprehensive in calculating EE and GHG emission is the IO table. IO table is method used by using a database on the economic sector [9]. Therefore, the calculation will relate to the total cost of the materials used in apartment construction.

2.1 IO Table Analysis

The input–output analysis developed by Nansai et al. [10] and modified by Surahman [11] is used to measure the embodied energy and embodied GHG emissions. The calculation method of energy intensity and GHG emission intensity estimation by IO analysis is carried out in six stages, starting from the determination of the import and domestic inverse matrix, estimation of fuel consumption, setting of the net contribution rate, setting of the net calorific value, setting of the calorific value, calculation of embodied energy intensity, and the estimation of the embodied GHG emissions intensity. The results from the calculation of the intensity from the IO table are then calculated to multiply the volume of material and the price of the material to find out the results of the value of embodied energy and embodied carbon. The calculations are explained through Eqs. 1 and 2.

$$EE = Q_p \cdot z_i \cdot EE_m \tag{1}$$

$$GWP = Q_p . z_i . GWP_m \tag{2}$$

where EE is embodied energy (GJ); EE_m is embodied energy intensity (GJ/ million IDR); GWP is embodied carbon (kg.CO); GWP_m is embodied carbon intensity (kg.CO/IDR); Q = material quantity ($m^3/m^2/m'$); z is material price (million IDR); i = material type alternative.

2.2 Case Study

This study selects three case studies of low-cost and middle-class apartment development in big cities in Indonesia, specifically apartments that have units with an area of less than 36 m^2, which consist of studio and two-bedroom types. The apartment chosen also has a similar type, with a total floor area of approximately 20 floors and an area of about 20,000 m^2.

To be comparable, the data in this study were analyzed based on their area. So the comparative analysis of the cases is determined using a building area of 1 m^2 as a functional unit of comparison in this study.

3 Result and Discussion

3.1 Embodied Energy and GHG of Architectural Works

Figure 1 summarizes the results of the total embodied energy values in the architectural work of each case. From the data of all cases, the largest number is in the screeding work, which is 0.168 GJ/m^2 in case B, 0.203 GJ/m^2 in case A, and 0.245 GJ/m^2 in case C. The second-largest value is ceramic work, which is 0.129 GJ/m^2 in case B, 0.136 GJ/m^2 in case C, and 0.152 GJ/m^2 in case A. Meanwhile, the third rank with the largest embodied energy value is lightweight brickwork, which is 0.098 GJ/m^2 in case B, which is 0.118 GJ/m^2 in case C, and 0.131 GJ/m^2 in case A.

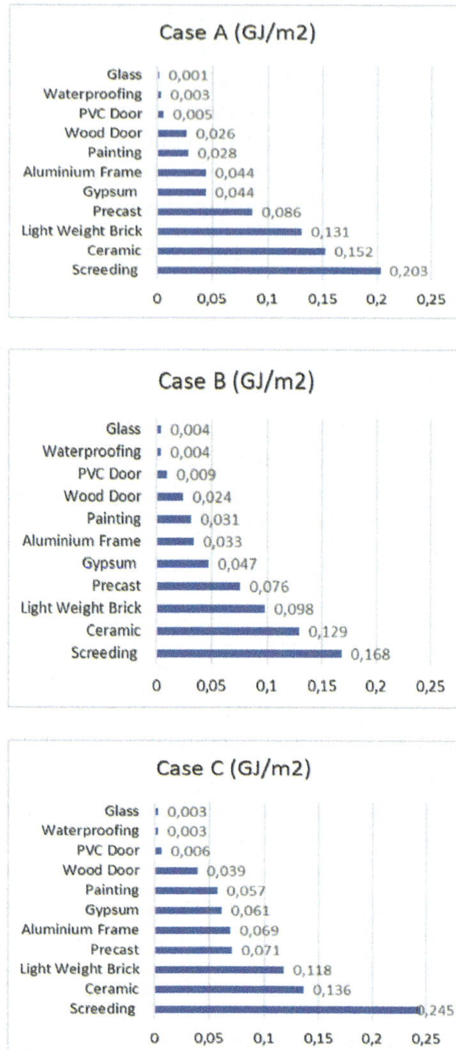

Fig. 1 The value of embodied energy in architectural works

As well as the embodied GHG data (Fig. 2), the EE values of all cases show similar trend. From the data of all cases, the greatest value is in the screeding work, which is 14.57 kg.CO/m^2 in case B, 17.78 kg.CO/m^2 in case A, and 21.11 kg.CO/m^2 in case C. The second-largest value is ceramic work, which was 11.5 kg.CO/m^2 in case B, 11.86 kg.CO/m^2 in case C, and 13.56 kg.CO/m^2 in case A. Meanwhile, the third-largest embodied energy value is the work of lightweight brick, which is 8.86 kg.CO/m^2 in case B, of 10.19 kg.CO/m^2 in case C, and 11.11 kg.CO/m^2 in case A.

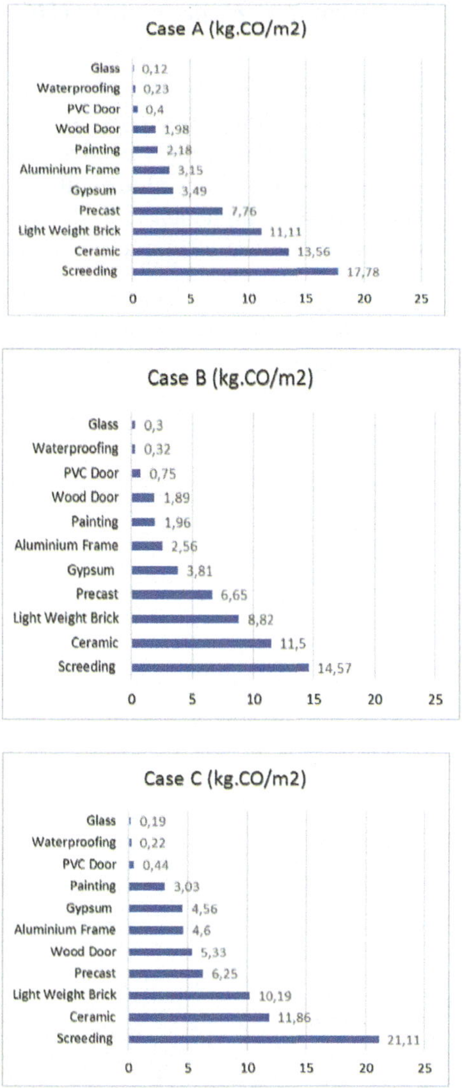

Fig. 2 The value of carbon emission in architectural works

From all the data, it can be seen that the embodied energy and GHG embodied values in the architectural work of low-cost apartment building have an identical pattern. This is because the volume on screeding work and lightweight brickwork is the largest volume of work in architectural work. As for ceramic work, this is due to the high intensity value of the building material.

3.2 Embodied Energy and GHG of Architectural Material

According to Figs. 3 and 4, it can be concluded that mortar or cement has the highest contribution to embodied energy and carbon emissions. It is followed by precast material and ceramic in the second and the third position with highest contribution of EE and GHG emission value of all apartments.

The result show high total value of the embodied energy and carbon emission was influenced by the use of mortar and precast. In the case of mortar in architectural

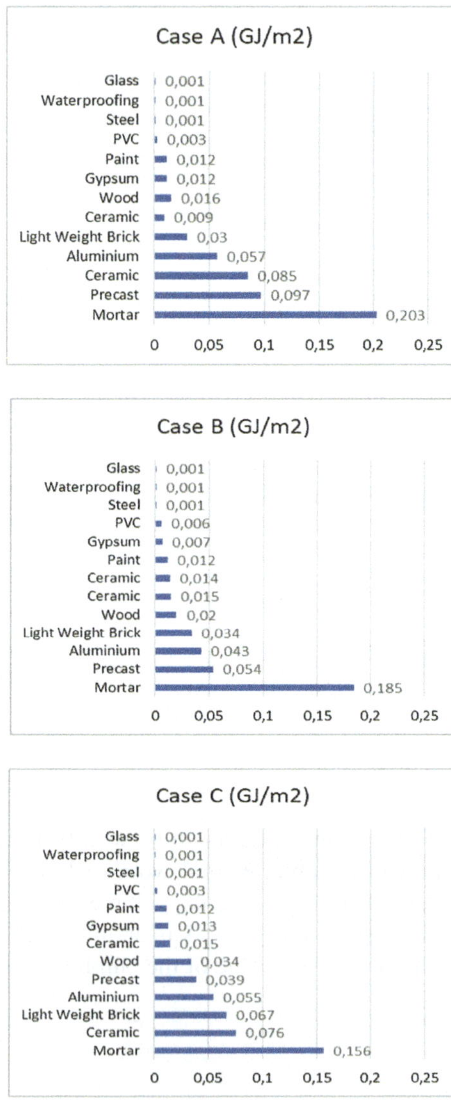

Fig. 3 The value of embodied energy of material in architectural works

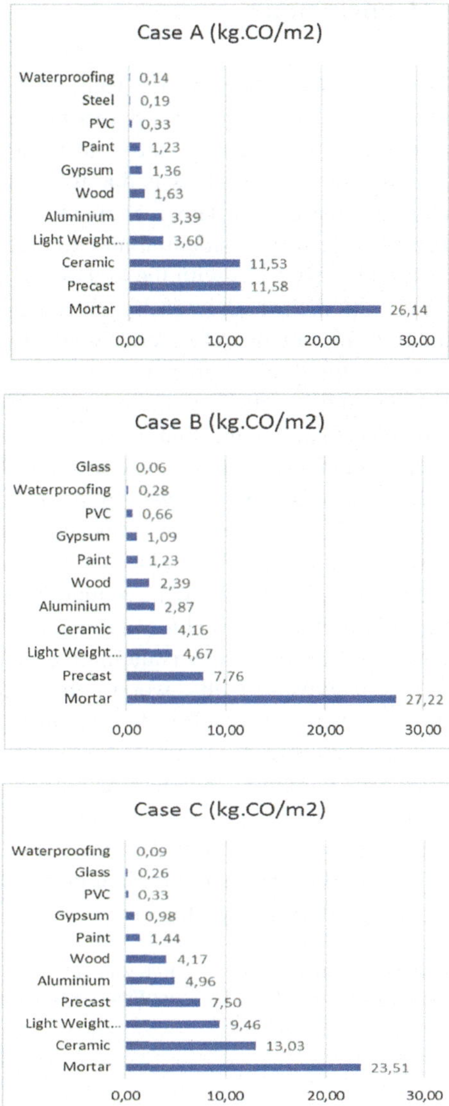

Fig. 4 The value of carbon emission of material in architectural works

work, the large volume in screeding and lightweight brickworks affects the volume of mortar, thus result in the high energy embodied value and carbon emissions. Meanwhile, in the case of precast work, the high intensity of the concrete influences the high energy embodied value and carbon emissions.

In the data of each work, the values of both EE and GHG emission of precast were below the value of EE and GHG emission of lightweight brick. However, in the data of each material, the values of both EE and GHG emission of precast wereabove the value of EE and GHG emission of lightweight brick.

4 Conclusion

The results show that the EE value and GHG emissions in low-cost apartments indicate an identical pattern. Screeding work and lightweight brickworks that use cement and concrete are contribute most to the value of EE and GHG emissions in architectural work. This is due to the large volume of cement material used in building construction. This is in line with the study done by Syngros et al. [12] and Shah et al. [13] on residential and commercial buildings who revealed that one of the building materials with the highest volume is concrete.

In addition, the high intensity of the material also affects the value of EE and GHG emissions in cement and concrete work. This is also proved by Tatsuo et al. [14] who states that the greatest energy intensity is in the steel material followed by cement material. In line with this, the Indonesia Ministry of Energy and Mineral Resources (ESDM) [15] also states that building materials such as cement and metals rank first and second in the manufacturing sector as the largest contributor to energy use in the manufacturing sector, followed by food and beverages.

Based on the patterns identified in the results that has been carried out, several alternatives are needed to reduce the value of EE and GHG emissions. Reducing the volume of high-intensity cement materials can be an option in reducing the energy and carbon embodied value. Lightweight brickworks that use mortar at most can be replaced by using other alternative materials that have lower intensity such as the use of gypsum walls, wood composites, bamboo, hollow concrete blocks or other materials [16–18]. Besides that, material installation techniques can be used as another alternative, for example by reducing the volume of mortar in wall finishing with the most optimal thickness design [18] and by designing materials with modular techniques [19, 20]. All in all, various design strategies in order to reduce embodied energy need to be developed so that they can become a reference for stakeholders involved in the apartment building process in order to realize low energy embodied and GHG embodied apartments.

Acknowledgements This research was supported by Productive and Innovative Research (Rispro) program of Indonesia Endowment Fund for Education (LPDP) 2021.

References

1. Ministry of National Development Planning (2014) Rencana Pembangunan Jangka Menengah Nasional (RPJMN) 2015–2019, Jakarta
2. Ministry of National Development Planning/National Development Planning Agency (2013) Indonesia Population Projection 2010–2035. Ministry of National Development Planning/National Development Planning Agency, Jakarta
3. Kumanayake R, Luo H, Paulusz N (2018) Assessment of material related embodied carbon of an office building in Sri Lanka. Energy Build 166:250–257
4. Ibn-Mohammed T, Greenough R, Taylor S, Ozawa-Meida L, Acquaye A (2013) Operational vs. embodied emissions in buildings—a review of current trends. Energy Build 66:232–245

5. IPCC (2008) 2006 IPCC guidelines for national greenhouse gas inventories—a primer
6. GlobalABC (2018) 2018 Global Status Report: Towards a zero-emission, efficient and resilient buildings and construction sector
7. British Standards Institution (2011) BS EN 15978:2011 Sustainability of construction works—assessment of environmental performance of buildings—calculation method
8. Dixit MK (2017) Life cycle embodied energy analysis of residential buildings: a review of literature to investigate embodied energy parameters. Renew Sustain Energy Rev 79:390–413
9. Onat NC, Kucukvar M, Tatari O (2014) Scope-based carbon footprint analysis of U.S. residential and commercial buildings: an input-output hybrid life cycle assessment approach. Build Environ 72:53–62
10. Nansai K, Moriguchi Y, Tohno S (2002) Embodied energy and emission intensity data for Japan using input-output tables (3EID)–inventory data for LCA. Natl Inst Environ Stud
11. Surahman U, Kubota T, Wijaya A (2016) Life cycle assessment of energy and CO2 emissions for residential buildings in Jakarta, Indonesia. In: IOP Conference Series: Materials Science and Engineering, vol 128, no 1
12. Syngros G, Balaras CA, Koubogiannis DG (2017) Embodied CO2 emissions in building construction materials of Hellenic dwellings. Procedia Environ Sci 38:500–508
13. Shah SMH, Junaid A, Hilal Khan R, Gardezi SSS (2019) Assessment of embodied carbon footprint of an educational building in Pakistan using Building Information Modelling (BIM). In: 11th International Conference, no. January, 2019
14. Tatsuo O, Yokoyama K, Takao S, Yokoo N, Yamamoto M (2014) Introduction of Annex 57—evaluation of embodied energy and carbon dioxide emissions for construction worldwide. World SB14 Conf., pp 26–32
15. Abdurrahman S (2018) Indonesia Energy Outlook 2018. Jakarta (2018)
16. Vidya A, Radha Ar (2008) Alternative low-cost building materials
17. Jurizat A, Ramadhan T (2020) Embodied Energy Pada Dinding Bambu Anyaman Dan Plester. J Arsit Zo 3(2):178–191
18. Jayasinghe C (2011) Embodied energy of alternative building materials and their impact on life cycle cost parameters. In: International conference on structural engineering and construction management, 16–18 December 2011, Kandy, Sri Lanka, no. January, pp 1–20
19. Hanifah Y, Dwika Reztrie N, Ramadhan T, Larasati D (2019) Evaluation of material selection on the initial embodied energy value of low-middle apartment in Indonesia. IOP Conf Ser Earth Environ Sci 294(1):012036
20. Reztrie ND, Larasati D (2019) Factors influence embodied energy and embodied carbon value at design phase of low middle class apartment in Indonesia. IOP Conf Ser Earth Environ Sci 294(1)

A BIM-Based Method for Building Energy Intensity (BEI) Evaluation

Taki Eddine Seghier and Lim Yaik Wah

Abstract Energy Usage Index (EUI) is used worldwide as a metric to evaluate and compare energy consumption of sustainable buildings. In Malaysia, EUI is referred as Building Energy Intensity (BEI) and it is adopted by different energy standards and green building rating systems. The preliminary evaluation of BEI in the Malaysian building sector is typically performed using an Excel-based tool known as Building Energy Intensity Tool (BEIT). This method has several limitations including but not limited to; (1) it supports only simplified building envelop configurations and (2) data input is manual and based on many assumptions. Due to the potential of Building Information Modelling (BIM), the opportunity for this method to adopt and benefit from BIM arises. Therefore, this research proposes a BIM-based approach for BEI assessment by integrating BEIT, BIM data and visual scripting to automatically extract the required data for BEI assessment and therefore overcome the current limitations. The applicability of the proposed system is validated in a BIM model of an office building. The proposed system provides a valuable decision support system for designers to analyze and compare the BEI of different design options.

Keywords Energy usage index (EUI) · Building energy intensity (BEI) · Energy efficiency · Building information modelling (BIM) · Visual programming

1 Introduction

In Malaysia, buildings consume a total of 48% of the electricity generated in the country [1]. Energy consumption metrics such as Energy Usage Index (EUI) and Building Energy Intensity (BEI) are developed to be used as guiding metrics to measure and compare the energy consumption of buildings. There are two major

T. E. Seghier (✉)
School of Architecture and Built Environment, Faculty of Engineering, Technology and Built Environment, UCSI University, Kuala Lumpur, Malaysia
e-mail: taki@ucsiuniversity.edu.my

L. Y. Wah
Faculty of Built Environment and Surveying, Universiti Teknologi Malaysia, Johor Bahru, Malaysia

T. Kang and Y. Lee (eds.), *Proceedings of 2021 4th International Conference on Civil Engineering and Architecture*, Lecture Notes in Civil Engineering 201,
https://doi.org/10.1007/978-981-16-6932-3_32

assessment models to evaluate the EUI/BEI of a building; assessment based on dynamic simulation and assessment based on static simulation. Dynamic simulation is usually performed using validated simulation engines such as EnegyPlus. However, static simulation is often performed using an Excel spreadsheet (template) based tool where the user inputs the required data manually to generate the simulation output. Both simulation models have advantages and disadvantages. However, one of the critical disadvantages of the static simulation approach is that data input is fully manual and based on assumptions in most cases.

In the Malaysian building industry, BEIT tool is widely used and recommended by green building guidelines and standards [2–4]. BEIT tool adopts a static simulation model in calculating the overall building energy consumption. Nonetheless, BEIT tool has several limitations which make it impractical and error-prone in various scenarios. Firstly, this tool was developed based on the concept of "Shoebox"; meaning buildings with complex forms should be simplified to regular parallelepiped forms before proceeding to the assessment. Secondly, BEIT tool supports only building envelops with maximum four elevations and only one material/window type per elevation. For instance, if the North elevation of a building contains two window types, the user is not be able to input the data of the second type because of tool's limitation. Thirdly, data input in BEIT tool is manual and based on assumptions in most cases. For example, lighting power and plug load in BEIT tool are assumed using a single input for all spaces regardless the number and type of spaces available in the project. In building performance simulation, it is very important to extract all this data from each space or at least using the averaged values of lighting power and plug loads of all spaces to ensure accuracy of the calculation. Furthermore, since data input is based on assumption in BEIT, it is believed that users must have expertise in building performance analysis to select the appropriate assumption.

On the other end of the spectrum, Building Information Modelling (BIM) provides a rich source of data that can be extracted and managed to benefit all phases of building lifecycle. BIM design process and technologies adoption has grown substantially worldwide. According to NBS [5] the awareness and usage of BIM has increased from 13 to 73% worldwide. Due to its enormous advantages, many scholars suggest adopting BIM-based process to solve issues related to green building evaluation [6, 7]. Based on the issues discussed earlier it is therefore important to link BEIT tool with rich database of information within the BIM model. Thus, in this study we propose a BIM-based method to improve one of the current widely used method in Malaysia for BEI assessment. The developed method is achieved by integrating BIM, visual scripting, and BEIT tool. The BIM model is used as a database for data extraction, visual scripting is implemented using Dynamo to develop scripts for automated data extraction from the BIM model, then the extracted data is processed and inputted in the BEIT tool to generate the BEI. Finally, the proposed BIM-BEI system is tested on a case study and the results are discussed.

2 Literature Review

Energy efficiency plays a pivotal role in sustainable building design. Building Energy Intensity (BEI) measures the intensity of energy consumption of a building per year. Most of design decisions taken by the project team have an impact on the end energy usage of the building. For instance, the size of windows, materials of building envelop, room volume, and others, all have an impact on energy consumption either directly or indirectly. There two main methods of BEI assessment; assessment based dynamic simulation and assessment based on static simulation. The latter is widely used method in Malaysia for BEI assessment, and it is often performed using the BEIT tool.

BEIT (Building Energy Intensity Tool) is a software tool developed to assess energy use in buildings and it is developed by the Association of Consulting Engineers Malaysia (ACEM) in 2010 [8]. The tool has gone through series of improvement where the last update was in 2018. According to the developers, this tool allows green building practitioners and energy auditors to perform easy BEI calculations, and to make quick and reliable predictions of energy savings that can be achieved for various new design and retrofit measures. BEIT is developed specifically for use under the equatorial climatic conditions found in Malaysia. However, it may also provide reasonable results in other hot and humid climate parts of the ASEAN region. The accuracy of BEIT tool output is dependent on the user's inputs. Unfortunately, all data input in BEIT are done manually and based on assumptions. Therefore, integrating BEIT with building information data and visual scripting could improve data input and increase the automation of BEI assessment.

According to Kensek [9], green building related design issues could benefit tremendously from BIM and visual programming integration. Many repetitive tasks and data extraction processes are being automated by developing visual scripts using software such as Dynamo. For instance, Khosakitchalert [10] developed a method which relies on visual scripting to improve the quality of material take-off for component elements. Seghier [11] used Dynamo and Revit to automate the calculation of Concrete Usage Index (CUI). Konis [12] developed a framework for building passive performance optimization in the early design stage.

3 Proposed BIM-BEI Prototype System

The BIM-BEI system is developed to overcome the current challenges and limitations in assessing BEI using BEIT tool. Additionally, it seeks to take advantage of BIM data during the assessment process to automate data input by relying on the data extracted from the BIM model. As shown in Fig. 1, the BIM-BEI system consists of three major parts: a BIM model and two modules developed for the purpose of this study; (1) OTTV and RETV module and (2) occupancy load, lighting power and plug load module. These three parts work together to extract all the necessary data from

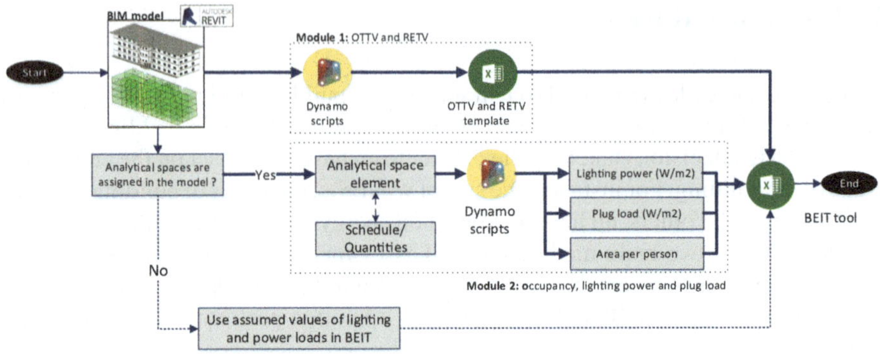

Fig. 1 BIM-BEI system prototype logic workflow

the BIM model which then is inputted in the BEIT tool to generate the assessment results of BEI.

3.1 *BIM Model and Analytical Spaces*

The BIM model consists of a database where all the necessary data is extracted from. For this reason, the modelling accuracy of the BIM model in Revit is highly important to ensure accurate data extraction. For the purpose of this study, only the architectural BIM model is required in addition to the analytical space elements which are often modelled by the mechanical engineer and therefore available in the BIM mechanical model of the building. The architectural model contains the building envelop elements such as wall, window and curtain wall, shading devices. This model consists the main source of data for Overall Thermal Transfer Value (OTTV) and Roof Thermal Transfer Value (RETV) calculation. On the other hand, the analytical spaces are used to calculate the averaged lighting power, plug loads and occupancy load of the building.

3.2 *Module 1: OTTV and RETV*

Heat gain through building envelop plays a fundamental role in BEI calculation. In Malaysia, OTTV and RETV are used as the main performance metric to assess the overall thermal transfer through building envelop and roof respectively (Department of Standards Malaysia 2014). Therefore, it is crucial step to automate the OTTV calculation to overcome the current limitation of BEIT tool which evaluates the OTTV in a very limited approach as discussed in the introduction. In this research OTTV module consists of a Revit template hosting the required shared parameters in addition to two Dynamo scripts developed to extract the required data for OTTV calculation from BIM model. These two scripts extract the thermal properties and physical parameters from building components namely walls, windows and roof. More details about these scripts and the developed Revit template are available in previous publications [13, 14].

3.3 *Module 2: Occupancy, Lighting Power and Plug Load*

Occupancy load, lighting power and plug load contribute significantly to the overall energy consumption of a building. As mentioned earlier in the introduction, these parameters are assumed in BEIT tool by the user instead of an accurate data input based on the building layout and space types. In this study, a Dynamo script is developed to extract the data related to these parameters from the BIM model. The data extraction process is performed through the following steps: (1) the script selects all the analytical spaces elements within the BIM model, (2) the script selects the relevant parameters hosted in the analytical spaces, (3) the selected parameters are processed to generate a real number (without a unit), and (4) the script calculates the averaged values of occupancy load, lighting power and plug load based on the unique values hosted in each space and the gross floor area of the building. The results are then exported to the BEIT tool.

4 Case Study

The BIM-BEI system is tested on a case study which consists of a typical office building with four levels, four elevations and 3473 m^2 of air-conditioned spaces (see Fig. 2). The physical and thermal parameters involved in BEI calculation are made

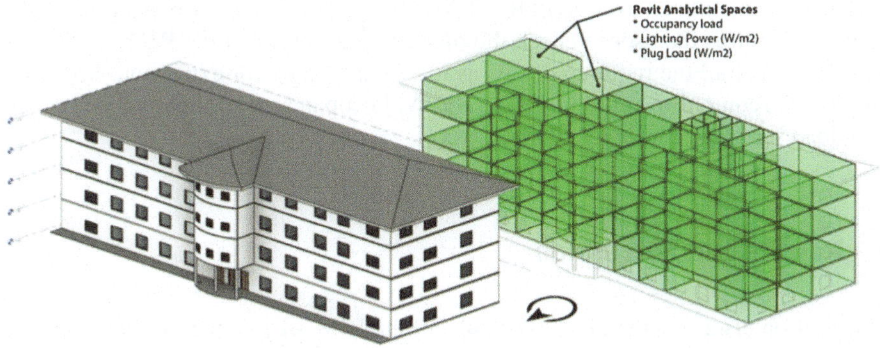

Fig. 2 BIM model of the case study (left) and assigned analytical spaces (right)

Table 1 Physical and thermal parameters of the case study

Building component	Type and parameters	Number of elements			
		N	S	E	W
Windows	**Type 1**: M-Fixed, large single glazed window, Area: 2.89 m^2, U-value: 5.6 W/(m^2.K), SC1: 0.9	32	36	4	4
	Type 2: M_Window-Double-Hang, No: 8, Area:2 m^2, U-value: 6.7 W/(m^2.K), SC1: 0.87	8	0	8	8
	Type 3: M-window, casement-double, No: 9, Area: 0.76 m^2, U-value: 3.1 W/(m^2.K), SC1: 0.87	9	0	0	0
	Type 4: M-Fixed, Plinkton glazed window, No: 3, Area: 0.55 m^2, U-value: 3.6 W/(m^2.K), SC1: 0.9	0	3	0	0
Shading devices	**None**	N/A			
Exterior walls	**Type 1**: Brick, common, U-value: 2.3 W/(m^2.K)	N/A			
Roof	**Type 1**: Warm roof concrete, U-value: 0.5 W/(m^2.K)	N/A			

available in the BIM model by selecting specific wall and window types using Revit. It is worth noting that each elevation of the building contains at least two different types of windows. The detailed data of the case study is given in Table 1.

The analytical spaces were added to the BIM model because all the data related to lighting power, plug loads and occupancy load is extracted from it. The data hosted in the analytical space elements in Revit are dependent on the space type parameter. Revit database contains a large number of analytical space types where the user can select from. Therefore, by selecting a specific space type (e.g., meeting room, enclosed office) other sub-parameters such as lighting power intensity will be loaded accordingly.

The implementation process of the BIM-BEI system follows two main steps. The first step requires the user to check the modelling accuracy of the BIM model, and make sure that all the required data is available and ready for extraction. Secondly, the user is required to run the OTTV, RETV, occupancy load, lighting power and plug load scripts accordingly.

5 Results and Discussion

This section discusses the obtained results of the BIM-BEI system during its implementation on the case study. The obtained results show that OTTV and RETV values generated by the proposed system are exactly equal to the values generate using BEIT tool alone. OTTV output is 30.12 W/m^2 while RETV output equals to 10 W/m^2 (see Fig. 3). However, the difference between the two methods resides in the data input process. When using the BIM-GBI system the data input is extracted automatically from the BIM model. On the other hand, by using the traditional method the user assumes the values before inputting them manually in the BEIT tool. As shown in Fig. 3, the data related to OTTV for example is extracted automatically from the BIM model using Dynamo then inputted to the BEIT tool. Additionally, the results obtained from the second developed module of BIM-BEI is 7.3 m^2/person for occupancy load, 8.5 W/m^2 for lighting power and 8.7 W/m^2 plug loads. Similarly, these values were extracted automatically and calculated based on the BIM model configuration. However, when using BEIT too alone the user selects predefined options from a drop list in the tool regardless of the building configuration. Unfortunately, this makes the assessment output dependent on assumptions given by the user instead of real and accurate data extracted from the building model. Finally, all the data generated using the BIM-BEI modules together with other BEI parameters (e.g., Air-Conditioning and Mechanical Ventilation (ACMV) system energy intensity) are inputted in the BEIT tool to generate the BEI of the building. As shown in Fig. 3, the BEI of the building is 131.1 kWh/m^2/year. It is worth noting that the table of BEIT tool shown in Fig. 3 is just a small portion of it.

6 Conclusion

This study investigated a method to support the assessment of Building Energy Intensity (BEI) by integrating the BIM data, visual scripting, and BEIT tool. The system prototype founded on this method relies on the data of the BIM model and two developed modules; module (1) OTTV and RETV and module, (2) occupancy

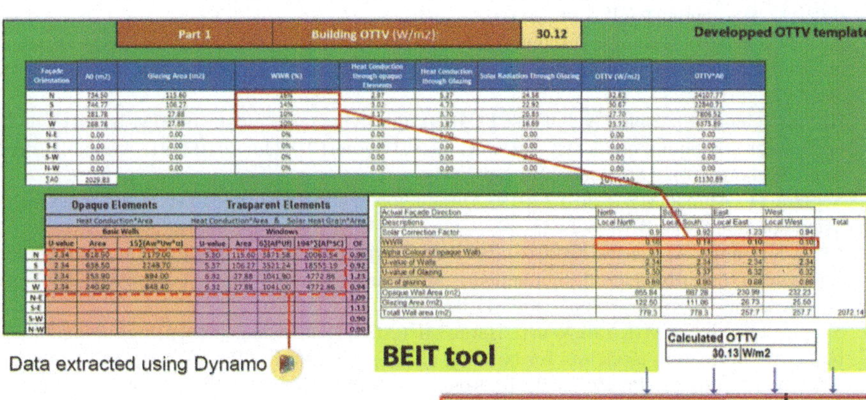

Fig. 3 OTTV data processing from the developed OTTV template to the BEIT tool

load, lighting power and plug load. The development process of the BIM-BEI system started by reviewing the limitations and challenges of BEIT tool. Then based on these limitations, the applicability of BIM data and visual scripting were investigated to develop scripts for data extraction. Finally, the developed system was tested on a case study building. The results have shown that the proposed BIM-BEI system is more advantageous compared to the current method of assessing BEI in term of automating data extraction and input to the BEIT tool.

Acknowledgements Authors wishing to acknowledge the financial support provided by UCSI University. Additionally, they would like to acknowledge the research funding from IJM Construction Sdn. Bhd. and Technology Depository Agency (TDA) Berhad through the Contract Research Grant, project no. 4C270, titled "Building Information Modelling (BIM) for Green Building Assessment".

References

1. Hassan JS, Zin RM, Majid MZA, Balubaid S, Hainin MR (2014) Building energy consumption in Malaysia: an overview. J Teknol 70(7):33–38
2. GBI (2013) GBI assessment criteria for residential new construction (RNC) 0–17
3. GreenRE (2015) Design reference guide 1–94
4. Lim Y, Shahsavari F, Sediadi E, Azli NF, Binti MN (2016) Building information modelling for building energy efficiency evaluation. Ace 42–48
5. NBS (2020) 10th Annual BIM Report. 2020, pp 1–39. Available from: https://www.thenbs.com/knowledge/national-bim-report-2020
6. Lim Y, Seghier TE, Harun MF, Ahmad MH, Samah AA, Majid A (2019) Computational BIM for building envelope sustainability optimization. 04001:2–6
7. Jrade A, Jalaei F (2013) Integrating building information modelling with sustainability to design building projects at the conceptual stage. Build Simul 6(4):429–444
8. GBI (2010) GBI RNC Residential Tool V1.2. Green Build Index Ref Guid Submiss Format Internet], pp 33–38. Available from: http://www.greenbuildingindex.org/Resources/GBITools/GBIRNCResidentialToolV2.0Final.pdf
9. Kensek K (2015) Visual programming for building information modeling: energy and shading analysis case studies. J Green Build [Internet]. 10(4):28–43. Available from: https://doi.org/10.3992/jgb.10.4.28
10. Khosakitchalert C, Yabuki N, Fukuda T (2019) Improving the accuracy of BIM-based quantity takeoff for compound elements. Autom Constr [Internet] 106(June):102891. Available from: https://doi.org/10.1016/j.autcon.2019.102891
11. Seghier TE, Ahmad MH, Lim Y (2019) Automation of concrete usage index (CUI) assessment using computational BIM. 6(1):23–30
12. Konis K, Gamas A, Kensek K (2016) Passive performance and building form: an optimization framework for early-stage design support. Sol Energy Internet 125:161–79. Available from https://doi.org/10.1016/j.solener.2015.12.020
13. Seghier TE, Ahmad MH, Yaik Wah L, Harun MF (2020) Data management using computational building information modeling for building envelope retrofitting BT—smart and sustainable cities and buildings. In: Roggema R, Roggema A (eds) Springer International Publishing, Cham, pp 205–216. Available from: https://doi.org/10.1007/978-3-030-37635-2_13
14. Seghier TE, Wah LY, Ahmad MH, Samuel WO (2017) Building envelope thermal performance assessment using visual programming and BIM, Based on ETTV requirement of Green Mark and GreenRE 4(3):227–235

Urban Space Planning and Development

Public Utilization of Rooftop Garden at Commercial Buildings in Hot and Humid Climate

Wardah Fatimah Mohammad Yusoff

Abstract Green roof is able to reduce the environmental issues such as the urban heat island. Besides that, the social benefits can also be achieved by the provision of green roof. Nevertheless, the application of it incurs additional cost such as the construction and maintenance costs. Therefore, it is pivotal to ensure the optimum utilization of green roof. This study aims to derive a preliminary idea regarding the public utilization of intensive green roof which is the rooftop garden. Two commercial buildings located in Malaysia had been selected for the study. The observation and the questionnaire survey methods had been executed to identify the characters of the rooftop gardens, as well as the public utilization of them. From the study, it is found that the design, vegetation types and other supporting spaces affect the utilization pattern and activities executed at the rooftop garden. These aspects also determine the purpose, frequency, time and duration of visiting the rooftop garden. The findings from this study are able to encourage further research of the appropriate green roof criteria, especially for the commercial buildings in hot and humid climate.

Keywords Rooftop garden · Commercial building · Hot humid climate

1 Introduction

The benefits of green roof to the environment have been discussed by many studies. Amongst the benefits are reducing the outdoor air temperature [1] thus mitigating the urban heat island effect [2–4], decreasing the cooling load energy for building [1, 5–7], enhancing stormwater retention [7], as well as improving the surrounding air quality [7–9].

Besides the environmental benefits, many studies are also interested in investigating the economic aspect of green roof such as the life-cycle cost of a green roof which includes the cost of installation, maintenance and many others. Examples of

W. F. Mohammad Yusoff (✉)
Department of Architecture and Built Environment, Faculty of Engineering and Built Environment, Universiti Kebangsaan Malaysia, Bangi, Malaysia
e-mail: wardahyusoff@ukm.edu.my

T. Kang and Y. Lee (eds.), *Proceedings of 2021 4th International Conference on Civil Engineering and Architecture*, Lecture Notes in Civil Engineering 201,
https://doi.org/10.1007/978-981-16-6932-3_33

such studies are by Bianchini and Hewage [10], Berto et al. [11], Niu et al. [12] and Mahdiyar et al. [13].

Other than environmental benefits, green roof also provides social benefits such as recreational use, as well as enhancing the aesthetic value of the building [14]. Some of the social benefit studies are also associated with the health benefits of the green roof such as by Triguero-Mas et al. [15] and Williams et al. [16]. Triguero-Mas et al. [15] investigated the social benefits of green roof that focused on the utilization of the rooftop for gardening activity. The study associated the rooftop gardening activity with the quality of life for people with intellectual disabilities. Meanwhile, Williams et al. [16] investigated the benefits of green roof in term of psychological impact. The focused groups in the study were the residents and workers who lived in the city. However, to date, the studies of social benefits of green roof are found to be lesser compared to the environmental benefits.

Each green roof project has its own uniqueness where it is characterized by the size of the roof, the building types, the local climate where it is situated, the life-cycle and many others [17]. Nevertheless, in general, the green roof can be classified into two types namely the extensive and intensive green roof. Though both types provide environmental benefits, the intensive green roof is found to be able to achieve greater reduction in air temperature and building cooling load energy, as well as enhancing thermal comfort compared to extensive green roof [1]. One of the examples of intensive green roof is rooftop garden. However, the intensive green roof has drawback of involving higher construction and maintenance costs compared to the extensive green roof [1]. The construction and maintenance costs are among the main hindrances that discourage the application of green roof in many buildings. Nevertheless, in the social aspect, the intensive green roof such as the rooftop garden provides more benefits as it can be accessed and utilized by people.

As the intensive green roof such as the rooftop garden incurs extra cost, it is essential that its benefits are fully gained for both, the environmental and social benefits. With the right approach of the green roof design, the environmental benefits are easily achieved. However, the social benefit such as the optimum usage of the intensive green roof by people is still questionable. Hence, this study investigates the public utilization of intensive green roof namely the rooftop garden. Focus is given to the rooftop garden at the commercial building compared to the other building typologies such as residential, institution and office as it provides more accessibility to the public.

2 Description of the Studied Rooftop Garden

The selected rooftop gardens for the study are located at Lot 10 Shopping Complex in Kuala Lumpur and 1 Utama Shopping Complex in Petaling Jaya, Selangor (Fig. 1). They were selected due to their locations which are at the urban area, where the urban heat island issue is inevitable.

(a) (b)

Fig. 1 The rooftop gardens at **a** Lot 10 Shopping Complex, and **b** 1 Utama Shopping Complex

Table 1 Description of rooftop gardens at Lot 10 and 1 Utama Shopping Complexes

Characters	Lot 10 Shopping Complex	1 Utama Shopping Complex
Type	Intensive green roof	Intensive green roof
Location	Rooftop (8th Floor)	Rooftop (7th floor)
Vegetation	Groundcovers, shrubs and trees	Groundcovers, shrubs and trees
Landscaping	Hardscape and softscape	Hardscape and softscape
Landscape layout	Diagonal design approach with open concept	Curvilinear design approach
Surrounding area	Services area, a gymnasium and an NGO's activities centre	Services area, futsal court and baseball court

The characters of both rooftop gardens are summarized in Table 1, while the layout plans are depicted in Fig. 2. In general, both of the rooftop gardens are situated at the roof level of the building, where the rooftop garden of Lot 10 Shopping Complex is located at 8th floor, while the rooftop garden of 1 Utama Shopping Complex is situated at the 7th floor. Hence, the height of both selected commercial buildings are almost similar. Both rooftop gardens are planted with the ground covers, shrubs and trees. They are also designed to have both, the hardscape and softscape. The main difference between them is the landscape layout where the rooftop garden of Lot 10 shopping complex has a diagonal design approach with open concept. On the contrary, the rooftop garden of 1 Utama Shopping complex is designed with curvilinear approach. The rooftop area of both shopping complexes is also found to accommodate other spaces besides the garden. At the rooftop area of Lot 10 shopping complex, there are also services area, gymnasium and spaces for the non-governmental organization (NGO) activities. Meanwhile, for the 1 Utama Shopping Complex, there are services area, futsal court and baseball court at the rooftop. More description of these two rooftop gardens are presented in Sect. 4.1.

(a) (b)

Fig. 2 The layout plans of rooftop gardens at **a** Lot 10 Shopping Complex and, **b** 1 Utama Shopping Complex. The green roof area is indicated by green colour

3 Research Methodology

The methods employed in examining the public utilization at the rooftop garden of both shopping complexes are observation and questionnaire survey. The purpose of observation is to identify the rooftop garden characters, while the objective of questionnaire survey is to determine the public utilization of the rooftop gardens. As the study is a preliminary investigation, the duration for the observation and questionnaire survey is seven days only for each rooftop garden. This duration includes the weekdays and the weekend.

Both methods were conducted simultaneously in the month of April 2019, from 10 am to 10 pm, which is in accordance to the operation time of the shopping complexes. There are 60 respondents for the rooftop garden of Lot 10 Shopping Complex and 100 respondents for the rooftop garden of 1 Utama Shopping Complex. The questionnaire survey form consists of two sections which the first section records the respondent's personal information, while the second section identifies the rooftop garden usage pattern of the respondent. In the second section, the purpose, frequency, time and duration of visit are documented.

4 Results and Discussion

The results are presented in two sub-sections namely the observation results and the questionnaire survey results.

4.1 Observation Results

From the observation conducted at both rooftop gardens, it is found that there are three factors that encourage the utilization of the rooftop garden. The first factor is the landscaping layout of the rooftop garden that is attractive and manage to provide various activities. The second factor is the vegetation types where they have to be the mixture of groundcovers, shrubs and trees. The shade elements are significant for hot and humid climate, which this can be achieved with the appropriate design and vegetation selection in the first and second factors, respectively. The third factor is the presence of other spaces around the rooftop garden, which can become attraction for more people to visit the rooftop area.

In term of the landscaping layout, it is found that each rooftop garden has elements that become the nodes of the users. The 1 Utama's rooftop garden has a pergola and a greenery tunnel that become the primary attractions to the users (Fig. 3a). These two elements are mainly used for relaxation and photoshoot. In addition, there are also water element and reflexology path at the garden which provide health benefit to the users. Meanwhile, for the Lot 10's rooftop garden, the node is mainly at the open field where many activities are executed there especially during the weekend. There are also vertical green walls that become the attraction for the users (Fig. 3b).

However, in term of vegetation types, it is found that the 1 Utama's rooftop garden has more trees that provide shades compared to the Lot 10's rooftop garden. This is in accordance to the garden's name which is known as Secret Garden. The landscaping layout that employs the curvilinear design approach provides surprise elements in the users' journey. In addition, the vegetation is also dense, with various types and species can be found at the garden. On the contrary, the Lot 10's rooftop garden layout is more towards the diagonal design approach with open concept. The vegetation there

(a) (b)

Fig. 3 **a** The greenery tunnel at the 1 Utama's rooftop garden, and **b** the vertical green wall at the Lot 10's rooftop garden

Fig. 4 **a** The fitness activity at the open area of Lot 10's rooftop garden, and **b** the visitors enjoy the greenery at the 1 Utama's rooftop garden

is also less dense and the types are found to be less variety compared to the 1 Utama's rooftop garden.

In term of the third factor, the rooftop area of both shopping complexes is found to accommodate other spaces besides the garden. At the rooftop area of Lot 10 shopping complex, there are a gymnasium and a centre for NGO's activities. There are many activities occur at the rooftop garden especially during the weekend such as the fitness activities which are held at the green open area (Fig. 4a). Meanwhile at the 1 Utama's rooftop area, there are futsal and baseball courts. All these supporting spaces have attracted many people to visit the rooftop area. The landscape layout of the 1 Utama's rooftop garden which is denser with vegetation compared to the Lot 10's roof top garden, has made this area been utilized most of the time. Visitors like to stroll along the provided pathway, while enjoying the variety of flora species planted at the garden (Fig. 4b). Hence, besides purposely visiting the facilities provided, they also take the opportunity to enjoy themselves at the rooftop garden.

4.2 Questionnaire Survey Results

The observation findings are then compared with the questionnaire survey results to understand further on the utilization of the rooftop garden by the public. The survey results of the visiting purpose (Fig. 5) indicate that many respondents visit to the rooftop garden for recreational activities. This is due to the rooftop garden layouts for both shopping complexes which provide attractive outdoor environment as well as areas for relaxation.

The results of the questionnaire survey indicate that majority of the visitors are the occasional users of the rooftop gardens (Fig. 6). The rooftop garden of Lot 10 Shopping Complex has recorded 39% of occasional visitors, while the rooftop garden

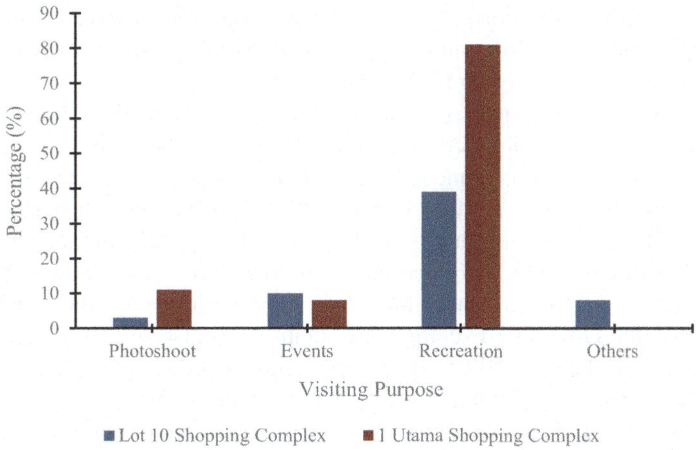

Fig. 5 The visiting purpose of the rooftop garden

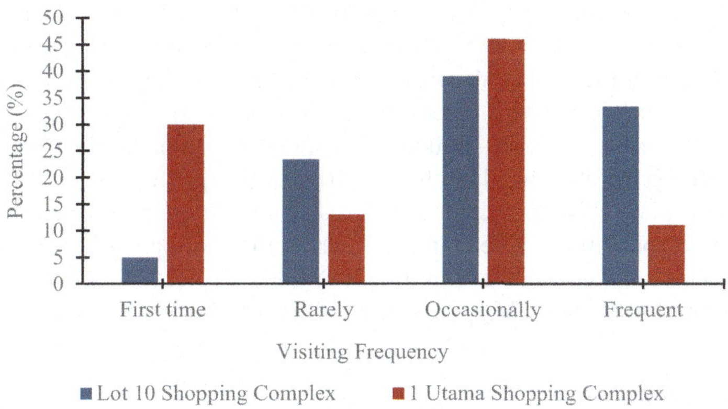

Fig. 6 The visiting frequency of the rooftop garden

of 1 Utama Shopping Complex has documented 46% occasional visitors. They are found to be amongst the shopping complex visitors as well as the workers. Figure 6 shows that the rooftop garden at the Lot 10 Shopping Complex manages to attract more frequent visitors which is around 33%, compared to the rooftop garden at 1 Utama Shopping Complex which is around 11% only. It demonstrates that those who have visited the garden once are attracted to repeat their visits. There are many factors that cause the revisiting of the users. These factors will be investigated further in future research to understand the criteria that aid or prevent the optimum usage of the intensive green roof. Nevertheless, Fig. 6 also demonstrates that there are many first time visitors at the rooftop garden of 1 Utama Shopping Complex, which is 30%, compared to the rooftop garden of Lot 10 Shopping Complex which has recorded 5% only. Hence, question arises on the visibility of the rooftop gardens among the public.

The visibility factor is important in ensuring that the public aware the presence of the rooftop garden at the shopping complex. The visibility aspect of rooftop garden will also be investigated further in the future research.

Besides the visiting frequency, investigation is also executed on the visiting time (Fig. 7), where it is in accordance to the operation time of the shopping complex that is from 10 am to 10 pm. It is found that both rooftop gardens are visited mainly at the evening hours. The rooftop garden of 1 Utama Shopping Complex has recorded 51% of visitors during the evening hours, while the rooftop garden of Lot 10 Shopping Complex has documented 39% visitors during the same hours. This is due to the normal activities pattern of Malaysian where evening hours are the usual time for outdoor activities. Other than evening hours, many users also visit the rooftop gardens during night time. The Lot 10's rooftop garden has recorded 28% visitors, while the percentage of visitors at 1 Utama's rooftop garden at night time is 17%. This is mainly due to the weather factor where the night time provides cooler outdoor environment compared to the daytime. Moreover, the visual effects created by the artificial lighting at the rooftop garden during night time also provides a scenic outdoor environment. In addition, the Lot 10's rooftop garden is found to attract more visitors at night compared to the 1 Utama's rooftop garden due to its open concept. The open concept of the garden may provide more safety condition to the visitors during the night time.

The duration of the visit is also examined in understanding the public utilization of the rooftop garden. Figure 8 demonstrates that for the 1 Utama's rooftop garden, majority of the users, which is around 54%, spend their time at the garden around 30 min to 1 h. On the other hand, for the Lot 10's rooftop garden, many users which is around 43%, spend shorter duration that is less than 30 min. Nevertheless, there are also users that spend time more than two hours at the rooftop garden of Lot 10. The long duration occurs during the night time due to the cooler outdoor environment at night compared to the day. Moreover, the artificial lighting effects in combination

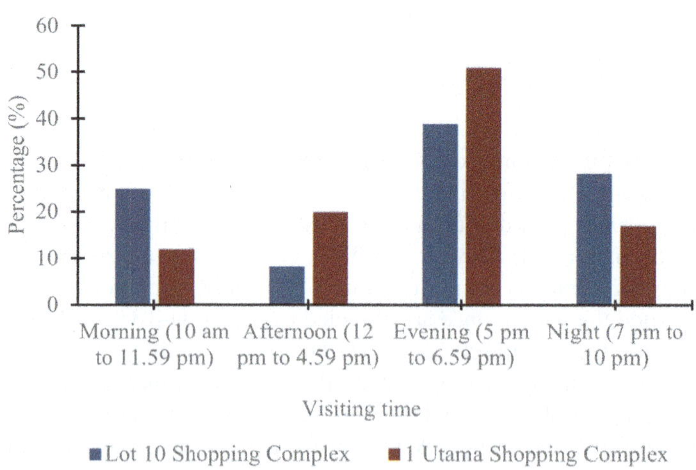

Fig. 7 The visiting time of the rooftop garden

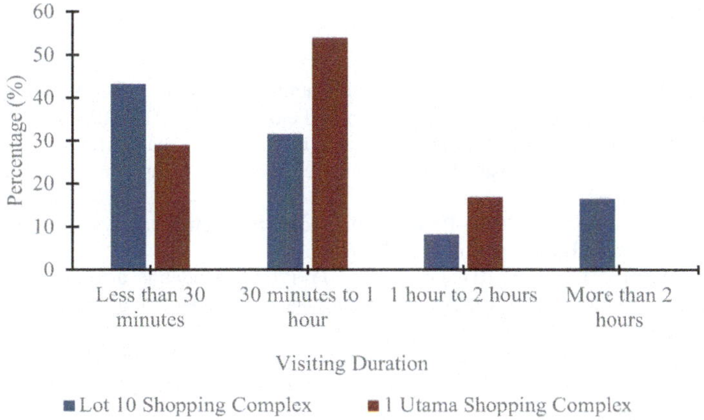

Fig. 8 The visiting duration of the rooftop garden

with the open concept of the Lot 10's rooftop garden have provided a scenic view during the night time. Nevertheless, there are also many other factors that influence the duration such as the user category (either shopping complex visitors or workers), the visiting purpose, the visiting frequency, the visiting time, as well as the rooftop garden outdoor condition.

5 Conclusion

In summary, the findings indicate that there are three factors that influence the public utilization of rooftop garden at commercial buildings in hot and humid climate, namely the landscaping layout, the selection of vegetation and the presence of other supporting spaces which able to attract people to visit the rooftop. These three factors also determine the purpose, frequency, time and duration of rooftop garden utilization by the public. The findings from this preliminary investigation also provide potential of embarking more studies on the utilization of rooftop garden at the commercial buildings. The understanding of public utilization is essential in ensuring that the rooftop garden is optimally used as the provision and maintenance of it incur certain costs.

Acknowledgements The author would like to acknowledge the financial support from the Ministry of Higher Education Malaysia through the Fundamental Research Grant Scheme (FRGS/1/2019/TK10/UKM/02/4-Air flow characteristics and behaviour around openings with various types of shading elements), and the National University of Malaysia (UKM) through the research grant GUP-2019-017. The author would also like to thank Nur Syazwani Abdul Majid who helped the execution of the questionnaire survey.

References

1. Aboelata A (2021) Assessment of green roof benefits on buildings' energy-saving by cooling outdoor spaces in different urban densities in arid cities. Energy 219:119514
2. Berto R, Stival CA, Rosato P (2018) Enhancing the environmental performance of industrial settlements: an economic evaluation of extensive green roof competitiveness. Build Environ 127(58–68):58–68
3. Bianchini F, Hewage K (2012) Probabilistic social cost-benefit analysis for green roofs: a lifecycle approach. Build Environ 58:152–162
4. Herath HMPIK, Halwatura RU, Jayasinghe GY (2018) Modeling a tropical urban context with green walls and green roofs as an urban heat island adaptation strategy. Procedia Eng 212:691–698
5. Kotzen B (2018) Green roofs social and aesthetic aspects. In: Pérez G, Perini K (eds) Nature based strategies for urban and building sustainability. Elsevier Inc., pp 273–281
6. Li Y, Qin X, Wu X, Chiew Y (2012) A green roof test bed for stormwater management and reduction of urban heat island effect in Singapore. Br J Environ Clim Change 2(4):410–420
7. Mahdiyar A, Tabatabaee S, Yahya K, Mohandes SR (2021) A probabilistic financial feasibility study on green roof installation from the private and social perspectives. Urban Forestry & Urban Greening 58:126893
8. McRae AM (2016) Case study: a conservative approach to green roof benefit quantification and valuation for public buildings. Eng Econ 61(3):190–206
9. Niu H, Clark C, Zhou J, Adriaens P (2010) Scaling of economic benefits from green roof implementation in Washington DC. Environ Sci Technol 44(11):4302–4308
10. Rowe DB (2011) Green roofs as a means of pollution abatement. Environ Pollut 159:2100–2110
11. Shafiquea M, Kima R, Rafiq M (2018) Green roof benefits, opportunities and challenges—a review. Renew Sustain Energy Rev 90:757–773
12. Sharma A, Woodruff S, Budhathoki M, Hamlet AF, Chen F, Fernando HJS (2018) Role of green roofs in reducing heat stress in vulnerable urban communities—a multidisciplinary approach. Environ Res Lett 13(9):094011
13. Triguero-Mas M, Anguelovski I, Cirac-Claveras J, Connolly J, Vazquez A, Urgell-Plaza F, Cole H (2020) Quality of life benefits of urban rooftop gardening for people with intellectual disabilities or mental health disorders. Prev Chronic Dis 17:200087
14. Williams KJH, Lee KE, Sargent L, Johnson KA, Rayner J, Farrell C, Williams NSG (2019) Appraising the psychological benefits of green roofs for city residents and workers. Urban Forestry & Urban Greening 44:126399
15. Wong NH, Cheong DKW, Yan H, Soh J, Ong CL, Sia A (2003) The effects of rooftop garden on energy consumption of a commercial building in Singapore. Energy Build 35(4):353–364
16. Yang J, Yu Q, Gong P (2008) Quantifying air pollution removal by green roofs in Chicago. Atmos Environ 42(31):7266–7273
17. Zinzi M, Agnoli S (2012) Cool and green roofs. An energy and comfort comparison between passive cooling and mitigation urban heat island techniques for residential buildings in the Mediterranean region. Energy Build 55:66–76

Chronological Development of Physical Planning of Hospital in Bangladesh

N. Afnan

Abstract The architectural design of hospitals in Bangladesh has undergone remarkable changes over the past few decades. To meet the evolving health needs of people and increasing demands for health services, the physical planning of hospitals has developed to a great extent. Rapid technological advancements have also affected the design of hospital. Newer medical technologies have significantly influenced the healthcare system. With the emergence of new kinds of chronic diseases, the conventional ways of treatment and care have changed a lot. These indispensable changes have had an impact on the design. Moreover, numerous components are being added to the design day by day and changes are being made accordingly. Architects are constantly challenged by the unforeseeable future needs and difficulties. They play an important role by ensuring flexible design solutions to overcome those unpredictable critical health challenges and creating a positive impact on hospital design. This paper highlights the chronological development of the physical planning of hospital in Bangladesh. This study can be used as a role model about how the physical design of hospitals has developed over the course of time in a developing country.

Keywords Hospital design · Medical technologies · Significant changes · Design guidelines

1 Historical Review

In ancient Bengal, the most popular form of treatment was Ayurvedic. The Muslim rulers developed It. During the colonial period, western medicines were recognized officially. Ayurvedic was also popular at that time. From the nineteenth century, the local authorities started to deliver health care facilities. Many dispensaries were established in that time and later converted to hospitals. For example, Pabna hospital, Shirajganj Hospital, etc. During the pre-independence period, 1200 dispensaries were established for the facilities of the rural poor population. By 1971, there were eight medical colleges and some specialized hospitals, such as Mental Hospital in

N. Afnan (✉)
Department of Architecture, Ministry of Housing and Public Works, Dhaka, Bangladesh

T. Kang and Y. Lee (eds.), *Proceedings of 2021 4th International Conference on Civil Engineering and Architecture*, Lecture Notes in Civil Engineering 201,
https://doi.org/10.1007/978-981-16-6932-3_34

Pabna, Seato Cholera Research Laboratory and Hospital, etc. Many sub-divisional and district hospitals were also established in this period. Development of the private sector had started during that time. For example, Islamia Eye Hospital, etc. After independence, more hospitals, such as Pongu Hospital, Shishu Hospital, Sohrawardi Hospital, etc., have been established. Some specialized hospitals, such as BIRDEM, N.I.C.V.D. etc., have also been established. The private sector has been developed to prevent people from going abroad to get better health facilities. For example, Square hospital, United Hospital Limited, Ibn Sina Specialized Hospital, LABAID Specialized Hospital, etc. More training facilities have been provided for the health care workers by the Government. To pursue postgraduate studies, research and specialization, new institutes have been established. Support has been provided to develop traditional medicine during this period [1].

2 Changing Needs

To meet the ever-changing needs, the architectural design of hospitals in Bangladesh has developed a lot over the past few decades [2]. Current trends are influencing the design guidelines. Sustainable strategies and other new trends, such as establishing a community setting in the hospital and demand for a homely atmosphere in the patient units have had an impact on the design. To accommodate the changes in the economy, politics, and society, hospital design has been changed [3]. Hospital design has also been developed to satisfy the needs of the people. In order to cope with the changing needs, different methods have been developed and applied to upgrade the design [4]. The arrangements of personal health services have been changing extensively in recent years. Moreover, there is an emergence of new challenges. In recent years, tremendous public health improvements have occurred due to social and economic evolution [5]. The increasing urbanization, globalization, fast-growing population and climate changes are changing the nature and scale of public health challenges. Furthermore, the changing expectations of our society have affected the transformation of the hospital design.

3 Advancement of Medical Science

Changes in hospital design were required due to the advancement of health sciences. The expectations of society regarding the hospital environment have also changed because of the newer medical technologies. The advancement in medicine has caused significant changes to the design of hospitals in Bangladesh over the years. For example, the advancement of analgesia techniques has resulted in the establishment of a short-stay unit for day-case surgery [3]. Rapid progress in surgical techniques has resulted in several changes to the design. Modifications have been done to the operating suites. Moreover, nanotechnology is being used for diagnosis purposes.

X-ray, CT scan and MRI room walls are being specially designed so that the rays do not go outside (Fig. 1).

Overall, medical technologies have had a direct influence on hospital design to support the development of new ways of treatment and care [7].

4 Technological Advancement

The discovery of new technologies has directly influenced the architectural design of hospitals in Bangladesh over the course of time. It is quite impossible for Modern hospitals to function effectively without technology. To render clinical excellence, technologies are required. The importance of technology is well understood during the time of clinical or non-clinical equipment failure. Unnecessary patient mortality may occur in case of equipment failure. Therefore, it is impossible to ignore the importance of technology in the daily running of hospitals. With the emergence of modern technologies, several changes have occurred in the hospital

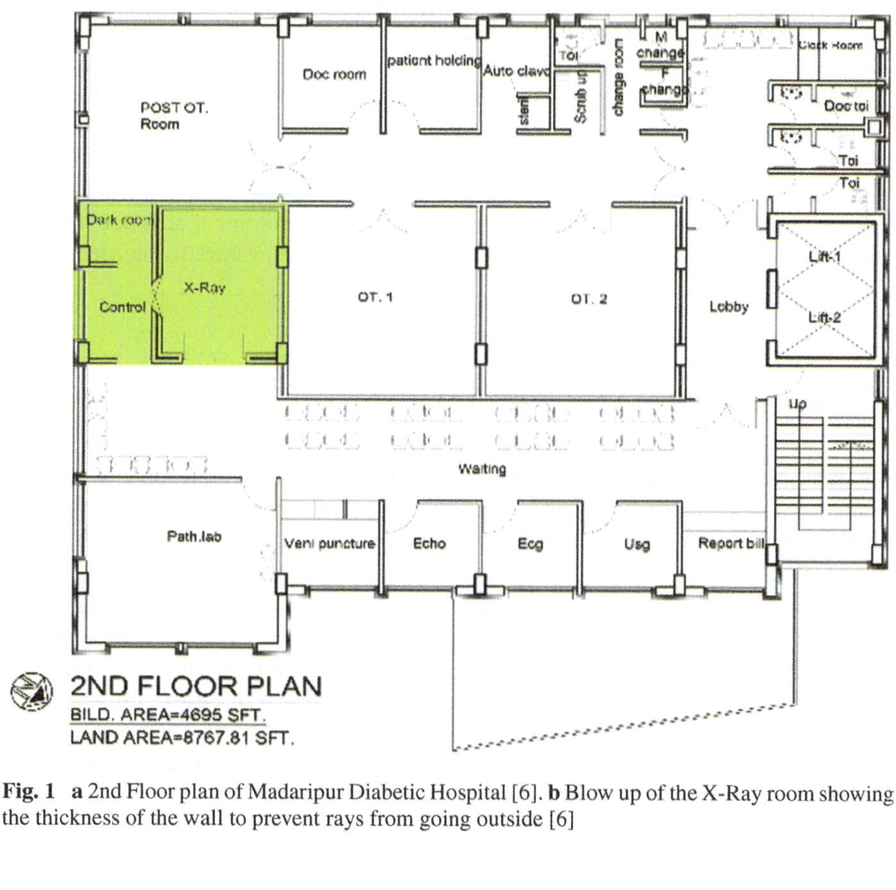

Fig. 1 **a** 2nd Floor plan of Madaripur Diabetic Hospital [6]. **b** Blow up of the X-Ray room showing the thickness of the wall to prevent rays from going outside [6]

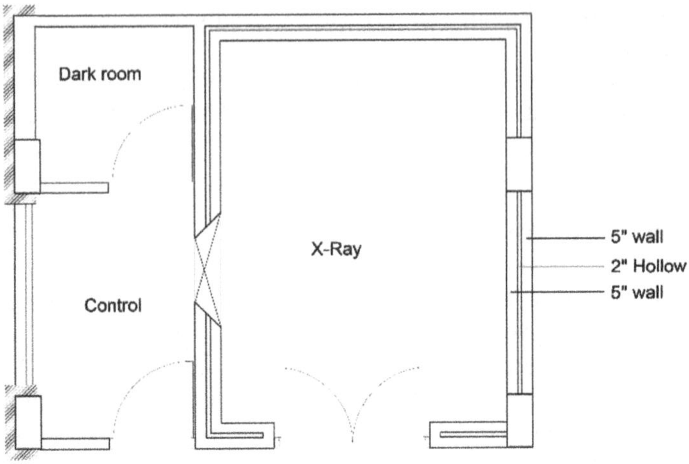

Fig. 1 (continued)

design of Bangladesh. For example, nowadays controlled temperature and humidity are provided in the operation theatre, labor room, delivery room, X-ray room and laboratory. Therefore, there is no need to add windows in the air-conditioned and light-controlled rooms. Cameras and computers have contributed to the replacement of Nightingale wards by cubicle ward layout. TV camera has been added to the operation block. Oxygen and other anesthetic gases are being supplied by pipes instead of cylinders. These pipes are influencing the wall and slab thickness. Mechanical vertical circulation was expensive and difficult. But technological advancement has made mechanical vertical circulation easier by adding elevators to the design [8] (Fig. 2).

5 Disease Related Planning of Hospital

The reasons behind the health problems faced by Bangladesh are poor water and sanitation infrastructure, malnutrition, rapid population growth, lack of hygiene, illiteracy, poverty, insufficient calorie intake etc. The increased rate of harmful use of alcohol and tobacco, air pollution and unhealthy diets are causing many chronic diseases. Poor environmental conditions are also responsible for many new health problems. There has been a rise of some non-communicable diseases, such as diabetes, cancer and heart diseases and some communicable diseases, such as HIV infections, tuberculosis and COVID-19 during the past few decades in this country [9]. Negligence towards following health guidelines is responsible for a high-level of diseases [8]. New ways of treatment and care are being introduced due to the significant changes in diseases. These disease-related challenges are responsible for the development of physical planning of hospitals in Bangladesh. For example, Corona

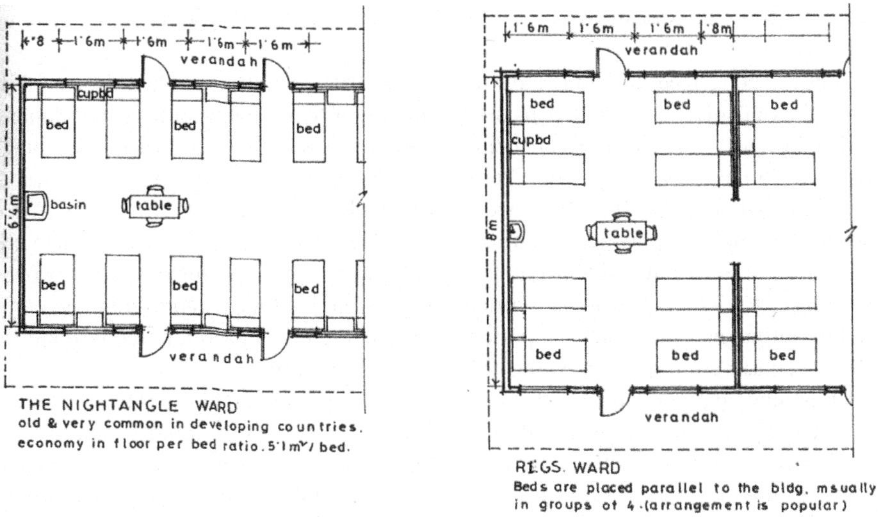

Fig. 2 Nightingale ward and Rig's ward [8]

unit and isolation unit have been added to the design after the COVID-19 outbreak. The laboratory used for the diagnosis of infectious particles such as SARS Cov-2 virus, tuberculosis, influenza etc. is being designed in such a way that the outside air cannot come inside and the infected inside air passes through the HEPA filter. This filter traps the harmful particles and purified air comes out. Hence, environmental pollution does not occur. Radiation therapy is used for the treatment of cancer. But during radiation therapy, the emission of rays can cause hazardous effects on the nursing personnel and the technical hands. To prevent this, the thicknesses of the wall and slab are specially designed and special door material is used. To serve the vast diversity of medical science new modern diagnostic technologies, surgical equipment, advanced machineries are impossible to allot in one hospital. Besides, different health care requires different facilities or environments.

As a result, the rapid growth of specialized hospitals, such as cancer, diabetic, cardiac, kidney, mental, tuberculosis, orthopedic etc. occurred during the past few decades. These hospitals are dedicated to particular diseases to provide improved quality of services. Depending on the segment type, different specialized hospitals also required different design criteria as well [10] (Fig. 3).

6 Morphological Analysis

According to previous experiences, low-storied hospitals that are horizontally expanding are more feasible and economical in Bangladesh [8]. But due to land shortage, vertical extensions are now being made instead of horizontal extensions (Fig. 4).

Highly centralized hospitals have shifted towards decentralized healthcare facilities [3]. A hospital generally consists of outpatient department, emergency department, laboratory, radiology department, inpatient department, operation theatre, post-operative room, labor room, dental unit, service department, administrative department etc. All of these functions are interdependent. In the large hospitals, there should be casualty department, traumatic department and physio-therapy department in addition to emergency department. Nowadays, there is no security in the hospital as before. To prevent unsolicited entrance, the hospital authorities prefer single entry though outpatient department, inpatient unit, emergency department, central sterile supply department require separate entries (Figs. 5 and 6).

Emergency patients may need to be taken to the operation theatre or labor room at any time. Therefore, emergency department, operation theatre, labor room, laboratory, X-ray, central sterilization unit should be close to each other. They can be situated on the ground floor. For example, Narayanganj hospital built by the Japanese where

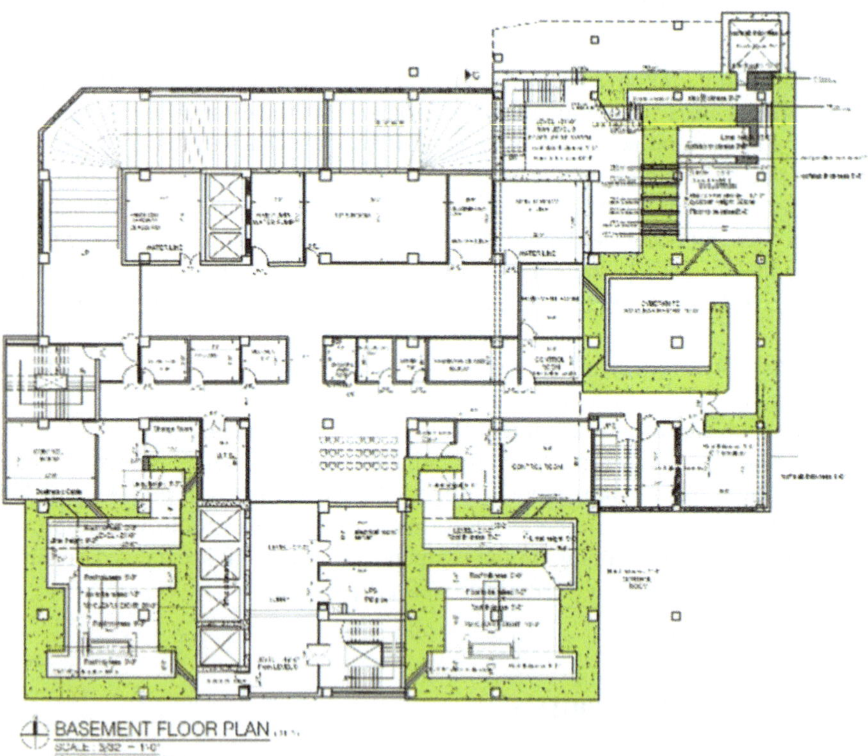

Fig. 3 **a** Basement Floor plan of Diagnostic & Oncology Department of BSMMU, Dhaka [6]. **b** Blow up of the radiation therapy room showing the thickness of the wall to prevent emission of rays [6]

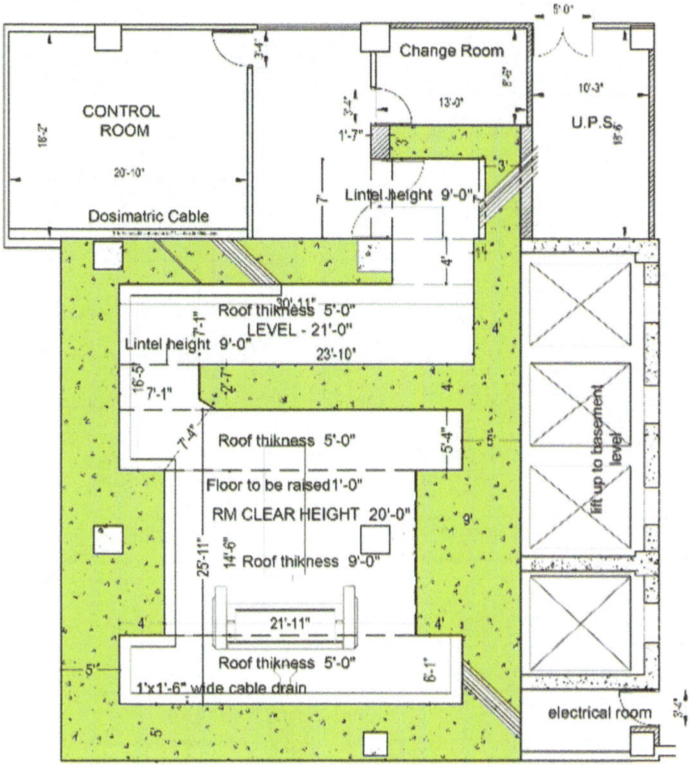

Fig. 3 (continued)

Fig. 4 High-rise hospitals in Bangladesh [11]

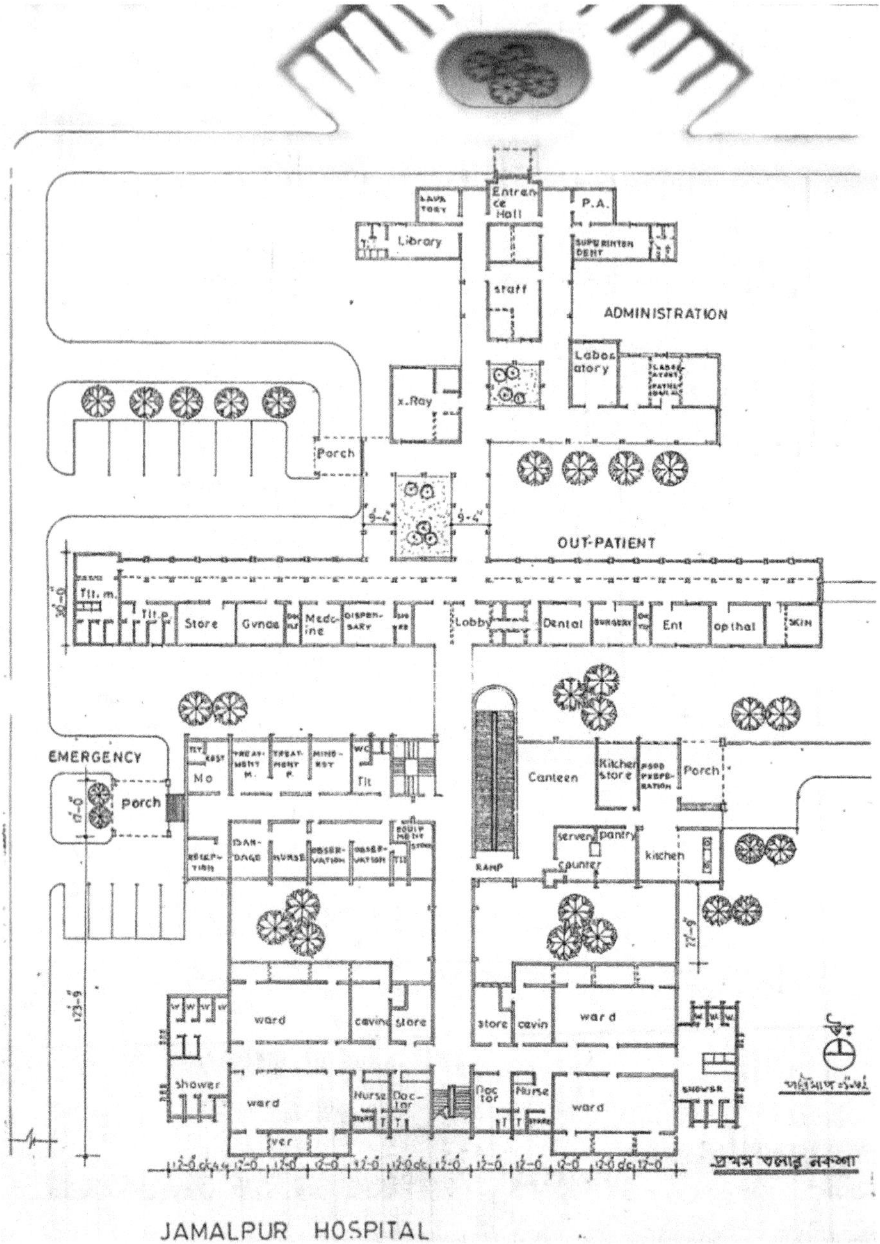

Fig. 5 Ground Floor plan of Jamalpur Hospital which is a horizontally expanding low-storied hospital [8]

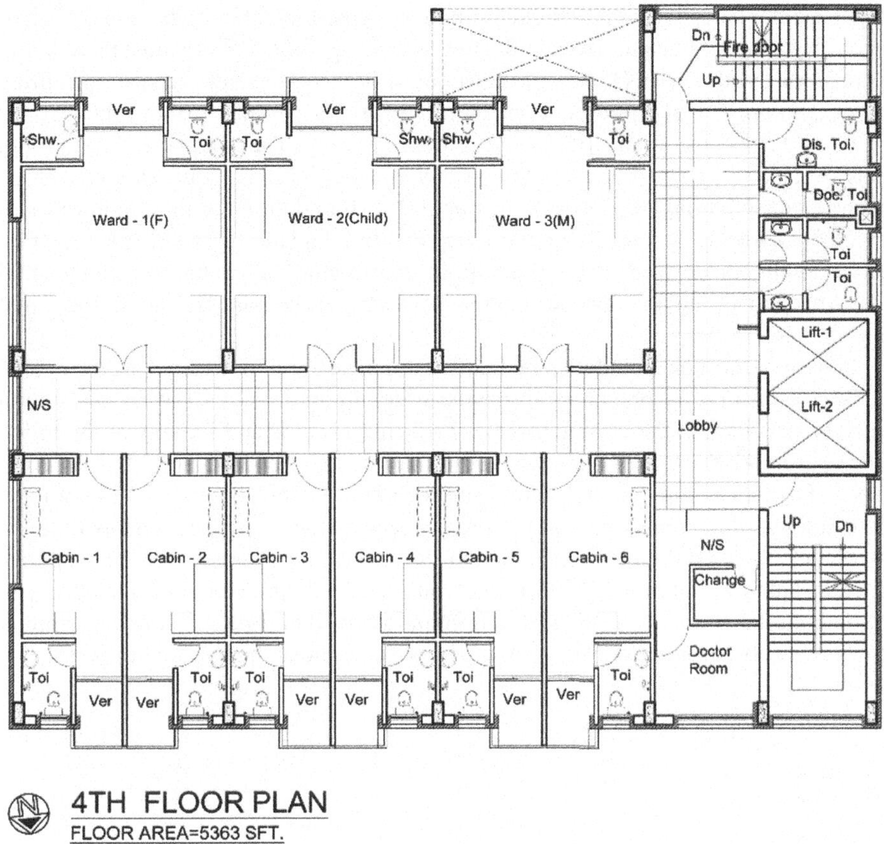

Fig. 6 4th Floor plan of Choddogram Diabetic Hospital, Cumilla where ward and cabin layout is shown [6]

these functions are situated on the ground floor. But the Bangladeshi surgeons do not prefer the operation theatre on the ground floor due to the dampness of the floor. According to them, septic tank is the reason for the filthy interior and exterior. Their opinion also differs regarding the plan of the operation theatre. In British system, sterilized equipment and linen are carried through 'Clean Corridor' and soiled equipment and linen through 'Dirty Corridor'. The operation theatre has separate entries for patient, doctor, staff and equipment in this system. In American system there is a single corridor and the operation theatre has one or two doors. Near the operation theatre, there are generally induction room, recovery room and rooms where OT instruments, linen, mobile X-ray machine and instruments for frozen section biopsy are kept. The outpatient department must be easily accessible. In large hospitals, there is generally a separate outpatient department connected with the main hospital by a corridor or bridge. Outpatient department depends on the diagnostic department (radiology, laboratory, etc.). Therefore, it is better to add a diagnostic department where there is a separate outpatient department. For better control, separate entry and exit must be provided to the consulting rooms. There must be an audio exam room with the ENT consulting room, a dark room with the eye consulting room, a minor OT and plaster room with the surgery consulting room, facility for ECG and Echocardiogram with the cardiologist's room. The dental consulting room must be large enough for the dental equipment. EEG room is required in neurology department. Nightingale wards were popular in Bangladesh as they required less space. Nowadays Rig's wards are more popular. In Bangladesh, a single nurse station serves a number of wards [8]. For example, Holy Family Hospital, Shishu Hospital etc. Attached veranda is needed in each ward so that the patients can sit there and dry clothes.

Playroom must be included in pediatric ward. Due to religious reasons, the people of Bangladesh prefer prayer room beside ward. In addition, east–west orientation must be provided in the arrangement of beds in the wards and fixtures of the toilet. The conservatism of the female ward is noticeable [8]. The laboratory has different types of technical units, such as sample collection room, biochemistry, immunology, hematology, blood bank, clinical pathology, bacteriology, cytology, histopathology, mycology, molecular biology, genetic lab etc. In the case of single-storied hospitals, kitchen, laundry, warehouse, pump house, substation, incinerator, mortuary, etc., can be built on the backside of the hospital. In multistoried hospitals, X-ray department, mortuary, substation, reservoir and parking are suitable at the basement [8] (Fig. 7).

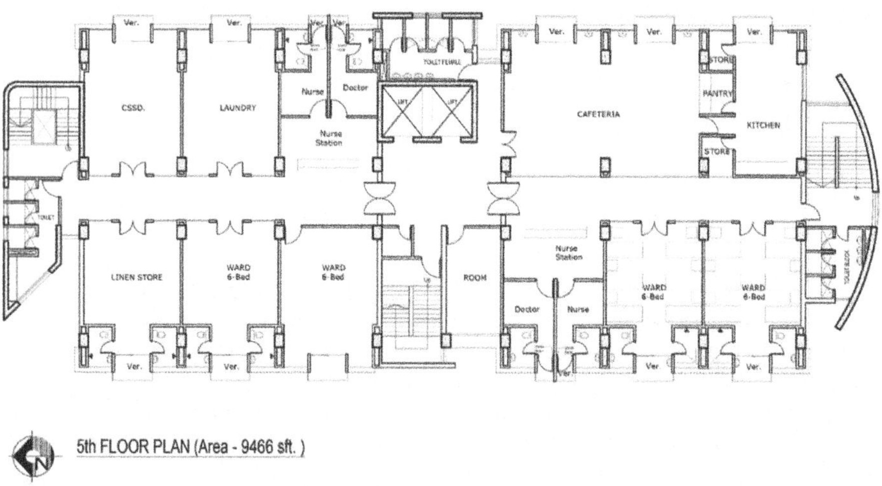

Fig. 7 5th Floor plan of Munshiganj Diabetic Hospital where cafeteria, kitchen, laundry, linen store, etc. services are shown [6]

7 Newer Components

Numerous components have been added to the health care system of Bangladesh day by day. After the development of new medical equipment over the years, new functions have been added to the hospital. Specialized wards have been increased in number. X-ray, ultrasonography, endoscopy, electrocardiogram etc. diagnostic tools have started being used for diagnosis. Later, other modern medical equipment, such as electromyography, angiography, CT scan, MRI etc. are added for pinpoint diagnosis. Consequently, CCU, Cath lab, etc., have been constructed. After the invention of the ventilator and other instruments of intensive care, ICUs have been established in hospitals. Biosafety cabinet has been included in the design to work with highly infectious viruses and bacteria. As a result, the infectious particles cannot come out and contaminate the working personnel and the outside environment. Kidney dialysis unit has been included for the treatment of end stage kidney failure. In some cases, short stays are required at the hospitals without admission such as taking injections that need several hours or as in the case of short-term observations. For these reasons, short-stay units have been established. Previously, dead bodies used to be kept on the morgue's table for days due to administrative complications and lack of refrigerators. This resulted in the emanation of a bad odor. For this reason, there was no mortuary inside the hospitals [8]. But nowadays, the dead bodies are kept in the mortuary refrigerator for the required time and hence the mortuaries are being constructed inside the hospital buildings. The vertical extension of hospital buildings introduced ramps as an alternative to the elevator to reduce the electricity cost. Fire stair has been added to the design to ensure fire safety. To accommodate commercial activities, pharmacies, coffee shops, and gift shops have been incorporated into hospitals [3]. Attached toilet has been provided to the doctors' room [8]. Moreover, accessible toilet has been introduced for the convenience of the differently abled people. The recent additions to the Bangladeshi hospitals are Corona unit for the isolation and treatment of COVID-19 patients. New PCR labs have also been established in many hospitals for the diagnosis of novel corona virus (Figs. 8, 9 and 10).

8 Role of an Architect

The role of an architect in hospital design is most important [12]. Architects are constantly faced with the problem of evaluating current effectiveness and developing alternatives to meet the health needs of people. This is undoubtedly a tough job due to the complexity of modern-day health care. To cope with the increasing complexity, there is a need to develop and apply methods [13]. One of the challenges that architects face is to design a flexible hospital to accommodate unforeseeable future demands. Anticipating the future needs is a difficult task as the changes take place rapidly. Architects need to plan in advance to accommodate future needs. Maximum flexibility and options for future expansion must be provided in the design. The design

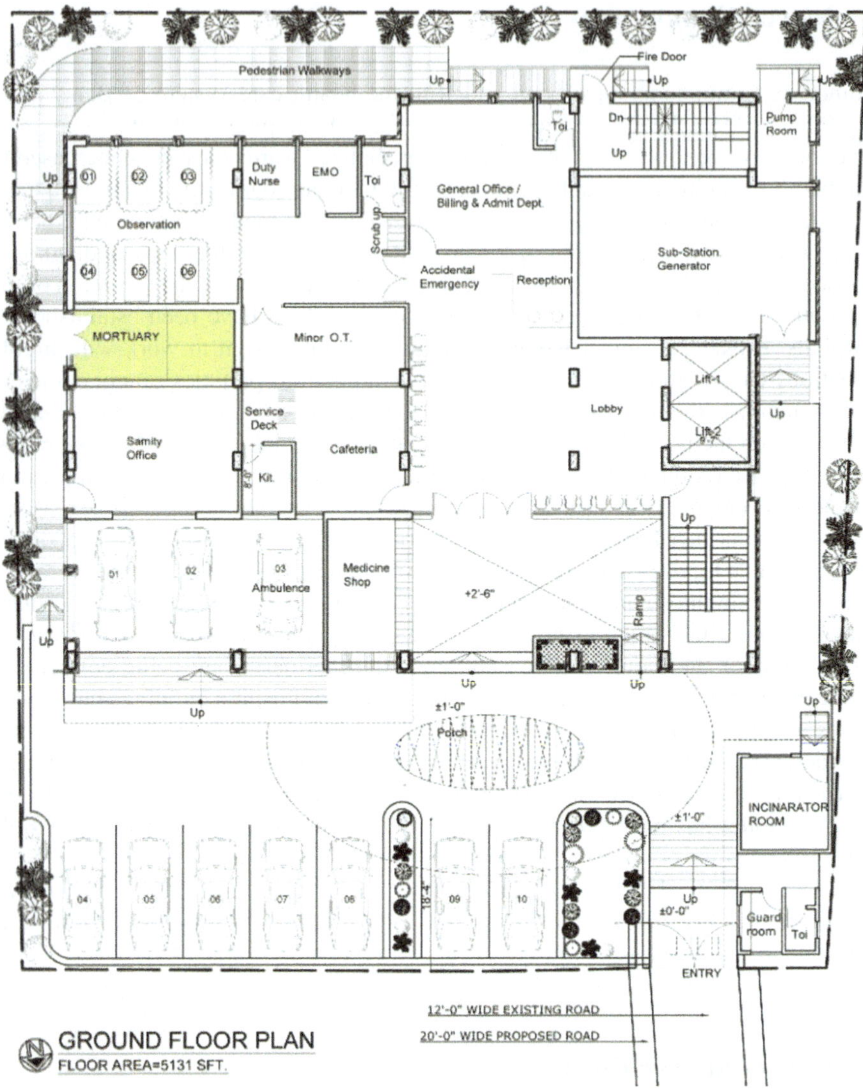

Fig. 8 Ground Floor plan of Choddogram Diabetic Hospital, Cumilla where the mortuary is constructed inside the hospital building [6]

process requires a lot of considerations. It is mandatory to have knowledge about the health sector. Architects need to focus on every aspect of a hospital while designing [12]. They must try to design in a way that creates a positive impact. The physical and emotional needs of the users need special attention [14]. Their satisfaction is a must. The architects need to ensure functionality while designing the hospital. They need to go beyond the traditional role and respond to the culture of the area. They

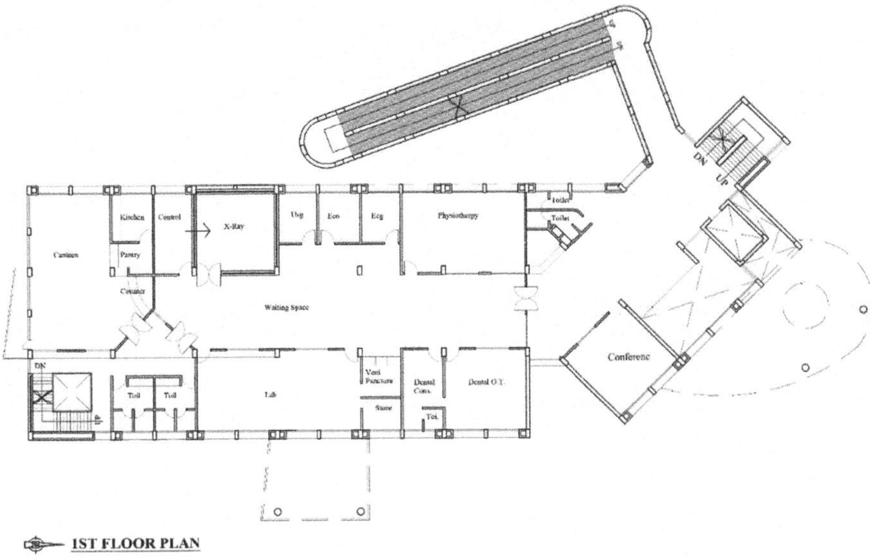

Fig. 9 1st Floor plan of Lalmonirhat Diabetic Hospital where ramp has been provided [6]

need to consider the constraints and try to overcome them. To recognize the vision of the hospital, it is important for the architects to participate in the planning process. The government guidelines are needed to be followed too [12].

9 Conclusion

Change over time is inevitable in the physical design of the hospital. Changes in different sectors necessitate frequent changes to the design. There were many limitations in health facilities previously. Technologies were not that advanced. Now due to the advancement of technology, design guidelines have changed significantly and numerous factors are needed to be considered while designing. To satisfy future demands, many more changes will occur in the future too. It is important to evaluate the previous designs, learn from them and apply that knowledge to the new designs.

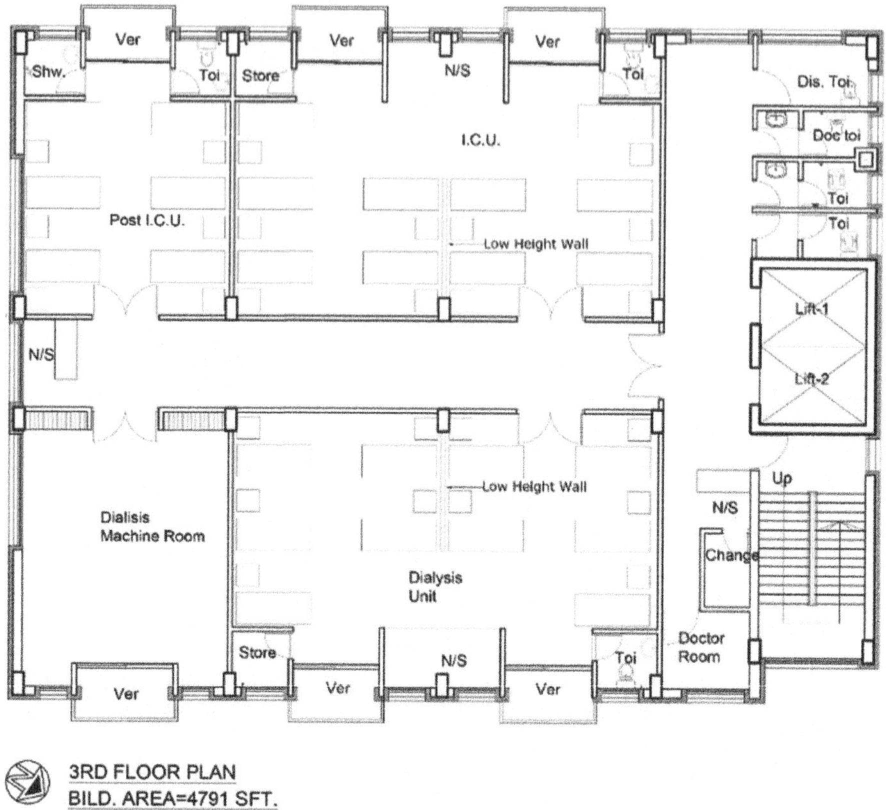

Fig. 10 3rd Floor plan of Madaripur Diabetic Hospital where ICU, Dialysis unit, Accessible Toilet have been provided [6]

References

1. Imran S (2021) History of Hospital and Bangladesh. SCRIBD, December 2010 [Online]. Available on: https://www.scribd.com/presentation/45219410/History-of-Hospital-and-Bangladesh. Last accessed: January 2021
2. Nagasawa Y (2019) Global Hospitals in 2050—a review of the historical development of hospital building studies from a global perspective, Architectural Institute of Japan, December 2019, Available on: https://doi.org/10.1002/2475-8876.12131. Last accessed: December 2020
3. Pilosof NP (2021) Planning for change: hospital design theories in practice. The American Institute of Architects, October 2005, Available on: https://www.brikbase.org/content/planning-change-hospital-design-theories-practice. Last accessed: January 2021
4. Snowdon A, Cohen J (2011) Strengthening health systems through innovation: lessons learned. International Centre for Health Innovation, November 2011 [E-Book]. Available on: https://www.ivey.uwo.ca/cmsmedia/3468017/global-health-systems-executive-summary1.pdf
5. Azzopardi-Muscat N, Brambilla A, Caracci F, Capolongo S (2020) Synergies in design and health. The role of architects and urban health planners in tackling key contemporary public health challenges. Acta Bio Med 91(3-S):9–20. https://doi.org/10.23750/abm.v91i3-S.9414

6. Drawings collected from "Department of Architecture". Ministry of Housing & Public Works, Bangladesh
7. Burke A (2021) Towards a new hospital architecture: an exploration of the relationship between hospital space and technology. PhD Thesis University of East London School of Architecture Computing and Engineering, January 2014, Available on: https://doi.org/10.15123/PUB.3895. Last accessed: January 2021
8. Shahidullah Md. (1932) Health and hospital architecture in Bangladesh. Department of Architecture, Ministry of Public works, Govt. of Bangladesh, October 1932 [Book]
9. Doshmangir L, Moshiri E, Farzadfar F (2020) Seven decades of primary healthcare during various development plans in Iran: a historical review. Archives of Iranian Medicine, Academy of Medical Sciences, I.R. Iran. https://doi.org/10.34172/aim.2020.24
10. Fralick PC (2020) Scarborough, Harold and Piercey, W. Douglas. "Hospital". Encyclopedia Britannica, April 2020. Available on: https://www.britannica.com/science/hospital. Last accessed: March 2021
11. Images collected from internet
12. Hospaccx Admin (2020) Role of healthcare architect in planning and designing hospitals. Hospaccx Healthcare Business Consultancy Pvt. ltd, May 2020 [Online]. Available on: https://hhbc.in/role-of-healthcare-architect-in-planning-and-designing-hospitals/. Last accessed: February 2021
13. Turley RE, Richardson WC, Hansen JV (2021) Morphological analysis for health care systems planning. Socio-Econ Plan Sci 9:83–88, April 1974, Available on: https://doi.org/10.1016/0038-0121(75)90013-0. Last accessed: February 2021
14. Pellitteri G, Belvedere F (2021) Characteristics of the hospital buildings: changes, process and quality. Architectural Research Centers Consortium (ARCC), June 2010, Available on: https://www.brikbase.org/content/characteristics-hospital-buildings-changes-processes-and-quality. Last accessed: January 2021

Reinterpretation of the Identity of East-European Cities' Riverside Territories (on the Example of the Historical Center of Uzhhorod, Ukraine)

O. I. Borshovskyi, H. P. Petryshyn, Y. O. Onufriv, and S. P. Tupis

Abstract The peculiarity of development of riverside territories of East-European cities compared to other European cities is their post-Socialist status, which demands a special approach. One of such cities is Uzhhorod that is situated on the border of Ukraine with Slovakia. The location of this city, as well as difficult historical transformations have formed today a multicultural environment which needs a proper treatment and preservation of its own identity. This and other peculiarities are elucidated in the context of re-interpretation of riverside territories' development of the central parts of Uzhhorod. According to the 'InterSave' method, a comprehensive analysis of the riverside areas of Uzhgorod is carried out, which includes an analysis of the functional structure, visual and landscape connections of the city with the river, accessibility of riverside areas for pedestrians and cultural value of riverside areas. During the research of the riverside territories of Uzhhorod several problematic situations were revealed. It is recommended to create a comfortable environment for people to stay in the riverside territory of Uzhhorod, using 12 criteria of public spaces quality developed by Jan Gale, with their adaptation to the conditions of the riverside spaces. It is determined that both for Uzhhorod and for other Eastern European historical cities, it is important to develop regulatory acts to preserve the identity of the urban landscape in the riverside zone.

Keywords Riverside territory · Historical city · Identity · Visual and cultural value · Functional structure · Pedestrian accessibility

1 Introduction

Historical cities of Eastern Europe, including Ukrainian cities, need a special approach, as for a long time they have been developing in the conditions of a totalitarian state. In the USSR cities developed as elements of state economical complex, with the stress on military industry development. Also, the communist ideology

---

O. I. Borshovskyi · H. P. Petryshyn · Y. O. Onufriv (✉) · S. P. Tupis
Department of Urban Planning and Design, Lviv Polytechnic National University, Bandery Str., 12, Lviv 79013, Ukraine
e-mail: yaryna.o.onufriv@lpnu.ua

T. Kang and Y. Lee (eds.), *Proceedings of 2021 4th International Conference on Civil Engineering and Architecture*, Lecture Notes in Civil Engineering 201,
https://doi.org/10.1007/978-981-16-6932-3_35

obscured the unique, regional features of cities. Today the processes of their development, reconstruction, building system, economic activity in the historical areal, defined on their territories, should be grounded on realization of the necessity of preservation and rational use of the objects of real estate cultural heritage, protection of traditional environment. The majority of historical cities developed near rivers. That is why an important problem of historical heritage of these cities' preservation in the context of their interconnections with the river, and namely: accessibility of riverside spaces for pedestrians; functional integrity of riverside territories, harmonious landscape and visual connections between the river and the city, preservation of the cultural identity of riverside spaces.

In the given research the problem of riverside territories' development is considered on the example of Uzhhorod, that is a city of regional subordination, administrative, economic and cultural center of Transcarpathian region. The city is the smallest regional center of Ukraine and is situated in the western part of the country on the border with Slovakia at the foot of the Carpathian Mountains. The river Uzh flows through the city. The area of the city is about 40 km^2; the population is about 115 thousand of people (at the end of 2020) [1]. The peculiarity of Uzhhorod is its multicultural landscape, like in other likely cities, for example Lviv (Ukraine) [2], an important task for architects is the organic inclusion and revival of other culture's elements into the spaces, and particularly, into riverside spaces.

The aim of the given research is specifying the actual peculiarities and perspective directions of development of riverside territories in the central part of Uzhhorod as a historical multicultural East-European city.

2 Materials and Methods

The methods of analysis 'InterSave' are used in the research of riverside territories of Uzhhorod. These methods were proposed by Danish scientists and have been used by the research of the cultural identity of cities of Ukraine, Poland, and Moldova [3]. Also, this methodology was interpreted by researching the identity of Szczecin (Republic Poland) in the context of identifying and preserving its riverside cultural landscapes [4]. The research is based on the criteria of specifying the identity of a place that are concluded from researching landmark public spaces of post-totalitarian East-European capitals [5].

For researching the mutual influence of the river Uzh and the architectural-planning structure of Uzhhorod there was taken the stripe of territory along the river with the width in average 500 m on the right bank and 500 m on the left bank of Uzh (Fig. 1). The width of the riverside stripe is different that is defined by the historical urban planning structure of the territory. In certain places the maximal width of the stripe reaches up to 1500 m. The defined stripe of the embankment does not limit and span the whole river valley of the river Uzh. The landscape of the river in the city near the natural elements that are characteristic for rivers contains also anthropogenic elements that belong to the urban planning frame of the territory. The

Fig. 1 Riverside territories alongside the river Uzh with the oldest historical center of the city of Uzhhorod (Ukraine): 1—Big industrial zone in the east, 2—old city, 3—little Halahov, 4—big Halahov, 5—district of the villa housing development, 6—borderline zone, 7—big industrial zone in the west, 8—Bozdoskyi park, 9—newly built residential area, 10—Svepomoch, 11—around the Petefi square, 12—Radvanka

combination of the natural and anthropogenic components forms the identity of the riverside spaces of the city.

Specifying the city identity that is based on the different forms of its interaction with the river demands the estimation of the riverside landscape. An important task in the research was singling out fragments of the riverside territory of the river on the right and the left banks that constitute integral landscape-space units. The division of the riverside territory into space units was made based on identifying the diverse character of the landscape in the riverside zone that is caused by the historical development of the city and physical peculiarities of the landscape. Individual space units are integral in their Urbanistic character that reflects their spatial identity. Based on the held analysis there were identified 12 landscape-spatial units, for which the accepted limits constitute clear landscape and structural limits, as well as functional barriers. The identified limits of the spatial units are the "limits" in the city structure from the point of view of the concept of the city image by Lynch [6]. The limits that delimitate the width of the spatial unit by the perpendicular line to the river are defined by the remoteness of the territory from the river. And the limits that specify the length of the spatial units are designated taking into account the likeness of the landscape of the given fragment of the riverside territory, heterogeneity of the city morphology, etc. The size of the identified spatial units has turned out to be very different depending on the type and degree of the given connections of the given plot with the river in the analyzed aspects (Figs. 2, 3, 4 and 5): functional-spatial, social, visual-composition, and cultural. Each spatial unit in a different way and with different force is "stretches" to the river, depending on physiographic peculiarities, traditions of a place, functional reclaiming of the territory, and the range of functional interaction of the center of the given spatial unit.

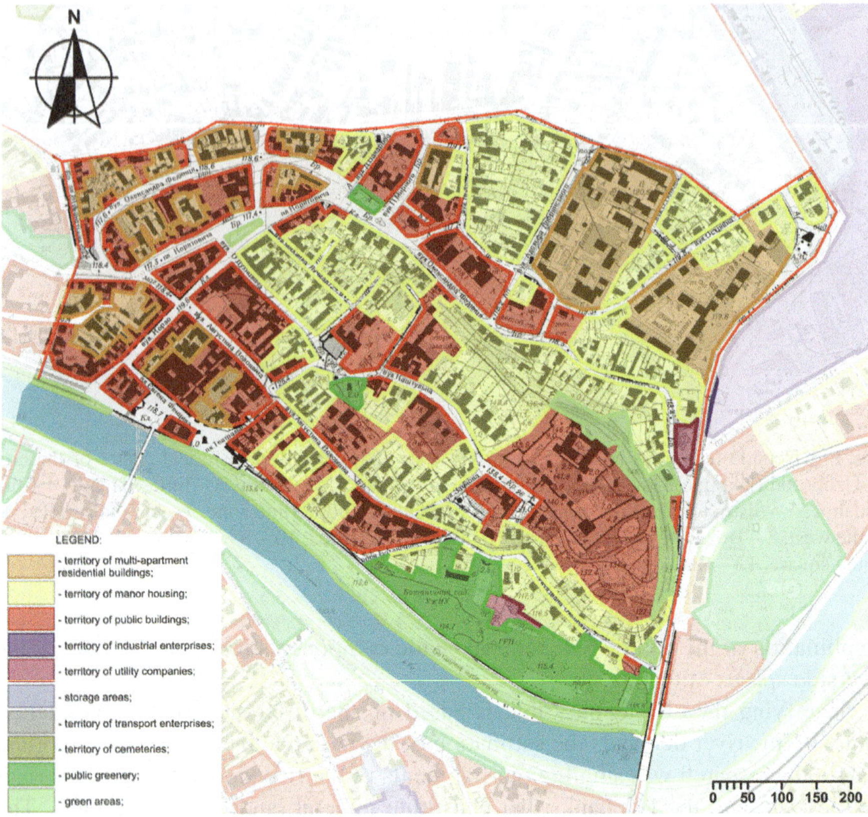

Fig. 2 Functional structure analysis

Every specified spatial unit has been given a name that signifies its identity (Fig. 1). The limits of the units are designated by the planning framework of the city, most often they run along the city streets which divide the city structure into significant historical elements, laid on different stages of the city development. Formulating the recommendations concerning making comfortable the riverside territories is based on the quality criteria, identified by Jan Gehl for the successful making comfortable of social spaces [7].

3 Results

According to the archeological data the city of Uzhhorod appeared on the territory of the modern Castle Hill as far back as at the end of the X—at the beginning of the XI c. Until the mid-XVI c., the city limits were limited by the Castle Hill and were lodged between two streambeds—of Big and Little Uzh. Until nowadays there have

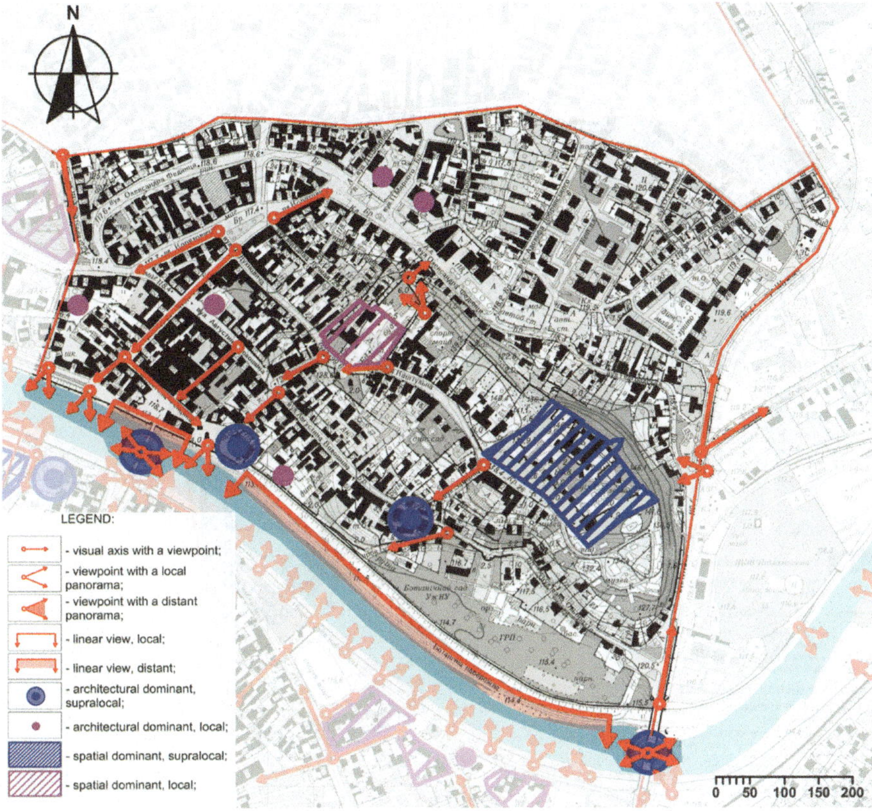

Fig. 3 Analysis of the visual-landscape connections of the city with the river

been preserved the outer defensive fortifications which were built in the period of the second half of the XVII c., being based on the northern-Italian defense doctrine that combines bricks and earth fortifications of rampart type with dry ditches. The Castle Hill, the basis of which is made of volcanic rocks of andesite creates a hill that is oblong from the north–west to the south–east. The south–east and south–west slopes of the hill are falling with sheer drops to the river valley. Beginning from the slopes of northern hills (the district of the modern Sobranetska Str.) the wide lowland stretches in the south–western direction up to the state border with Slovakia; from the southern side, this territory smoothly changes into the valley of the river Uzh [8].

The outlined above methodology is illustrated on the example of the detailed analysis of landscape-space unit No. 2 "Old city" on the right bank of the river Uzh, that is the oldest part of the city, and namely the territory around the Castle hill, that is limited from the north by the streets of Pidhirna and Karmeliuk, from the west by the Zhupanatskyi street, from the east—by Shumna street, and from the south the district flanks the river Uzh (Fig. 1).

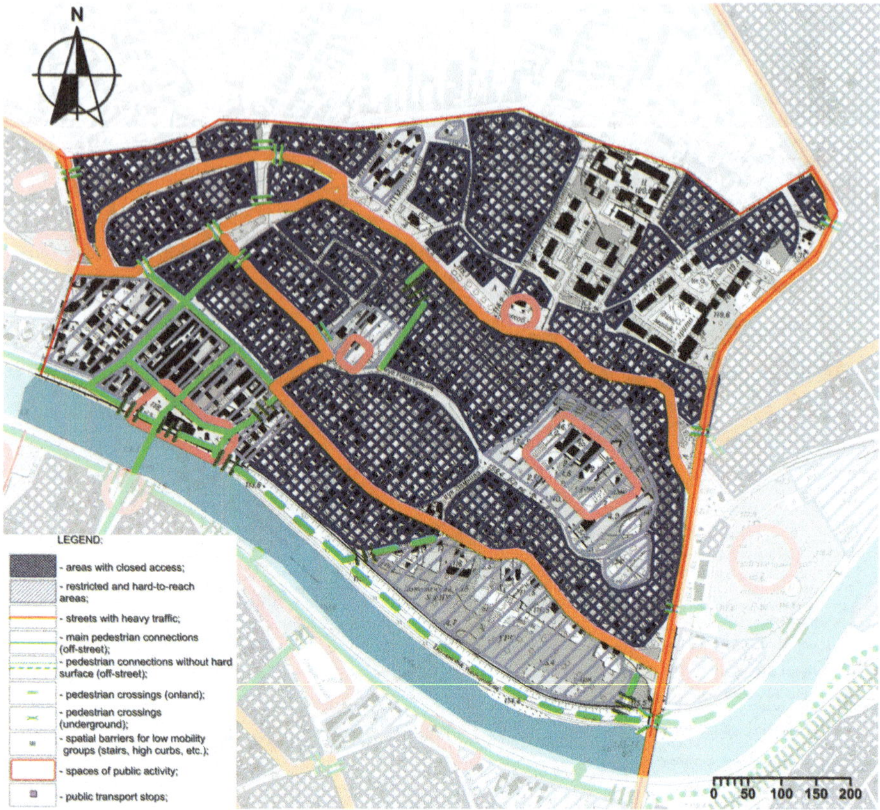

Fig. 4 Analysis of the accessibility of riverside areas for pedestrians

3.1 Functional Structure Analysis

Functional structure of the district (Fig. 2) is analyzed from the point of view of the specifics of using the riverside territories. In the analysis of the functional structure there are separated the dominating functions on the riverside territories:

- Residential territories;
- Territories of public development (public foodservice, educational establishments, theatre and philharmonic);
- Public services;
- Recreation zone (Botanical garden, railway line for children).

In the given case the territories of the public development, and namely summer places of restaurants and cafes, as well as the square in front of the theatre with the foot bridge and the public garden in front of the philharmonic are the main catalysts of the spatial development of this district in the riverside zone. Here the biggest activities of the city inhabitants can be observed and a big concentration of

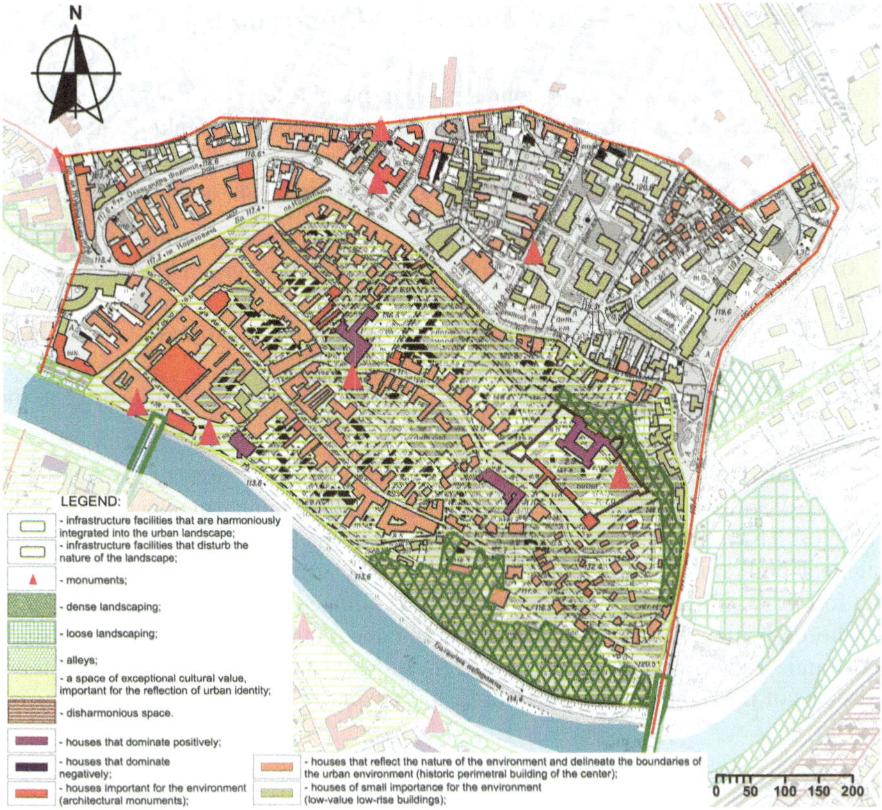

Fig. 5 Analysis of the cultural value of riverside areas

tourists. These squares are elevated above the river and strengthened by the high wall. Whereas the Botanical embankment is on the level of the river line of the Uzh River, it is unmanned and has a more natural character (Fig. 6).

Fig. 6 View on the right bank of the Uzh river with the botanical embankment in the central part of the city of Uzhhorod (Photo of 25.09.2020)

3.2 Analysis of the Visual-Landscape Connections

Research of the visual-landscape connections of riverside territories with the river and city dominating places foresees the assessment of their qualities in the composition and symbolic dimensions. (Fig. 3). In particular, there were defined the elements of the landscape presentation of the active and passive character, including:

- The existing viewing manifestations, axes and panoramic views, that are opening from the banks on the river itself or on the river valley and delineate its spatial range;
- The objects dominating in the landscape both in the local and over the local context: point like and spatial ones;
- Important visual axes that are joining the important objects, dominating in the riverside landscape.

The main landscape dominant element of the analyzed territory of the "Old city" is the Castle hill together with the architectural-spatial dominant element—the Castle complex. The main architectural dominating elements here are the Transcarpathian regional philharmonic, Uzhhorod Greek-Catholic Cathedral and the building of the former Basilian monastery (today there is situated the physical department of Uzhhorod National university). From the Botanical embankment there are opened panoramic views of the left bank, and from it there can be seen the main dominant elements of the right bank of the river Uzh (Fig. 7).

Fig. 7 View on the embankment with the building of the philharmonic—the former synagogue of Uzhhorod. On the background there can be seen the Cathedral (Photo of 25.09.2020)

3.3 Analysis of the Territory's Accessibility

Analysis of the space's accessibility characterizes the quality of public spaces on the riverside territories, communication connections between the public spaces and the river (Fig. 4). The analysis of connections is fulfilled from the point of view of the accessibility of spaces for pedestrians. Riverside territories are divided into three categories according to their accessibility:

- open public places;
- hard for access or of the limited access;
- closed—inaccessible.

This part of the city has close connections with the river presented by the pedestrian streets (Korzo street), arcades (arcade Bati) and the foot-bridge (Fig. 8). The main barrier for the pedestrian movement from the city in the direction of the river is the steep terrain of the Castle hill, where the pedestrian communications are often solved by the flights of stairs, which make the movements for the groups of population with restricted mobility more difficult. Also, the access to the Botanical embankment is barred by railway line for children and is limited by the closed private spaces of villa development that are situated within the river valley.

Fig. 8 Foot-bridge in the central part of Uzhhorod (Photo of 25.09.2020)

3.4 Cultural Value Analysis

The research of the cultural values combines in itself composition, architectural, historical and symbolic qualities of the city riverside landscape during whole its historic development in the close interconnection with the river (Fig. 5).

The analysis of the cultural value according to this methodology is somewhat changed according to the research tasks. According to the conducted analysis, the most valuable architectural objects on the defined area are the defensive complex of the castle on the Castle hill, Uzhhorod Greek-Catholic Cathedral, Transcarpathian regional philharmonic (in the past it was a synagogue), the physical department of Uzhhorod National university (the building of the former Basilian monastery) and others. There distinguish themselves also such important for the forming of the city space character objects of greenery: the chestnut alley alongside the puppet theatre and the Ye. Fentsyk garden; a separate tree (botanical monument)—Masaryk's ash tree; forested plots of the Botanical garden along the river. Also, places of memory have symbolic meaning, one of which is situated on the embankment near the former synagogue (monumental sign to the Holocaust victims).

4 Discussion

Based on the conducted above complex analysis of the riverside territory on the given area we can conclude that the riverside zone is divided into two fragments: very much changed urbanized embankment and almost unchanged natural embankment. Each of these fragments is unique and valuable in its way, that is why it is expedient to preserve its essence.

The physical access to the river for the pedestrians in this area should be planned, and there should be held some urban planning measures which would improve the accessibility for the groups of population with the restricted mobility (designing rampant, stairlifts, etc.).

In the context of preserving the visual character of the landscape on the analyzed small part of the city, there is fragmentarily observed the visual worsening of the space, and also there is complicated access to the water in some places because of building the one-family households of the doubtable aesthetics directly on the riverbank.

The catalysts of the public activity in the riverside zone are the open pedestrian spaces (squares and cities) in front of the public objects and the alleys on the embankments developed for the passive rest and the passage of the bicycle transport.

City identity, except for the development, system of greenery and open spaces, is created also by the architectural details, such as signboards, small architectural forms (kiosks, sculptures, street furniture, etc.), and the lighting. In particular, in Uzhhorod, as well as in some East-European cities (for example, in Wroclaw), there was made broadly used the small sculpture culture, in particular on the embankment there are situated some miniatures, which strengthens the city's cultural identity and attracts city inhabitants and tourists to the river. Also, diverse artistic-cultural arrangements that are associated with the local traditions (the festival of cheese and wine, city's day, etc.) and other manifestations of the city acupuncture have an important significance for the city identity [9].

5 Conclusion

For the sake of forming a comfortable environment for the stay of people in the riverside zones of cities important is the provision of basic needs: protection from the road traffic (pedestrian embankments), protection from crime (lively public life during the day and night, quality lighting in the dark time of the day), protection from the possibility of flooding and negative atmosphere influences (flood control measures), the possibility to walk on foot (no obstacles, accessibility for everyone), possibility to stop and observe the river (sitting or standing), the possibility to see (comfortable visual distances, lighting in the dark time, open spaces to the water), possibility to communicate (low level of noise, places for sitting), the possibility to train physically and play games (interactive places which encourage creativity on the embankments), taking into account the human scale, the possibility to enjoy the advantages of climate, aesthetics of the city design elements.

For Uzhhorod, as well as for other East-European historical cities, it is important to develop the regulatory acts for preserving the identity of the city landscape in the riverside zone. One can meet the examples of attentive treatment of preserving the cultural landscapes in many countries, like the USA, Canada, the EU countries, where strict rules have been introduced, which pertain even to the smallest architectural details, such as the range of colors, construction material, the height of fences, etc. [10].

References

1. State statistics service of Ukraine. http://www.uz.ukrstat.gov.ua/statinfo/dem/2020/nasel_0112_2020.pdf. Accessed 01/2021
2. Cherkes B, Linda S (2019) Rebirth of multicultural identity in public spaces of Lviv. IOP Conf Ser: Mater Sci Eng 471:072019 (2019)
3. Larsson B, Skoog G (2009) Multiethnic built heritage in west Ukraine and Moldova—a challenge for urban planning and development, Asplund-biblioteket, Lund, p 114
4. Petryshyn H, Czałczynska-Podolska M, Sochacka-Sutkowska EA, Pilarczyk A (2012) Budowa tożsamości szczecina w oparciu o wykorzystanie kulturowych wartości krajobrazu nadrzecznego miasta, Wydawnictwo Hogben, Szczecin, p 498
5. Tscherkes B (2014) Identität, architektur und rekonstruktion der stadt, Lit Verlag, Germany, p 272
6. Lynch K (1960) The image of the city. The MIT Press, USA
7. Gehl J (2010) Cities for people, Island Press, USA, p 285

8. Ministry of Culture of Ukraine (2014) Research institute of monument protection and research.In: 2014 historical and architectural key plan of Uzhhorod, Zakarpattia region, defining the boundaries and modes of monuments and historical areas protection zones use, Kyiv, Ukraine, p 126
9. Lerner J (2014) Urban acupuncture, Island Press, USA, p 139
10. Lozynskyi R (2020) Cultural landscapes of Galicia: the ethics of human interaction with nature, Lviv, Ladeks, p 224

Classifying Essential Morphological Characteristics to Understanding Urbanity in the Old Town of Chiang Mai

Chompoonut Kongphunphin and Manat Srivanit

Abstract The aims of the study are to achieve an understanding of urban morphology and classify essential morphological characteristics, and to measure the degree of urbanity. Chiang Mai's old town and its surrounding radius one-kilometer buffer were used to be the study area. In this study, the analysis of urban morphology properties including street network configuration and the degree of accessibility, density of buildings and open space, and patterns of land use and building, which investigated with tools called space syntax, space-matrix, and mixed-use index, respectively. All three analytical tools were combined through the Geographical Information System (GIS) to classify the morphological characteristics. The characteristic results show that there are quite high degrees of accessibility in the old town and main roads in its buffer. All building types in the old town are low rise, most of them are mixed functions between living and doing small business. The combination results reveal that there are mixed of various degrees of urbanity in the old town; highly Urban, in-between (high), in-between (middle), low-Urban and suburban. The results of the study can be used to develop strategies and design planning for the old town of Chiang Mai city and applied to be an approach of the urban planning study for other Thailand's cities.

Keywords Morphological characteristics · Urbanity · The old town of Chiang Mai

1 Introduction

The concept of urban morphology is a study of human settlement in order to understand the spatial structure and character of shape and form and the process transform of urban. Studying morphology covers various scales of urban elements included individual buildings, lots, and cities [1–4]. Analysis of urban morphology can be

C. Kongphunphin (✉) · M. Srivanit
Faculty of Architecture and Planning, Thammasat University, Pathumthani, Thailand

M. Srivanit
Center of Excellent in Urban Mobility Research and Innovation (CoE-UMRI), Thammasat University, Pathumthani, Thailand

T. Kang and Y. Lee (eds.), *Proceedings of 2021 4th International Conference on Civil Engineering and Architecture*, Lecture Notes in Civil Engineering 201,
https://doi.org/10.1007/978-981-16-6932-3_36

considered from the street systems, plots, buildings, the result of the historical evolution, urban tissues, and building types that can be adapted to contemporary structures, the relationship of public spaces built to support social practice which is a way to support the livelihood of the city [4, 5]. Moreover, the study of urban morphology is a process of understanding the shape and form of cultural landscapes caused by the physical structure, realization and social action, which refers to the urban characteristic [1–3]. The study of urban morphology is an analysis of the urban structure and spatial composition of the pattern of access network, building density, and land and buildings utilization, which is consistent with the elements in the study of morphology that consisted of three main components: (1) road system, (2) building system, and (3) building use pattern [5]. Each element is summarized as a concept that in terms of access network pattern, the traffic networks in urban centers are often concentrated in access to the area in all directions. The hub area is made up of axial lines connected at angles (approximately 90° or more), those lines swerve and loop-back with a path connecting them in all directions. The transportation network in the cities (hub) consists of short routes that are often linked together as a grid, which is facilitated a widespread traffic network within the area [6]. In cases of building density patterns and open space, urban centers are densely concentrated with a mass of small buildings. The allocated forms of the building mass are quite structured. The scales of building mass are various mixed sizes. And the proportions of open space are quite low compared to the building mass. Moreover, the proportion of the roads are usually small because they were provided to the area of the building mass, most are for commercial and service activity [7]. For building use patterns, most of the building uses in urban centers are commercial and service or other activities that related to the economy of the area, such as shopping malls, shops banks, hotels, and markets. Those economic buildings are concentrated in the area caused intense activity than the surrounding area. Morphological characteristics classifying is included three analytical methods for analyzing essential components [3, 8, 9]. Space syntax is a tool that widely used for calculating the connection degree of roads in terms of topological, geometric, and metric distances [10]. Space-matrix is a method used for measuring the degree of building density [11]. Mixed-use index is a method used for computing the mixing degree of a functional mixture of buildings [3, 9]. In terms of measure the degree of urbanity, it is a quantitative morphological analysis that applied the grid analysis of GIS to integrate space syntax, space-matrix, and mixed-use index for classifying urban forms [3, 5, 9, 12].

Chiang Mai is the largest and fastest-growing city in the northern region of Thailand. Chiang Mai's old covered an area of approximately 2.72 km^2. It is surrounded and shaped to square by historic walls and moats. There are five city gates on the historic walls for entering the old town, two gates on the south and three gates separated on the north, east, and west. Within the old town, it is known as a historical site of Chiang Mai city and neighborhood areas, a seven-hundred-year-old community, where is a tourist attraction. There are many old communities and their temples, and road networks. The old city is dense with buildings and varied activities [13, 14]. Chiang Mai's morphology is formed by the combination of urban planning design on beliefs in axis and direction of the city according to the beliefs of the indigenous people and actual use that is still visible from the past to nowadays [14].

This study is the classifying essential morphological characteristics of the old town of Chiang Mai by analyzing street network, which analyzes the access in the district and urban levels of the study area, studying building density and open space by analyzing the relationship between open space and building, and studying land and building use by using percentage mix of land and buildings use among housing, working, and amenities, as the degree of integration of the study area. The results of this study can lead to the realization of the complexity and essential elements of urban fabrics. That can be used as a guideline for planning Chiang Mai city and other cities.

2 Research Methodology

This study used Chiang Mai, the largest and fastest-growing city in the northern region in Thailand, to be the study area. The boundary of the analysis focused on the old town of Chiang Mai and its neighboring areas with a one-kilometer radius surrounding as shown in Fig. 1. In order to get the morphological characteristics, there are two steps of the analytics; (1) urban morphological analysis and (2) urban morphological classification as shown in Fig. 2. The first step, space syntax, space-matrix, and mixed-use index are methods of the spatial analysis. The second step is a combination of the spatial analysis that shows morphological characteristics of the old town of Chiang Mai. All spatial methods are analyzed on Geographical Information System (GIS) on raster system. A cell raster size used in the study is 150×150 m per cell, the size of the cells is dependent on the context and the power of the computers used [15].

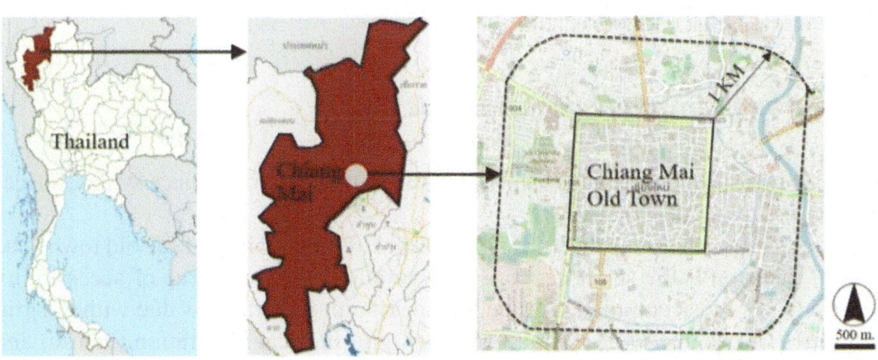

Fig. 1 The location of study area

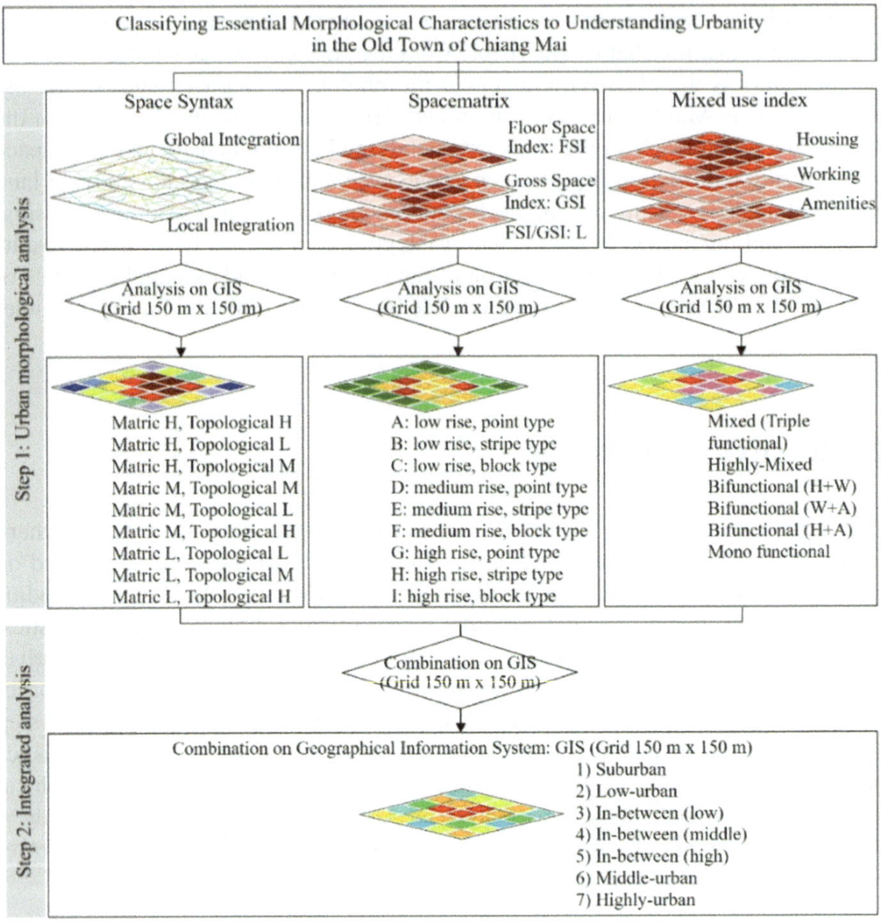

Fig. 2 The analytical framework of the study

2.1 *Urban Morphological Analysis*

This step measures degree of urbanity by three spatial methods; space syntax, spacematrix, and mixed-use index.

Space syntax measures the accessibility level of the mobility of the old town. The gain of this step is the street network configuration and the degree of accessibility that evaluates the accumulation value, the maximum integration value within a grid cell [15]. The degree of accessibility in each grid of local integration (metric) and global integration (topological) is classified to high: H, medium: M, and low: L level. The crossing analysis of both is shown nine types; metric H, topological H; metric H, topological L; metric H, topological M; metric M, topological H; metric M, topological M; metric M, topological L; metric L, topological H; metric L, topological M and metric L, topological L.

Space-matrix considers the pattern, density of buildings, and open space of the old town. That computes the maximum value, the sum of gross floor area, and footprint per area of a grid cell [9, 11]. The Obtain of the step is building density. Floor Space Index (FSI), Ground Space Index (GSI), and L are calculated as Eqs. (1), (2), and (3), respectively.

$$FSI_f = F/A_f \tag{1}$$

where F is the gross floor area in meter square, and A_f is the area of the urban fabric in meter square.

$$GSI_f = B/A_f \tag{2}$$

where B is the built-up surface or building footprint in meter square, and A_f is the area of the urban fabric in meter square.

The average number of floors (L) is derived from the basic indicators FSI and GSI can be calculated as follows:

$$L = FSI_f/GSI_f \tag{3}$$

The overlay of FSI, GSI, and L has presented nine types composing of A: low rise, point type; B: low rise, stripe type; C: low rise, block type; D: medium rise, point type; E: medium rise, stripe type; F: medium rise, block type; G: high rise, point type; H: high rise, stripe type and I: high rise, block type.

Mixed-Use index examines the patterns of land and building use of the old town focusing on residents, workplaces, and amenities. It is quantified the weighted average value, percentage of the three functions in a grid cell (measured in gross floor area). The overlay of housing, working and amenities use displays six minor of three major types of mixed-use, there are highly mixed, mixed (Triple-functional), Bi-functional (housing and working), bi-functional (amenities and working), bi-functional (housing and amenities), and mono-functional [3, 9].

2.2 *Urban Morphological Classification*

This step is the combination of the final results of each urban morphological analysis together into one map by using GIS. The overlay of space syntax, space-matrix, and mixed-use index represented the level of urbanity to seven classes of morphological characteristics of each grid cell of the study area composing of highly urban, middle-urban, in-between (high), in-between (middle), in-between (low), low-urban and suburban.

3 Results and Discussion

3.1 Urban Morphological Characteristics

The result of street network configuration analysis from space syntax shown the global integration and local integration of the old town of Chiang Mai and its one-kilometer radius surrounding as shown in Fig. 3a and b. The integration values of both global integration and local integration were classified to nine classes from −1 to 1.651 and −1 to 4.158, respectively. From the results, it can be implied that the streets surrounding the old town were the highest intensity of connection. Comparing between in the old town and its surrounding, the streets in the old town were higher intensity of connection than the outer surrounding.

The overlay of global integration and local integration from space syntax given eight of nine types of degree of accessibility as shown in Fig. 4a. The metric H, topological L type was absent in this study. Metric H, topological H and metric L, topological L types were the highest ratio of the study area (22.28%). There were four types that appeared in the old town. Metric H, topological H type was the highest ratio (66.66%) followed by metric H, topological M, metric M, topological M and metric M, topological L. Their percentiles were 18.12, 6.61, and 6.61, respectively.

The integration among FSI, GSI, and L from space-matrix of the study area shown nine types of the building density patterns as shown in Fig. 4b. Type C (low rise, block type) was the highest ratio of the study area (48.50%). There were three types

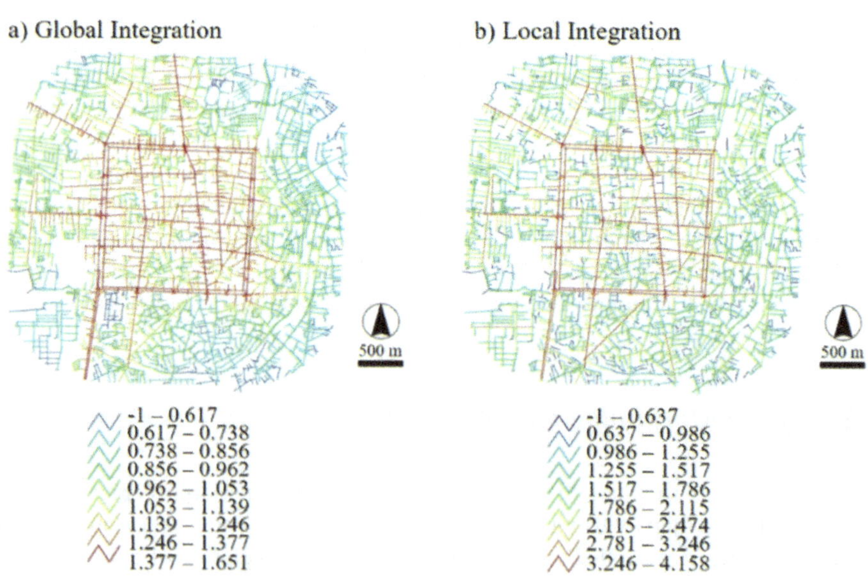

Fig. 3 Global Integration **a** and local Integration **b** of the old town of Chiang Mai and its one-kilometer radius areas

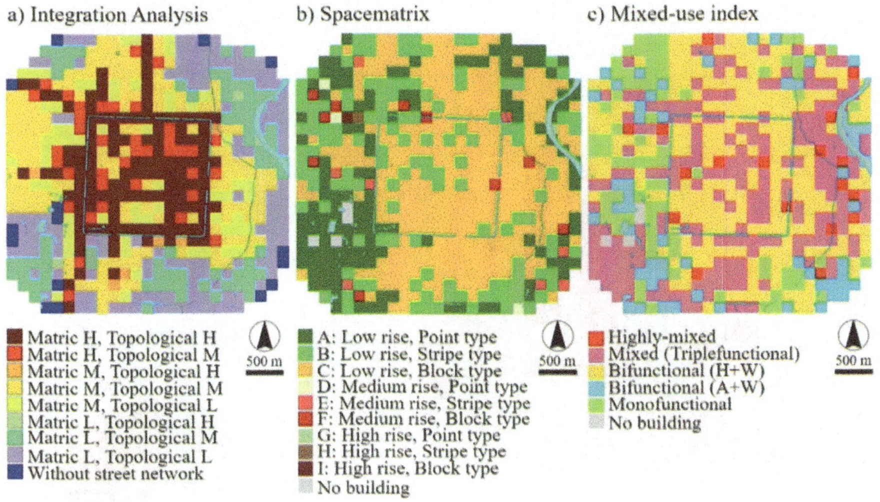

Fig. 4 The results of the urban morphological analysis; integration analysis (space syntax) (**a**), space-matrix (**b**) and mixed-use index (**c**)

that appeared in the old town. C type (low rise, block type) was the highest ratio (80.17%) followed by B (low rise, stripe type) and F (medium rise, block type). Their percentiles were 18.18, and 1.65, respectively.

The integration among housing, working, and amenities from mixed use index of the study area shown six of seven types of the degree of land use diversity as shown in Fig. 4c. Bi-functional (housing and amenities) was missing. Bi-functional (housing and working) type was the highest ratio of the study area (37.37%). There were four types that appeared in the old town. Bi-functional (housing and working) type was the highest ratio (56.86%) followed by mixed (Triple-functional), highly—mixed, and mono-functional, their percentiles were 37.19, 5.12, and 0.83, respectively.

3.2 *Result of the Spatial Integration Analyze*

The combined result from space syntax, space-matrix, and mixed used index of the old town of Chiang Mai and its one-kilometer radius areas were classified to seven classes as shown in Fig. 5a. According the distribution proportions of each category in the study area as shown in Fig. 5a and b, suburban type was the highest ratio (24.56%) followed by in-between (middle), low-urban, in-between (low), in-between (high), middle-urban, and highly urban, their percentiles were 21.58, 20.00, 15.61, 12.98, 1.40, and 0.88, respectively. In terms of the old town, there were five types of morphological characteristics. In-between (middle) was the highest ratio (47.11%) followed by in-between (high), low-urban, highly urban, and suburban,

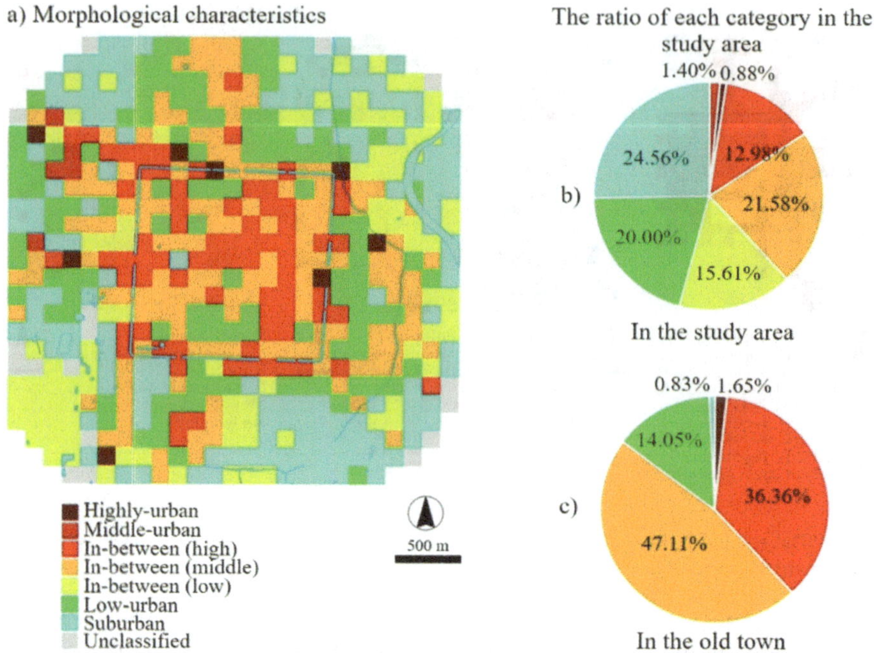

Fig. 5 Various categories shown morphological characteristics (**a**) and the distribution proportions of each category in the study area (**b**) and old town of Chiang Mai (**c**)

their percentiles were 36.36, 14.05, 1.65, and 0.83, respectively. the distribution proportions of each category in old town of Chiang Mai was shown in Fig. 5a and c.

Implying from the results, in the old town of Chiang Mai was higher degree of urbanity more than one-kilometer radius surrounding. The areas located farther from the old town, the lower urbanity. The areas located evenly between in and out of the old town of Chiang Mai were the highest degree of urbanity. Morphological characteristics of the old town of Chiang Mai was mixed, high–low of urbanity. The one-kilometer radius surrounding of the old town was mixed categories from the lowest to the highest of urbanity. There were small blocks of the highest urbanity widely located on the main roads of the surrounding areas.

Considered with urban form and urban image of each category as shown in Fig. 6, highly-urban, middle-urban, and in-between (high) types were located on the main roads, dense with quite tall buildings, three floors and over, for commercial and service. In-between (middle), in-between (low), and low-urban types were located on secondary and alley roads. Their locations were lower density of buildings. Most of their function and pattern of buildings were small scale for living. For the suburban type, they were located in the suburb areas with low accessibility and density of buildings. Most of their function and pattern of buildings were small scale for living, there were large open spaces or green spaces between each residential building.

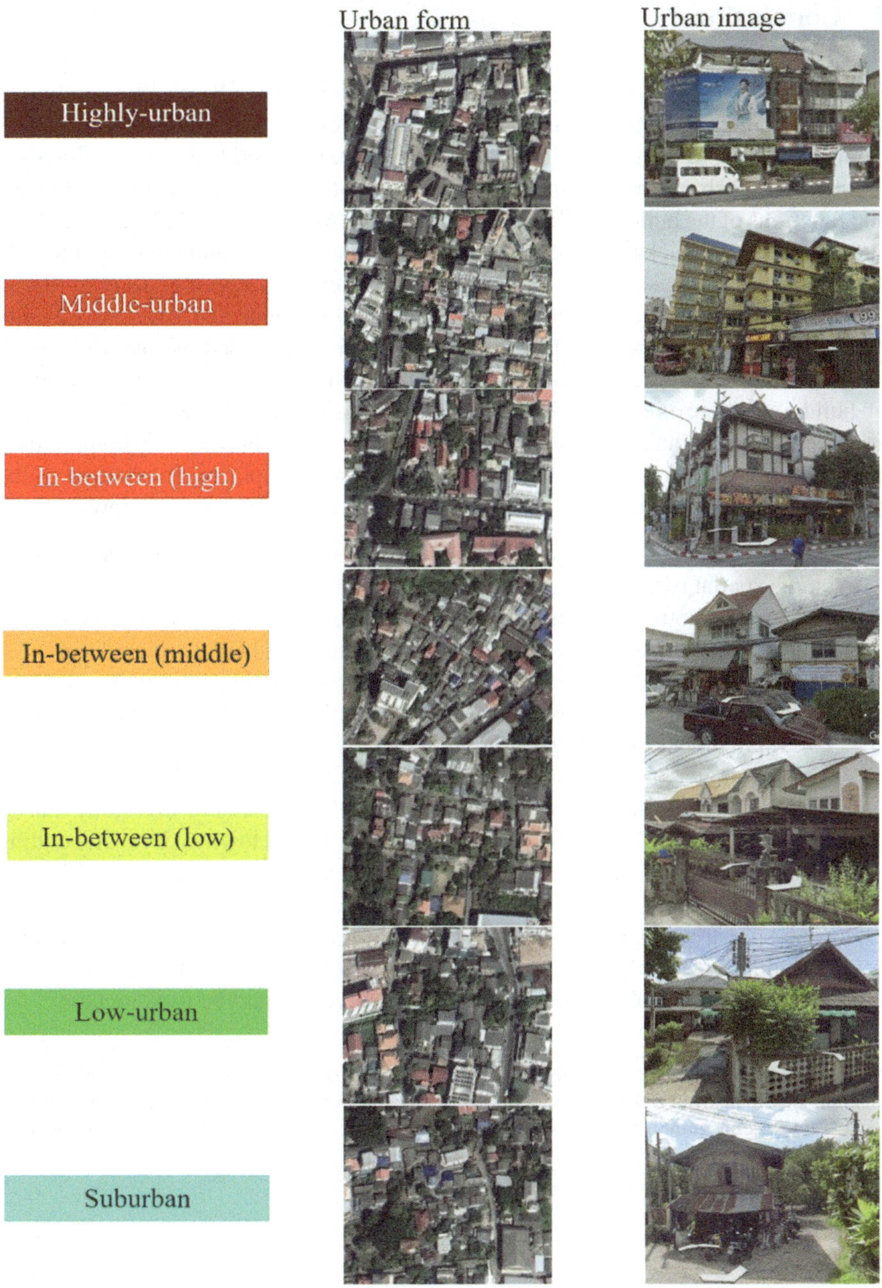

Fig. 6 Urban form and urban image of each category of morphological characteristics of the old town of Chiang Mai

4 Conclusion

Chiang Mai's old town is an old community of Chiang Mai where is outstanding in socio-economic. Its degree of urbanity is medium-level mixed of various categories from the highest to the lowest. It has a unique context derived from the combination of the essential characteristics included street networks, building densities and types, and functions of the buildings. The related results between the morphological analysis and the study area background create the clearer characteristic and identity of the old town. In terms of the street network configuration and the degree of accessibility, it has many lines of road and streets, and they are highly connected, especially evenly between in and out of the old town of Chiang Mai, and secondary and alley roads in the old town. In cases of the building density patterns and the mixed functions of buildings, there is a high density of buildings in the old town. Most of their architectures are small-scale for living and doing small business at home called shop house, mixed-function between residences and workplaces. In addition to the research results, there are considerations from research synthesis as follows:

(1) The spatial analysis results of the study different from the previous study, some spatial types, the metric H, topological L type from space syntax and bi-functional (housing and amenities) from mixed use index, are absented in the study. That represented the characteristics of the city where there are differences in geographic features, populations, and socio-economic conditions.
(2) The results of the study represent the urban expansion of urbanization. There are small blocks of the highest urbanity widely located on the main roads of the one-kilometer buffer areas. That can be implied that urban areas of Chiang Mai are expanded from the old town to the outer. The results of this study can be used to guide for balancing planning between the old town and its surrounding and applied to be an approach of the urban planning study for Chiang Mai city and other cities.
(3) The spatial secondary data obtained from the relevant departments of town planning is the only primary type of data used in this study. To make the study clearer, the data obtained from the actual survey should be used together.

Acknowledgements The authors thank the department of public works and town and country planning of Chiang Mai (DPT) for share their data and supported this study. This research did not receive any specific grant from funding agencies in the public, commercial, or not-for-profit sectors.

References

1. Oliveira V (2013) Morpho: a methodology for assessing urban form. Urban Morphol 17(1):21–33
2. Oliveira V (2016) Urban morphology an introduction to the study of the physical form of cities. Springer, Cham

3. Ye Y, van Nes A (2014) Quantitative tools in urban morphology: combining space syntax, space-matrix, and Mixed-use Index in a GIS framework. Urban Morphol 18(2):97–118
4. Sangsehanat S (2018) Morphological identity of Bangkok. Thailand Research Fund, Bangkok
5. Conzen MP (2010) The elusive common denominator in understanding urban form. Urban Morphol 14(1):55–58
6. Paksookcharoen K (2011) Urban discourses through morphological structures. Acad J Architect 63–76
7. Hillier B, Hanson J (1984) The social logic of space. Cambridge University Press, Cambridge, U.K.
8. Marshall S, Caliskan O (2011) A joint framework for urban morphology and design. Built Environ 37(4):409–426
9. Ye Y, van Nes A (2013) Measuring urban maturation processes in Dutch and Chinese new towns: combining street network configuration with building density and degree of land use diversification through GIS. J Space Syntax 4(1):17–37
10. Hillier B (1999) Centrality as a process: accounting for attraction inequalities in deformed grids. Urban Des Int 4(3–4):107–127
11. Berghauser-Pont M, Haupt P (2007) The relation between urban form and density. Urban Morphol 11(1):62–63
12. Marcus L (2010) Spatial capital. J Space Syntax 1(1):30–40
13. DPT (2020) Chiang Mai old town. Department of public works and town and country planning of Chiang Mai (DPT), Chiang Mai
14. Rangsirot S, Dankittikul C, Kasemsuk A (2011) Urban morphological analysis for Chiang Mai walking street space utilization improvement. Art Archit J Naresuan Univ 1(2):54–67
15. ESRI (2001) ArcGIS spatial analyst: advanced GIS spatial analysis using raster and vector data. ESRI White Paper, New York

Integrated Space in Low-Income Housing in Kuala Lumpur

Aida Syafiqah Suryadi, Hazrina Haja Bava Mohidin, Asrul Mahjuddin Ressang, Ainoriza Mohd Aini, Sharifah Salwa Syed Mahdzar, and Mahmud Mohd Jusan

Abstract The rapid urbanization and population growth in Malaysia cause increased rural migrants and poverty in the city centre, particularly Kuala Lumpur. Therefore, the needs for affordable housing for the low-income group are crucial. However, low-income housing design is subject to guidelines and standards set by the Ministry of Housing that cause the housing unit to be fundamental and lack understanding towards the occupant's lifestyle. Past studies have shown comparisons of spatial studies in various low-income housing layouts; however, none has identified the most integrated space utilized. Therefore, this paper investigates the value of spaces in Low-income housing in Kuala Lumpur. Using Space Syntax, this study identifies the most dominant space by looking at the integration of space. The finding of this study will determine the parameters in which improving the future Low-income housing design. This study found that entrance hall is the most dominant space followed with kitchen and living room. This paper will benefit policymakers, low-income house developers and consultants, researchers in the architecture field, and society.

Keywords Low-income housing · Space integration · Space syntax

1 Introduction

Malaysia is in Southeast Asia and is one of the fastest developing countries in this region. Greater Kuala Lumpur, usually referred to as the Klang Valley, has over seven

A. S. Suryadi · H. H. B. Mohidin (✉) · A. M. Ressang
Department of Architecture, Faculty of Built Environment, University of Malaya, Kuala Lumpur, Malaysia
e-mail: hazrinahaja@um.edu.my

A. M. Aini
Department of Real Estate, Faculty of Built Environment, University of Malaya, Kuala Lumpur, Malaysia

S. S. S. Mahdzar · M. M. Jusan
School of Architecture, Faculty of Built Environment, Universiti Teknologi Malaysia, Johor Bahru, Malaysia

T. Kang and Y. Lee (eds.), *Proceedings of 2021 4th International Conference on Civil Engineering and Architecture*, Lecture Notes in Civil Engineering 201,
https://doi.org/10.1007/978-981-16-6932-3_37

million residents. This growing urban population has become pronounced since the 1980s. People from all over Malaysia flock to the capital city, searching for jobs to earn a better living.

Consequently, the rapid urbanization causes the increasing number of settlers and slum areas in the city. Social housing schemes such as the People's Housing Project (PPR) scheme have been one of the government's approaches to cope with a housing shortage in the cities. The housing scheme aims to improve the quality of life, eradicate poverty among low-income groups, and resettle the urban squatter. The units' spatial quality needs to be considered significant in providing a good living for low-income people. Most of the studies in the literature mentioned a lack of space consideration given towards the low-cost housing.

This research study aims to identify the most dominant spaces of the low-income housing units through space syntax analysis of the space integration within the unit's layout. A study was conducted with 3-bedroom PPR units of similar spatial arrangement to explore space integration in the design. Integration is a variable that shows how space relates to other spaces in its surroundings [1]. This study is the key factor in the understanding of the relationships between users and the spaces. Hence, this study will benefit developers and consultants, researchers in the architecture field, in determining the parameters to improve low-income housing's spatial quality.

2 Literature Review

2.1 Low-Income Housing

The provision of low-cost housing units in Malaysia contributes significantly to ensure more houses built to relocate slums and provide a good home for the low-income people since the 1960s [2]. The program specifies that each low-cost shelter must have a minimum built-up space, including bedrooms, a kitchen, and a bathroom.

Over the years, the low-income housing design has faced multiple revisions with providing an adequate home for low-income people. Starting from 1999 and onwards, the three-bedroom flat unit introduced with a fitted living hall, bedrooms, bathrooms, kitchen, and drying area. However, despite the changes, the houses often fail to deliver humanistic factors' basic needs [3]. This is probably because of the financial constraint and cost-saving aspect in constructing and developing low-income housing [4].

Precedent studies concerning the low-income housing units can divide into (a) one that analyzed the spatial layout by the comparison of the low-income housing units and traditional houses (b) ones relating to physical factors such as privacy, thermal comfort, and safety [5] ones relating to lifestyle and culture. If we look at the researchers analyzing the spatial structure of traditional versus modern houses of low-income housing in interpreting its characteristics and implication, one of a recent study [6] mentioned that the modern low-cost housing units' spatial structure are segregated while the traditional Malay houses more towards the integrated spatial structure. Other recent studies investigating the unit's spatial structure by looking into the thermal comfort parameters and suggesting that the traditional Malay houses spatial structure is an ideal model and low-cost housing units design have not yet achieved an ideal thermal comfort for its occupants [5].

Other studies by Seo et al. [7] investigate changes in occupants' lifestyle by looking into the spatial structure of the modern apartment in Malaysia and found that the modern apartment in Malaysia has little consideration towards its occupants' vernacular lifestyle. Other studies [8], analyzing the spatial layout concerning the occupants' cultural value of the Low-income Housing and found that the current design of low-income housing has not met its occupants' adequate cultural needs. This led to the modification of spaces and abandoned low-cost housing units. Most of the occupants of the low-income housing renovated their houses to fit their everyday needs and activities. Therefore, the study aims to identify the space's integration concerning the domestic space used. Hillier and Hanson [9] mentioned that highly utilized spaces tend to have high integration values. This study is significant to determine the design parameters in improving the future Low-income housing design.

3 Methodology

Floor plans of 3-bedroom PPR Flats units were selected and classified based on the two phases; the first generation is the PPR Flats built between 1999 and 2010, and the second generation is the PPR Flats built in year (2011–2018). These two types of unit plans were analyzed using space syntax methods to obtain basic syntactic measures and difference factors [9].

This space syntax analysis will represent various spatial structures such as connectivity, control value, visual privacy, etc. Nevertheless, as the study's purpose, integration value is selected as the analysis index in examining the spatial properties. The integration index presents the degree of 'deep' and 'shallow' of the space within the units' plan [10]. In other words, the most integrated and segregated spaces within the units' plan can be identified using the integration value of each of the spaces in the housing unit. Spaces with a high level of social function and activity tend to have high integration values.

For this study, sets of floor plans of the two-generation PPR flats units were drawn using Autodesk CAD and analysed using UCL Depth map V.8. In this software, the units' plan is represented in a series of convex spaces where one convex space indicates one unit of space. This is done by developing a convex map. The convex space is the rooms that are defined by walls such as entrances, living hall, kitchen, bedrooms, dry area, and bathrooms. After running the analysis, integration data for each unit space were exported into an Excel spreadsheet. Next, the analysis results were statistically processed and compared.

4 Analysis and Discussion

4.1 Description of Basic Terminologies

The syntactic properties of spatial structure can be related to embedding the degree of space integration; the relationship can be applied to the other types of space layouts. Consider, low-income housing in Malaysia has been addressed for lack of space consideration. The low-income housing design is significant to analyse in addressing a new parameter for improvement in future development references.

Figure 1 shows a configuration of two spaces, A and B, with an outside space of C. the spaces lack openings. Figure 2 shows two different possibilities for the among spaces A, B, and C. The spaces are denoted as a dot "O", the access among them is denoted as lines "____", and each of the settings can be represented by a graph called a justified graph in terms of syntactical analysis. This graph is significant because it provides information regardless of space's size or relative location [9, 11].

Figure 1(i) shows that spaces A, B, and C have equal access to one another, forming a ring round type of layout as illustrated in the justified graph (Fig. 1(ii)). The same layout but with termination access within spaces B and C (Fig. 1(iii)). Thus, a linear type of layout is identified (Fig. 1(iv)).

The book the social Logic of Space [9] described several concepts to analyze these space layout settings in numerical terms. The first term is 'depth', which calculates the required number of spaces to pass before reaching another space from a certain point. Figure 1(i) shows that one can directly access A or B from point C when the depth value of points A and B is 1. Meanwhile, the depth value of point C is 0. As shown in Fig. 1(iii), the depth value of c is 0, and the depth values of A and B are 1 and 2, respectively. This showed that one must pass through A to reach space B, which makes the depth values of B higher than A.

The mean depth [3] is calculated based on the depth values gained. The calculation of mean depth can be expressed as [11].

$$\boldsymbol{MeanDepth(MD)} = \frac{thevalueofdepthofspacesfromtherootpoint}{thetotalnumberofspacesinthesystem} \tag{1}$$

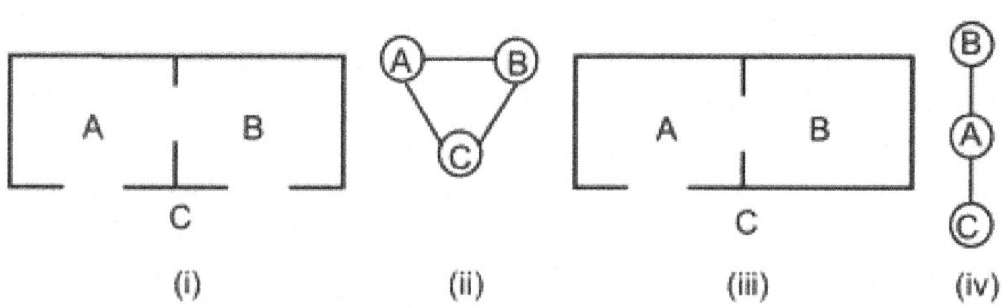

Fig. 1 A diagram taken from Tim Stonor's lecture (space syntax 101, 2021) of the configurational relationships between spaces and its justified graph

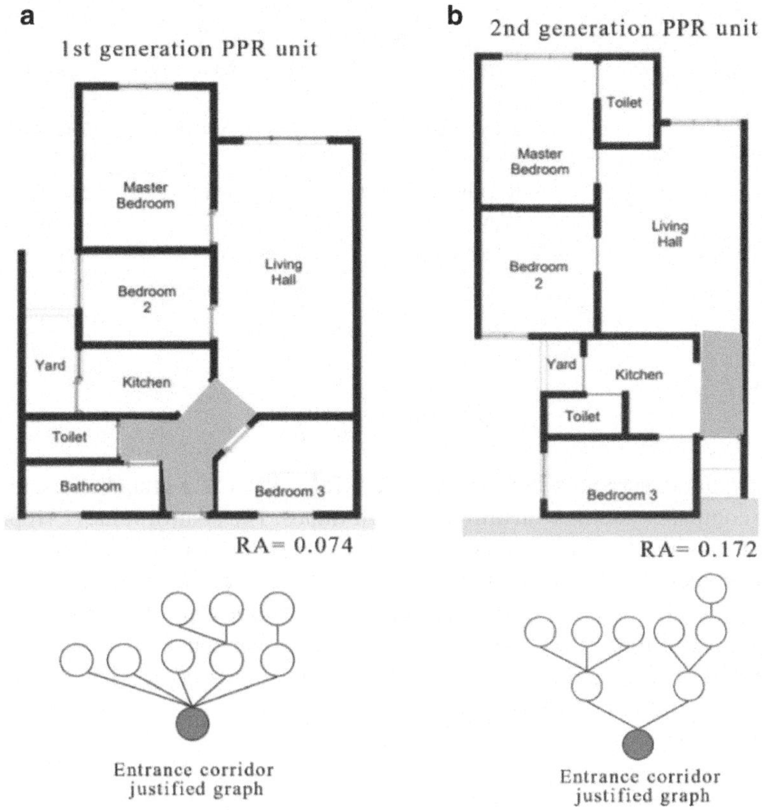

Fig. 2 Relative asymmetry (RA) of entrance corridor in both PPR units

The 'deep' or 'shallow' of a space from a root point can be identified with the MD values. The lowest value of depth indicates that the system is the shallowest, for instance, Fig. 1(ii). Meanwhile, the highest value of depth is when the spaces are connected in a linear sequence (Fig. 1(iv)). Therefore, the notion of relative asymmetry (RA) can be obtained by comparing the depth of space from a root point with the theoretical relation of depth and shallowness. In other words, this calculation of values will represent the integration of a space with its system from a root point. To calculate RA, the following equation employed

$$\boldsymbol{RelativeAsymetry(RA)} = \frac{2(\boldsymbol{MD} - 1)}{\boldsymbol{thetotalnumberofspacesinthesystem}} \quad (2)$$

Hillier and Hanson [9] mentioned the inverse of RA values could indicate the degree of integration of the spaces. This shows that a higher RA value of space represents a lower degree of integration with its system and a lower RA value indicates high integration.

Table 1 The terminologies and their implication used in the syntactical analysis

Terminologies	Higher value	Lower value
Depth	Far from the origin point	Near the origin point
Mean depth	'Deep' system	'Shallow' system
Relative asymmetry (RA)	Segregated	Integrated
Integration	Integrated	Segregated
Control value (E)	Strong connectivity	Less connectivity

$$\boldsymbol{Integration} = \frac{1}{\boldsymbol{RA}} \tag{3}$$

Lastly, another related term is real RA (RRA). RRA is a global measure that is used to compare systems with different depth values. The control value (E) represents the connectivity of a particular space to other neighbor spaces. Each system has a certain number (n) of neighbor spaces. The number (n) referred to as the node count. Therefore, each neighbor spaces 1/n summed up to measure the control value for each space. The higher the control value, indicates higher connectivity and vice versa. Table 1 describes the terminologies and their implication used in syntactical analysis proposed by Hillier and Hanson [9].

4.2 'Integration' of Rooms in the Floor Plan

Figures 2 show the selected PPR housing unit plans in Kuala Lumpur with their entrance indication. These units' plan schemes have different spatial compositions with a similar number of rooms and have only one entrance—both PPR units are attached with a living hall, kitchen, two bathrooms, and three bedrooms. However, most changes and modifications of spaces can be seen with the kitchen area's enlargement [12]. The modification happens by replacing the dry area by adding a wet kitchen.

Integration of spaces in the PPR housing unit is measured by the relative asymmetry (RA) [11]. It is a value between zero and one computed for each node in the justified graph by considering the distance between the nodes. RA equal to zero means the highest possible integration, whereas one defines the lowest integrated space from other rooms, which indicates the highest segregation value. A highly integrated space is a space from which other rooms within the units can easily be reached.

For example, in Fig. 2a, the entrance corridor has a relative asymmetry equal to 0.074; space is the second-lowest RA in the flat units (the kitchen and living room have 0.393 and 0.123, respectively).

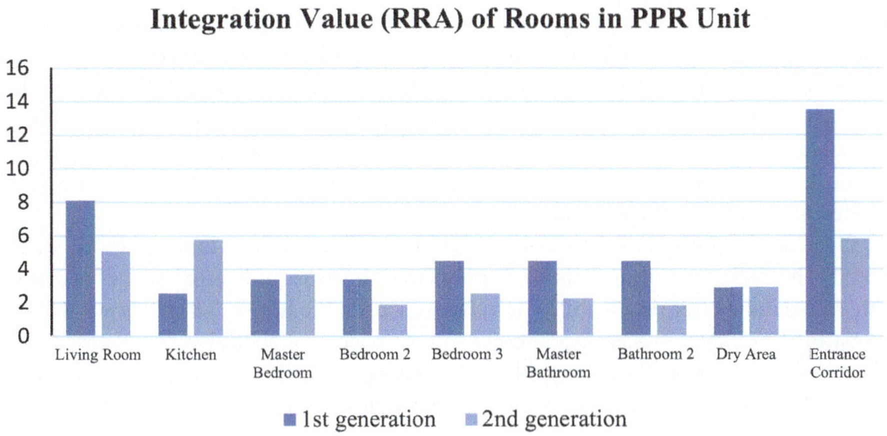

Fig. 3 Integration values of rooms in both PPR units

It is interesting to note; the same space remains with the highest integration value in first and second-generation units. The data's most striking result is that the entrance corridor's integration value of first and second-generation unit layout is 13.15 and 5.81, respectively. The high integration value shows that this space is the most permeable as it connects the servant and the houses' serve area. However, in the second generation PPR unit, the kitchen presents the second most integrated space; meanwhile living hall is the second most integrated space in the first generation PPR units. The kitchen in the 2nd generation PPR units controls the accessibility of three spaces: the 2nd bathroom, Yard, and 3rd bedroom. Thus, the kitchen has more integration value than the living room in the unit.

Overall, the analysis of the justified graphs for both generations of PPR units revealed that most are relatively deep and asymmetric. The MD for the first generation PPR units is 1.63 and has increased to 2.61 for the second generation PPR unit. This represents that the first generation appears in more symmetric order whereby spaces are directly connected to the root point. The second-generation PPR unit is much more asymmetrical and has deeper depth. Interestingly, the RRA for the first generation PPR unit is 0.13, while the second generation PPR unit indicates a higher value with 1.45. This signifies that the spatial configuration of the 2nd generation PPR unit is more integrated compared to first-generation. Thus, connectivity and accessibility within the rooms have improved (Fig. 3).

5 Conclusion

This paper intended to report the initial findings of ongoing research that explores the spatial configuration of low-income house units and their parameters. Therefore, this study selected the two units of PPR flats in Kuala Lumpur with different spatial layouts. Both units' floor plans were compared and quantitively analysed. The floor plans were investigated for their morphological spatial configuration. It discovered that spaces with the highest RRA are the same in both generation PPR units. In this case, the entrance. However, the kitchen has the second highest RRA value after entrance corridor in 2nd generation units, while the second highest RRA value is in the 1st generation units. Therefore, the living hall and kitchen are the dominant spaces after entrance hall in both housing units.

Overall, the 1st generation units have higher RRA values and are more segregated. The 2nd generation units have a lower RRA value, indicating spaces are more integrated and better permeability to other spaces in the house. These shows that the latest PPR unit design has improved terms of spaces connectivity and accessibility. However, it is still early to conclude the findings. Further investigation needed in terms of the relationship between spatial configuration and households' activities.

Acknowledgements This study is part of research work 'Improving Indoor Space quality of B40 Housing' under Ministry of Higher Education FRGS grant FRGS/1/2019/SSI11/UM/02/2.

References

1. Dettlaff W (2014) Space syntax analysis–methodology of understanding the space. Ph. D. Interdiscip J 283–91
2. Plan TM (2012) 2010 Tenth Malaysia plan 2011–2015. In: Speech presented by the prime minister introducing the motion to table the tenth Malaysia plan at the Dewan Rakyat, Malaysia, vol 3
3. Hisham NN, Amin NDM (2020) Housing design for urban poor in Kuala Lumpur: a literature review. IOP Conf Ser: Earth Environ Sci 012083
4. Zainal NR, Kaur G, Ahmad NA, Khalili JM (2012) Housing conditions and quality of life of the urban poor in Malaysia. Proc Soc Behav Sci 50:827–838
5. Mohd Sahabuddin MF, Gonzalez-Longo C (2015) Traditional values and their adaptation in social housing design: towards a new typology and establishment of air house standard in Malaysia. ArchNet-IJAR: Int J Archit Res 9:31–44
6. Yahya ASB (2019) Comparative pattern of spatial visibility and stationary activity in Malay traditional and low-income houses
7. Seo KW, Ghani MZA, Sarkom Y (2021) Relocating home activities: spatial experiments in Malaysian apartment houses to accommodate the vernacular lifestyle. J Asian Archit Build Eng 1–15
8. Mohit MA, Ibrahim M, Rashid YR (2010) Assessment of residential satisfaction in newly designed public low-cost housing in Kuala Lumpur, Malaysia. Habitat Int 34:18–27
9. Hillier B, Hanson J (1989) The social logic of space, Cambridge university press
10. Hillier B (2007) Space is the machine: a configurational theory of architecture, Space Syntax
11. Bafna S (2003) Space syntax: a brief introduction to its logic and analytical techniques. Environ Behav 35:17–29
12. Goh AT, Yahaya A (2011) Public low-cost housing in Malaysia: case studies on PPR low-cost flats in Kuala Lumpur. J Des Built Environ 8

Application of UAV-Based Oblique Photography in Architectural Design: The Case of Mengyuan Resort Hotel in Yunnan, China

Guiye Lin and Kun Sang

Abstract With the development of remote sensing and images techniques, the oblique aerial photography (OAP), which relies on the unmanned aerial vehicles (UAV), has become increasingly significant, showing advantages of high efficiency, low cost, and high spatial resolution in various disciplines. The OAP technology provides an efficient method for the acquisition of spatial information, site surveying, and mapping. It can also be applied to improve the scientific process of architectural design and planning. This article, therefore, introduces a methodology using point clouds obtained by the OAP and takes the Mengyuan resort hotel in China as a case to discuss the meaning and possibility of employing the OAP in architectural design. This research process includes five main steps: site investigation, drone route planning, data acquisition, model construction, and the application of 3D models. The design project of the Mengyuan hotel is finally fulfilled based on the built 3D point clouds, which can be a reference for future architectural analysis, planning, and research, combining with multiple spatial and remote sensing methods.

Keywords Oblique aerial photography · UAV · Architectural design · Remote sensing

1 Introduction

Nowadays, along with computer and internet technologies, the advances of photogrammetry have driven the multiple uses of UAV in more and more disciplines and practical cases. And an increasing number of scholars are focusing on the research on the flight performance of drone, the quality of its photographs, as well

G. Lin
Department of Civil, Environmental and Architectural Engineering, University of Padua, 35131 Padua, Italy

K. Sang (✉)
Department of Advertising, Xiamen University Malaysia, 43900 Sepang, Selangor, Malaysia
e-mail: kun.sang@xmu.edu.my

T. Kang and Y. Lee (eds.), *Proceedings of 2021 4th International Conference on Civil Engineering and Architecture*, Lecture Notes in Civil Engineering 201,
https://doi.org/10.1007/978-981-16-6932-3_38

as the applications of UAVs, especially in the inspection works for military, industrial and civil purposes [1]. Meanwhile, UAV has also been integrated with other methodologies, such as the Geographic Information System (GIS), Territorial Laser Scanning (TLS) and Building Information Modelling (BIM), and so forth [2, 3].

Among these, the oblique aerial photography (OAP) is an innovative and useful photogrammetry and UAV technology discussed frequently in recent years. It challenges the limitations of the traditional surveying method, which can only photograph from vertical angles and directions. The OAP utilizes several sensors to cooperatively collect data from multiple angles and obtains massive information more efficiently. Through modeling and calculations, it can reflect the spatial positions, geometry, and texture of the surveyed objects. Besides, 3D real scene models with higher reliability and accuracy can be constructed with the help of OAP, to meet the needs to access the research fields [4]. Currently, the building of 3D models requires a combination of 3D scanning and OAP technology, which are widely used in the surveying and monitoring of land and environment.

In the field of architecture, the creative uses of OAP based on the functions of investigation and documentation attract more attention in academia, including the protection and visualization of historic buildings [5], modeling of smart city, the detection of defects in heritage structures [6], and the preliminary survey and mapping for architectural construction [7]. But its application in the field of architectural design has been seldom discussed. A professional drone systems or technical standard for architecture design have not been formed yet. Thus, how to establish an effective application system with OAP to improve the scientific architectural planning and design is of great significance, and it is still a challenge waiting for further discussions.

Based on the OAP technology, this paper establishes a real-life 3D spatial information system, and conducts a case analysis of a resort hotel in Yunnan, China, which can set a reference for other similar design and planning projects in architecture. With the built 3D models, architects and designers can accurately and freely express their creative minds through the software, like Sketch-up, 3D Max, etc. However, facing a real project site, a huge workload of surveying, mapping, and modeling may cause a loss of objective information due to the errors in the surveying process, resulting in unexpected problems in design process [8]. Especially for those projects in remote areas, it is impossible to acquire the comprehensive site data and information immediately, such as the terrain condition, surrounding environment, building layout and forms, materials, and other decorative details. Thus, introducing the OAP into the process of architectural design can effectively deal with the difficulties mentioned above [9]. More importantly, the built 3D models provide architects with comprehensive information in time, solving the time-consuming problems and errors in traditional surveying methods.

2 Methodology

In this paper, the OAP method is mainly utilized for the whole project, which includes four steps to assist architectural design projects. The specific operational process is shown in Fig. 1. During the site preparation, the design purposes are clarified according to the needs of the clients, as well as the survey requirements of the real site and the characteristics of the scanner. The buildings in the study area can be marked on the satellite map. Then, a preliminary survey plan is made according to the shape of the site and the requirement of accuracy. After on-site surveys, combined with the collected site information and the constraints, such as cost, time, and human resources, a detailed route plan for the drone is designed to build the 3D models, which will be applied for more design purposes. Some factors are also considered in this process, such as the shooting range, image accuracy, working period, quality of images, etc.

The planning of aerial survey routes is the primary process in the whole project. Because the images obtained from the OAP are superimposed on some important feature points, but one drone flight usually takes a few hours to complete. The environmental conditions, such as light and shadow always change in the daytime. Thus, the site to be sensed should be photographed under the same conditions (minimum light changes) to ensure the continuity of color and texture of the photography. Then, a large number of obtained images usually lead to a higher rate of overlapping between images and a higher efficiency of data post-processing. Besides, other requirements need also to be considered, such as a windless weather condition, equipment costs, etc.

After data collection, the point clouds are obtained, namely a cluster of points with coordinate systems, which need to be further pre-processed, such as removing the noises and integrating all the images according to the feature points. Furthermore, in the process of model construction, there are many options of software in the market, such as Smart3D, Pix4D, Agisoft Photoscan, etc., which can process the remote sensing photographs automatically, and output the data of digital elevation model (DEM), Digital Surface Model (DSM), Digital Orthophoto Map (DOM), and other real scene models. In order to meet the needs of high accuracy of 3D models, the Pix4D (www.pix4d.com) is selected as the platform to do the data processing

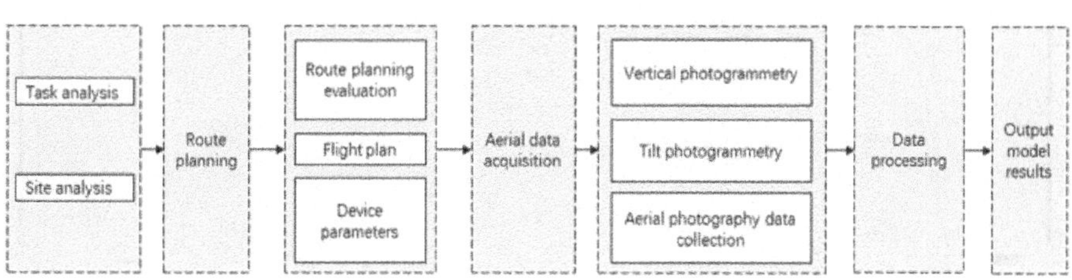

Fig. 1 The workflow of this paper: preliminary work preparation; UAV aerial photography route design and data acquisition; model construction; model applications

and model construction [10]. The models built by Pix4D can be also employed and combined in other modeling software, such as Revit, 3D Max, Rhino, and Recap for multiple uses, whose results are helpful for further architectural planning and analysis, building landscape models, as well as information displaying and interpreting [11].

3 Application

3.1 Site Condition

Based on the discussion above, this project focuses on a resort hotel located in the Mengyuan scenic area in Xishuangbanna, Yunnan Province in the southwest of China. It covers an area of 1000 m^2. This area is famous for its tropical landscape, high biodiversity, and Thai culture, which is a special area in China, bordering Myanmar and Laos. And the Scenic area of Xishuangbanna is one of the top touristic destinations in China, being popular among visitors. However, the county Mengyuan locates in the border of the scenic area, waiting for further development and touristic planning (see Fig. 2). The built hotel will be the first one in this region and will be an important touristic facility for this area in the future. The design site is located in

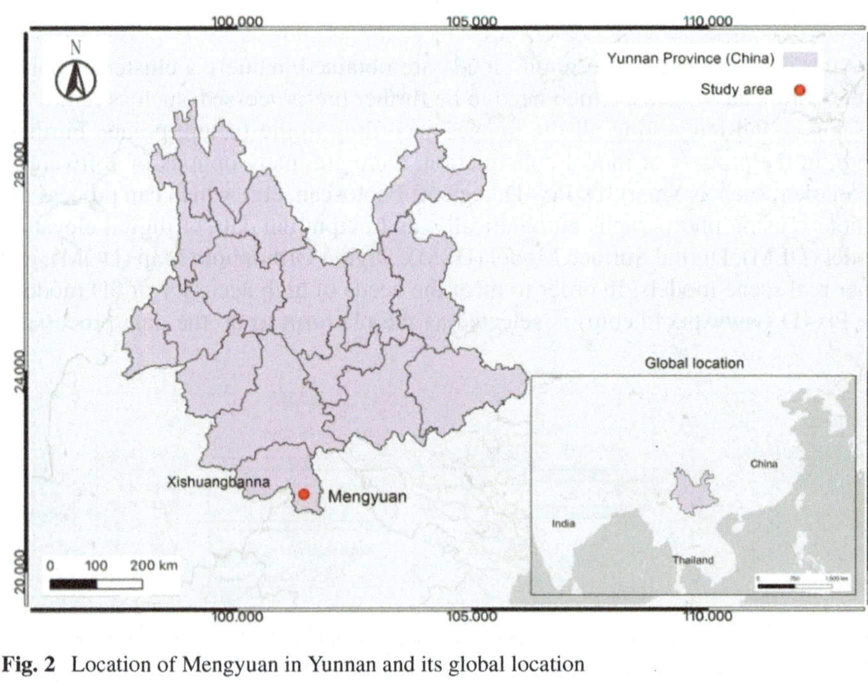

Fig. 2 Location of Mengyuan in Yunnan and its global location

Fig. 3 Photos of the site condition

a mountainous area, which makes the traditional surveying methods hard to access (see Fig. 3). As is seen, the OAP method owns its advantages in this kind of project.

3.2 Equipment Selection and Route Design

According to relevant design requirements, the "DJI Mavic Pro" is the drone tool selected for the OAP (see Fig. 4). It is a rotorcraft which can fly at an altitude (lower than 100 m) for aerial surveys and scanning, without any indoor or outdoor limits. Because of the internal positioning and sensing system, it can automatically avoid obstructions to ensure the safety of flight. Then, the flight route is planned as is shown in Fig. 5, which is a bow-shaped route suitable for the site shape and topography. In this project, drones are used for multiple flight photography, and four times of flights are planned, including one time for vertical photography and three times for 45° photography. The red route represents the flight paths with an angle of 45°, and the white lines indicate the shots at a vertical angle [12].

Fig. 4 DJI Mavic 2 Pro UAV platform

Fig. 5 Planning the flight route

3.3 *Image Processing*

Then, the images obtained from UAV are imported into the software Pix4D, and the attribute information is also assigned immediately, including their geographic coordinates, image parameters, etc. The Pix4D can automatically extract, match the feature points, and calculate the position of photos through aerial triangulation to generate the point clouds (a set of data points in space). The irregular triangulation network is constructed according to the point cloud to produce the digital surface model. In addition, the territorial texture is also mapped to produce the real scene model, mainly including six steps: importing data into the Pix4D, applying Pix4D for an aerial triangulation, make the schematic diagram of local site, generating triangular mesh surface, projecting the site restoration, and presenting the projected result (see Fig. 6). Some errors may be generated in the process of model construction. Thus, the Recap software (www.autodesk.com/products/recap) is a useful tool to remove the errors and noises to build accurate point cloud models [13].

Fig. 6 Point clouds construction (from generating triangular mesh surface to projecting the site restoration)

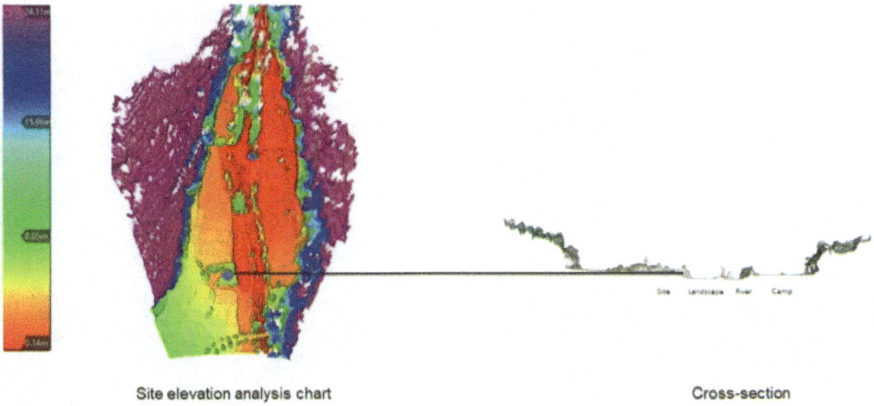

Fig. 7 Height analysis based on different colors; a cross section is made based on the elevation analysis

3.4 Data Analyzing

After building the point clouds, the site can be analyzed through other software. For example, the point clouds can be utilized to analyze the drainage slopes, site suitability, building heights, public and green areas, etc. With the simulations in Recap software, other building conditions (sunshine and ventilation) can be also analyzed. Furthermore, through color analysis of the model, the building materials, style, and texture of the site can be evaluated. For example, the analysis of height difference is calculated in this paper. As is shown in Fig. 7, different colors represent different heights, facilitating the understanding of the slopes of the site and visualization of architectural spaces [14].

3.5 Modeling and Design Results

As a result, the point clouds are linked to the software "3Dmax" (www.autodesk.com/products/3ds-max), so that the architects can design the project within a virtual environment (see Fig. 8). A final design example of the hotel is shown in Fig. 9. In this process, the architects understood better the environmental conditions, especially the elevation of the site based upon the 3D model, at the same time, the structure and design process of architecture are all processed in the software with the help of the built models, which means a higher accuracy and efficiency during the design process.

Finally, the 3D model of the hotel is fulfilled on a real 3D scene, so that the design project can be shared among professionals immediately to discuss the design idea and image rendering. The clients and future tourists can also easily access to the 3D

Fig. 8 Conceptual design based on the 3D max platform

Fig. 9 Examples of the design result—Mengyuan hotel

model to know better the idea of the designer, as well as the inner structure of the building, and many other details expressed by the 3D model.

4 Conclusion

OAP is an effective and useful tool for site sensing and mapping. It can also be applied in the architectural projects to support the design process and to improve the accuracy and flexibility. After reviewing the character and advantages of OAP, this article discusses the method of applying the 3D point cloud models obtained by the OAP to the site investigation and the design process during the architectural design process, taking the design project of Mengyuan resort hotel as a case study. The results demonstrate the feasibility and reliability of the OAP. This method not only solves the problem of missing data for some design projects but also deals with the limitations of traditional graphic design. It saves greatly workload and costs, and it is worthy of promotion in future architectural design work, as well as in other related research.

However, in this study, the architectural design process with the help of OAP considers more the natural and environmental conditions of the site in Mengyuan, but the cultural aspects of the site were ignored, such as the local culture, identity of local people, and the expectation of the future tourists. Thus, a combination of using UAV/OAP to communicate with the future clients, related authorities, and other stakeholders can be an independent issue, which will facilitate the process of participatory design in architecture, improve the communication efficiency and a mutual understanding among all the stakeholders [15].

Besides, further study can be done focusing on its combination with other advanced sensing and mapping tools, such as Geographic Information System (GIS), and Augmented/Virtual Reality (AR/VR), to support other urban planning and touristic activities.

Acknowledgements The project was funded by the China Scholarship Council (CSC), Grant Number: 201708440230.

References

1. García-Pulido LJ, Jaramillo JR, Dorado MIA (2017) Heritage survey and scientific analysis of the watchtowers that defended the last Islamic kingdom in the Iberian Peninsula (thirteen to fifteenth century). Int Arch Photogramm Remote Sens Spatial Inf Sci XLII-2(W5):259–265
2. Ivosevic B, Han YG, Kwon O (2017) Calculating coniferous tree coverage using unmanned aerial vehicle photogrammetry. J. Ecol Environ 41(10):1–8
3. Ivošević B, Han YG, Cho Y, Kwon O (2015) The use of conservation drones in ecology and wildlife research. J. Ecol. Environ 38(1):113–118
4. Qu Y, Yang Z, Yao Q (2019) The application of UAV oblique photography technology in architectural design. In: DEStech transactions on engineering and technology research, DEStech Publications, Changsha, China, pp 123–127
5. Püschel H, Sauerbier M, Eisenbeiss H (2008) A 3D model of Castle Landenberg (CH) from combined photogrammetric processing of terrestrial and UAV-based images. Int Arch Photogramm Remote Sens Spatial Inf Sci 37(B6b):93–98
6. Aydin CC (2014) Designing building façades for the urban rebuilt environment with integration of digital close-range photogrammetry and geographical information systems. Autom Constr 43:38–48
7. Aicardi I, Chiabrando F, Grasso N, Lingua AM, Noardo F, Spanò A (2016) UAV photogrammetry with oblique images: first analysis on data acquisition and processing. Int Arch Photogramm Remote Sens Spatial Inf Sci 41:834–842
8. Kang C, Cheng Y, Wang F, Zong M, Luo J, Lei J (2020) The application of UAV oblique photogrammetry in smart tourism: a case study of Longji terraced scenic SPOT in Guangxi province. Int Arch Photogramm Remote Sens Spat Inf Sci 42:575–580
9. Liu S, Tong X, Chen J, Liu X, Sun W, Xie H, Ye Z (2016) A linear feature-based approach for the registration of unmanned aerial vehicle remotely sensed images and airborne LiDAR data. Remote Sens 8(2):82

10. Gupta L, Jain R, Vaszkun G (2015) Survey of important issues in UAV communication networks. IEEE Commun Surv Tutorials 18(2):1123–1152
11. Maxa JA, Mahmoud MSB, Larrieu N (2016) Joint model-driven design and real experiment-based validation for a secure UAV and HOC network routing protocol. In: Integrated communications navigation and surveillance (ICNS), IEEE, Herdon, USA, pp 1E2–1
12. Landau M, Van-Delden S (2017) A system architecture for hands-free UAV drone control using intuitive voice commands. In: Proceedings of the companion of the 2017 ACM/IEEE international conference on human-robot interaction, Association for Computing Machinery, Vienna, Austria, pp 181–182
13. Tisdale J, Ryan A, Zennaro M, Xiao X, Caveney D, Rathinam S, Sengupta R (2006) The software architecture of the Berkeley UAV platform. In: 2006 IEEE conference on computer aided control system design, IEEE, Munich, Germany, pp 1420–1425
14. Zhou S, Gheisari M (2018) Unmanned aerial system applications in construction: a systematic review. Constr Innov 18:453–468
15. Lin C, He D, Kumar N, Choo KKR, Vinel A, Huang X (2018) Security and privacy for the internet of drones: challenges and solutions. IEEE Commun Mag 56(1):64–69

A Classification Urban Precinct Ventilation Zones Using Key Indicators of Spatial Form: Case Study in Bangkok

Manat Srivanit, Daranee Jareemit, and Jiying Liu

Abstract This study focuses on developing a geospatial clustering-based methodology using the k-means method, which could be applied by city planners to effectively classify key urban form indicators concerning the ventilation environment in Bangkok city, Thailand. All indicators were calculated with standard GIS-based tools commonly utilized by urban planners. The results showed that the inner core of the city's forms is heterogeneous, including three precinct ventilation zones. In this study, the precinct ventilation map could be used to propose primary investigations for delimiting the main problem areas in terms of natural ventilation performance during the initial stage for urban heat mitigation and air quality management planning.

Keywords Urban ventilation · Precinct ventilation zone · GIS-based analysis · Urban form indicators

1 Introduction

The rapid urbanization of Bangkok, the capital city of Thailand, an increase of urban vehicles leads to traffic congestion and overloaded road. This traffic is a significant cause of air pollution problem in the inner areas with high dense land used. According to the US AQI report [1], Bangkok is ranked in the 12th list of the cities with the worst outdoor air quality in the world. Such high pollution levels could cause respiratory disease and risk the elderly, children, and pregnant women. Policy-makers have made

M. Srivanit (✉)
Center of Excellent in Urban Mobility Research and Innovation (CoE-UMRI), Ahmedabad, India

D. Jareemit
Thammasat University Research Unit in Architecture for Sustainable Living and Environment, Pathumthani, Thailand

M. Srivanit · D. Jareemit
Faculty of Architecture and Planning, Thammasat University, Pathumthani, Thailand

J. Liu
School of Thermal Engineering, Shandong Jianzhu University, Jinan, China

T. Kang and Y. Lee (eds.), *Proceedings of 2021 4th International Conference on Civil Engineering and Architecture*, Lecture Notes in Civil Engineering 201,
https://doi.org/10.1007/978-981-16-6932-3_39

an effort to controls urban traffic congestion and reduce vehicle-induced pollutant dispersion.

Besides traffic control, urban ventilation is another key parameter that could dilute air pollution in the city. The effect of urban morphology on air ventilation and pollutant dispersion has been investigated in different scales. Street canyon design, including a ratio of street width to building height (H/W) [2–4], building roof shape [3, 5], building block length [4, 5], has been widely investigated in a local context or district area. Regarding urban scale, urban sprawl with complex urban forms, relevant to open space ratio, frontal area index, variations of building height, and street patterns, could block the wind flow in the city. These factors result in a high accumulation of air pollutants at street level [6–8]. In high density and poor ventilation areas, the high pollutant concentration is serious because the residence time of the local air is longer than those of low-density areas. The precinct ventilation zone classification has been commonly used to identify the poorly ventilated areas resulting from air quality problems. This measure is affected by external meteorological conditions (synoptic wind, sea/river-land breeze, and local wind environment) and morphological characteristics, as shown in Fig. 1.

Bangkok has a morphological identity characteristic with complex urban form and variations in building height, leading to a poor wind environment. This effect relates to the urban heat island effect and air quality problem. Previous works have been classified the urban morphology using local climate zone classification and its impact on urban heat effect and outdoor thermal comfort [9, 10]. Reference [11] investigated the effect of different residential-cluster forms on street wind velocity in Bangkok suburbs. Lack of study uses urban ventilation indicators to classify the zones in Bangkok. In this study, to understand ventilation performance in urban scale, this study selected the most five-key urban-form, including (1) building coverage ratio

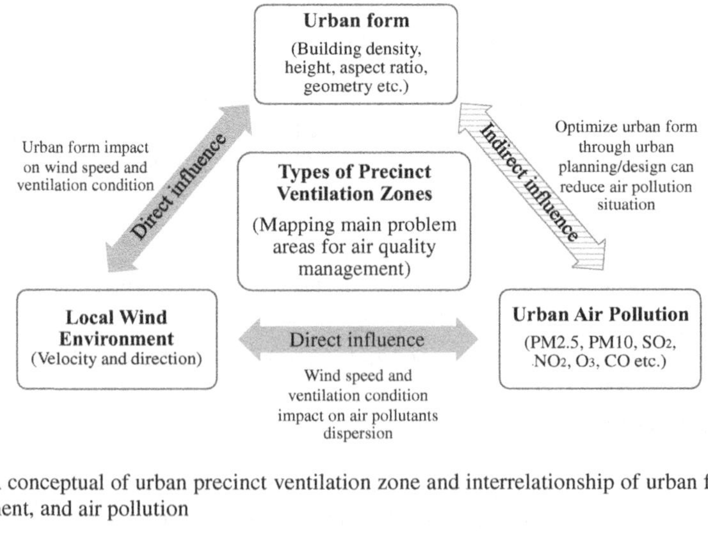

Fig. 1 A conceptual of urban precinct ventilation zone and interrelationship of urban form, wind environment, and air pollution

(BCR), (2) floor area ratio (FAR), (3) Sky View Factor (SVF), (4) average of building height (AvgH), and (5) standard deviation of building height (StedH), which were commonly used to assess the urban ventilation performance [7, 12–14] and air quality studies for the local scale [12, 15–17]. The classification will provide designers, and urban planners have an understanding of the precinct ventilation performance. In addition, it can be further investigated to enforce policies or solutions for a specific zone, which has critical air ventilation and pollution problems.

2 Research Methodology

The methodological approach of the precinct ventilation zones (PVZs) classification proposed in this study is presented in Fig. 2. This study uses the GIS-based method to classify the groups of urban morphology in Bangkok city using five key indicators, which are summarized in Table 1.

The studied location used in this study is the inner core of the Bangkok Metropolitan Area (BMA). This area is defined as the high land-use intensity and the most densely populated area in BMA. It covers an area of 125 km^2, which includes eight districts. Typically, those locations are in the modern central business districts of BMA (Fig. 3a). The geospatial datasets were retrieved from the Department of City Planning, Bangkok Metropolitan Administration, and Open-source QGIS V.3.12.1 software was used in this study. In order to delimit different PVZs with standard features, a grid size of 100 × 100 m which was overlaid with 12,519 grid cells, was applied to calculate each urban form indicator (Fig. 3b–f). The view of the data's distribution using histograms can be seen in Fig. 4. Pearson's correlation analysis was used to assess the effective urban form indicators. In this study, classifying PVZs was performed by Cluster Analysis following the Ward method by running the k-means in statistics software, and the Elbow method was used for determining the optimal number of clusters [18].

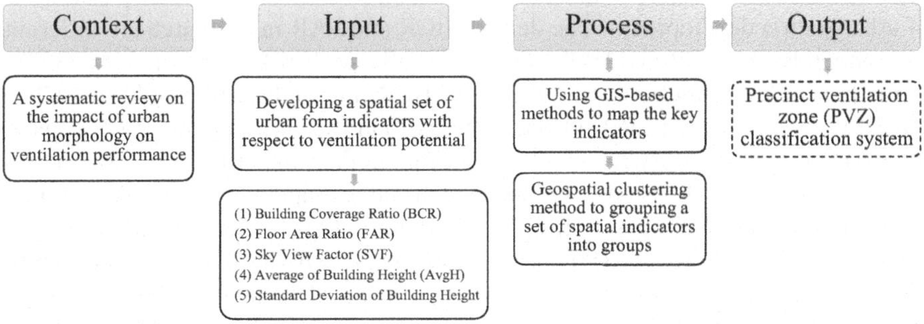

Fig. 2 The methodological approach was used in this study

Table 1 A spatial set of urban form indicators with respect to ventilation potential

Factors	Equation	References
Building coverage ratio (BCR)	$BCR(\%) = \frac{\sum_{i=1}^{n} building\ area_i}{grid\ area} \times 100$	Zhou et al. [15]
Floor area ratio (FAR)	$FAR(\%) = \frac{\sum_{i=1}^{n} total\ floor\ area_i}{grid\ area} \times 100$	
Sky view factor (SVF)	$SVF = \sum_{i=1}^{n} S \frac{2}{2\pi} \sin\left(\frac{\pi}{180}\right) \sin\left(\frac{\pi(2\alpha i-1)}{2n}\right) \frac{360}{\theta i}$	Lindberg and Grimmond [17]
Average of building height (AvgH)	$AvgH = \frac{\sum_{i=1}^{n} building\ height_i}{n}$	
Standard deviation of building height (StdH)	$StdH = \sqrt{\frac{\sum(building\ height - \mu)^2}{n}}$	Zheng et al. [16]

3 Results

Statistically, Pearson's correlation measures the linear correlation between two variables and reflects this correlation with a value between +1 and −1. In the analysis, 1 shows a positive linear correlation, and −1 shows a negative linear correlation. The correlation matrix, presented in Table 2, could help examine multi-collinearity issues between the predictor variables. If the variables are highly correlated with the correlation coefficient, the value will close to 1. It is extremely hard to evaluate the true effects of the predictor variables. According to the calculation, it was found that the urban form indicator has the expected signs with each of the five indicators. All show positive correlation coefficients at 0.01 level. Table 3 presents the resulting centroid table where we can see the average value of each indicator in each cluster.

The characteristics of each urban form indicator could be classified into three clusters: low, moderate, and high values, as summarized in Fig. 5. The BMA has planned via land-use controls and zoning ordinances to increase various possibilities of urbanization development. The defined BCR and FAR in each area are different. Consequently, the BCR and FAR values show a large distribution range. According to the box-plot distributions, in the case of BCR, the highest threshold of BCR is 20–71.3%, representing the mid-rise zone as found in the standard PVZ 2, while the lowest threshold of BCR of 0–26.4% occurs in the standard PVZ 1, which is an open low-rise zone. This study used the median value as an indicator for determining the PVZ. For example, densely high-rise buildings (Fig. 6) were classified into the standard PVZ 3. According to Fig. 5, this zone shows the high values in floor area ratio (FAR), an average of building height (AvgH), a standard deviation of building height (StdH). At the same time, the building coverage ratio (BCR) and sky view factor (SVF) in this zone show moderate and low values, respectively. The high value of FAR (>5.5%), AvgH (>25 m), and StdH (>27.3 m) could increase the precinct surface

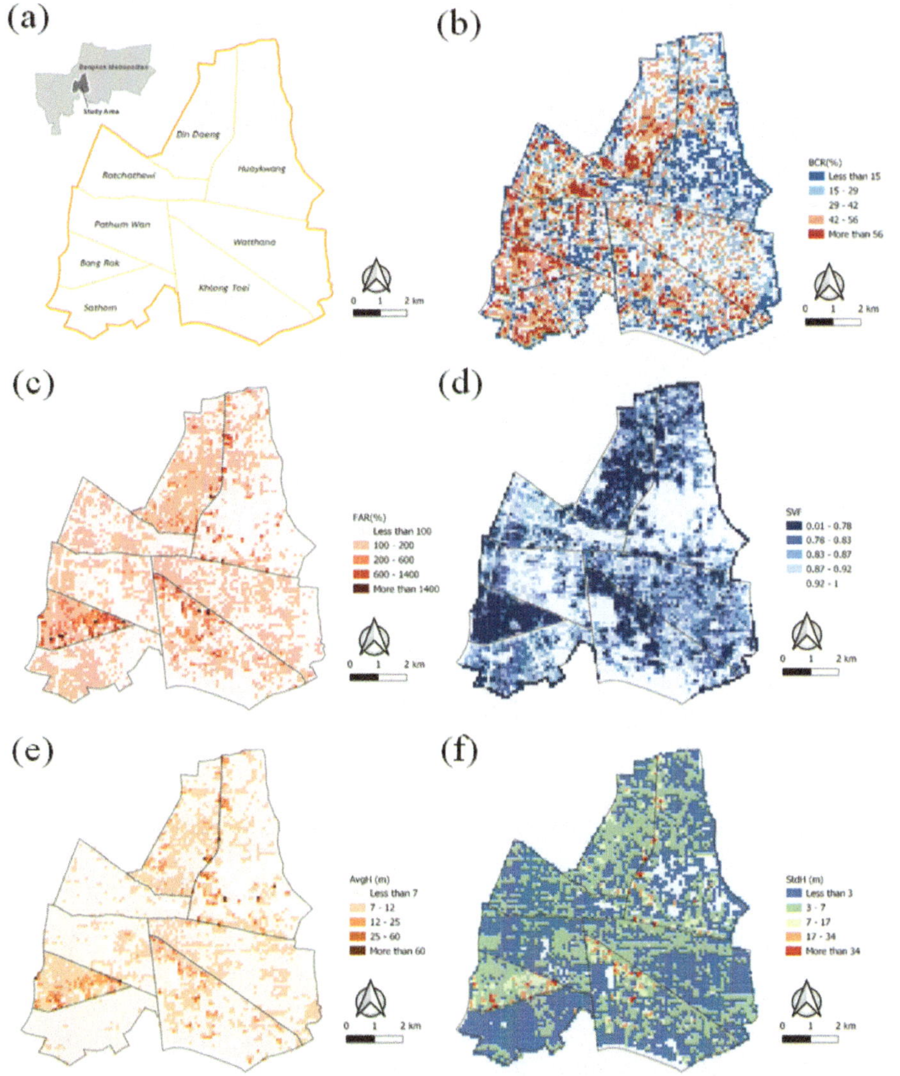

Fig. 3 The location of the study area (**a**) and key indicator maps for precinct ventilation performance classification (**b–f**)

structure. These design factors could block the parallel wind causing a reduction in wind speed while increasing turbulence.

From the GIS-based calculation method, the PVZ 3 covers 176 grid cells, which the area accounts for 2.4% of the total studied area. The spatial pattern of PVZ 3 is along a linear of the mass rapid transit corridor, especially in the four districts include; Bang Rak, Khlong Toei, Huai Khwang, and Din Daeng, respectively (Fig. 7). It is

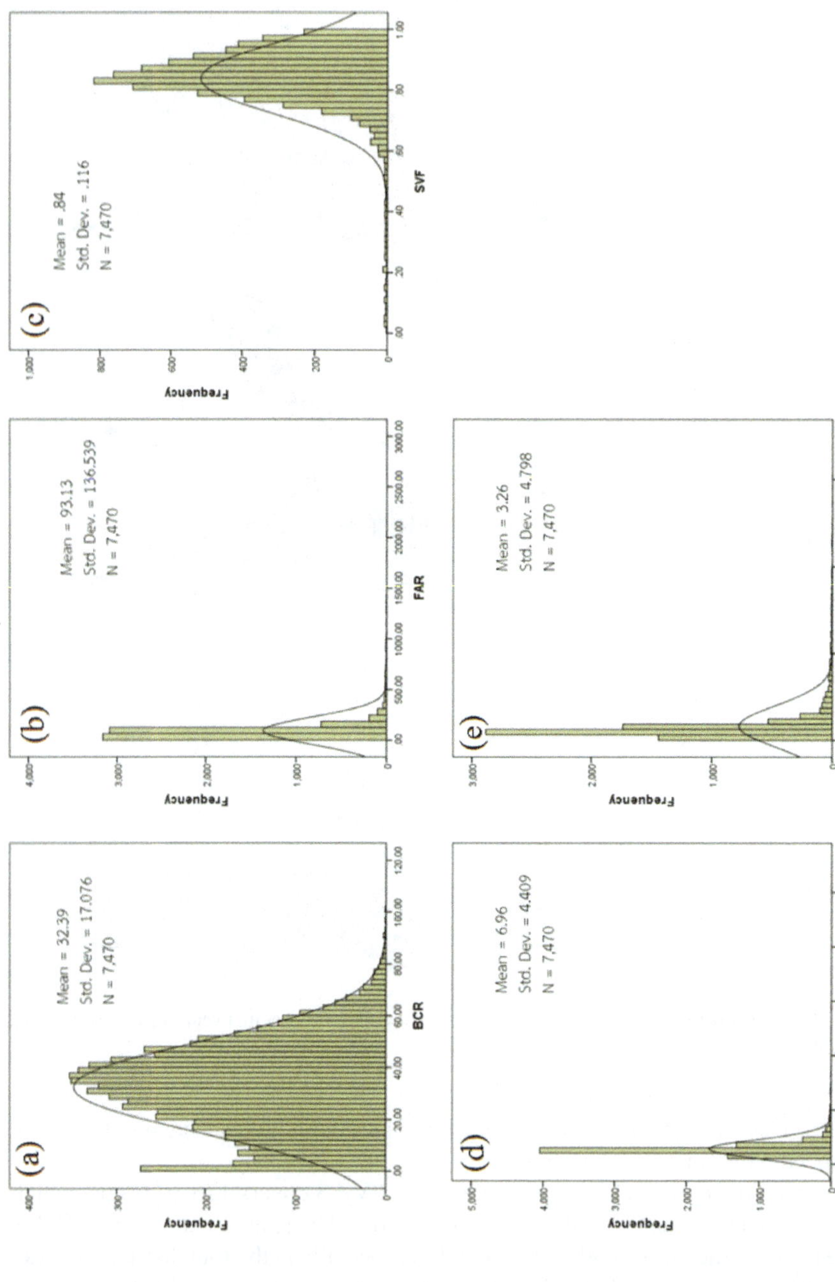

Fig. 4 Histogram and summary statistics of five-key urban-form indicators (**a**–**e**)

Table 2 Correlation matrix of Pearson correlation coefficients of five key indicators

		BCR	FAR	SVF	AvgH	StdH
BCR	Pearson correlation	1	0.538 **	0.545 **	0.568 **	0.339 **
	P-value		0.000	0.000	0.000	0.000
FAR	Pearson correlation	0.538 **	1	0.238 **	0.613 **	0.698 **
	P-value	0.000		0.000	0.000	0.000
SVF	Pearson correlation	0.545 **	0.238 **	1	0.491 **	0.244 **
	P-value	0.000	0.000		0.000	0.000
AvgH	Pearson correlation	0.568 **	0.613 **	0.491 **	1	0.703 **
	P-value	0.000	0.000	0.000		0.000
StdH	Pearson correlation	0.339 **	0.698 **	0.244 **	0.703 **	1
	P-value	0.000	0.000	0.000	0.000	

Note N = 12,519, (**) = Correlation is significant at the 0.01 level

Table 3 Final cluster centers of k-means cluster analysis

	Cluster 1	Cluster 2	Cluster 3
BCR	Low	High	Moderate
	(3.24)	(40.24)	(31.70)
FAR	Low	Moderate	High
	(0.07)	(1.02)	(5.50)
SVF	High	Moderate	Low
	(0.83)	(0.74)	(0.41)
AvgH	Low	Moderate	High
	(1.73)	(6.70)	(25.03)
StdH	Low	Moderate	High
	(0.60)	(2.88)	(27.26)

because of the effects of the transit-oriented development (TOD) on Bangkok's urban land-use planning,

Meanwhile, a compact low-rise zone with the low BCR, FAR, AvgH, and StdH occur in the standard PVZ 1 (see Fig. 5c). Such low values result in a high SVF. This zone is commonly designated for low-rise residential buildings and mid-rise commercial buildings to maintain the residential and natural structures.

In this study, the precinct ventilation classification system begins with proper protocols that fully cover the morphological characteristics closely related to the wind environment. As shown in Fig. 1, the simplification from the complex morphological characteristics at distinct scales to the protocol of precinct ventilation zone follows the five key indicators. There are high-rise and even tall buildings for Bangkok's central business district areas, few open spaces, tending to be compact. In comparison, Bangkok's suburbs, exhibiting the characteristics of the low-density area, inclusively consist of small and low-rise buildings and many open spaces. However, in this area

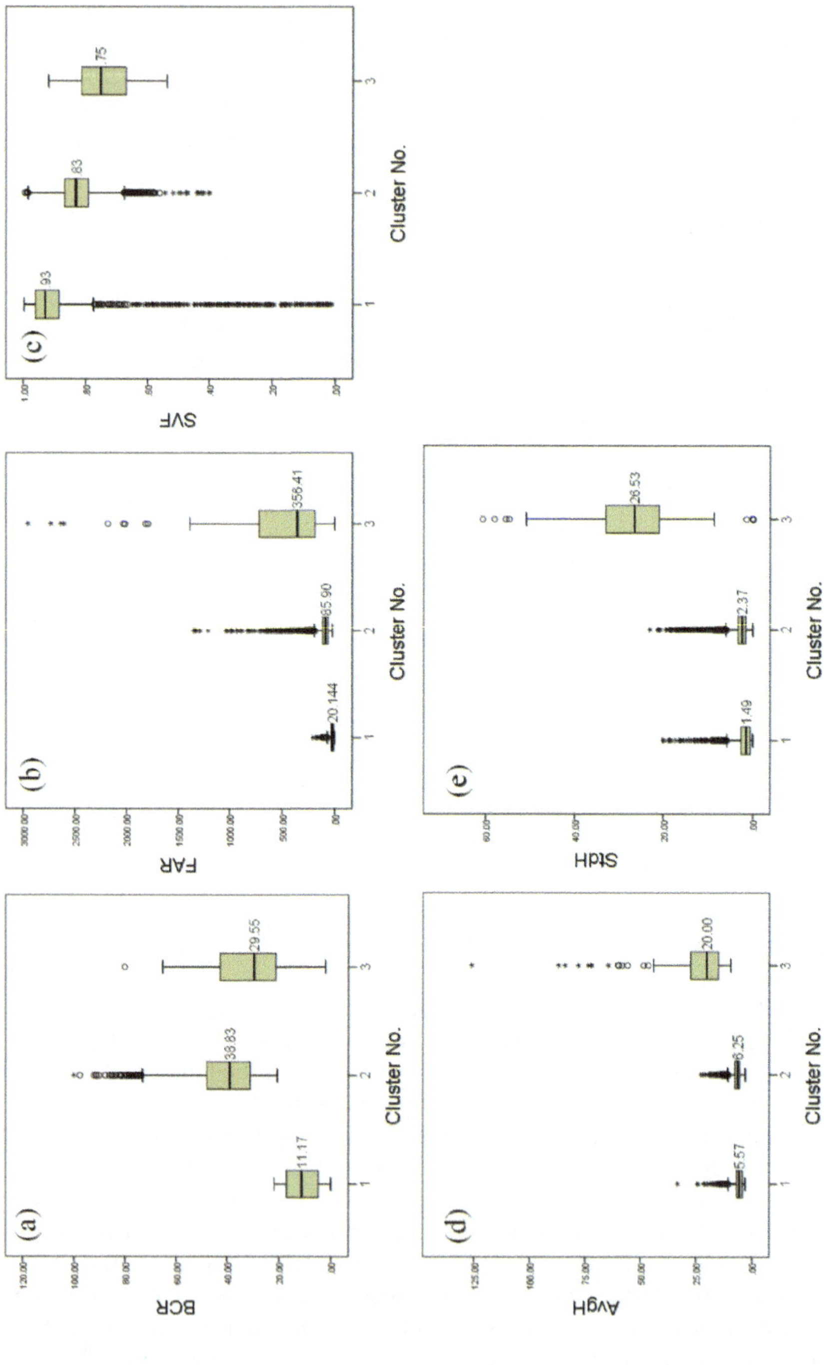

Fig. 5 Box-plot representations of the median values (thick black line) for each indicator (**a–e**)

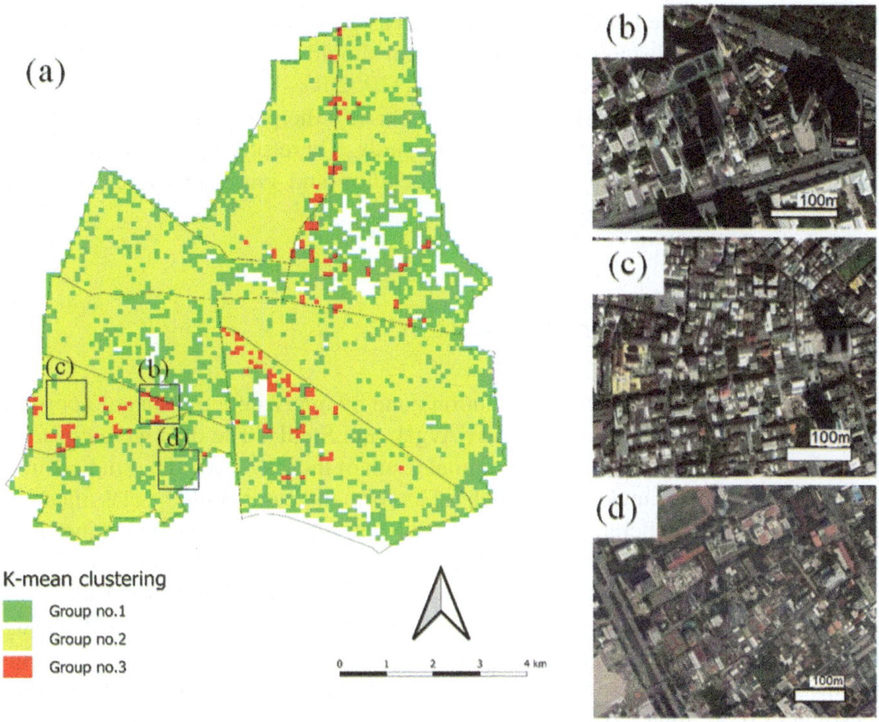

Fig. 6 Mapping the distribution of precinct ventilation zones (PVZs) on 100 m × 100m grid cell resolution in the study area

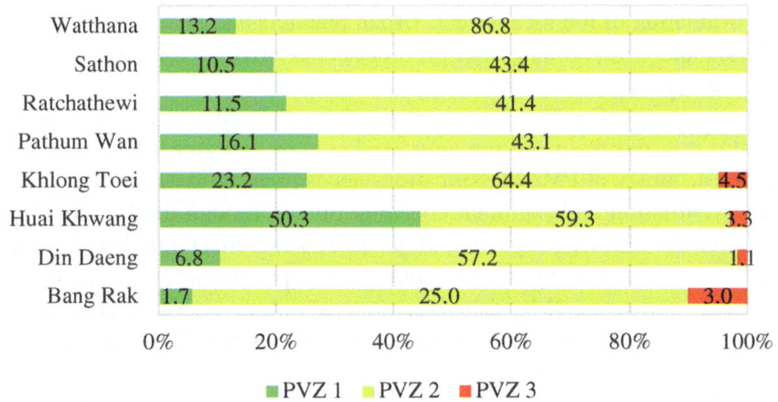

Fig. 7 Types of precinct ventilation zones (PVZs) and their district-level distribution of proportion within the study area

with the densely planted trees, the trees may act as the wind barriers to reduce the local air ventilation. The design of windbreaks form should be considered for promoting urban ventilation in the densely planted trees [19, 20].

However, the five key indicators used to classify the precinct ventilation do not account for the tree planting effects in this study. Therefore, further studies might be needed to investigate the effects of vegetation and green structures on urban ventilation and pollution dispersion.

4 Discussion and Conclusion

This study classified a precinct ventilation zone by applying a geospatial approach using five indicators: BCR, FAR, SVF, AvgH, and StedH. The data were clustered using the k-means method to get an optimal result of the PVZ map. The studied areas are the inner core of Bangkok city. Each urban form indicator could be classified into three clusters: low, moderate, and high values. It was found that most of the inner areas in Bangkok are in the standard PVZ 2, representing mid-rise buildings, whereas the PVZ 3 (densely high-rise buildings) accounts for 2.4% of the total studied area. This zone covers four central business districts: Bang Rak, Khlong Toei, Huai Khwang, and Din Daeng, with the highest precinct surface structure, it cloud block the wind and increase wind turbulence. The precinct ventilation classification method could constitute a preliminary step to assess further the risk of poor ventilation in near-surface pollutant concentrations in different areas. The research results provided a classification methodology and development for PVZ map and spatial urban form datasets. Then, future studies should use the local-scale climate modeling for an in-depth evaluation of the local effect. Urban planners and architects could identify the methods and enforcements that promote better urban ventilation and pollutant dispersion for specific ventilation zone.

In this study, the five key indicators used for the precinct ventilation classification were referred from the previous studies. For applying in Thailand's urban context, the accuracy of PVZ classifications needs to be validated with the field data of airflow ventilation and pollutant concentration. Although the low-density areas (PVZ 1) have low surface roughness, urban green structures might have a significant impacts on local air ventilation and air pollutant dispersion for low-density areas. In contrast, pollution removal by urban trees could reduce high concentration levels. These factors should be investigated in particular typologies and other related studied aspects.

Acknowledgements This project is supported by a corporate research fund from the National Science and Technology Development Agency and Thammasat University (research grant number: JRA-CO-2563-13105-TH). We especially want to thanks Danaipat Prasitreak and Titi Chuaytong for data collection and GIS analysis in this research project.

References

1. The Nation Thailand, Bangkok ranks 12th on list of cities with the worst air, https://www.nationthailand.com. Last accessed 2021/02/23
2. Chen L, Hang J, Sandberg M, Claesson L, Di Sabatino S, Wigo H (2017) The impacts of building height variations and building packing densities on flow adjustment and city breathability in idealized urban models. Build Environ 118:344–361
3. Yazid AWM, Sidik NAC, Salim SM, Saqr KM (2014) A review on the flow structure and pollutant dispersion in urban street canyons for urban planning strategies. SIMULATION 90(8):892–916
4. Vardoulakis S, Fisher BEA, Pericleous K, Gonzalez-Flesca N (2003) Modelling air quality in street canyons: a review. Atmos Environ 37(2):155–182
5. Wen H, Malki-Epshtein L (2018) A parametric study of the effect of roof height and morphology on air pollution dispersion in street canyons. J Wind Eng Ind Aerodyn 175:328–341
6. Yang J, Shi B, Shi Y, Marvin S, Zheng Y, Xia G (2020) Air pollution dispersal in high density urban areas: research on the triadic relation of wind, air pollution, and urban form. Sustain Cities Soc 54:101941
7. He B-J, Ding L, Prasad D (2019) Enhancing urban ventilation performance through the development of precinct ventilation zones: a case study based on the Greater Sydney, Australia. Sustain Cities Soc 47:101472
8. Yuan C, Ng E (2012) Building porosity for better urban ventilation in high-density cities—a computational parametric study. Build Environ 50:176–189
9. Khamchiangta D, Dhakal S (2019) Physical and non-physical factors driving urban heat island: case of Bangkok Metropolitan Administration, Thailand. J Environ Manage 248:109285
10. Srivanit M, Hokao K, Iamtrakul P (2014) Classifying thermal climate zones for urban and environmental planning in Bangkok Metropolitan Area. J Arch Plan Res Stud 11(1):73–92
11. Jareemit D, Srivanit M (2019) Effect of street canyon configurations and orientations on urban wind velocity in Bangkok Suburb Areas. IOP Conf Ser: Mater Sci Eng 690:012006
12. Yang J, Shi B, Zheng Y, Shi Y, Xia G (2020) Urban form and air pollution disperse: key indexes and mitigation strategies. Sustain Cities Soc 57:101955
13. Yang R, Zhang J, Shen S, Li X, Chen J (2007) Numerical investigation of the impact of different configurations and aspect ratios on dense gas dispersion in Urban Street Canyons. Tsinghua Sci Technol 12(3):345–351
14. Ng W-Y, Chau C-K (2014) A modeling investigation of the impact of street and building configurations on personal air pollutant exposure in isolated deep urban canyons. Sci Total Environ 468–469:429–448
15. Zhou X, Okaze T, Ren C, Cai M, Ishida Y, Mochida A (2020) Mapping local climate zones for a Japanese large city by an extended workflow of WUDAPT Level 0 method. Urban Clim 33:100660
16. Zheng Y, Ren C, Xu Y, Wang R, Ho J, Lau K, Ng E, Zheng Y (2018) GIS-based mapping of local climate zone in the high-density city of Hong Kong. Urban Clim 24:419–448
17. Lindberg F, Grimmond CSB (2010) Continuous sky view factor maps from high resolution urban digital elevation models. Clim Res 42:177–183
18. Thorndike RL (1953) Who belongs in the family? Psychometrika 18:267–276
19. Gromke C, Ruck B (2007) Influence of trees on the dispersion of pollutants in an urban street canyon—experimental investigation of the flow and concentration field. Atmos Environ 41(16):3287–3302
20. Zhu J-J, Jiang F-Q, Takeshi M (2002) Spacing interval between principal tree windbreaks-based on the relationship between windbreak structure and wind reduction. J For Res 13(2):83–90

Updating Healthy City Variables for the Post-pandemic Era Using Delphi Method

Maria Cacique, Takazi Okuda, Ana Areces, and Ou Sheng-Jung

Abstract The purpose of this paper is to update variables used to measure Healthy City by using Delphi Method since the current Pandemic has made it necessary to re-think and re-evaluate the entire complex notion of Healthy City. The assessment of variables under this new perspective is analyzed through a panel of experts in the field's knowledge. For this first exploratory analysis, Mexican multidisciplinary specialists were selected to weight a group of variables collected from a wide bibliographic review. Surveys were conducted by online platforms and experts were able to rate the selected variables through two different rounds of questions with cross feedback. The obtained result is an updated and renewed list with the most significant variables weighted and agreed by different experts. The Delphi Method allowed the intellectual interaction between experts and reopens the debate about what a healthy city should be in the XXI century. This paper also identifies further specific uses of this method to deepen scientific research on Healthy City and may be adapted for additional analysis in various countries with specific characteristics as well as different scales of study.

Keywords Healthy city · Delphi method · Healthy cities variables · Post-pandemic cities · Expert consensus

M. Cacique · T. Okuda
Department of Architecture, Chaoyang University of Technology, No. 168, Jifeng East Road, Wufeng District, Taichung City 413, Taiwan

A. Areces
Universidad Politécnica de Madrid, Ctra.Valencia Km 7. Bloque 7, 28031 Madrid, Spain

O. Sheng-Jung (✉)
Department of Landscape and Urban Design, Chaoyang University of Technology, No. 168, Jifeng East Road, Wufeng District, Taichung City 413, Taiwan
e-mail: sjou@cyut.tw

T. Kang and Y. Lee (eds.), *Proceedings of 2021 4th International Conference on Civil Engineering and Architecture*, Lecture Notes in Civil Engineering 201,
https://doi.org/10.1007/978-981-16-6932-3_40

1 Introduction

The current COVID-19 pandemic has opened a new urban debate about how cities should address a healthier development for future generations. Healthy Cities for a Post-Pandemic Era are expected and the need of a comprehensive review of its meaning and characteristics is the first step of the process. Assessing the challenge to create healthier cities through the re-evaluation of the variables involved first and then the dynamic relations between them may provide a deeper understanding of the system.

For the creation of a healthy city model, this paper focuses on the instruments used to choose the variables and their specific weigh in the Healthy City Index. Since healthy city indicators are expected to play a crucial role in the dynamic process of healthy city progress, the establishment, monitoring, and testing of the indicators are very important [1], thus the methodology approach used in their selection should respond to this complex changing global reality. There is a growing awareness among city leaders and policy-makers of the impact of the urban environment on health outcomes and inequalities. Increasingly, practitioners in built environment city departments, such as housing, planning, transport and regeneration, seek new tools and guidance to understand how their respective policies and decisions can support the creation of healthier cities [2]. In this sense, the World Health Organization aims to create a health-supportive environment, to achieve a good quality of life and states that a healthy city is one that is continually creating and improving those physical and social environments and expanding those community resources which enable people to mutually support each other in performing all the functions of life and developing to their maximum potential [3]. And ever since their beginning in 1986, Healthy Cities projects all over the world have been confronted with the issue of evaluation. Nevertheless, after 20 years, many key dilemmas constantly reappear; people are often looking for a kind of 'magic' list of universally applicable indicators to evaluate these initiatives [4]. Efforts have been made since the end of the 80s, to implement different programs and projects in which the public, private and social sectors have been involved in the creation of healthier cities, but the measurement of the results is unclear; "these indicators are usually not standardized, consistent or comparable (over time or across cities), nor do they have sufficient endorsement to be used as on-going benchmarks [5]."

The article points out this problem and seeks a methodological approach with a different perspective. The exploratory research of a new suitable methodology to measure a Healthy City that this study aims brought the investigation towards the Delphi Method (also known as Estimate-Talk-Estimate or ETE). This Method was developed at the beginning of the Cold War by The Project RAND to avoid the shortcomings of traditional forecasting methods; it was devised in order to obtain the most reliable opinion consensus of a group of experts by subjecting them to a series of questionnaires in depth interspersed with controlled opinion feedback [6]. Since its inception in the studies, the RAND Corporation has conducted Delphi studies focusing on various areas of knowledge, sectors, regions, and countries [7].

Nowadays, Delphi method is common in the natural sciences and various medical subfields, applied studies aimed at predicting future developments in various areas, finding practical solutions, in policymaking [8], and its value for this research is based on this technique, of a qualitative nature, is recommended when there is insufficient information available for decision making [9]. Therefore, in this work, the usefulness of the Delphi method to update the variables used to measure the Healthy City is tested through the generation of a consensus that has allowed updating and weighting the group of variables evaluated.

2 Research Process

As a way to demonstrate the worth of the Delphi Method for the Healthy City Index, the research combines qualitative and quantitative analysis that goes from the bibliographic research to collect the existing variables that will be taken as the basis of the process, to the subsequent steps that include the implementation of Delphi.

The research process began in identifying and selecting Organizations/Institutions worldwide in charge to evaluate Healthy Cities and the variables employed. Then, the study reviewed, classified and regrouped them to filter out those that were repetitive. The study also evaluated variables with same characteristic or if were conceptualized in the same way as another one to get a new group of variables for the investigation. The new organized group of variables was, with the selection of experts, the foundation to apply the Delphi Method. In order to conduct the experiment it was decided to find Mexican experts on Healthy Cities of multiple backgrounds and adapt the Delphi Method to the specific research and global pandemic situation.

Delphi Method is usually a process that is carried out in 2–4 rounds of surveys depending on the consensus reached after each round, usually involving between 6 and 30 experts. All participants are experts without direct communication among them. A controller will provide feedback to the respondents after each round and analyse and filter the group's results overviewing the process. The general process consists of 4 phases [9]: Definition, Conformation of the informants group, execution of consultation rounds and results.

These phases were adapted for this research as follows:

- Definition

It starts with the definition not only of the method but also of the dimensions and variables to be explored. The specific process for the present investigation starts updating the Healthy Cities Variables considering the experience and knowledge gained during the Pandemic. To define the initial list, an exhaustive bibliographic review is carried out identifying the most important Organizations involved worldwide in this matter: World Urban Campaign- UN Habitat, World Health Organization through the World Healthy Cities Project, Building Research Establishment and Alliance for Healthy Cities.

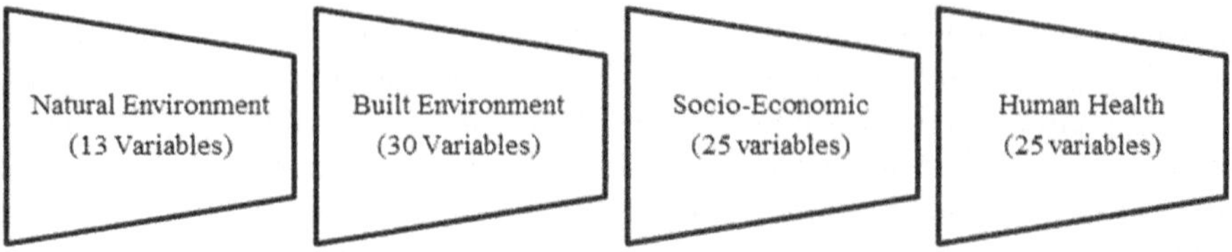

Fig. 1 Dimensions to evaluate and variables included

Fig. 2 Panel of Experts conformation

As a result, a total of 212 variables were obtained. A second step is to filter and compile the variables to discard those that were evaluating the same characteristic. At the end of this process, the resulting variables were 93 and are grouped into 4 dimensions identified (Fig. 1): Natural Environment, Built Environment, Socio-Economics and Human Health.

- Conformation of the informants group

An e-mail invitation was sent to potential participants with a detailed explanation of the procedure and the dates for its execution. All those who were contacted accepted to participate. The Method was conducted with a panel of 13 Mexican experts in 4 specific fields (Fig. 2): Government, Educational, Healthcare, and Urban Planner. They were Architects, Engineers, Doctors, and Urban Planners who are in charge of decision-making in the different fields of Healthy Cities care.

- Execution of consultation rounds.

There were 2 questionnaires submitted to the pool expert within one week of each other. A total of 2 rounds were held. Due to the current Pandemic confinement and international locations of experts, an online platform was required. Several software options were tested in order to select a suitable one. Several pilot tests were design and tested to identify the best option for the surveys. Google Forms was the platform selected for the first and second round given the amount of data, an Excel Sheets with information was also design needed and e-mailed to each of the experts. The following is a condensed procedures register:

- First Round (Questionnaire 1)

A first questionnaire was sent to the experts through the Google Forms online platform. The survey Design mainly focused on weighting the group of 93 variables regarding to the evaluation of Healthy Cities. The experts had to rate each variable from 1 to 5, where 1 was not important and 5 was most important. There was also an option for respondents to suggest new variables to enhance the list based on their experience and knowledge. The 13 experts sent their surveys responses and the controller analyzed the results to obtain an average rating of each of the variables as well as new proposals.

- Second Round (Questionnaire 2)

The second survey was meant to readjust the weight of the variables result of the first round. The Google Form and the Excel Sheet were designed with more information: the 93 variables plus the extra ones proposed by experts all of which were briefly described highlighting their relationship to Healthy Cities, the average rating assigned by the group to each variable and their individual rating of each variable. The group of experts was asked to reevaluate their answer to the first survey with the new information to obtain a verified rating of every variable. Each expert was anonymously confronted with the average rating of the group in each variable and this second evaluation would define if consensus were achieved.

3 Results

The results of the two surveys were gathered and organized in order to understand the level of consensus that the pool of experts had on each variable. The rating criteria from 1 to 5 given by the experts, 1 meaning not important and 5 the most important, made clear that variables rated 4 and 5 were the ones most significant to achieve the Healthy City Index. Therefore, the selection criteria would be and average rating of 3.75 or more to guarantee sufficient consensus among the experts on each variable (75% of the scale from 1 to 5), the variables with less rating than 3.75 would be discarded (see Table 1 for a better understanding).

Table 1 Final group of variables

Rating	Interpretation	%
1	Not Important	25
2	Less Important	
3	Neutral	
4	Important	75
5	Most Important	

Average Rating	Percentage
5.00	100%
3.75	75%

Criteria Examples

Rating	Times of that rating	%
1	0	0.00
2	1	7.69
3	1	7.69
4	6	46.15
5	5	38.46
	13 experts	100

84.62% (75% or more by adding the last 2 ratings together = CONSENSUS to include that variable in the Final List)

Variable	Average Rating	Percentage
Air Quality	4.92	98.40 (more than 75% of **CONSENSUS**)

On the first survey, as Fig. 3 shows, the most important variables were: Natural Environment 10 (% of land with green space, air quality, water quality, noise pollution, soil pollution, amount of living space, recycling, percentage of water pollutants removed from total sewage produced, river quality, environmental health services), Built Environment 22 (ventilation in buildings, public transit and safety, adapting to support pandemic needs, adequacy of housing facility, % of dwellings, m^2 of public indoor and outdoor space per capita, housing affordability, building quality control index, percentage of energy derived from renewables, electricity consumption per year, street connectivity: average number of three or more-way intersections per square kilometer, pedestrian city design, network for cycling in cities, public transport network cover, community engagement participation, economic resilience, institutional policy planning, disaster management and pandemics, % of population with potable water supply service, % of population with authorized electricity service, % of population served by wastewater collection, % of population living in slums), Socio-Economic 20 (culture, total population, age and sex breakdown, ethnic distribution, assessment of impact of economy on health, sources of social stress, description of social support mechanisms/network, hospitals and community health facilities, schools, community centers, percentage of population living in substandard accommodation, unemployment rate, percentage of people earning less than the mean per capita income, percentage of disabled persons employed, crime occurrence rate, urban density, standard of monitoring enforcement, rates of homelessness, elderly people living alone receiving care, social welfare expenditures) and Human Health 16 (disabilities, waste management, food hygiene, perceptions of health and well-being, immunization rate, nutrition, alcohol and drugs consumption, domestic

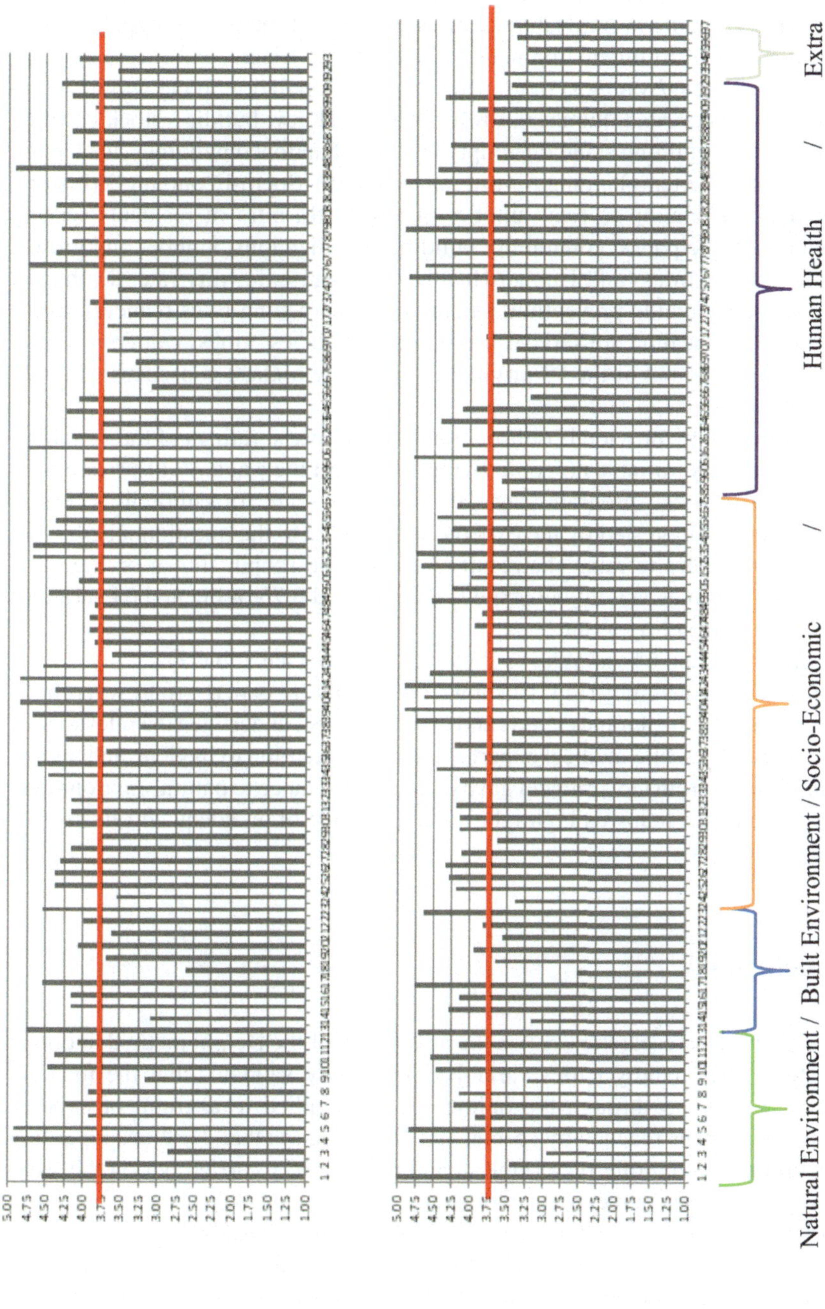

Fig. 3 Variables average rating 1st round and 2nd round respectively

violence, access to safe drinking water, cause of death, number of inhabitants per nurse, percentage of population covered by health insurance, rate of smoking in adults, population that exercise regularly, existence of a health education program). The discarded variables were: Natural Environment 3 (climate, topography, insects and rodent control), Built Environment 8 (land tenure, freight and cargo, description of urban planning/zoning system, percentage of population living within 500 m of highway or major road, number of refurbished and new building stock certified with a sustainable building standard, empty lots and abandoned buildings, leadership for building back better, planning for migrant workers), Socio-Economic 5 (history, abortion rates in relation to total number of live birth, ratio of residents serving as volunteers, participation community building organizations, number and growth rate of cultural facilities) and Human Health 9 (birth rate, fertility rate, death rate, injures/accidents, suicide prevention, suicide rates, screening rates (cancer), availability of primary health care services in foreign languages, fitness centers). A total of 68 variable remained of the 93 original list with the highest consensus while 25 were discarded.

The second survey provided confirmation of the experts' ratings of the variables. At this point, 4 new variables (people living alone and their age, % of private non-profit funding in social programs, availability of day-care for children, access to sustainable food), suggested by the experts, were added to the questionnaire having a total of 97 variables to rate. The result of the second round, shown on Fig. 3, is that of the 97 variables, 33 did not achieve the required 3.75 rating and 64 variables remained as the most important.

The final list of the 64 variables is shown as follows (see Table 2).

The following line chart (Fig. 4) shows the results by expert group wherein as could be expected, educational experts and urban planners underlined the relevance of territorial variables while medical experts paid more attention to hygiene and human health and experts from Government gave greater importance to natural environment. This is relevant as it helps us highlight the importance of an adequate selection of the panel of experts.

4 Conclusions

This paper reports a research based on the adaptation of the Delphi method to obtain the update of the Healthy Cities variables. The study validates that the final set of variables, obtained through the online surveys applied to a group of experts, reflects the consensus reached by specialists. This suggests that decisions regarding the selection and weighting of variables can be effectively carried out on the basis of a survey of professional experts in the field.

In order to accomplish the final list of variables to create the Healthy Cities Index, the first step taken was to reassert the number and significance of the early list of variables. Out of the 93 variables, the expert's consensus discarded 33 keeping 64 variables founded to be most important for the final Index. It was noticed at the

Table 2 Final group of variables

Natural environment 10	Built environment 23	Socio economic 18	Human health 13
% of land with green space, air quality, water quality, noise pollution, soil pollution, amount of living space, recycling, percentage of water pollutants removed from total sewage produced, river quality, environmental health services (water, noise, air, soil)	Ventilation in buildings, public transit and safety, adapting to support pandemic needs, adequacy of housing facility, % of dwellings, m^2 of public indoor and outdoor space per capita, housing affordability, building quality control index, % of energy derived from renewals, electricity consumption per year, pedestrian city designs, network for cycling in cities, public transit network cover, community engagement participation, economic resilience, leadership for building back better, institutional policy planning, planning for migrant workers, disaster management and pandemics, % of population with potable water supply service, % of population with authorized electricity service, % of population served by wastewater collection,% of population living in slums	Total population, age and sex breakdown, ethnic distribution, assessment of impact of economy on health, sources of social stress, description of social support mechanism/ network, hospitals and community health facilities, schools, community centers, % of population living in substandard accommodation, unemployment rate, % of people earning less than the mean per capita income, crime occurrence rate, urban density, standard of monitoring enforcement, rates of homelessness, elderly people living alone receiving care, social welfare expenditures	Death rate, waste management, food hygiene, perception on health and wellbeing, immunization rate, nutrition, alcohol and drugs consumption, domestic violence, access to safe drinking water, cause of death, percentage of population covered by health insurance, population that exercise regularly, existence of health education program

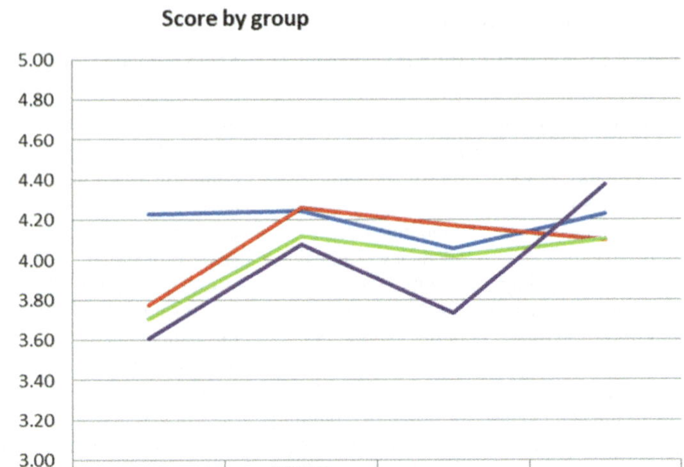

	GOVERNMENT	URBAN PLANNER	EDUCATIONAL	HEALTH CARE
NATURAL ENVIRONMENT	4.23	4.24	4.06	4.23
BUILT ENVIRONMENT	3.78	4.26	4.17	4.10
SOCIO-ECONOMIC	3.71	4.12	4.02	4.10
HUMAN HEALTH	3.61	4.08	3.73	4.37

Fig. 4 Average score for each group of experts

first round of the survey the necessity of redefining the variables due to the different backgrounds of the experts. Each specialist had a slightly different outcome to each of the concepts that leading the study to a higher level of complexity. The second round of the survey took this into consideration and added a definition of the variables so that expand the knowledge of each of the concepts. The findings of this adaptation of the methodology were significant due to the interactions of the different position of the experts. The final outtake in the modified methodology made it clear that even though the experts came from different fields of knowledge the results integrated each of their points of view to the final group of variables making the outcome a successful complex Index.

It could be considered that the method was partial due to the selection and number of experts, time limits and global condition at the time of the study; however, there is value not only in the first approach to the integration of the index but also in the use of the Delphi Method. Through the research, some challenges were found. The first one was the selection of the appropriate variables for the construction of the index. In order to create a valid index, a systemic vision upon the variables and their relationship with each other is a must. It is also needed to integrate measurable and controllable indicators, easily adaptable to each case study. The systemic Index should be able to adapt to different scales and local conditions where they may be applied and should reflect different possible Healthy Cities approach showed by the experts on this paper. The weighing of each the variables, the relative importance to each other and even the type and characteristics of the measurement must be relative to local conditions. The variables system must me cohesive and flexible to adapt to deferent scenarios. The present research showed that some variables used in different local index were too general or ambiguous to be applied while others were too specific, making it difficult is to evaluate them under different scenarios.

A second challenge was the conformation of the panel of experts. The group was chosen in order to have the widest and deepest approach to the matter. The expert's profiles were balanced between academic and professional backgrounds in Mexico in the fields of Public Health, Urban Planning and city administration among others. Finding the right expert to achieve a cohesive group was crucial to the overall discussion and results. The amount of participants in the group of study was found to be relevant although the original Delphi Method only suggests it to be a group from 6 to 30 members. A group too small wouldn´t consider many different points of view while a group too wide could lead to an endpoint in the discussion. The right amount and variety of experts was found to be decisive for the results and led to a possible deeper analysis with the methodology (adapting a third round of the questionnaire). Regrouping variables in separated systems to be evaluated and then reestablishing the complete variable scenario are possible further investigation methods.

The paper has cast significant results to establish new methodologies for the evaluation of Healthy Cities. The final set of selected and weighted variables is well documented and has a well-established consensus among the experts but further studies should be conducted to create a systemic and measurable index for Healthy Cities, this new integrated system reached has to have the potential to measure and weight each variable depending on the conditions existing in every case study and still being able to have cohesive results. A subsequent research not only is relevant but necessary and the Delphi Method, as this work shows, would allow a successful methodology to find consensus among experts under specific conditions or scales of study.

References

1. Bureau of Health Promotion, Department of Health, Taiwan (2010). Healthy cities in Taiwan. Available: https://my.ksu.edu.tw/file/bbcode/w3Pages/40488/6265ff57-cc46-4cfb-8e8e-b63fff18f659/Healthy%20Cities%20in%20Taiwan%E5%81%A5%E5%BA%B7%E5%9F%8E%E5%B8%82%E5%9C%A8%E8%87%BA%E7%81%A3(English%E8%8B%B1%E6%96%87%E7%89%88).pdf
2. Pineo H, Zimmermann N, Cosgrave E, Aldridge RW, Acuto M, Rutter H (2018) Promoting a healthy cities agenda through indicators: development of a global urban environment and health index. Cities Health. https://doi.org/10.1080/23748834.2018.142918
3. World Health Organization (1998) Health Promotion Glossary. Healthy City Definition. Available: https://www.who.int/healthpromotion/healthy-cities/en/
4. O'Neill M, Simard P (2006) Choosing indicators to evaluate healthy cities projects: a political task? Health Promot Int 21(2):145–152. https://doi.org/10.1093/heapro/dal006
5. World Bank (2007) City Indicators: Now to Nanjing. Available: http://documents1.worldbank.org/curated/en/707781468263936449/pdf/wps4114.pdf

6. Dalkey N, Helmer O (1962) An experimental application of the Delphi method to the use of experts Memorandum RM-727/1-ABRIDGED, pp 11–16. Available: https://www.rand.org/content/dam/rand/pubs/research_memoranda/2009/RM727.1.pdf
7. Zartha JW. Manrique J, Montes JM, Palacio JC, Orozco GL (2020) Delphi Method in Emerging Technologies. Available: https://www.researchgate.net/publication/346099992_Delphi_Method_in_Emerging_Technologies
8. Fink-Hafner D, Dagen T, Doušak M, Novak M, Hafner-Fink M (2019) Delphi method: strengths and weaknesses. Method Zv 16(2):1–19. Available: https://www.researchgate.net/publication/337570516_Delphi_Method_Strengths_and_Weaknesses/citations#fullTextFileContent
9. Reguant-Álvarez M, Torrado-Fonseca M (2016) The Delphi method. University of Barcelona. Institute of Educational Sciences, REIRE, vol 9, Núm 1, pp 6–12. ISSN: 2013-2255. 10.1344. Available: http://diposit.ub.edu/dspace/bitstream/2445/110707/1/654735.pdf

The Association of Neighbourhood Aesthetics with Physical Activity in Gated and Non-gated Neighbourhoods in Developing Countries

Yasmeen Gul, Gul Ahmed Jokhio, Tehmina Bibi, and Syeda Maria Zaidi

Abstract Neighborhood Aesthetics are considered important for enhancing the physical activity among people and gated neighborhoods have more aesthetical features than non-gated. The objective of this study is to compare the association of perceived aesthetics and perception of physical activity (PoPA) in gated and non-gated neighborhoods in developing countries. Sixteen neighborhoods and 1042 participants were selected from Karachi, Pakistan. Results show that both the perceived aesthetics and PoPA are significantly higher and dependent on each other in gated neighborhoods than non-gated neighborhoods of developing countries. Therefore, it can be concluded that the future neighborhoods should have aesthetical feature for encouraging PA.

Keywords Neighborhood environment · Physical activity · Gated communities · Developing countries · Aesthetics · Walking

1 Introduction

It is agreed by several researchers that the physical activity is an important attribute to keep people away from diseases. There are different guidelines for individuals to be physically active in a week such as the guidelines of World Health Organization are that one should get minimum 150 min of walking or 75 min of sports activities

---

Y. Gul (✉)
Alghurair University, Dubai, UAE
e-mail: yasmeen.gul@agu.ac.ae

G. A. Jokhio
British University in Dubai, Dubai, UAE

T. Bibi
Azad and Jammu Kashmir University, Muzaffarabad, Pakistan

S. M. Zaidi
Balochistan University of Engineering and Technology, Khuzdar, Pakistan

T. Kang and Y. Lee (eds.), *Proceedings of 2021 4th International Conference on Civil Engineering and Architecture*, Lecture Notes in Civil Engineering 201,
https://doi.org/10.1007/978-981-16-6932-3_41

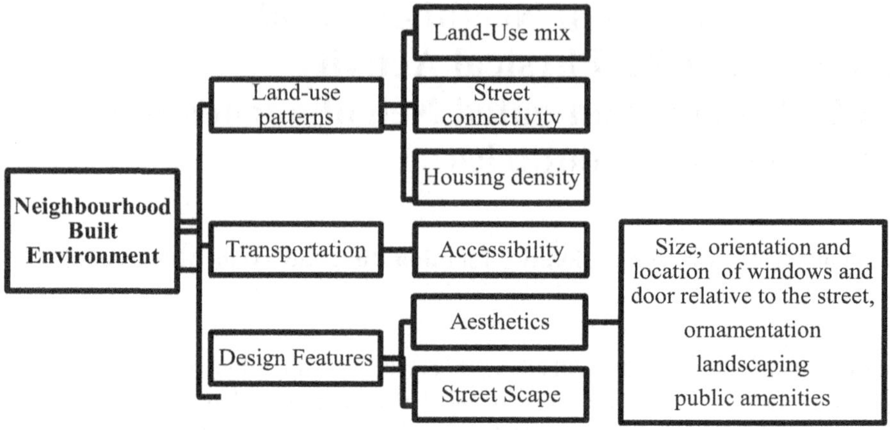

Fig. 1 Hierarchy of neighborhood built environment indicators

or any combination to get 600 MET[1] in a week [1]. Neighbourhood environment is considered very important to enhance physical activity. The neighbourhood-built environment has three important parts that include land-use patterns; accessibility and design features [2–4], which can affect physical activity. The design feature is either the streetscape [5] or aesthetical features at neighbourhood level. Aesthetical features are the qualities that contribute to the attractiveness of neighborhoods such as the design of buildings, including the size and orientation of windows, the location of the door relative to the street, decoration, ornamentation; landscaping, particularly trees and the shade and the availability of public amenities such as street benches and lighting. Neighborhood with desirable qualities provides a strong sense of place and a clear identity [2, 6]. The Hierarchy of neighborhood-built environment indicators are given below in Fig. 1.

Much research has been done on this topic to calculate the effects of Aesthetics on physical activity reported that neighborhood aesthetics such as sidewalk cafes, street furniture and trees have positive effect on walking for transport which means people who live near sidewalk cafes are 10% more physically active because of walking for transport than others [7]. It is reported in one study that residents of high-walkability neighborhoods with high residential density, land use mix, street connectivity, environmental aesthetics, and safety had 70 more minutes of physical activity than those who didn't have these things in their neighborhoods [3, 8]. In another study, it is reported that accessibility, opportunities, and aesthetics had significant associations with physical activity at neighborhood level [9]. Neighborhood environmental attributes are important for physical activity. It is reported that

[1] The Metabolic Equivalent of Task (MET), or simply metabolic equivalent, is a physiological measure expressing the energy cost of physical activities and is defined as the ratio of metabolic rate (and therefore the rate of energy consumption) during a specific physical activity to a reference metabolic rate, set by convention to 3.5 ml O_2 kg^{-1} min^{-1} or equivalently:

$$1\,\text{MET} = 1\,\frac{\text{kcal}}{\text{kg}*\text{h}} = 4.184\,\frac{\text{kJ}}{\text{kg}*\text{h}}$$

neighborhoods were more physically active when there were aesthetics with other important features such as the accessibility of shops, the social environment and recreational facilities [10, 11]. In another study, the objectively collected data was analyzed through environmental audits of all street segments. Neighborhood characteristics included the land use environment, transportation environment, recreational facilities, aesthetics, and social environment and it was reported that recreational activity was positively associated with perceived aesthetics [12]. It is reported in another study that male and female both were less physically active when there were no pleasing and aesthetical features in neighborhoods [13, 14]. Green spaces such as lawns, flower beds, trees are part of aesthetics and have positive association with physical activity. One study investigated the impacts of green spaces on mental health and physical activity and reported that the greenest neighborhoods had the lowest risks of poor mental health. Overall, levels of physical activity were higher in greener neighborhoods [15]. More broadly, the "physical environment" refers not just to the built environment but also to the aesthetics that have the potential to influence choices about physical activity [2]. It can be concluded from reviewing the literature that there is a relationship between aesthetics of neighborhoods and physical activity levels. At the same time there has emerged a new form of neighborhoods which is walled and gated from the rest of the nearby neighborhoods in a city called gated communities and neighborhoods (GC's). These neighborhoods have recently emerged and gained popularity around the world among consumers [16]. The trend of gating is growing very fast in developing countries as well and it is considered that the GC's in developing countries [17–19]. It is also reported in a study of developing countries that over physical activity especially vigorous physical activity is higher in gated neighborhood than non-gated neighborhoods [20] and this is also reported that the vigorous physical activity is higher in gated neighborhoods in Karachi Pakistan because of having more physical activity facilities [11] but there is no difference in physical activity due to perception of crime in developing countries [21]. There is a gap in literature has been found to compare the aesthetics and its effects on physical activity behavior in gated and non-gated neighborhoods with special reference to the developing countries. Therefore, the objectives of this study are to compare the preceived aesthetics and perception of physical activity (PoPA) in gated and non-gated neighbourhoods and also to investigate the association between these two. The remaining of this paper is as follows: at first stage, the methods of data collection and analysis are presented. Stage two presents the results; stage three discusses the results with existing literature and provides conclusions for this study.

2 Methods

A metropolitan city Karachi from Pakistan has been selected as study area because Karachi has all types of neighborhoods such as low-rise, medium rise and high-rise in both gated and non-gated categories. The selection criteria is discussed in detail somewhere else [11, 20]

The data for neighborhood aesthetics and physical activity of people have been collected through the questionnaire (perceived method). For perceived aesthetics people were provided 5 choices from strongly agree to strongly disagree if they consider their neighborhoods to have aesthetical features. They were provided options such as do they have public amenities (i.e. street benches, street lights and paths ways), landscaping, attractive building facades with respect door and window opening and ornamentations. If they tick more 80% options then their answer were considered strongly agree, from 60 to 80% their answers were considered Agree, while it was considered neutral if they have ticked 40–60% of given options, 20–40% tick marks showed disagree and less than 20% ticks showed strongly disagree with the aesthetical features in their neighborhoods. Similarly for calculating perception of propel for physical activity, the people were asked, do they feel that these aesthetical features have attracted people in their neighborhoods to do physical activity such as walking, cycling and sports activities at neighborhood level. They were provided 5 choices of strongly agree to strongly disagree for PoPA as well. Strongly agree was number 5 and strongly disagree was numbered 1 for calculations. Statistical analysis was done in two parts using SPSS statistics. Part 1 explained the group statistics and independent sample t-test to check the difference of perceived aesthetics and PoPA of people in gated and non-gated neighborhoods, while at second level, cross-tabulation and gamma tests were computed to check the association between these two variables also in gated and non-gated neighborhoods separately.

3 Results

3.1 Comparison of Perceived Aesthetics and PoPA in Gated and Non-gated Neighbourhoods

An independent sample t-test was conducted to compare perceived aesthetics of neighborhoods and PoPA of people in gated and non-gated neighborhoods. There is a significant difference in the perceived aesthetics, for gated neighborhoods (M-3.3 and SD-1.1) and for non-gated (M-3.0SD-1.5); conditions $(t = -1040) = 3.67$ and p-0.0001. Similarly there was a significant difference between PoPA of people, for gated (M-3.38 SD 1.1) and for non-gated (M-3.4 SD-1.3) condition (t-1040) $= 4.1$ and p-0.000). These results suggest that both the perceived aesthetics and PoPA of people are significantly higher in gated neighborhoods than in non-gated. This shows that people of gated neighborhoods feel that their neighborhoods have more aesthetical features and they also feel that these features have encouraged people to be physically active through walking, cycling and sports activities in their neighborhoods. The results of group statistics and independent sample t-test are presented in Tables 1 and 2.

Table 1 Group statistics for perceived aesthetics and PoPA in gated and non-gated neighbourhoods

	Neighbourhoods	N	Mean	Std. deviation	Std. error mean
Perceived aesthetics	Gated	499	3.3206	1.19629	0.05355
	Non-gated	543	3.0037	1.54990	0.06651
PoPA	Gated	499	3.8036	1.11466	0.04990
	Non-gated	543	3.4788	1.36205	0.05845

Table 2 Independent sample t test results for perceived aesthetics and PoPA in gated and non-gated neighbourhoods

		Levene's test for equality of variances	t-test for equality of means					Remarks
		Sig	t	df	Sig. (2-tailed)	95% Confidence interval of the difference		
						Lower	Upper	
Perceived Aesthetics	Equal variances assumed	0.000	3.672	1040	0.000	0.14759	0.48632	Perceived aesthetics is significantly higher in gated neighbourhoods
	Equal variances not assumed		3.712	1010.377	0.000	0.14939	0.48453	
PoPA	Equal variances assumed	0.000	4.191	1040	0.000	0.17271	0.47686	PoPA of People is significantly higher in Gated Neighbourhoods
	Equal variances not assumed		4.226	1026.501	0.000	0.17398	0.47559	

3.2 *Association Between Perceived Aesthetics and PoPA in Gated and Non-gated Neighbourhood*

The results of Cross-Tabulation in Table 3 show that there is an association between perceived aesthetics and PoPA of people in both gated and non-gated neighborhoods. The results show that in Non-gated neighborhoods 31.5% agree that there is an association between perceived aesthetics and PoPA and 28.4% strongly agree with this association. Similarly in non-gated neighborhoods 31.5% people agree with the association and 32.9% strongly agree with this association. 11.2% people in non-gated neighborhoods were neutral while they were 23.2% in gated neighborhoods. The percentages of participants who strongly disagreed or disagreed with this association were 11.4% and 17.5% respectively in non-gated neighborhood. On the

Table 3 Cross-tabulation between perceived aesthetics and PoPA in gated and non-gated neighbourhoods

Neighbourhoods				PoPA					Total
				Strongly disagree	Disagree	Neutral	Agree	Strongly agree	
Non-gated	Perceived aesthetics	Strongly Disagree	Count	23	20	23	44	27	137
			% Within aesthetics	16.8%	14.6%	16.8%	32.1%	19.7%	100.0%
		Disagree	Count	15	35	13	35	18	116
			% Within aesthetics	12.9%	30.2%	11.2%	30.2%	15.5%	100.0%
		Neutral	Count	1	3	9	6	2	21
			% Within aesthetics	4.8%	14.3%	42.9%	28.6%	9.5%	100.0%
		Agree	Count	10	27	11	61	37	146
			% Within aesthetics	6.8%	18.5%	7.5%	41.8%	25.3%	100.0%
		Strongly Agree	Count	13	10	5	25	70	123
			% Within aesthetics	10.6%	8.1%	4.1%	20.3%	56.9%	100.0%
	Total		Count	62	95	61	171	154	543
			% Within aesthetics	11.4%	17.5%	11.2%	31.5%	28.4%	100.0%
Gated	Perceived Aesthetics	Strongly Disagree	Count	5	5	7	9	15	41
			% Within aesthetics	12.2%	12.2%	17.1%	22.0%	36.6%	100.0%
		Disagree	Count	5	12	7	28	42	94
			% Within aesthetics	5.3%	12.8%	7.4%	29.8%	44.7%	100.0%
		Neutral	Count	0	3	82	17	11	113
			% Within aesthetics	0.0%	2.7%	72.6%	15.0%	9.7%	100.0%
		Agree	Count	7	14	11	78	56	166
			% Within aesthetics	4.2%	8.4%	6.6%	47.0%	33.7%	100.0%

(continued)

Table 3 (continued)

Neighbourhoods				PoPA					Total
				Strongly disagree	Disagree	Neutral	Agree	Strongly agree	
		Strongly Agree	Count	5	6	9	25	40	85
			% Within aesthetics	5.9%	7.1%	10.6%	29.4%	47.1%	100.0%
	Total		Count	22	40	116	157	164	499
			% Within aesthetics	4.4%	8.0%	23.2%	31.5%	32.9%	100.0%
Total	Perceived Aesthetics	Strongly Disagree	Count	28	25	30	53	42	178
			% Within aesthetics	15.7%	14.0%	16.9%	29.8%	23.6%	100.0%
		Disagree	Count	20	47	20	63	60	210
			% Within aesthetics	9.5%	22.4%	9.5%	30.0%	28.6%	100.0%
		Neutral	Count	1	6	91	23	13	134
			% Within aesthetics	0.7%	4.5%	67.9%	17.2%	9.7%	100.0%
		Agree	Count	17	41	22	139	93	312
			% Within aesthetics	5.4%	13.1%	7.1%	44.6%	29.8%	100.0%
		Strongly Agree	Count	18	16	14	50	110	208
			% Within aesthetics	8.7%	7.7%	6.7%	24.0%	52.9%	100.0%
	Total		Count	84	135	177	328	318	1042
			% Within aesthetics	8.1%	13.0%	17.0%	31.5%	30.5%	100.0%

Table 4 Gamma test for perceived aesthetics and PoPA of people in gated and non-gated neighbourhoods

Neighbourhoods				Value	Asymp. Std. error [a]	Approx. T [b]	Approx. Sig.
Non-gated	Ordinal by ordinal	Gamma	Zero-order	0.149	0.046	6.097	0.000
	N of valid cases			543			
Gated	Ordinal by ordinal	Gamma	Zero-order	0.286	0.052	2.837	0.005
	N of valid cases			499			
Total	Ordinal by ordinal	Gamma	Zero-order	0.242	0.034	7.062	0.000
			First-order partial	0.224			
	N of valid cases			1042			

other hand the percentages of participants who strongly disagreed or disagreed were 4.4% and 8% respectively in gated neighborhoods. The results highlighted that the number of people who disagreed with this association were less in numbers in gated neighborhoods. The results show that people in gated neighborhood feel that their neighborhoods have more aesthetical features and they also feel that people used to do more physical activity such as walking, cycling and sports activities there in their neighborhoods due to aesthetical features.

The gamma test results in Table 4 show that the R value of the association between perceived aesthetics and PoPA in non-gated neighborhoods is 14.9% while it is 28.6% for gated neighborhoods. The overall R values for association between perceived aesthetics and PoPA is 24.2% which shows that there is small but significant association between these two variables in both types of neighborhoods.

The two types of results show that there is a significant difference between perceived aesthetics and PoPA in both types of neighborhoods. The residents of gated neighbourhoods feel that they have more neighbourhood aesthetics in terms of landscaping, ornamentations, attractive building facades, public amenities such as street benches, lights and paths in their neighborhoods and they also have perception that due to these facilities people are more physically actively in terms of walking and sports activities.

4 Discussion and Conclusions

This study has highlighted gated neighborhoods have more aesthetical features; therefore, people are more physically active than the non-gated neighborhoods. This result is consistent with the previous studies of Blakely and Snyder [16] that gated neighborhoods have more Aesthetical features than non-gated, and therefore, the behavior of people changes towards physical activity and they become more physically active as compared to the people of non-gated neighborhoods. This result is consistent to some extent with the previous study of Saelens et al. [8] where the researchers reported that people who live in high walkable neighborhoods with more land-use mix, street connectivity, residential density and neighborhood aesthetics are physically more active than the people who don't have these things in their neighborhoods. The results of this study are also consistent with the results of Hoehner et al. [12] that there is a positive relationship between perceived aesthetics and PoPA. Those researchers also used questionnaire method for data collection and the results show that along with other attributes, the attractive features were found positively associated with physical activity. This study is consistent to some extent with the results of Ball et al. [13]. They investigated the relationships between aesthetics, and walking companions and walking for exercise or recreation of Australian adults. They concluded that neighborhood aesthetics, and walking companions are important correlates of walking for exercise among urban Australians. The results of this study are also consistent with the previous study of Lovasi et al. [7] that there is relationship between neighborhood aesthetics and walking. They evaluated whether neighborhoods with more aesthetic amenities (sidewalk, cafes. street trees, and clean sidewalk) are associated with active transportation (walk for transport).

Strengths of this study are that it is the first study in Karachi, Pakistan with large study sample size, which makes it one of the largest studies so far in developing countries which has calculated the perception of people for aesthetical features of their neighborhoods and their importance for physical activity. This result can a be a guideline for policy makers that neighborhood aesthetics are also important which not only give pleasing effects to the residents but also attracts people to have good health for being physically active. Limitations of this study are that this is a cross-sectional study and uses subjective method. It was also difficult to obtain access to the gated neighborhoods for conducting this study. These issues may limit generalizability of the findings of this study. This study has calculated the neighborhood aesthetics through perceived method so the future neighborhoods can calculate these features by using some standard instruments. Future studies can also calculate the effects of aesthetical features on social behaviors of people as well.

The comparison between gated and non-gated neighborhoods shows that there is more aesthetics; therefore, people are more physically active in gated neighborhoods than non-gated neighborhoods. It can be recommended that the future neighborhoods should have aesthetical features such as public amenities, attractive facades and landscaping for enhancing physical activity at neighborhood level.

References

1. World Health Organization (2010) Global recommendations on physical activity for health. WHO Library Cataloguing-in-Publication Data
2. Handy S (2004) Community design and physical activity: what do we know? And what don't we know? Obesity and the built environment: improving public health through community design, pp 1–7

3. Gul Y, Sultan Z, Jokhio GA (2020) Contribution to the environmental sustainability by improving the walking behaviour through neighbourhoods' design with special reference to developing countries, E3S Web Conf 02002
4. Gul Y (2021) The effectiveness of walkability index for social sustainability in developing countries using geographical information system. J Des Sci Appl Arts 364–384
5. Gul Y, Jokhio GA, Bibi T (2020) Walk towards sustainability: improved neighbourhood street connectivity helps. E3S Web Conf 211
6. Gul Y (2017) Effects of neighborhood environment on physical activities in gated and non-gated neighborhoods in Karachi, Pakistan. University Technology Malaysia, Johar Bahru
7. Lovasi GS, Schwarts-Soicher O, Neckerman K, Konty K, Kerker B, Quinn J, Rundle A (2013) Aesthetic amenities and safety hazards associated with walking and bicycling for transportation in New York City, Ann Behav Med, pp 76–85
8. Saelens BE, Sallis JF, Black JB, Chen D (2003) Neighborhood-based differences in physical activity: an environment scale evaluation. Am J Public Health 1552–1558
9. Humpel N, Owen N, Leslie E (2002) Environmental factors associated with adults' participation in physical activity—a review. Am J Preventive Med 188–199
10. Mota J, Almeida M, Santos P, Ribeiro JC (2005) Perceived neighborhood environments and physical activity in adolescents. Preventive Med 834–836
11. Gul Y, Sultan Z, Moeinaddini M, Jokhio G (2018) The effects of physical activity facilities on vigorous physical activity in gated and non-gated neighborhoods. Land Use Policy 77:155–162
12. Hoehner CM, Ramirez LKB, Elliott MB (2005) Perceived and objective environmental measures and physical activity among urban adults. Am J Preventive Med 105–116
13. Ball K, Bauman A, Leslie E, Owen N (2001) Perceived environmental aesthetics and convenience and company are associated with walking for exercise among Australian adults. Am J Preventive Med 434–440
14. Gul Y, Sultan Z, Moeinaddini M, Jokhio GA (2019) The effects of socio-demographic factors on physical activity in gated and non-gated neighbourhoods in Karachi, Pakistan. Sport Soc 1225–1239
15. Richardson EA, Pearce J, Mitchell R, Kinghamc S (2013) Role of physical activity in the relationship between urban green space and health. Public Health 318–324
16. Blakely EJ, Snyder MG (1997) Divided we fall: gated and walled communities in the United States. In: Ellin N (ed) Architecture of fear. Princeton Architectural Press, New York
17. Breitung W (2012) Enclave urbanism in China: attitudes towards gated communities in Guangzhou. Urban Geogr 33(2):278–294
18. Leisch H (2002) Gated communities in Indonesia. Cities 19(5):341–350
19. Gul Y, Sultan Z, Johar F (2016) Effects of neighborhood's built environment on physical activities in gated communities: a review. Int J Built Environ Sustain 60–69
20. Gul Y, Sultan Z, Moeinaddini M, Jokhio GA (2018) Measuring the differences of neighborhood environment and physical activity in gated and non-gated neighborhoods in Karachi, Pakistan. Urban Design 24(3):494–513
21. Gul Y, Sultan Z, Jokhio GA (2018) The association between the perception of crime and walking in gated and non-gated neighbourhoods of Asian developing countries. Heliyon e00715

Urban Transformation Processes in Post-industrial Agglomerations (on the Example of Drohobych Agglomeration in Ukraine)

Y. Bardyn and Y. Onufriv

Abstract Drohobych agglomeration was created based on the joint industrial use of natural resources. Therefore, when planning the further development of agglomeration territories, it is necessary to analyze: the actual geological and natural potential, and the environmental consequences of the previous use of the territories. The strategy of development of the Drohobych agglomeration is the further use of the basic territorial resource—post-industrial territories, and joint solution of environmental problems. By analogy to the term "post-industrial territories", the article deals with the concept of "post-industrial agglomeration", which was based on the extraction and processing of minerals. The de-industrialization process causes the relocation of industrial sites and the reclamation of mining landscapes to meet the new needs of agglomeration development. These processes are connected with the transformation of the planning and landscape-spatial structure of the industrial centers of agglomeration—Drohobych, Boryslav, and Stebnyk.

Keywords Urban agglomerations · Post-industrial territories · Urban transformation

1 Introduction

The Drohobych agglomeration consists of the cities of Drohobych, Boryslav, Truskavets, Stebnyk, Skhidnytsya, and neighboring settlements of Drohobych district. They form a group of neighboring settlements that have intense economic, cultural, and household ties at the base of the common labor market, mental similarity, and integrated history. These cities form a polycentric agglomeration, where Drohobych is the central city, Stebnyk, and Boryslav are industrial centers, Truskavets and Skhidnytsia are recreational [1]. In this article, the problems of functioning and development of cities—industrial centers of Drohobych agglomeration are discussed.

Y. Bardyn · Y. Onufriv (✉)
Urban Planning & Design Department, Lviv Polytechnic National University, Bandery Str. 12, Lviv 79013, Ukraine
e-mail: yaryna.o.onufriv@lpnu.ua

T. Kang and Y. Lee (eds.), *Proceedings of 2021 4th International Conference on Civil Engineering and Architecture*, Lecture Notes in Civil Engineering 201,
https://doi.org/10.1007/978-981-16-6932-3_42

The earliest settlements in the territory of the modern Drohobych agglomeration arose based on rich mineral resources—rock and potassium salt, oil, and its derivatives [2]. The peculiarity of the formation of these cities within the agglomeration, namely Drohobych, Boryslav, and Stebnyk, is that the mining was the main city-forming function of these cities. The geological structure of the earth's surface, namely the occurrence of deposits of rock and potassium salts, oil, gas, and ozokerite in cities and adjacent territories, caused in the past the development of the mining and processing industry, technological processes in which required the optimization of industrial objects interconnections. This was the main stimulating factor in the development of a system of agglomeration communications—highways of territorial importance (T-14-02 Skhidnytsya-Pisochna (47 km), T-14-15 Mostyska-Sambir-Boryslav (19 km), T-14-18 Nyzhankovychi-Sambir-Drohobych-Stryi (17 km), as well as railway tracks, in this connection a specific planning structure of each of these cities with large areas of mining zones was formed (Fig. 1).

At present, the formed transport links between cities are relevant only partially due to the overall decline of the mining industry within the agglomeration. Problems with the functioning and, as a result, suspension of the activity of some enterprises, such as Private Joint Stock Company Stebnyk Mining and Chemical Enterprise "Polymineral", led to the accelerated degradation of these territories, as well as to the threatening environmental and economic situation in cities. So, today there is an

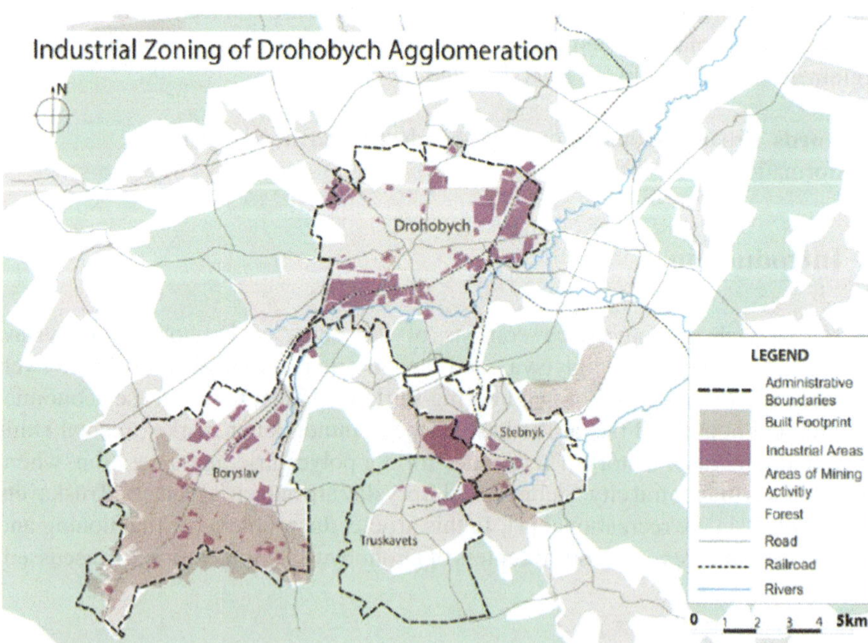

Fig. 1 Industrial territories and mining zones in the planning structure of Drohobych agglomeration (Lviv region, Ukraine). *Source* Bardyn Y

urgent need to re-profile and determine the directions of reclamation of large areas of mining landscapes in the planning structure of cities of Drohobych, Boryslav, and Stebnyk—in the past industrial centers of Drohobych agglomeration.

Today Drohobych agglomeration is in ecological risk and extremely low stability of the geological environment [3], as confirmed by numerous studies in the field of geography, geology, ecology, and chemistry, the main of which are the works of Ukrainian scientists: Rudko, Ivanov, and Kovalchuk [4], Bilonizhka and Dyakiv [5], Gaidyn [6], Yatchyshyn [7], Pavlyuk [8], Bondarenko [9] and others. Similar geo-ecological problems related to the consequences of the activity of similar mining enterprises also exist in other regions of the world and Ukraine, as evidenced by separate scientific works: Soroka, Novodvorskaya, Volkova, Gavrilov, & Tsmyg [10]; Żurek, Diakiv, Szarek-Gwiazda, Kosiba, & Wojtal [11], etc.

Due to the existing problems and the need to re-profile mining enterprises, as well as to determine the directions of the most optimal use of the post-industrial territories of Drohobych agglomeration, there is a need for the rehabilitation of disturbed landscapes. Many developed countries already have a long-standing experience of reclamation of such lands, and accordingly, there are many scientific works in this direction: Kuter [12]; Mborah, Bansah, and Boateng [13]; Ostręga and Uberman [14]; Wolkersdorfer and Bowell [15]; Madziarz, Mizera, and Dębkowski [16]; Rostanski [17]; Kasprzyk [18] et al. The reviewed works concern both general approaches to the reclamation of mining territories, the definition of evaluation criteria and areas for reclamation of territories, and the analysis of specific successful examples of reclamation in the world.

Existing studies of the geo-ecological situation in Drohobych agglomeration, scientific approaches to the reclamation of disturbed landscapes in Ukraine and worldwide, as well as available design materials and concepts related to the future development of Drohobych agglomeration cities [19], Drohobych City Council [20], Ukrainian State Institute for City Planning "Mistoproekt" [21], Truskavets City Council [22], and others), can be the basis for the formation of scientifically weighted recommendations for the recultivation of the mining landscapes of Drohobych, Boryslav, and Stebnyk. To determine the common priority direction for the development of cities of Drohobych agglomeration, it is necessary to have a comprehensive analysis of their functional and landscape-spatial organization and to determine the potential and "strengths" in the unification of territories of settlements as post-industrial agglomeration.

2 Material and Research Methods

The following research methods are used in the article: a content analysis of scientific publications on the topic, historical analysis of the study area, statistical method, comparative analysis of existing design developments, method of generalization and synthesis. The research methodology is to find "strengths" in the unification of territories of settlements as agglomeration for new activities that will complement the existing structure of the economy.

The sequence of the study is set out in the following steps:

- analysis of statistics concerning the existing resource base and economic situation in the post-industrial cities of Drohobych agglomeration;
- analysis of existing project proposals and approaches to solving such problems, and determining the possibility of adaptation of these approaches to the study area;
- formulation of priority development direction for cities of Drohobych agglomeration.

3 Results

Today Drohobych agglomeration has many problems and needs to identify new directions of development. Against this background, it is important to identify the negative tendencies in the urban planning structure and its natural basis organization.

Modern Drohobych is a developed natural-economic complex, the second-largest industrial city in the region. At the same time, Drohobych, like Boryslav, and Stebnyk - is a historical settlement with preserved historical habitat and traditional character of this environment. Nowadays, the main industrial branch of the city is mechanical engineering and oil refining. Most businesses are active. The site of the former salt cellar, measuring 5.8 hectares, which now houses the museum and still produces salt according to old technology, needs reorganization. Today, there are already conceptual proposals for the revitalization of this territory with the creation of public space with different functions: museum-production, commercial, recreational, cultural, artistic, and entertaining [23].

The peculiarity of Boryslav is that it is the only city in the world that is located on industrial oil and gas and ozokerite deposits with numerous sources of mineral and medicinal waters. As of 2004, Boryslav was one of the six greenest cities in Ukraine. The area of green space per inhabitant of the city here was 69.8 m^2 (according to international standards, this figure should be at least 20 m^2). At the same time, the city is suffering from environmental pollution from oil and its refining products. Not only did the oil get into the air, but it also concentrated in wells, gullies, groundwater. Water has become not only poisonous but also fire hazardous. Also, the high level of gas pollution of the surface layer of the atmosphere—that is, the presence in the air of production zones and habitats of gaseous hydrocarbons of natural origin (methane, ethane, propane, butane, and other methane hydrocarbons) in significant concentrations is one of the contemporary environmental problems of Boryslav [24]. The city will become a safer place to live if the problem of pollution would be eliminated. Another problem for the functioning of the city is the presence in the structure of the urban landscape of large areas of post-mining landscape complexes, such as oil wells with sanitary protection zones, as well as the territory of the former

Open Joint Stock Company "Boryslav ozokerite", which are used today irrationally. In particular, there is a problem in the city with old oil wells, which were closed after the development was completed, and today in the places of some former wells the sinkholes are formed. The locations of these wells are unknown because they are randomly distributed throughout the city and invisible on the surface of the earth. It is necessary to research and identify the location of such wells and take precautionary measures to eliminate the danger of sinkholes. In case of elimination of the above-mentioned negative phenomena and possible threats, post-industrial territories can be a resource for the development of the city.

The town of Stebnyk is known in Ukraine for its environmental problems. These problems are related first of all to the activity of Polymineral State Enterprise, which was engaged in the production of potassium salt and production of potassium-magnesium fertilizers and belongs to environmentally hazardous objects of the Lviv region. Over the last few decades, several environmental catastrophes have occurred in the area related to the mining industry: from the breakthrough of a salt-brine tailings dam into the aquifer and rivers, to floods of mines and uncontrolled formation of karst, which occur till now [25, 26]. Today, no more is being mined. The production activity of Polymineral State Enterprise was discontinued in 1988, and the territory that was included in the mining zone is now in no way used in connection with the danger of sinkholes and subsidence of the earth's surface. There are scientific developments concerning the elimination of the danger of karst formation and conservation of the Stebnyk State Polymineral Plant [27]. However, the necessary remediation measures require substantial investment and have not yet been fully implemented. In the case of anthropogenic danger elimination, further biological reclamation and reorientation of this territory to other functions will be possible, which can be a significant impetus to Stebnyk's development in the structure of Drohobych agglomeration.

These cities have many common characteristics, but they also have differences in development (Table 1). The highest demographic potential is present in the city of Drohobych, which has twice the population of Boryslav and more than three times more than in Stebnyk. However, in Drohobych, as well as in Boryslav, there is a negative indicator of population migration, partly due to a lack of jobs. Urban development potential is different in each of the cities considered. The area occupied by the territory of the city is four times larger in Drohobych and Boryslav compared to Stebnyk. According to the master plans of cities in Drohobych, Boryslav and Stebnyk there is territorial potential for housing construction and public services, industry, and recreation. The largest area of post-industrial territories is present in Boryslav, which includes landscaped areas of general and limited use around oil wells, which are mosaically located almost throughout the city, as well as the territory of the Boryslav Ozokerite Plant. Due to these green areas, the city has such a high degree of landscaping. However, most of them, except some more complete ones, will be difficult to refocus in the future due to their patchwork structure and integration into the residential development area. In Stebnyk there is also a likelihood of re-profiling the post-industrial territories of the State Polymineral Enterprise, with an estimated area of 90 ha, for other functions. Drohobych has the greatest transport potential, but other cities also have developed road and rail connections. Investigated cities

Table 1 Characteristics of the potential of Drohobych agglomeration cities [28–30]

Potential	Units of measurement	Drohobych	Boryslav	Stebnyk
Demographic				
Population	Thousand people	76	34	21
Population of working age	%	60	57	60 [a]
Population older than working age	%	20	20	20 [a]
Migration movement of the population	Person	−298	−88	ND
Urban planning				
Area of the settlement	ha	4450	3700	890
Area of unoccupied territories Including accommodation:	ha	583 [b]	221 [b]	165 [a]
• Housing and objects of public services		392 [b]	154 [b]	88 [a]
• Industry		135 [b]	18 [b]	44 [a]
• Recreation		54 [b]	48 [b]	32 [a]
Area of post-industrial territories	ha	5.8 [a]	1000 [a]	88 [a]
Main Bus / Railway station	Units	1/–	– /–	– /–
Small bus/train station	Units	1/1	1/1	1/–
Local highways	Presence	T-14-18	T-14-15	T-14-02
The length of railway tracks	kм	9.1	5.0	5.9
Cemeteries	ha	29,7	15,6	12,3
Solid household landfill	ha	–	3.0	–
Nature and landscape				
Area of forests, forest parks	ha	281	90	57
Area of water reservoirs		115	36	13
Nature Reserve Fund Including:		154.95	2.03	ND
• Botanical monuments of nature		0.66	–	ND
• Geological nature monument		–	2.03	ND
• Parks-monuments of landscape architecture		154.29	–	ND
Engineering and technical				
Security of construction with central water supply	%	86	73	70 [a]
Drainage		86	73	70 [a]
Electricity supply		100	100	95 [a]
Gas supply		97	93	80 [a]
Hot water supply		77	51	40 [a]

(continued)

Table 1 (continued)

Potential	Units of measurement	Drohobych	Boryslav	Stebnyk
Historical and cultural				
Monuments of architecture of national importance	units	4	3 [c]	–
Monument of architecture of local importance		191	2 [c]	2
Monuments of monumental art		10	9 [c]	–
Monuments of history		14	35 [c]	–
Muzeums		5	2	–
Theaters and cinemas		5	1	–
Houses of culture		7	3	2
Churches		7	4	6
Public service				
Shopping malls, markets	units	5	3	1
SUpermarkets		9	3	1
Hotels		2	1	–
Sanatoriums and camp sites		4	1	–
Number of hotel rooms		69 [d]	632 [d]	–
Scientific and educational				
Higher education institutions	units	10	4	1
Schools and pre-schools		32	6	8
Scientific institutions		6	–	–
Libraries		7	8	1
Medical care				
Hospitals, polyclinics	units	19	6	2
Number of hospital beds	units	740	210	50 [a]
Economic and industrial				
Foreign Direct Investment in 2018	ths. dollars. USA	18,313.7	1734.1	ND
Capital investment for 2018	thousand UAH	296,871	227,046	ND
Number of enterprises	units	539	195	13 [a]
Area occupied by enterprises	ha	543 [a]	220 [a]	93 [a]
Financial result (balance)	thousand UAH	84,193.6	−4229.0	ND
Profitable enterprises	%	76.3	67.0	ND
Number of employees in enterprises	units	8323	2690	ND
Number of full-time employees	units	14,904	5558	ND

(continued)

Table 1 (continued)

Potential	Units of measurement	Drohobych	Boryslav	Stebnyk
Average monthly nominal wage (average per employee)	UAH	6291.12	7415.15	ND
Formal employment in the registered labor market	%	36.8	33.1	15.3 [e]
Registered unemployment		0.8	3.2	1.3 [e]
Economically active population in the unregistered labor market		35.9	25.8	38.7 [e]
Economically inactive population		26.5	37.9	44.7 [e]
Other factors				
Emissions of pollutants (2018) into atmospheric air from stationary sources of pollution	T	898	335	ND
Waste generated (for 2018) I-IV hazard classes		14,641	1455	ND
Waste disposal (for 2018) of I-IV hazard classes		14	141	ND

Notes [a] Approximate data taken
[b] According to the current town planning documentation
[c] Within the Boryslav City Council
[d] Institutions of legal entities and individual entrepreneurs are taken into account
[e] General data on Drohobych district; ND—no data

are provided approximately equally in engineering and technical terms. Drohobych has a significantly larger area of city parks. Stebnyk lacks nature reserve assets. The historical and cultural potential is also the highest in Drohobych, including 191 architectural monuments of local significance and 4 of national significance. Drohobych also has considerable scientific and educational potential. The relatively larger number of rooms in hotel-type establishments of the developed private sector could be seen in Boryslav. The economic and industrial potential is currently the lowest in Stebnyk. The environmental situation in Stebnyk and Boryslav is quite threatening. However, Drohobych is the main air pollutant and producer of industrial waste. The problem concerns the recycling of industrial waste as well as solid household waste since there is no landfill in the city.

Still, Drohobych remains an industrial center of agglomeration. There are more than 500 businesses in the city, more than half of which are profitable. There are also territorial reserves for industrial development. However, the industrial sector still needs investment and partial reorientation to other sectors of the economy. Promising for tourist purposes is the post-industrial territory of Drohobych saltworks, which confirms successfully implemented projects based on salt mines in the world, such as the Wieliczka salt mine near Krakow (Poland) with a museum, a medical center, a lecture hall, a concert hall, and a sports hall; Klodawa Salt Mine

(Poland)—tourist center, museum, operating enterprise; Lion Salt Works in the UK—a museum, venue for festivals, concerts, and art events; Saline Luisenhall in Grone (Germany)—museum, operating company, spa; Galle Salt Farm and Museum and Museum (Germany); Jacobsen Salt Co in Portland (USA) and others. Nowadays the tourist route "Salt Way", which includes the Drohobych Salt Plant, is becoming more popular. The significant historical and cultural potential of the city and the availability of interesting industrial sites will contribute to the development of local and industrial tourism in the city, which in turn will require an increase in the number of hotels, restaurants, and other tourist facilities.

Boryslav also has an inexhaustible industrial potential. Despite this, many environmental problems have arisen in the city today, related to the formation of associated gases and sinkholes at the site of old oil wells. Exploration and production of oil from deeper horizons, which, according to scientists, contains about 300 million tons of oil, as well as the processing of related gases from the oil industry for electricity and heat in the future, with investments, can improve the environmental and economic situation in the city. Boryslav ozokerite deposit is also promising. In particular, there are proposals to create a recreation area within the northern part of the mining outlet of Open Joint Stock Company "Boryslav ozokerite". The plant's territory is empty today and no production is taking place. However, with the presence of the investor, it is possible to continue the production of ozokerite with the construction of a new mine and the transformation of the old mine into a museum, which would be unique, since ozokerite is extracted in only thirty countries. Part of the post-industrial territory of the enterprise could be used for recreation by residents. Today the city already has a museum of the oil industry, and there are also significant territorial reserves with preserved oil wells and old oil pumps, around which a Skansen—a museum of the open-air oil industry, could be organized, which would become part of the Carpathian-Galician oil industry path and transboundary oil trail extending from Gorlice, where a similar Skansen oil industry "Magdalena" has already been created, through Jaslo, Krosno, Sanok, and Ustishki Dolishny to Boryslav, Drohobych, and Lviv. Today, Boryslav is a garden city, as the degree of landscaping in the city is very high. However, this landscaping is large of limited use, as it is a sanitary protection zone from oil wells. These green areas are integrated into residential areas, which can be a "strong point" for the development of private mini-hotels and provide a base for overnight visitors of the historic industrial sites. Thus, industrial tourism can become a complementary branch of the city's economy.

Modern Stebnyk is a post-industrial city where there is no longer any prospect of resuming the production of potassium-magnesium salts [31]. The city has a large area of post-mining territories, which in the long term, after carrying out all the necessary remediation activities and additional research can be redeveloped into recreational territories, namely: on the site of the Salina saltworks, where the remains of the building have been preserved there is an opportunity to create a salt-making museum; based on the Stebnyk State Plant "Polymineral" of mine № 1 it is possible to create a speleological hospital [32]; in open areas within the mining area, you can create campsites with outdoor sports fields, trails for extreme sports, or a field for music festivals. The development of beekeeping in the city as an ecological alternative for restoring Stebnyk's natural historical landscapes can be promising.

Therefore, Stebnyk should develop industrial and green tourism, as well as recreation to improve the respiratory tract. This is confirmed by the already existing successful practice in the world of caving in the former potash salt mines. The proximity of Stebnyk to Truskavets—a well-developed balneological resort, is also an additional enhancing factor for the development of tourism in Stebnyk in the future.

4 Conclusions

Based on a comprehensive analysis of the potential of Drohobych, Boryslav, and Stebnyk, it can be confirmed that industry is no longer the core function of these cities, and therefore they need new complementary functions, and in the case of Stebnyk, reorientation to another branch of the economy. Therefore, the use of free territories in cities should be aimed at enhancing economic benefits and at the same time improving the environmental situation in cities. Post-industrial territories are an important resource for urban development, which, in the full or partial change of the functional designation of these sites, will in the future lead to an improvement of the ecological status of disturbed landscapes. The architectural and planning organization of the post-industrial territories in cities, when transforming them into public spaces, should be carried out taking into account historical continuity and dynamic adaptation with the use of energy-saving technologies [33]. In the context of the set goals, we consider the complementary direction of development for the cities of Drohobych agglomeration—the development of tourism infrastructure based on both industrial sites and natural landscape resources with recreational and therapeutic potential. The results of the study can be used in the creation of a development strategy for Drohobych agglomeration, as well as in the development of projects for the restoration of post-mining landscapes.

References

1. Baranetskyi T (2017) Analysis of opportunities and needs of cooperation of territorial communities within the Drohobychina agglomeration. NGO "European Dialogue" p 13
2. Bila KO (2011) Socio-economic reforms in the context of Ukraine's integration choice. Mat VII Internat Res Prac Conf Econ and Soc Iss of today p 136
3. Ukrainian State Institute for City Planning "Dipromisto" (2009) Lviv region. Scheme of territory planning. Graphic part: Scheme of environmental protection (Kyiv)
4. Rudko GI, Ivanov EA, Kovalchuk IP (2019) Mining geosystems of the Western region of Ukraine (Kiev–Chernivtsi: Bukrek), p 376
5. Bilonizhka P, Diakiv V (2009) Chemical and mineral composition of waste of potash ore enrichment of Stebnyk deposit and their impact on the environment. Bull I. Franko Nat Uni of Lviv Geol Ser 23:162–174

6. Gaidyn AM (2013) Environmental problems of potassium salt and sulfur extraction areas. Ecol Environ Manag 16:127–134
7. Yatchyshyn YY (2007) Improvement of ecological safety in waste-free technologies of processing of multicomponent salt-containing materials (Lviv: Lviv Polytech. Nat. Uni.), p 19
8. Pavliuk V (2012) Negative consequences of uncontrolled flooding of the Stebnyk potash mine (Precarpathian, Ukraine). Geol Geochem Combust Min **1–2**:91–101
9. Bondarenko MD (2001) Dynamics and prediction of the geological environment of the salt and sulfur deposits of the Precarpathian region (Lviv: I. Franko Nat. Uni. of Lviv), p 18
10. Soroka AV, Novodvorskaya MA, Volkova OA, Gavrilov DA, Tsmyh MS (2018) Ecological consequences of mining (potassium salts) in the Soligorsk district. Mat. IV. Sci. Conf. Kazan, pp 19–21
11. Żurek R, Diakiv V, Szarek-Gwiazda E, Kosiba J, Wojtal AZ (2018) Unique Pit Lake created in an opencast potassium salt mine (Dombrovska Pit Lake in Kalush, Ukraine). Mine Wat Environ 37:456
12. Kuter N (2013) Reclamation of degraded landscapes due to opencast mining. In: Advances in landscape architecture. Namık Kemal University, Turkey
13. Mborah C, Bansah KJ, Boateng MK (2016) Evaluating alternate post-mining land-uses: a review. Environ Poll 5(**1**):14–22
14. Ostręga A, Uberman R (2010) Kierunki rekultywacji i zagospodarowania – sposób wyboru, klasyfikacja i przykłady. Górn. i Geoinż. 34(4):445–461
15. Wolkersdorfer C, Bowell R (2005) Contemporary reviews of mine water studies in Europe, Part 2. Mine Water Environ 24:2–37
16. Madziarz M, Mizera A, Dębkowski R (2012) Projekt "Rekultywacja obszarów zdegradowanych działalnością górniczą na terenie gminy mirsk z utworzeniem ścieżki śladami dawnego górnictwa Kruszców" jako koncepcja kompleksowych działań w zakresie ochrony i wykorzystania dziedzictwa górniczego dolnego Śląska. Dzieje górnic. – elem. Europ. Dziedz. Kult. 4:273–289
17. Rostanski A (2000) Rekultywacja i zagospodarowanie nieuzytkow popremyslowych – rozwiazania alternatywne. Inż Ekol 1:81–86
18. Kasprzyk P (2009) Kierunki rekultywacji w górnictwie odkrywkowym. Probl Eko Kraj 14:7–15
19. Lafler et al (2012) Drohobych profile: demography, economy, budget (REOP project)
20. Drohobych City Council (2020) Urban planning documentation and historical and architectural plan: https://drohobych-rada.gov.ua/investor/misto-docs/ (accessed: 03.2020)
21. Ukrainian State Institute for City Planning "Mistoproekt" (2011) Project stage of the Master plan "Adjustment of the Master Plan of Boryslav city in Lviv Region" (Lviv), p 123
22. Truskavets City Council (2020) Town-planning activity. Master plan. Zoning: http://www.tmr.gov.ua/home/mistobudivna-diialnist (accessed: 04.2020)
23. Department of Restoration of Architectural and Artistic Heritage of Lviv Polytechnic National University 2019. Drohobych salt plant: https://www.facebook.com/Drohobychsaltplant/ (accessed: 03.2020)
24. Scientific Center for Earth Aerospace Research of the Geological Sciences Institute of the Ukrainian National Academy of Sciences 2002. Geochemical control of the contamination of air pollution in the city of Boryslav with the use of Earth remote sensing materials: Report on the creation of NTP (final) (Kyiv), p 100
25. Vlasiuk A (2016) Technogenic earthquake near Truskavets (map): http://www.drogobych.com/2016/09/blog-post.html (accessed: 04.2020)
26. Dudar O (2020) Two failures in two and a half years: whether landslide near Stebnyk could be prevented: https://dyvys.info/2020/03/16/dva-provallya-za-dva-z-polovynoyu-roky-chy-mozhna-bulo-zapobigty-zsuvu-gruntu-bilya-stebnyka/ (accessed: 04.2020)
27. Diakiv V, Tsar H (2010) A model of leaching, solidification and self-isolation of easily soluble salts from near-surface salt-clay deposits in the tailings pond of potash deposits of the Precarpathian region. Min Collect 60(2):136–147

28. Central Statistics Office in Lviv region (2020) Passport of the Lviv region: https://www.lv.ukrstat.gov.ua/eng/bank/about_statregion.php (accessed: 04.2020)
29. Institute for Regional Studies of the National Academy of Sciences of Ukraine (2016) Conducting an analytical study to determine the dynamics and structure of the unemployed economically active population of Lviv region: https://loda.gov.ua/upload/users_files/22/upload/Struktura_nezaynyatogo_ekonomichno_aktYvnogo_naselennya.pdf (accessed: 04.2020)
30. Public cadastral map of Ukraine: https://map.land.gov.ua/?cc=3461340.1719504707,6177585.367221659&z=6.5&l=kadastr&bl=ortho10k_all (accessed: 04.2020)
31. Varyvoda ZV (2014) Will there be an environmental catastrophe in Stebnik: https://protruskavets.org.ua/chy-taky-bude-ekolohichna-katastrofa-v-stebnyku/ (accessed: 04.2020)
32. Rudnytska ND et al (2000) Prospects for speleotherapy development at the Truskavets resort. Proc. of the Inter. Cong. Probl. of Inf. of Recr. and Tour. Act. in Ukr.: Prosp. for Cult. and Econ. Dev., pp 255–257
33. Kravchenko OV (2015) Principles of architectural and planning organization of open urban spaces with disturbed territories. National Aviation University, Kyiv, p 237

Management of Flood Risk with Spatial Planning in the Sri Lankan Context

H. M. M. Herath, N. T. S. Wijesekera, and R. L. H. L. Rajapakse

Abstract Characterization, assessment, and management of the complexities and uncertainties incorporated with flood risk are important for the operationalization of flood risk management. Flood risk is required to be managed by the integration of objectives all stakeholders who are associated with it. Therefore, the operationalization of flood risk management is determined by the conceptualization and assessment of flood risk in the planning process, as well as the discourse of the stakeholders in the governance context, and the available interaction mechanisms for participatory decision making. This research paper reviews the level of operationalization of flood resilience of the current practice of flood risk management considering the Sri Lankan context. Panadura urban area has been selected as the case study and it is a coastal urban area, which located in the downstream of the Bolgoda basin. Bolgoda basin which is in the wet zone of Sri Lanka is affected by frequent floods particular during the southwest monsoons. The case study has highlighted that flood risk and land uses are planned and managed separately in the current practice. This study has further highlighted, that at present, spatial planning is not able to provide a common guideline for stakeholders' integration in the face of increasing flood risk. There is an obstruction in the practice of Flood Risk Management due to the absence of a common framework for risk assessment and management of all the stakeholders. Therefore, this study has highlighted that spatial planning practices are yet to adapt to the notion of resilience for structuring the uncertainties in flood risk management.

Keywords Flood · Risk · Uncertainty · Stakeholders · Spatial planning · Sri Lanka

---

H. M. M. Herath (✉) · N. T. S. Wijesekera · R. L. H. L. Rajapakse
Department of Civil Engineering, University of Moratuwa, Moratuwa, Sri Lanka
e-mail: malani@uom.lk

T. Kang and Y. Lee (eds.), *Proceedings of 2021 4th International Conference on Civil Engineering and Architecture*, Lecture Notes in Civil Engineering 201,
https://doi.org/10.1007/978-981-16-6932-3_43

1 Introduction

1.1 Increasing of Flood Damage with the Intensification of Urban Development in Sri Lanka

Sri Lanka has a tropical climate characterized by two monsoon seasons. Sri Lanka is routinely affected by recurring floods. Floods have become a significant disaster particularly in urban areas mainly distributed along the western coastal plain of Sri Lanka. Approximately, one-third of the total population of the country is living in the western region of Sri Lanka. Flood damages are tending to increase, particularly in coastal urban areas in the western region where the elevation is prominently less than six meters from the mean sea level. The main drainage basins of this region that are highly affected by recurrent flooding are the Ja Ela basin, Kalu Oya basin, Greater Colombo basin, and Bolgoda basin. Each government in the past has allocated significant amounts of public funds for flood risk management. However, flood risk is still a significant problem particularly in urban areas of Colombo, Panadura, Kaduwela, etc., despite the flood risk management practices implemented so far.

1.2 Flood Management Towards Flood Risk

In the recent years, it has been recognized that the traditional flood control structural measures are inadequate to cope up with extreme flood occurrences [1]. Further, it has been intensively discussed that the traditional flood risk management practices have created a new set of problems in rapidly urbanizing regions in the face of climate change [2, 3]. The concept of socio-ecological systems suggests that resilience is the key to manage complex systems and reduce vulnerability that is a result of the inherent uncertainty of flood risk management [4]. In view of that, flood risk management has been defined as a multidimensional concept [5, 6]. Decisions on traditional flood risk management were usually based on the stationarity assumptions of flood occurrence [7]. Recently, it has been recognized that the flood risk cannot be estimated based on probability models due to its non-linearity and the non-stationarity [8]. Moreover, it has been further recognized that flood risk management with the command and control practices is maladaptive due to its dynamic nature [9]. The transition of flood management towards flood risk management practices [10] has highlighted the requirement of review the flood risk assessment, methods of flood risk management, and methods of operationalization during the past years [11, 12]. Accordingly, it has been highlighted that the flood damage assessment and vulnerability assessment have limitations in decision making in the face of the increasing complexity of flood risk due to urban development and climate change. Furthermore, the assessment and management of uncertainty have become the main tasks of the decision-making process of flood risk management.

2 Research Objectives and Methodology

The decision-making of flood risk management is a complex process. The decision-making process is characterized with many aspects such as characterization of flood risk, flood risk assessment, evaluation of the merits of flood risk management options, identification of the flood risk management strategies, evaluation of the merits of objectives, strategies, implementation of guidelines, evaluation of the achievement of the long term objectives, etc. Accordingly, defining a common objective for all stakeholders for evaluating the alternative options of flood risk management strategies is a challenging task in flood risk management. Therefore, the operationalization of flood risk management is governed by the planning processes as well as the governance structure of the context. It is, therefore, necessary to structure the complexity of the planning process for the implementation of flood risk management. Moreover, at present, the planning process does not provide a platform for a strategic assessment of flood risk in the face of increasing urban development and climate change. There is no common guideline to structure the complex planning process and evaluation of flood risk management options. The Analytic Hierarchy Process (AHP) is a multi-criteria decision-making technique which permits the relative assessment and prioritization of alternative options [13]. The Delphi technique is a systematic and interactive method that depends on a panel of independent professionals. Delphi technique permits to reach a consensus, through the collection of experts' opinions on a given issue during successive stages of questionnaire and feedback. Analytical Hierarchy Process (AHP) and Delphi technique are several analytical techniques which can be used to structure complex decision-making process considering the real ground situations.

First, the conceptual framework of the decision model, including the criteria and sub-criteria of the framework, is developed on the basis of a systematic review of the literature. The criteria and sub-criteria of the framework are selected and prioritized using Integrated Analytical Hierarchical Process (AHP) and Delphi technology. Later on this developed framework was used to review flood risk management practices in the selected case study area. Panadura urban area was selected for this research due its location on the downstream of the Bolgoda basin. Field observation, questionnaires and interviews with key experts and stakeholders were used to collect the information related to the case study area. A systematic survey was conducted with stakeholders for gathering the data, information and opinions. After that, the data and information were analyse with reference the developed framework. It has analysed the existing flood risk management practices in order to identify the opportunities and the challenges for operationalization of the flood risk management.

3 Development of the Framework for Operationalization of Flood Risk Management

The operationalization of flood risk management is determined by four main aspects of the planning process and the governance context. These factors include conceptualisation and assessment of the planning process the policy discourses and the available participatory mechanisms of the governance context. It is required to conceptualize the non-linear interactions of flood risk during the planning stage to identify the stakeholders of the decision-making process. The flood risk assessment and the evaluation of alternative management measures are required to consider the multiple objectives of the stakeholders at multiple scales, multiple scenarios, and the wellbeing of the society in long term. Therefore, the decision making is required to be a continuous and integrated process to manage the uncertainty. This is achieved by creating integration among multiple objectives of multiple stakeholders at different scales while considering the interdependencies of stakeholders. The level of integration among the stakeholders is determined by the governance structure of each context which includes the discourse of the stakeholders and the available participatory decision-making mechanism. The discourse of the stakeholders has been recognized to be of utmost importance in the participatory decision-making and implementation process in the operationalization of flood risk management. Participatory governance mechanism means the availability of an integrated interaction mechanism among stakeholders to share power and resources for collaborative decision making, and implementation of flood risk management goals, strategies, and guidelines. Moreover, a participatory governance mechanism is able to facilitate the legitimacy of the participation of all stakeholders in the decision-making and operationalization of common objectives.

3.1 Selection of Case Study Area

The Bolgoda basin can be broadly categorized into three regions as northern, southern, and north-eastern. There are two large lakes named North Bolgoda Lake and South Bolgoda Lake and both lakes are interconnected by the Bolgoda River (Ganga) (Fig. 1). The Bolgoda North Lake discharges to the sea through the Panadura estuary while the South Bolgoda Lake is connected to the sea via Thalpitiya Canal (Ela). Bolgoda South Lake is also connected to the Kalu Ganga basin through a manmade canal. The northern region of the basin is covered by highly urbanized areas while the southern and the northeastern regions of the Bolgoda basin are covered by urbanized and semi-urbanized human settlements. Panadura area is a coastal urban centre which is located downstream of the Bolgoda basin. Water which flows off the Bolgoda basin connect to the sea at the two locations of Panadura town (Fig. 2). The annual average rainfall for the Bolgoda basin is about 2500 mm. The Study on Storm Water Drainage Plan for the Colombo Metropolitan Region which is conducted by the Japan International Cooperation Agency has estimated probable Bolgoda basin average 24-h

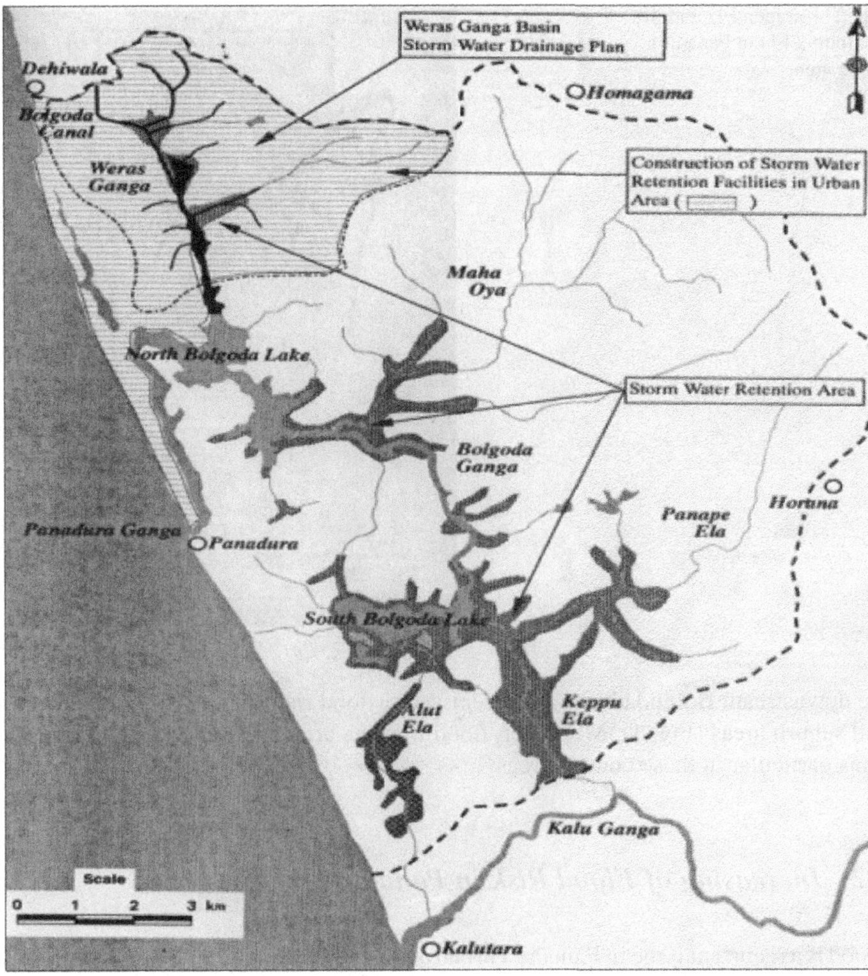

Fig. 1 Panadura urban area in the Bolgoda Basin. https://openjicareport.jica.go.jp/pdf/11716586_10.pdf PDF file ((jica.go.jp)

Rainfall for the return periods of 2, 5, 10, 25, and 50 years are 103 mm, 137 mm, 160 mm, 188 mm and 210 mm, respectively. The Projected rainfall patterns have suggested more intense rainfall episodes in the future [1].

The total population of the Panadura urban council area in 2012 was 45,136 and the population density was 7339 per square kilometre. The population growth rate was 0.70% during 1981–2001. The population growth rate has been increased up to 2.41% during the years of 2001–2012. The main reasons for the population growth in this area are the increase of direct access to the main city of Colombo, development of infrastructure facilities, the establishment of the Modarawila industrial estate, etc. During recent years, flooding can be seen as a very common disaster in urban areas of

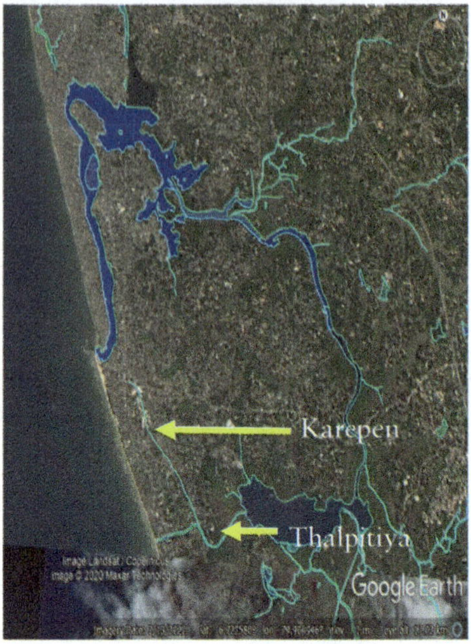

Fig. 2 Karapen Ela and the Thalpitiya Ela in Panadura urban area

the downstream Bolgoda basin. In recent years, flood risk is increasing in Panadura and suburb areas (Fig. 1). Moreover, flood impacts are triggered with the monsoon rains particular in these coastal areas.

3.2 Increasing of Flood Risk in Panadura Urban Area

Flood is a recurrent issue in Panadura urban area after 2000 particularly in the vicinity of the Karepen ela, town centre, etc. Kerapen Ela starts from the Modarwila industrial area and drains into the Thalpitiya Ela. The length of this canal is about 3.5 km (Fig. 2). 'Moderawila' was a small pond, distributed over 90 acres in extent, and it was reclaimed by the Sri Lanka Land Development Corporation (SLLDC, previously known as SLLRDC) in 1979 for industrial development. During the monsoon periods, water flows from Panadura town center and left-hand side town area to this canal. As a result, Karapen Ela has been experiencing floods even after a minor shower in recent years. Before 2000, no flood occurrences have been recorded in these areas. However, in the aftermath of 2000, this area is affected by recurrent floods. Water levels are varied from 1 foot –5 feet and usually remained for more than three days at very close to the canal. The elevation of Thalpitiya Ela is higher than that of Karapen Ela. Therefore, water is not drained from Karapen Ela into Thalpitiya Ela during the monsoon periods. As a result, the surrounding areas of the Karapen Ela gets inundated. The elevation of the Thalpitiya canal is also higher than the mean sea level and it is locked by a heavy sand bar at the sea outfall. There is no continuous water flowing into the sea because of the sand bar formation at the sea outfall of the Thalpitiya canal. There are bottlenecks for water flowing during the southwest monsoon as well. In recent years, a flood control project was proposed considering the flood situation in Pandura urban area. The drainage improvement project at Panadura was initiated during the 2012–2014 period to manage the flood risk in this area.

3.3 Current Flood Risk Management Practices

Irrigation Department of Sri Lanka has been authorised to control river floods according to the Flood Ordinance No. 22 of 1955. Sri Lanka Land Reclamation and Development Corporation (SLLRDC) was established in 1968 under Sri Lanka Land Reclamation and Development Corporation Act No. 15 of 1968. This act was amended as Act No. 35 of 2006 enabling SLLRDC to take legal actions against unauthorized land reclamation and water pollution. SLLRDC was re-designated as "Sri Lanka Land Development Corporation" with the amending Act No. 52 of 1982 to the Colombo District (Low Lying Areas) Development Board Act No. 15 of 1968. The Sri Lanka Land Development Corporation is responsible for planning and management of urban drainage systems. Sri Lanka Land Development Authority has designed and constructed a drainage system to overcome the flood risk in urban areas. Therefore, at present, Sri Lanka Land Development Corporation (SLLDC) is the main agency that is engaging in the planning and implementation of the regulations related to the development and maintenance of wetlands and urban drainage systems. A boundary of 60 m has been declared as Bolgoda Environmental Protection Area by Central Environmental Authority (CEA) under the gazette notification No. 1634/23 on 30th December 2009 under the National Environmental Act no. 47 of 1980. The rest of the canals drains, and waterways are adhering to the special 04 gazette notice No. 1662/17 dated 14th July 2010 published by Sri Lanka Land development authority. The Sri Lanka Land Development Corporation (SLLDC) is the main stakeholder in flood control in urban areas too. The Urban Development Authority (UDA) defines the floodplains in urban areas following the giving guidelines mainly by the Irrigation Department and the Sri Lanka Land Development Corporation. Flood risk management is currently being operationalized under the direction of technical experts. Technical experts from flood risk management authorities reserve a stronger position in the decision-making process.

4 Development of the Framework for Operationalization of Flood Risk Management

It has been reviewed the current flood risk management practices in Panadura using the developed framework [2–4, 14].

4.1 Conceptualization of Flood Risk in the Planning Process

There are many causative factors for the gradual increase in flood risk in Panadaura urban area. The main causative factors for flood occurrence in the downstream Bolgoda basin is the lack of proper stormwater drainage facilities and obstructed natural drainage system. Natural drainage patterns of the Bolgoda basin have been changed significantly due to rapid land-use transformation due to the infrastructural development in recent years. Reclamation of wetlands for urban development activities is a common practice of the local communities due to the higher demand of the lands, particularly for housing. After the 1980s, many of the low lying paddy lands and marshy lands of this region were converted to commercial, residential, or industrial uses. Unauthorized construction encroaching on water bodies have hence reduced the capacity of waterways. This unplanned land use transformation has led to an increase in flood hazard as well as flood vulnerability in these areas. Figure 3 illustrates the land use transformation in the catchment area of Karapen Ela during the last years. Ill management of land use transformation, has led inadequate hydrological connections between the primary, secondary and tertiary canals and the floodplains. In addition, the majority of water canals in the area are blocked with municipal solid waste during the monsoon periods due to the non-availability of proper solid waste management systems.

Identification and estimation of flood risk is the crucial element for the decision making process of flood risk management. Hence, a decision making process is necessary to appraise the probability of flood occurrence and the impact of flooding

Fig. 3 Land use transformation in the catchment area of the Karepen ela during 1994, 2004 and 2020, respectively

and then understand the causative factors to increase flood risk. Flood damage assessments are used to estimate the impacts associated with floods that might occur in the future. Flood damage assessment provides information for decision making to flood risk managers. At present, the occurrence of flood hazards is considered to arise from a probability distribution pattern. Flood frequency models considers the occurrence of extreme events to arise from a serially independent time series with a probability distribution pattern. The hydrological parameters of flood events are considered fixed over the design life of a project. In the current practice, Flood management structural measures are designed to control only a certain, predefined design flood. Climate change has eliminated assumptions of the past, relatively stationary hazard occurrence to a non-stationary condition. Therefore, a new dimension has been added to the uncertainty associated with flood hazard predictions. Furthermore, the worsening of floods due to climate change will bring an additional burden to inland coastal communities.

Present, flood risk assessment and flood risk management, are still the territories of hydrologists. Hydrologists design the capacity of the flood control structural measures considering the historical rainfall intensities and the peak runoff analysis of the catchment. In the current practice, flood risk is conceptualized mainly based on climatological and hydrological inputs. Alternative flood risk management measures are evaluated based on the Expected Annual Damage (EAD) assessment. Designing capacities of flood management measures for the mainstreams very often consider resisting a relatively broader probable range of climatic change in the 50-year recurrence storms. But designing the capacities for urban drainages with small catchments consider resisting with a shorter range of climatic scenario of 5–10 year recurrence storms. The estimation of flood damages also depends on several assumptions. For example, the hydrological parameters are also considered homogeneously distributed over the basin.

The components of the flood risk are subjected to change over the period of time due to multiple reasons which are influenced at different spatial scales such as local, regional, and global. For example, the population growth has increased in Panadura urban area in recent decades mainly due to the proximity to the city of Colombo. Floodplain land use development can be changed in the future due to many factors related to the area such as the level of urban development in Panadura. Therefore, it is a challenging task to predict the socio-economic conditions and associated land-use changes over the catchment and the floodplains in the future. Flood risk estimations based on the hydrologic models are usually associated with significant uncertainty [4, 5]. Decision making is based on the probability models which are used to simulate flood flows on floodplains for few climatic scenarios rather than the multiple scenarios of climatic change. Also, it does not consider the worst-case scenarios of climate change and urban development. The alteration of water flow patterns due to future land-use transformation is quite difficult to predict in decision making. The inhabitants of the floodplains have complex reasons for living with this flood risk, such as land scarcity, employment, etc. Flood risk is a product of complex interactions among hydrology, meteorology, and socio-economy of a region which are linked among themselves in a nonlinear manner with many feedback processes.

The stakeholders' interpretation of the complex independencies between urban landuse development and hydrological systems have not been considered in the existing decision-making process. The current planning process pays a significantly less attention to the modification of hydrological flow patterns and flood risk due to reactive and proactive human responses such as land reclamation, rainwater harvesting, etc. For example, urban development along the floodplains is a vital component of flood exposure, and changes made in this component can lead to variations in flood risk. Illegal developments in canal reservations cannot be controlled by the existing land use regulations and have caused difficulties in canal maintenance. Accordingly, the uncertainty incorporated with social feedback is also not considered in the current flood risk assessment. Therefore, uncertainty poses a significant challenge to the implementation of flood risk management practices and the achievement of the effectiveness of flood risk management measures. However, currently, there is a high expectation of structural flood risk management measures by the public community. Moreover, the narrow framing of flood risk could not recognize the complex interdependencies of the hydrological and social system and the spatial and temporal dynamics of flood risk. As a result, the planning process could not able to recognize the role of each stakeholder and could not able to define the common objectives of the flood risk management system. Furthermore, there is an ill coordination between relevant agencies such as local authority, Urban Development Authority, Sri Lanka Land Development Cooperation, etc.

4.2 Flood Risk Assessment

The current decision-making process is characterized by technically competent and economically concerned factors. Current flood risk management and land use planning practices are mainly aimed to maximize socio-economic development from the flood-prone areas. Further, huge investments are made in structural flood control measures. As explained above, the alternative flood risk management measures, design capacities, etc. are selected mainly based on economic evaluation. The evaluation process compares costs of management measures (cost of investment, maintenance, and management) for increasing the safety against flood events with the decrease in expected flood damage cost. In general, reduction of flood damage, improvement of lands of the floodplain etc. are considered as the benefits of flood risk management. However, predetermined safety factors would not be sufficient to protect land uses along with the floodplains such as low-income settlements, particularly in areas where the rapid urban development has taken place. Protection of low-income settlement are also not considered in the selection of the safety levels of existing flood management measures.

Flood control structural measures encourages urban development in flood hazard-prone areas. In addition, the recovery and relief mechanisms might also provide a disincentive for community preparedness for extreme flood events. Further, current cost-benefit analysis considers that the cost and the benefits of flood risk management measures are distributed evenly among all the stakeholders. Hence, the existing evaluation process does not help to address the social injustice of the local communities. However, the fundamental principle of cost-benefit analysis requires to increase social wellbeing rather than economic wellbeing [5]. On the other hand, increasing safety levels of the structural measures will bring negative impacts to the local fishermen communities, farmers, as well as the residents.

Implementation of structural flood control measures adversely impacts the ecosystem-dependent livelihoods and economy. Flood control structural measures would not improve the living conditions, housing quality, environmental quality, and economy of the local community. For example, the objective of the Karapen ela development project is mainly to discharge the rainwater to the adjoining waterways as soon as possible during the monsoon seasons. There is no maintenance of the Karepen ela canal during the dry periods. Natural waterways and floodplains provide several benefits to society. For example, floodplains provide many goods (agriculture, fish) and services (flood control, water quality regulation, and food chain support) to society. But appreciation of these functions in deciding the common objectives of the flood risk management system depends on the perceptions of the interdependencies of the hydrological system, land uses by decision-makers and the local community.

A fundamental factor of any flood risk management system is to smoothen the function of natural water flow patterns considering the regional and local hydrological connectivity. Bolgoda Basin is characterized by difficulties in natural draining of stormwater. This situation is further challenged by climate change and the reduction of water retention areas due to the rapid urban development of this region recently. Accordingly, the flood control structural measures create new environmental risks such as flooding, pollution, and biodiversity decrease, reduction of fishery activities, and increase of water pollution, waterlogging and abandoning of agricultural activities. For example, currently, many of the paddy lands are abandoned downstream of the Bolgoda basin. The evaluation of flood management measures requires the consideration of the possible outcomes in the long term of the projects rather than an assessment of only economic efficiency. Flood risk changes spatially and temporally due to increased urban development over the floodplains and therefore, the planning process is necessary to focus on manage flood risk rather than control flood events.

Evaluation of flood risk management alternative measures in the decision-making process is required to consider gaining of social benefits in the long run. The importance of estimating the multiple impacts considering the whole systems approach making has been highlighted by decision-makers recently [6, 7].

It is not possible to completely provide protection for the lands from flood occurrence. The current practice of flood risk management is more localized, reactive, and isolated rather than being proactive and hydrologically integrated. Unfortunately, at present, reclamation is conducted legally as well as illegally throughout the basin due to the absence of a common framework for land development. At present, the main flood risk management policies keep on stressing the importance of the necessity of both non-structural and structural options to deal with uncertainties. If the flood risk management system mainly focuses only on channel improvements it requires a huge cost. Further, the impacts of operationalization of structural flood management measures do not bring more social and environmental benefits. Therefore, it is important to maintain the existing marshes, lowlands, floodplains, and paddy lands as stormwater retention areas. It is necessary to consider entire hydrological interactions and floodplains as components of a single ecosystem. Reconnecting the retention areas of floodplains will be economically feasible to face the stronger storms in the future. Therefore, the planning process is required to pay attention to explore the potentials to conserve the hydrological connectivity of the Bolgoda river basin. Floodplains are required to be allocated for multi-functional land uses considering the trade-off of diverse objectives of stakeholders such as urban development, flood risk management, etc. Accordingly, the objective of the planning process is to explore the potential ways to improve the ecological stability of the floodplains while maintaining ecological functions and supporting urban development. The objectives of Flood risk management are required to be broad and mainly aimed to reduce potential negative impacts on society, economy, and ecology.

4.3 Flood Risk Management Policies

Operationalization of flood risk management is required a common framework to strategically manage the complexity and the unpredictability of flood risk. Moreover, floodplains are required to be adjusted by the objectives of diverse stakeholders. Flood risk management is a collaborative decision-making process that includes multiple stakeholders representing many different sectors of development. Therefore, one of the key roles of spatial planning is coordinating different stakeholders, interpreting policy directions, and monitoring and implementation decisions to balance the natural landscape. It is necessary to periodically monitor the achievement of the objectives. Legalize Local community participation and negotiation and consensus-building among stakeholders encourage the practising of preparedness strategies.

Disaster management Act, no. 13 of 2005 is demanding for comprehensive risk-based flood management framework. Disaster management act aims to transform flood risk management practices away from emergency response toward preparedness. The National Disaster Management Coordination Committee was established in the year 2007 for the improvement of the Hyogo framework. Furthermore, in the year of 2010, the National Disaster Management policy was established to perform as the major governing mechanism for the management of disasters. It includes important concepts such as the participation of local-level authorities, supporting multi-stakeholder approaches and determining standards for response and relief. Sendai Framework for Disaster Risk Reduction 2015–2030 has also highlighted the importance of multilevel governance for flood risk management. The Sendai Framework stresses the importance of stakeholder participation and sharing of local knowledge systems in the planning process. Local community involvement in decision making is required for effective risk reduction or preparedness activities. Following the Sendai framework, the government of Sri Lanka has begun developing the National Disaster

Risk Management Plan for 2018–2030. This new plan has highlighted the participation of stakeholders representing various sectors related to disaster as well as development. In addition, the Urban Development Authority Act No. 41 of 1978 and amendments have provided a flat form for operationalization of integrated planning and implementation of economic, social, and physical development in declared urban areas.

However, the Flood Ordinance No. 22 of 1955 does not highlight the practice of any non-structural flood risk management measures and does not provide incentives for strategies to improve preparedness. Both local participation and adaptive capacities of local communities are also not recognized in the current planning process. The current practice of flood management is predominantly centred on controlling floods through engineering structural measures, mainly channelling the floodwater. At present, state departments and agencies often act independently in the decision-making process of flood risk management. Therefore, it can be seen that the flood risk management system is mainly governed by a centralized system. The existing standalone approaches in flood management and land use planning reduces the effectiveness of investments in flood risk management. However, the involvement of multiple stakeholders in the decision-making process raises many questions. These include the identification of legitimate stakeholders, their roles, etc. moreover, legitimacy becomes increasingly complex in the context of current flood risk management, as the scope is broad and diffuse. Further, land use planning is not recognized as a tool for flood risk management. The Sendai framework is not yet set into operationalization in the Sri Lankan context.

4.4 Participatory Tools for Land and Water Integration

In the current practice, spatial planning and flood risk management are functioning as two separate domains. Spatial planning is not provided with a common framework for all stakeholders to incorporate the flood risk into urban development. Moreover, there is no guideline to urban land use transformation managing the long term impacts on the water resources at regional and the local scales. As a result, water-related issues such as flooding is not among the key priorities of spatial planning practices. Further, the responsibility sharing, power-sharing, resource sharing related to Disaster Risk Reduction (DRR) is not clear even at the national level. There are no interactive mechanism, tools, etc. for spatial integration among stakeholders to increase socio-ecological functions of regions, and local areas.

5 Conclusion

Conceptualization and assessment of flood risk and governance mechanics and the discourse of the stakeholders are the main criteria that can be used to review the existing flood risk management practices. Flood risk assessment is more grounded on quantifying the likelihood of flooding. The safety levels of the flood management structural measures are determined based on the linear rational models. The case study has revealed that the existing practices are more focused on flood management rather than flood risk management. Flood risk is increasing in Panadura and suburb areas during recent years, mainly due to the absence of integrated decision making. However, many of the stakeholders including the local communities have not yet recognized the importance of risk-based flood management. Hence, the still sectoral approach is dominated by the decision-making process. Therefore, the objectives of present flood management policies and the practices are not integrated with other connected development activities such as urban development, agriculture, etc. The case study has highlighted that spatial planning does not play a significant role in flood risk management in the current practice. Increasing flood risk indicates the gap between flood risk management and spatial planning.

Acknowledgements Authors would like to acknowledge support by the Senate Research Committee (SRC) Conference & Publishing Support Grant University of Moratuwa, Sri Lanka.

References

1. Alahacoon N, Matheswaran K, Pani P, Amarnath G (2018) A decadal historical satellite data and rainfall trend analysis (2001–2016) for flood hazard mapping in Sri Lanka. Remote Sens 10(3):448
2. Herath HMM, Wijesekera NTS (2020) Development of a conceptual framework to operationalize the flood risk management. IOP Conf Ser: Earth Environ Sci 563(1):012020. IOP Publishing
3. Herath HMM, Wijesekera NTS (2020) Transformation of flood risk management with evolutionary resilience. E3S Web Conf 158:06005. EDP Sciences
4. Herath HMM, Wijesekera NTS (2021) A methodological framework for operationalization of flood risk management. IOP Conf Ser Earth Environ Sci 691:012018
5. Jonkman SN, Brinkhuis-Jak M, Kok M (2004) Cost benefit analysis and flood damage mitigation in the Netherlands. Heron 49(1)
6. Thaler T, Nordbeck R, Löschner L, Seher W, Thaler T, Nordbeck R, Löschner (2020) Coopßeration in flood risk management: understanding the role of strategic planning in two Austrian policy instruments. Environ Sci Policy 114:170–177
7. van Herk S, Rijke J, Zevenbergen C, Ashley R (2015) Understanding the transition to integrated flood risk management in the Netherlands. Environ Innov Soc Transit 15:84–100
8. Koutsoyiannis D, Montanari A (2007) Statistical analysis of hydroclimatic time series: Uncertainty and insights. Water Resour Res 43(5)
9. Apel H, Thieken AH, Merz B, Blöschl G (2004) Flood risk assessment and associated uncertainty. Nat Hazard 4(2):295–308
10. Kundzewicz ZW (2002) Non-structural flood protection and sustainability. Water Int 27(1):3–13
11. Jongman B, Ward PJ, Aerts JC (2012) Global exposure to river and coastal flooding: long term trends and changes. Glob Environ Chang 22(4):823–835
12. Gunderson L (2010) Ecological and human community resilience in response to natural disasters. Ecol Soc 15(2)

13. Aerts JCJH, Botzen WJ, Clarke KC, Cutter SL, Hall JW, Merz B, Michel-Kerjan E, Mysiak J, Surminski S, Kunreuther H (2018) Integrating human behaviour dynamics into flood disaster risk assessment. Nat Clim Change 8(3):193–199
14. Herath HMM, Wijesekera NTS (2021) A methodological framework for operationalization of flood risk management. IOP Conf Ser: Earth Environ Sci 691(1):012018. IOP Publishing

Project Construction and Management

VPL Prototypes for Determining Disassembly Parameters of BIM Models

B. Sanchez, C. Rausch, C. Haas, and T. Hartmann

Abstract Reuse of building components has the potential of maximizing the residual utility for the building materials in the End-of-Life (EoL) stage of a construction project. In this matter selective disassembly is an efficient method to retrieve reusable components from building assets. Unfortunately, the implementation of selective disassembly planning for buildings is still limited due to the challenge of generating accurate and correct analytical disassembly models for building archetypes. Therefore, the aim of this paper is to develop algorithms for the automatic determination of the necessary information for disassembly building models. The approach presented in this study uses BIM as the main platform and interface to implement the computational model. First, the necessary information for building disassembly models is investigated according to the Sequential Disassembly Planning for Buildings (SDPB) approach and the BIM parameters for disassembly models are suggested. Then, two algorithms are developed and tested in the Visual Programming Language (VPL) Dynamo to automatically populate the missing information in the BIM model. Finally, the results of the enriched BIM model are verified according to the analytical solution for disassembly models for buildings.

Keywords Disassembly model · Building information modeling · Disassembly planning

B. Sanchez (✉)
Escuela de Ingenieria y Ciencias, Tecnologico de Monterrey, Via Atlixcayotl 2301, 72453 Puebla, México
e-mail: benjamin.sanchez@tec.mx

C. Rausch · C. Haas
Civil and Environmental Engineering Department, University of Waterloo, 200 University Ave. W, Waterloo, Ontario N2L 3G1, Canada

T. Hartmann
Department of Civil and Building Systems, Technical University of Berlin, Gustav-Meyer-Allee 25, 13156 Berlin, Germany

T. Kang and Y. Lee (eds.), *Proceedings of 2021 4th International Conference on Civil Engineering and Architecture*, Lecture Notes in Civil Engineering 201,
https://doi.org/10.1007/978-981-16-6932-3_44

1 Introduction

Reuse of building components has become a very important topic because year by year an immense amount of building materials ends up as trash in landfills [1]. To overcome this problematic the construction industry has implemented technological advancements such as Building Information Modeling (BIM), material passports, and Life Cycle Assessment (LCA) [2], in order to decrease the amount of waste production through reduce, reuse, and recycle. However, the implementation for product recovery management is still scarce due to the lack of research about reclamation protocols for reusable materials [3]. In this matter, selective disassembly planning is an advantageous strategy for the recovery of building components and materials for their future reuse or recycle [4, 5]. Selective disassembly planning has the objective of optimizing the dismantling process of products compounded by different components and connections [6].

Unfortunately, selective disassembly planning for buildings is still underdeveloped in comparison to other industries such as manufacturing, automobile, and electronic [4, 6]. According to Rausch et al. [5] disassembly planning for buildings is still underdeveloped because its implementation depends on the preprocessing of a considerable amount of data for a given disassembly model. According to previous investigations a major challenge is the manual arbitration of key topological interrelations between building parts. These limitations make the process inefficient, time consuming, and prone to error [5].

2 VPL Prototypes for Determining Disassembly Parameters

BIM is arguably the most important technology used in the construction industry for planning and monitoring building processes along all the life cycle of a building project [7]. BIM is a highly organized 3D model-based graphical interface for the efficient planning, designing, constructing, and management of buildings assets. In the las three decades, a plenty of investigations have explored the benefits of implementing BIM in different life cycle stages of building projects, however its implementation for the EoL stage is still underdeveloped [4, 6]. The application of VPL design tools in BIM environments, such as Dynamo, have increased the scope for developing assessments in a highly interactive and user-friendly environment. VPLs offers a descriptive workspace for defining, arranging, and correlating computational functions [8]. To the knowledge of the authors this is the first study that uses a VPL for the determination of BIM parameters for disassembly models.

In previous studies, the authors presented, developed, and demonstrated a novel selective disassembly planning for buildings named SDPB [4, 6]. The SDPB approach optimizes disassembly sequences for retrieving components from building disassembly models. This approach is based on the disassembly graph theory, where a disassembly model of a product must be integrated as an analytical model, or Disassembly Graph (DG) model, to perform the computational analysis. The DG model is a set of five constraint matrices (*CC*, *MC*, *PC*, *HC*, and *LC*). The constraint matrices contain the information of the local physical restrictions (components c_n and fasteners f_n) for components extraction of a given disassembly model (see [4, 6]). Without an automated way for populating the information in the DG model, all the physical restrictions must be determined by visual inspection of the disassembly model. As it was discussed before, this process is prone to error due to the high amount of information to retrieve and for physical human impossibilities for visualizing all the parts of the assembly. Therefore, it is necessary to implement a systematic way to determine the missing parameters for disassembly models.

Based on the case studies presented in previous studies [4, 6], as well as literature review about disassembly graphs theory [9, 10, 11], five information parameters for a complete disassembly model for buildings were identified. The disassembly parameters are: (1) *Disassembly part type*, (2) *Disassembly ID*, (3) *Connected components*, (4) *Connected fasteners*, and (5) *Motion constraints*. According to the DG theory, any element that is meant to be part of the same disassembly model must be categorized as a component, or a fastener. This attribute is defined as *Disassembly part type* in this study. The *Disassembly ID* parameter refers to the unique identifier for components c_n and fasteners f_n inside the same disassembly model. The parameters *Connected components* and *Connected fasteners* registers the building elements membership and interrelationship between building parts. These parameters are explained in more detail in the case study as well as the *Motion constraints* attribute.

3 Case Study—BIM Elements' Parameters for Selective Disassembly

The two VPL Dynamo prototypes are presented with a functional demonstration in the BIM software Autodesk Revit. The building assembly archetype used in these examples comes from a previous study [6]. Figure 1 shows the BIM disassembly model.

3.1 Example 1 - Disassembly Part Type, Disassembly ID, Connected Components, and Connected Fasteners

Figure 2 shows a group of four shared parameters for any BIM element of the disassembly model. The proposed shared parameters are *Part type*, *Disassembly part ID*, *Hosted components*, and *Connection hosted*. This panel is the user interface display, where the designer can interact with the Dynamo prototype and the BIM model. The Dynamo graph consists of six main parts where each part has groups of nodes

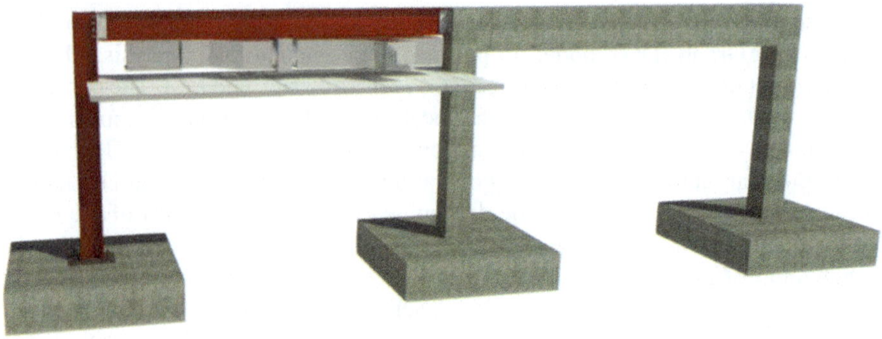

Fig. 1 BIM disassembly model

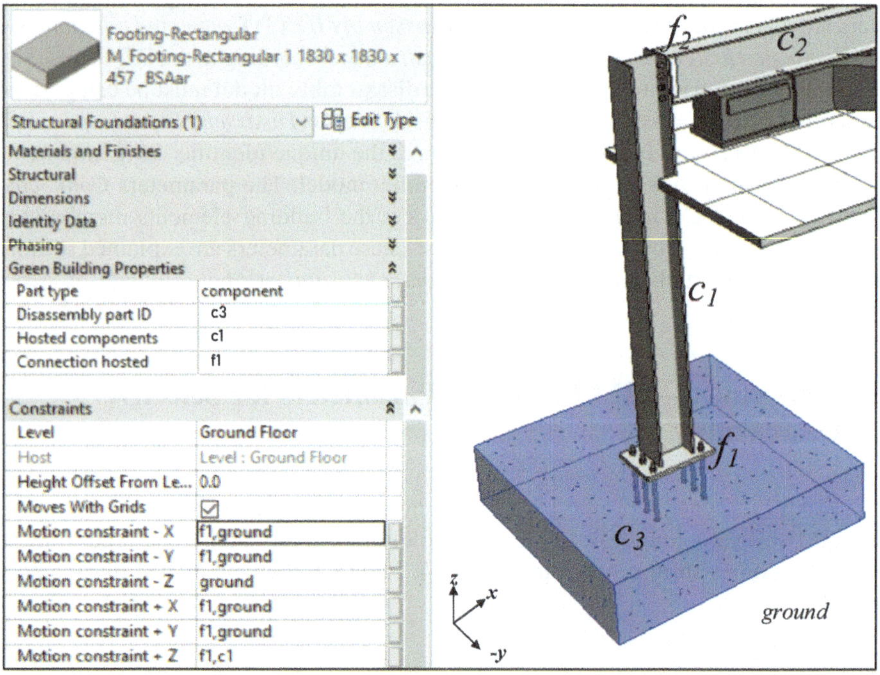

Fig. 2 Shared parameter values for selective disassembly

to perform the required task. The Dynamo prototype's six main parts are described as follows. Part 1, "select all model elements" contains only one automated node. Part 2, "sorting elements" consists of two groups of nodes. The first group provides a detailed identification of the categories of the BIM elements and the second one sorts the BIM elements by part type. A BIM element of the disassembly model, or BIM part, can be a component or a fastener. The final output is a list with the BIM elements identified as components and another list for fasteners. Part 3, "disassembly ID for components" consists of two groups of nodes used to establish the disassembly ID for components. The first group retrieves the bi-univocal BIM element ID for each component. Then, a unique disassembly part ID name for each BIM element is associated (e.g., $c_{\#}$). The second group of nodes inserts the disassembly components ID's into BIM project information. Part 4, "Disassembly ID for fasteners" is homologue the previous part 3 but applied for fasteners. In Part 5 and 6, "Selecting hosted fasteners and components" the information related to the hosted components and fasteners is retrieved directly from the BIM model.

For the disassembly model displayed in Fig. 1, the Dynamo prototype found 19 BIM elements, 10 components, and 9 fasteners. Figure 2 shows the completed parameters for selective disassembly of the BIM element c_1. The component c_1 is the host component of component c_2 (hosted component—steel beam) through fastener f_2 (hosted fastener—double angle connection).

3.2 Example 2 – Motion Constraints

Figure 2 shows a group of six shared parameters for any BIM element of the disassembly model which are added to the constraint properties. The shared parameters are the motion constraints in the generic six principal extraction directions: *Motion constraint* $+x$, *Motion constraint* $-x$, *Motion constraint* $+y$, *Motion constraint* $-y$, *Motion constraint* $+z$, and *Motion constraint* $-z$. This information does not exist in BIM models, however, it can be automatically inferred and extracted from the disassembly model. The six principal extraction directions are fundamental information to be able to estimate disassembly sequences for targeted components to retrieve from a disassembly model.

The selective-disassembly-parameters Dynamo prototype for example two is displayed in Fig. 3. The Dynamo graph consists of five main parts. To determine the motion constraints, the Dynamo prototype simulates the virtual projection of a selected model element in each of the principal extraction directions and inside of a working space of 1.5 m [6]. Then, the Dynamo prototype records all the constraining elements from the host model that intersects with the virtual projection of the selected model element. In the end, the motion constraints are loaded into the BIM model. The Dynamo prototype is aligned to the workflow for BIM elements interference check. The Dynamo prototype parts are described as follows.

Part 1—selecting BIM elements. This part contains a couple of nodes. In this part, the user must select the model element to analyze and the host model. According to the BIM interference check workflow, the model element to analyze is the BIM element that will be examined to look for interferences. The host model is the group of BIM elements that will be part of the physical constraints for the interference assessments. This node automatically runs the assessment element-by-element in a loop until finishing with the whole disassembly model.

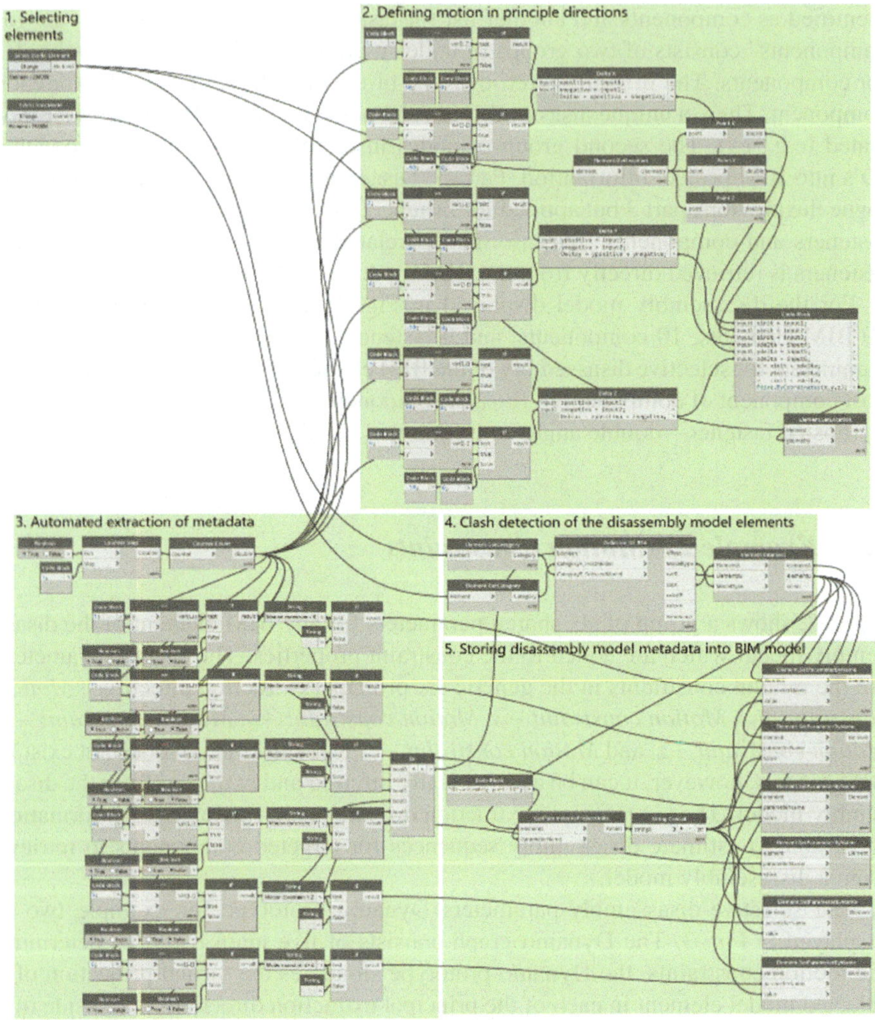

Fig. 3 Selective-disassembly-parameters Dynamo prototype full graph – example 2

Part 2—defining motion in principle directions. Part 2 of the Dynamo graph consists of three groups of nodes. The first group provides the 3D location of the model element to analyze. This location is the point of reference before the simulation of the virtual projection in each of the principal extraction directions and inside of a working space. The second group defines the parameters for the virtual projection in each principal extraction direction (+x, −x, +y, −y, +z, and −z) according to the coordinates system in the BIM environment. The parameters are incremental offsets until reaching the boundary of the working space of 1.5 m. This can be easily adjusted to fit the application. The third group of nodes sets the new location of the model element to run an interference check analysis.

Part 3—automated generation of metadata. Part 3 of the Dynamo graph consists of two groups of nodes. The first group automatically generates the periodic pulse to run the Dynamo prototype. Each pulse represents one simulation of the virtual projection of the model element to analyze. The algorithm ends when all the possible positions for each of the principal extraction directions have been evaluated. The second group automatically assigns the intersected elements to the selective-disassembly-parameters of the model element under analysis.

Part 4—clash detection of the disassembly model elements. Part 4 of the Dynamo graph consists of one group of nodes. This is a customized node from a Dynamo tool kit for clash detection between BIM elements. This node locates invalid intersections between the selected elements in the disassembly model. For example, the model element to analyze is the isolated foundation (footing) c_3 and the host model is integrated by the BIM elements ground,f_1, and c_1. The principal extraction directions under study are marked in red as well as the elements that collide with the projection of the model element to analyze c_3.

Part 5—storing disassembly model metadata into BIM model. This part consists of two groups of nodes. The first group temporarily stores the queue of intersected elements along a principal extraction direction. The intersected elements are identified according to their disassembly ID part. The second group of nodes populate shared parameter values in the BIM model for selective disassembly constraints properties. For the disassembly model displayed in Fig. 1, the Dynamo prototype found all the motion constraints for the BIM elements. Figure 2 shows the completed BIM parameters for selective disassembly of the element c_3.

4 Conclusions

The results of this study demonstrate that through this kind of automated tools it is possible to bridge the gap for the implementation of sophisticated disassembly planning techniques for building models. This study provides a comprehensive method to systematically analyse BIM models with the objective of generating metadata that is necessary for disassembly models. The prototypes presented in this study demonstrated to be efficient, affordable, and user friendly for the assessment of building subsystems with a considerable number of building parts. Future research should be done to optimize the architecture of the programming of the Dynamo prototypes as well as demonstrating their applicability for more building subsystems.

Acknowledgements Authors would like to acknowledge the financial and the technical support of Writing Lab, TecLabs, Tecnologico de Monterrey, in the production of this work.

References

1. Badi S, Murtagh N (2019) Green supply chain management in construction: a systematic literature review and future research agenda. J Clean Prod 223:312–322
2. Jalaei F (2015) Integrate building information modeling (BIM) and sustainable design at the conceptual stage of building projects
3. Mahpour A (2018) Prioritizing barriers to adopt circular economy in construction and demolition waste management. Resour Conserv Recycl 134:216–227
4. Sanchez B, Rausch C, Haas C (2019) Deconstruction programming for adaptive reuse of buildings. Autom Constr 107:102921
5. Rausch C, Sanchez B, Haas C (2019) Spatial parameterization of non-semantic CAD elements for supporting automated disassembly planning, Banff, AB, Canada, pp 108–115
6. Sanchez B, Haas C (2018) A novel selective disassembly sequence planning method for adaptive reuse of buildings. J Clean Prod 183:998–1010
7. Mattern H, König M (2018) BIM-based modeling and management of design options at early planning phases. Adv Eng Inform 38:316–329
8. Marmo R, Polverino F, Nicolella M, Tibaut A (2020) Building performance and maintenance information model based on IFC schema. Autom Constr 118:103275
9. Smith S, Hung P (2015) A novel selective parallel disassembly planning method for green design. J Eng Des 26:283–301
10. Smith S, Smith G, Chen W (2012) Disassembly sequence structure graphs: an optimal approach for multiple-target selective disassembly sequence planning. Adv Eng Inf 26:306–316
11. Smith SS, Chen W (2011) Rule-based recursive selective disassembly sequence planning for green design. Adv Eng Inform 25:77–87

Accuracy of Digital Photogrammetric Survey from Aerial Images for Architectural Documentation in a Monuments: A Case Study of Wat Sing Sam Khok, Thailand

M. L. Varodom Suksawasdi

Abstract This study is a part of research to describe how to obtain various photogrammetric products from aerial photographs using a drone. The article aims to know the accuracy tolerance in the process of using different measure instruments. The study compared each method's pros and cons for a case study of the architecture documentation in Wat Sing Samkhok Pathum Thani Province, Thailand. The first section describes the physical condition of the case study, such as location, the scope of the study area, the Character of the building, and equation concepts using Agisoft PhotoScan software.

Keywords Virtual tourism · Architecture documentation · Photogrammetry

1 Introduction

1.1 Site Existing Context

Wat Sing is an ancient temple since the late Ayutthaya period where an architectural style reflects the prosperity of local art, which is now almost difficult to find. Therefore, providing information and knowledge of architecture for conservation and virtual tourism should offer opportunities to the general public to access diversity. The threatening is caused by destroying the resource's value, which often occurs until a tourist destination is popular. (Not restrict the access of large group of people.) The concept of the Image of the City was to analyze the physical Identity of the city [1]. There are five essential elements as Edge, Path, District, Node, and Landmark in a case study. EDGE consists of the natural border of the Chao Phraya River and highway 3111 road, representing the border. PATH is consists of highway 3111 road

M. L. Varodom Suksawasdi (✉)
Faculty of Architecture and Planning, Thammasat University Rangsit Campus, Pathumthani, Thailand
e-mail: varodom@ap.tu.ac.th

Research Cluster in Livable Environment and Architectural Design Laboratory (LEAD-Lab), Pathumthani, Thailand

T. Kang and Y. Lee (eds.), *Proceedings of 2021 4th International Conference on Civil Engineering and Architecture*, Lecture Notes in Civil Engineering 201,
https://doi.org/10.1007/978-981-16-6932-3_45

will have a small alley perpendicular to the Chao Phraya River line connecting to Bangna-Bang Toey Road. It looks like a line of road grids. DISTRICT is a village with an agricultural area with factories inserted in some areas. The highlight is the old Mon community and Mon temples lined along the river. NODE is located at ING NAM Sam Khok market. There is a variety of traffic transit service systems. LANDMARK is the temple that lines along the river, probably the highlight of the area because besides being a destination, the eye destination is a guiding signal.

The historical tourism in Thailand, Focus on Thammasat University Rangsit Center, is located in Pathumthani Province, governed by the local area. In the past, that can be travel to the center of prosperity. The route must be pass along the riverside paths on both sides of the Chao Phraya River. Both of these have the potential to be seen as of importance. The study of the architecture of the riverside temple to be promoted tourism in this area. A case study will give the best benefit of looking at the overall dimensions of the culture (see Fig. 1).

However, a person cannot enter the temple inside if he genuinely thinks they will do activities within that temple. In addition, the concept of urban conservation has its own Identity in the meaning of the cultural landscape and a review of related research such as the study of cultural tourism resources of the Mon community [2, 3] and potential as a Buddhist tourist attraction of temples in Pathumthani Province and a study of the cultural tourism potential of temples in Pathumthani Province [4]. As a principle, all there. The spatial features are also intertwined. The temple itself or the abbot is more an image of the city than an architectural feature [5] on the other hand, historical data is truly unique in the community area (see Fig. 2).

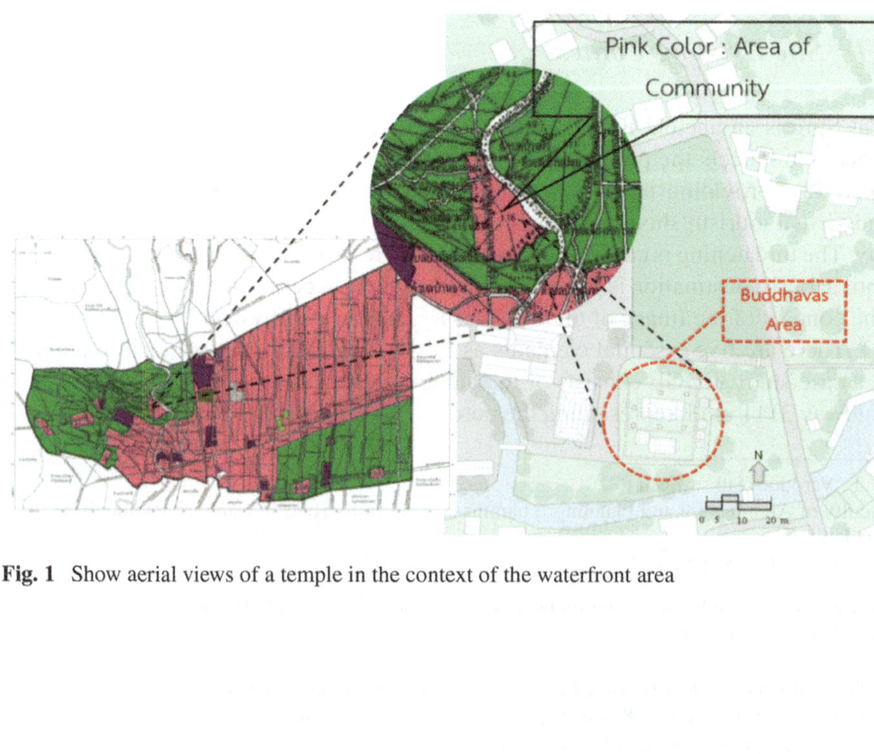

Fig. 1 Show aerial views of a temple in the context of the waterfront area

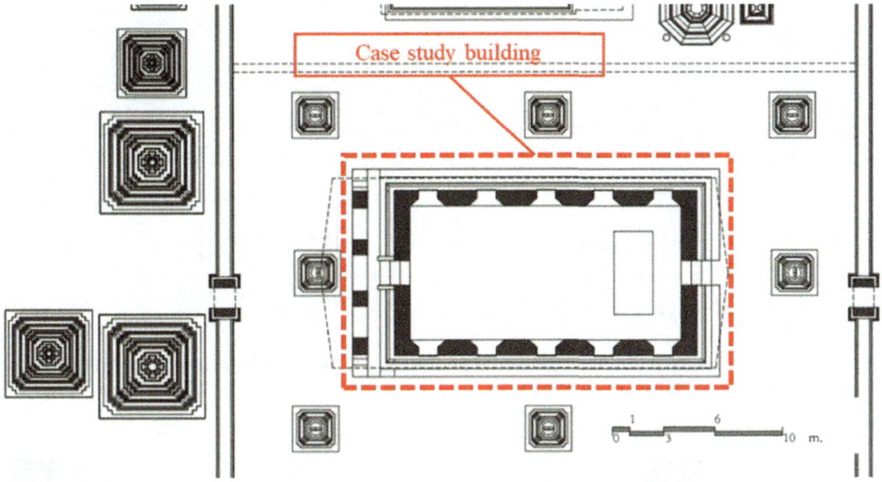

Fig. 2 Show a plan of building in the temple

2 Research Methodology

In this study, three types of measurement instruments are used to collect distance and architectural elements of Thai monuments. Firstly, a simple tape measure with a maximum length of 5 m combines three measuring scales, including meters, centimeters, and millimeters. This tool is certified by Thai measurement standards, The Weights and Measures Act, B.E. 2466 [6]; this study used four decimal places for the distance measurement.

The laser distance meter system proved to be an easy-to-use, highly versatile instrument. While initial calibration required time and several attempts before reaching acceptable calibration levels, the spatial data collection itself was fast and highly efficient, and similar to the total station. It was only less so, in bright sunlight conditions, when it was difficult to see the laser beam for targets beyond ~10 m, although laser enhancement glasses help mitigate this. Data was recorded automatically in digital format and could be viewed in real-time on a tablet was another significant advantage of this system. Exporting and loading the data to a computer was also easy and intuitive [7].

Secondly, the laser distance meter is an electronic surveying instrument that incorporates electronic distance measurement (EDM) to measure 3D locations with a high degree of accuracy and precision [8]. It has a function to calculate the area and height from the Pythagorean equation (see Fig. 3).

Finally, using the photogrammetric process products consisting of digital tools, the result is a digital file, Raster Image to Pixies file, calculating values through image dot color. Agisoft PhotoScan software is created 3D Modeling (see Fig. 4).

This Agisoft PhotoScan software uses linear vector-based equations to calculate the following concepts, Collinearity of Bundle from Image Ray Condition; This

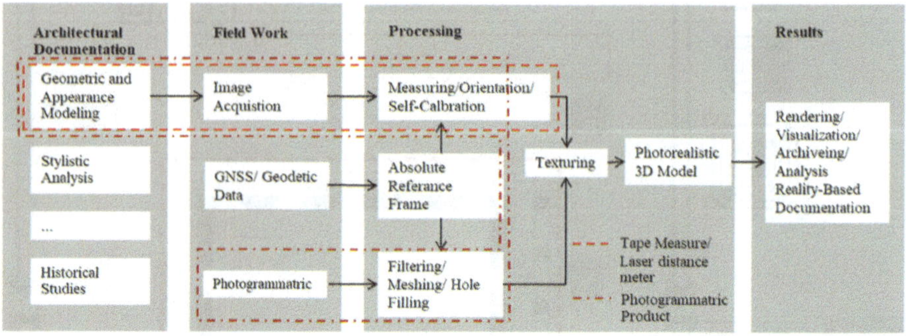

Fig. 3 Show the general pipeline to delivered photorealistic 3D models from image sensors, using Drone, Agisoft Image Software [9]

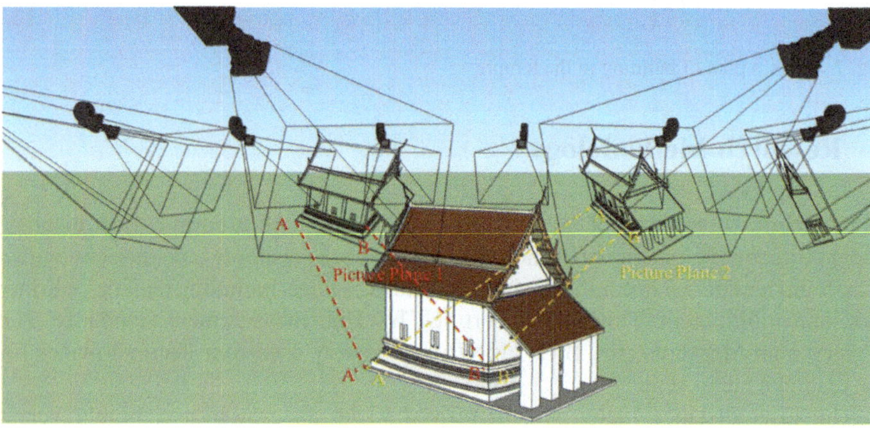

Fig. 4 Show the results that can tell the correct proportion (true shape)

makes it possible to analyze the relationship between the origin photo, the point on the subject, and the point on the photograph (see Fig. 5).

The Collinearity Condition Equations are

$$x_a = x_0 - f\left[\frac{m_{11}(x_A - x_0) + m_{12}(y_A - y_0) + m_{13}(z_A - z_0)}{m_{31}(x_A - x_0) + m_{32}(y_A - y_0) + m_{33}(z_A - z_0)}\right]$$

$$y_a = y_0 - f\left[\frac{m_{21}(x_A - x_0) + m_{22}(y_A - y_0) + m_{23}(z_A - z_0)}{m_{31}(x_A - x_0) + m_{32}(y_A - y_0) + m_{33}(z_A - z_0)}\right]$$

$M = f(\omega, \emptyset, k)$ Represents the rotation matrix, the xyz axis system is the x′y′z′ system.

The collected data from flying with DJI Mavic 2 Pro drone with a sensor 1inch CMOS, FOV Lens about 77°, maximum flight time (no wind condition) about 31 min,

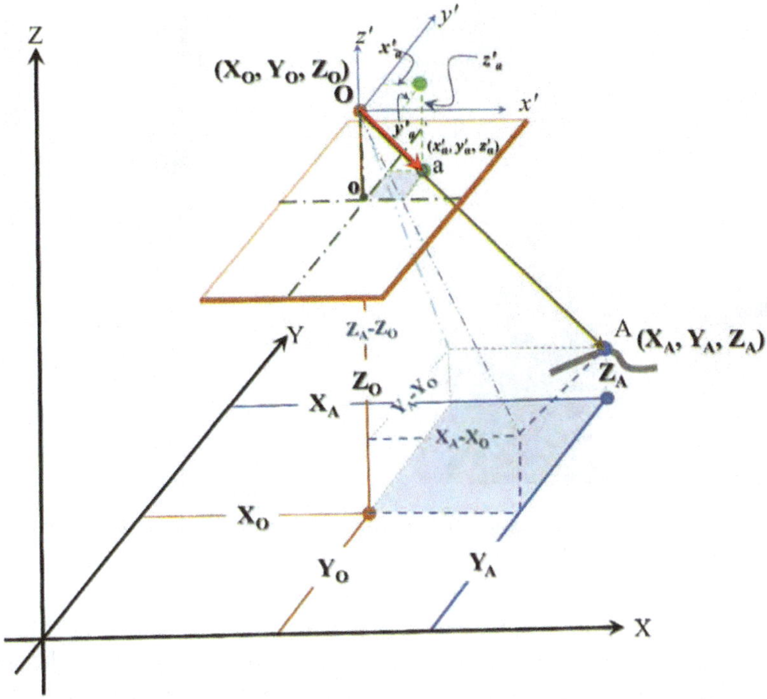

Fig. 5 Show diagram of collinearity condition equations [10]

and suitable flight procedures. By flying around the object to collect images in each of the images circling and assessments to use this system. The desired result is the point cloud formats.

3 Results

The result was compared the value of the differences between the selected architectural elements such as Roof height, Base length, Column width. A baseline is an index contour that uses a contour line start level 0 m above the reference line (see Fig. 6).

The results of data collected from different Instruments were analyzed by Root mean square error (RMSE), which is a statistical measure of the magnitude of a varying quantity. It is a generalized standard deviation and is useful when differences between subgroups or relationships between variables are looked at value. The name comes from the fact that it is the square root of the mean of the values. In order to achieve the objective of this analysis, a selection of a random sample of points was made [11]. The deviation of the data is not implicit. The calculating the coefficient

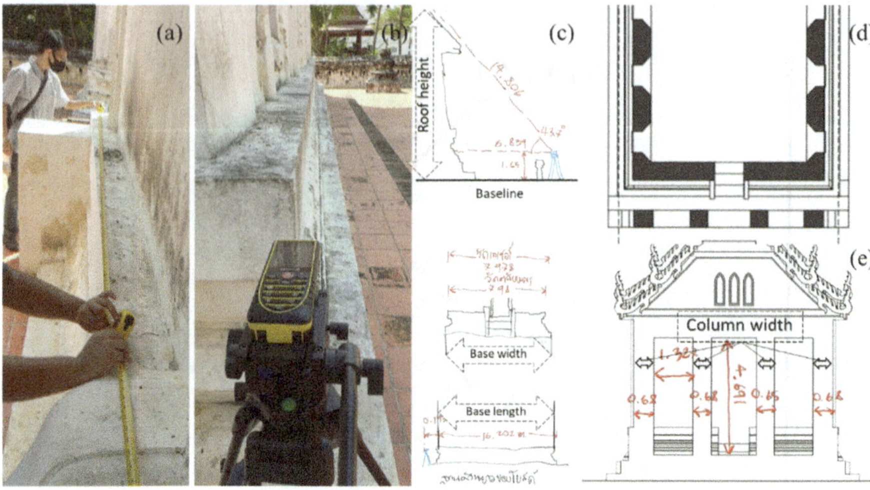

Fig. 6 Show following **a** tape measure, **b** laser distance meter, **c** the survey data, **d** ubosot plan, **e** ubosot elevation

of variation as a percentage of the roof height was a slight data deviation. Measuring height with a tape measure tool is impossible because the building that holds the data is a conservation building. To set up a frame to be used for the position to perform the measurement. It could lead to damage to the building. Therefore, In this case, the height measurement of the roof to ignored. The problem encountered in the collecting data is the Accessories component. The Tri-pod mount for the Laser distance meter camera makes it difficult to set it against the edge of the line.

The photogrammetric procedure is as follows: Take series of images obtained from a high angle shot. In which the series of images is a continuous pattern revolving around the building [12, 13], then imported into Agisoft Photo Scan software, which the image correlation program to created point cloud and create a surface mesh to create 3D Modeling (see Fig. 7).

Images were flying around the object. The program calculates a group of Point-cloud, which is processing the relationships of 45 images in order. It references both the distance, direction and color of programmatically calculated plane elements. The next step will be to select the region of Pointcloud's clusters so that the program will compute a plane to define both interconnected planes when the relationship of many interconnected planes can create 3D Modeling.

Comparison of three datasets is matching each pair of instruments and analyzing the RMSE values. Table 1 shows the comparison between Tape measure and Laser distance meter, Table 2 between Tape measure and Photogrammetry, and Table 3 between Laser distance meter and Photogrammetry. When calculating values like MAD, MSE, MAPE, RMSE should be lower, the better, assuming that the value is 0, it is proven accurate x is 100% correct. In practice, the probability of error = 0

Fig. 7 Show agisoft photo scan software and surface mess of 3D modeling

is complicated to prove. The table can lead to overfitting where the datasets are too close to each other. When analyzing, the proof of accuracy is unreliable.

4 Conclusions

This section has presented a comparative instruments approach for Architectural Documentation for accurate 3D Modeling. However, the analysis from the condition of work, size of the job, amount of time required, human resources, steps that result in errors. Reduce the hassle. The results will be the same. But the algorithm and the learning curve are not the same as the data collection process.

Proving that each tool has the best accuracy using the matching toolkit of the result data, it appears (see Table 4), RMSE is evident that the tape measure and Laser distance meter has the lowest value, so it is assumed that the tool, when used together, will have similar accuracy. However, considering that the unequal number of n resulted in the lowest mean selection between Laser distance meter and Photogrammetry in each set. Using this tool is probably the best answer.

The advantage of Tape measure tools is that they are cheap, easy to learn to use. However, the measurement accuracy is lower than the other instruments, with extremely high tool durability. In this situation, this tool ensures highly accurate data acquisition. The advantages of a Laser distance meter are moderately priced. Nowadays, they are much lower in price and also feature a wide variety of feature functions. Some functions can recognize the measured data uploaded to digital information; some require batteries that are sometimes difficult to replace. Perhaps, some types of batteries are unique that to find complicated in the market. The operation is not complicated. It can reduce the steps of manual labor. The last part is Photogrammetry High fidelity, good storage in a wide area but requires knowledge and work

Table 1 Data between the tape measure and laser distance meter comparative

Instrument	Tape measure (T_e)	Laser distance meter (L_e)	Error ($T_e - L_e$)	Absolute values of error	Square of error	Absolute value of error
	x (m)	y (m)	$x-y$	$\lvert x-y \rvert$	$(x-y)^2$	$\lvert x-y/x \rvert$
Base width	7.9800	7.9780	0.0020	0.0020	0.0000	0.0003
Base length	16.0000	16.0090	−0.0090	0.0090	0.0001	0.0006
Column width	0.6700	0.6810	−0.0110	0.0110	0.0001	0.0164
Roof height	None	11.8830	None	None	None	None
		Total	−0.0180	0.0220	0.0002	0.0172
		n	3.0000			
Mean absolute deviation		MAD	0.0073			
Mean square error		MSE	0.0001			
Root mean square error		RMSE	0.0083			
Mean absolute percentage		MAPE	0.5744			

Table 2 Data between the tape measure and photogrammetry comparative

Instrument	Tape measure (T_e)	Photogrammetry (P_e)	error (T_e-P_e)	Absolute values of error	Square of error	Absolute value of error
	x (m)	y (m)	x−y	|x−y|	$(x-y)^2$	|x−y/x|
Base width	7.9800	7.9800	0.0000	0.0000	0.0000	0.0000
Base length	16.0000	16.1200	−0.1200	0.1200	0.0144	0.0075
Column width	0.6700	0.6760	−0.0060	0.0060	0.0000	0.0090
Roof height	None	11.6500	None	None	None	None
		Total	−0.1260	0.1260	0.0144	0.0165
		n	3.0000			
Mean absolute deviation		MAD	0.0420			
Mean square error		MSE	0.0048			
Root mean square error		RMSE	0.0694			
Mean absolute percentage		MAPE	0.5485			

Table 3 Data between laser distance meter and photogrammetry comparative

Instrument	Laser distance meter (L_e)	Photogrammetry (P_e)	error (L_e-P_e)	Absolute values of error	Square of error	Absolute value of error
	x (m)	y (m)	x−y	|x−y|	$(x-y)^2$	|x−y/x|
Base width	7.9780	7.9800	−0.0020	0.0020	0.0000	0.0003
Base length	16.0090	16.1200	−0.1110	0.1110	0.0123	0.0069
Column width	0.6810	0.6760	0.0050	0.0050	0.0000	0.0073
Roof height	11.8830	11.6500	0.2330	0.2330	0.0543	0.0196
		Total	0.1250	0.3510	0.0666	0.0341
		n	4.0000			
Mean absolute deviation		MAD	0.0878			
Mean square error		MSE	0.0167			
Root mean square error		RMSE	0.1291			
Mean absolute percentage		MAPE	0.8534			

Table 4 Data various deviation between tape measure, laser distance meter, and photogrammetry comparative

		Tape measure (T_e) vs Laser distance meter (L_e)	Tape measure (T_e) versus photogrammetry (P_e)	Laser distance meter (L_e) versus photogrammetry (P_e)
Mean absolute deviation	MAD	0.0073	0.0420	0.0878
Mean square error	MSE	0.0001	0.0048	0.0167
Root mean square error	RMSE	0.0083	0.0694	0.1291
Mean absolute percentage	MAPE	0.5744	0.5485	0.8534

Table 5 Varies measure instruments comparative for architectural documentation of a monument

Varies	Tape measure	Laser distance meter	Photogrammetry
Pros	Low price	Mid price	High price
	Low learning curve	Mid performance over time and Mid effort	High performance, less time, and less effort
Cons	Low accuracy	High accuracy	High learning curve
	Durable tool	Moderate process	High price, sensitive tools
	Low performance	Need for a battery	High process

procedures that follow. There is also a price for both the tools and the software that can be used to do the work as well (see Table 5).

There are scattered parts of the archaeological site, such as laterite stone. In this situation, the area must be measured with a building. Using the Tape Measure method is almost impossible to obtain accurate data. Therefore, choosing to be accurate instruments will make data collection the fastest and most efficient. Further of this study has been helping decision between suitable several measurement instruments in architectural conservation.

Acknowledgements The authors gratefully acknowledge the financial support provided by Thammasat University Research Fund under the TU research Scholar, Contract No. 125/2561, and this research has been approved by a research ethics committee of Thammasat University.

References

1. Lynch K (1960) The image of the city. MIT Press, Cambridge
2. Kitcharoenpaisal P (2010) The study of mon community cultural tourism resources to develop ecotourism in Pathumthani Province. Ecotourism Planning and Management. Bangkok, Srinakharinwirot University. Master of Science
3. Wisedsind N et al (2017) The potential of buddhist tourismdestination:temples in pathumthani province. J Graduate Stud Valaya Alongkron Rajabhat Univ 11(3):109
4. Suvannin W et al (2014) A study of cultural tourism potentiality of buddhist templesin pathum thani province. Journal of graduate studies valaya alongkron rajabhat university 8(3):297
5. O'connor RA (1983) A theory of indigenous Southeast Asian urbanism. Inst Southeast Asian Stud 47(4):132
6. Central Bureau of Weights and Measures. Sara Na Ru Chang Twng wad. Retrieved from http://www.cbwmthai.org/Activity_Detail.aspx?id=34. (2013). Last accessed 2020/05/10
7. Almeida-Warren K, Braun DR, Carvalho S (2021) The DistoX2: a methodological solution to archaeological mapping in poorly accessible environments. J Archaeol Sci Rep 35:102688
8. Jalloh Y et al (2017) Conventional total station versus digital photogrammetry in land development applications. J Environ Anal Toxicol 07
9. David CC, Charles M, George G, Georgina LB, Thomas W, Caroline N (2017) UAVs in context: archaeological airborne recording in a national body of survey and record. drones2010002 2(1)
10. Waisurasingha C (2020) Photo survey Photogrammetry, Se-Education Plc
11. Charnnarong S (2012) Photogrammetry in architectural conservation: literature review and possible applications for Thai traditional architecture. NAJUA 9
12. Waisurasingha C (2013) Investigated study of close range photogrammetry for high precision measurement. Eng Appl Sci Res 31(4):377–394
13. Mouget A, Lucet G (2014) Photogrammetric archaeological survey with UAV. ISPRS Ann Photogramm Remote Sens Spatial Inf Sci II-5:258

A BIM-Based Method to Automate Material and Resources Assessment for the Green Building Index (GBI) Criteria

Taki Eddine Seghier, Chavanont Khosakitchalert, and Yaik-Wah Lim

Abstract Sustainable or green building rating systems were developed in many countries. Green Building Index (GBI) is the rating system developed in Malaysia, which consists of six main criteria. Material and Resources (MR) assessment is one of the main criteria that require the quantification of building materials to calculate the cost of materials and giving GBI points. Conventionally, the assessment process is done manually, which is a time-consuming and error-prone process. Building Information Modelling (BIM) technology gives a new possibility to extracted material quantities instantly from a model, which can support the process of MR assessment. Therefore, this research aims to develop a BIM-MR integration method that integrates the material quantities extracted from BIM with the developed templates and scripts according to the GBI to form automatic assessment results. The prototype system was developed using Autodesk Revit and Dynamo extension, and the validation is done using a case study. With this system, the MR assessment report can be done automatically, which will save time and cost to produce the assessment documents.

Keywords Building information modeling · Green building index · Material and resources assessment · Assessment · Green building rating systems

1 Introduction

The high demand for a sustainable or green building design in recent years leads to the initiation of various rating systems for sustainable building assessment in

---

T. E. Seghier
School of Architecture and Built Environment, Faculty of Engineering, Technology & Built Environment, UCSI University, Kuala Lumpur, Malaysia

C. Khosakitchalert (✉)
Department of Architecture, Faculty of Architecture, Chulalongkorn University, Bangkok, Thailand
e-mail: chavanont.k@chula.ac.th

Y.-W. Lim
Faculty of Built Environment and Surveying, Universiti Teknologi Malaysia, Johor Bahru, Malaysia

T. Kang and Y. Lee (eds.), *Proceedings of 2021 4th International Conference on Civil Engineering and Architecture*, Lecture Notes in Civil Engineering 201,
https://doi.org/10.1007/978-981-16-6932-3_46

many countries. Green Building Index (GBI) is a sustainable building rating system utilized in Malaysia [1]. It consists of six main criteria, including energy efficiency, indoor environment quality, sustainable site planning and management, materials and resources, water efficiency, and innovation.

This study focuses on the material and resources (MR) criterion. This criterion is divided into 7 sections including materials reuse and selection, recycled content materials, regional materials, sustainable timber, storage and collection of recyclables, construction waste management, and refrigerant and clean agents. The first four sections calculate the points based on the quantification of building materials. For example, if the project has reused materials over 5% of the project's total material cost, it will be given 2 points for the material reuse and selection sections [1]. Currently, the process of quantification of building materials and calculation of the points is done manually, which is a time-consuming and error-prone process.

Building Information Modeling (BIM), a technology that integrates a digital model representing a building with information [2], is an efficient tool that can be used to assist the sustainable building assessment because all information of the building can be extracted directly from the BIM model [3, 4]. Moreover, BIM has introduced a new method for quantity takeoff (MTO) called BIM-based quantity takeoff. It is proven that this method is the fastest and most reliable if the input information is complete and correct [5–8]. Therefore, BIM-based quantity takeoff can be used to automate the MR assessment process.

The objective of this research is to develop a BIM-MR integration method that automatically evaluates the quantifiable criteria under MR of a building design based on GBI criteria. The prototype system was developed using Autodesk Revit and Dynamo. The proposed method is validated using a case study. The results are then discussed, and the conclusions are summarized.

2 Literature Review

There are many sustainable building rating systems in the world, such as the Leadership in Energy and Environmental Design (LEED), Green Star, Comprehensive Assessment System for Built Environment Efficiency (CASBEE), and Thai's Rating of Energy and Environmental Sustainability (TREES) [9, 10]. GBI is one of the sustainable building rating systems developed specifically for the environmental, cultural, and social contexts in Malaysia [1]. The assessment process is similar to other rating systems. It is done by evaluating building design documents and supporting documents with the criteria and then give the points. If the total points surpass the minimum requirement, the building will get the certificate.

Many studies have attempted to integrate these sustainable building rating systems with BIM. Azhar et al. [9] proposed a conceptual framework to integrate BIM with the LEED rating process. They found that BIM can generate results very quickly compared to a manual method, but only 17 LEED credits and 2 prerequisites can be generated from BIM. Wong and Kuan [11] applied the use of parameters and schedules in Autodesk Revit with their developed template for BEAM-Plus assessment, which is a sustainable rating system in Hongkong. Barnes and Castro-Lacouture [12] developed a LEED toolbar in Autodesk Revit. Biswas et al. [13] developed an Autodesk Revit add-in that can calculate energy consumption, quantify embodied carbon, and evaluate output based on LEED requirements. Khoshdelnezamiha et al. [14] and Seghier et al. [15, 16] developed prototype systems to automatically calculate and optimize the Overall Thermal Transfer Value (OTTV) and Envelope Thermal Transfer Value (ETTV) based on GBI assessment and Green Mark criteria respectively using Autodesk Revit and Dynamo. Lim et al. [17] proposed a process of energy efficiency assessment in GBI criteria using Autodesk Revit. Ilhan and Yaman [18] developed a green building assessment tool that evaluates BIM models according to a green material database.

Recently, the usage of Visual Programming (VP) tools (e.g., Dynamo) in BIM-related research to automate different tasks has increased drastically. For example, Seghier et al. [16] automated the evaluation of the Concrete Usage Index (CUI) by developing a Dynamo script able to extract the necessary information from the BIM model and perform the required calculation according to CUI requirements. Nik-Bakht et al. [19] investigated the possibility of developing a BIM-based calculation method using Dynamo scripting to estimate the acoustical properties of buildings (Reverberation Time) with acceptable accuracy and details. Abbasi et al. [20] combined a multi-objective optimization algorithm with BIM and LCA and VP to determine the trade-off between embodied and operation energy with a focus on renewable energy use.

However, the assessment of MR of a building using BIM has not yet been proposed. In a conventional method, the cost of MR of a building must be calculated manually and then evaluate with the GBI criteria. BIM-based quantity takeoff, which is the faster and more reliable tool [5–8], can improve the assessment process by integrating the material quantities extracted from BIM with the developed templates to form automatic assessment results.

3 Proposed BIM-MR Integration Prototype System

The BIM-MR integration method is developed to automate the assessment of the MR criteria under GBI. As shown in Fig. 1, data processing in the BIM-MR integration prototype system is performed through three major stages. The first stage requires having a BIM model which is modeled with a level of development (LOD) equals to 300 and contains all material data including cost. The second stage relies on the developed Revit template and its new shared parameters. In this stage, the properties related to sustainable materials or product usage in the project are identified by the user. The final stage consists of using a developed Dynamo script to automatically extract martial data and generate MR results according to GBI requirements.

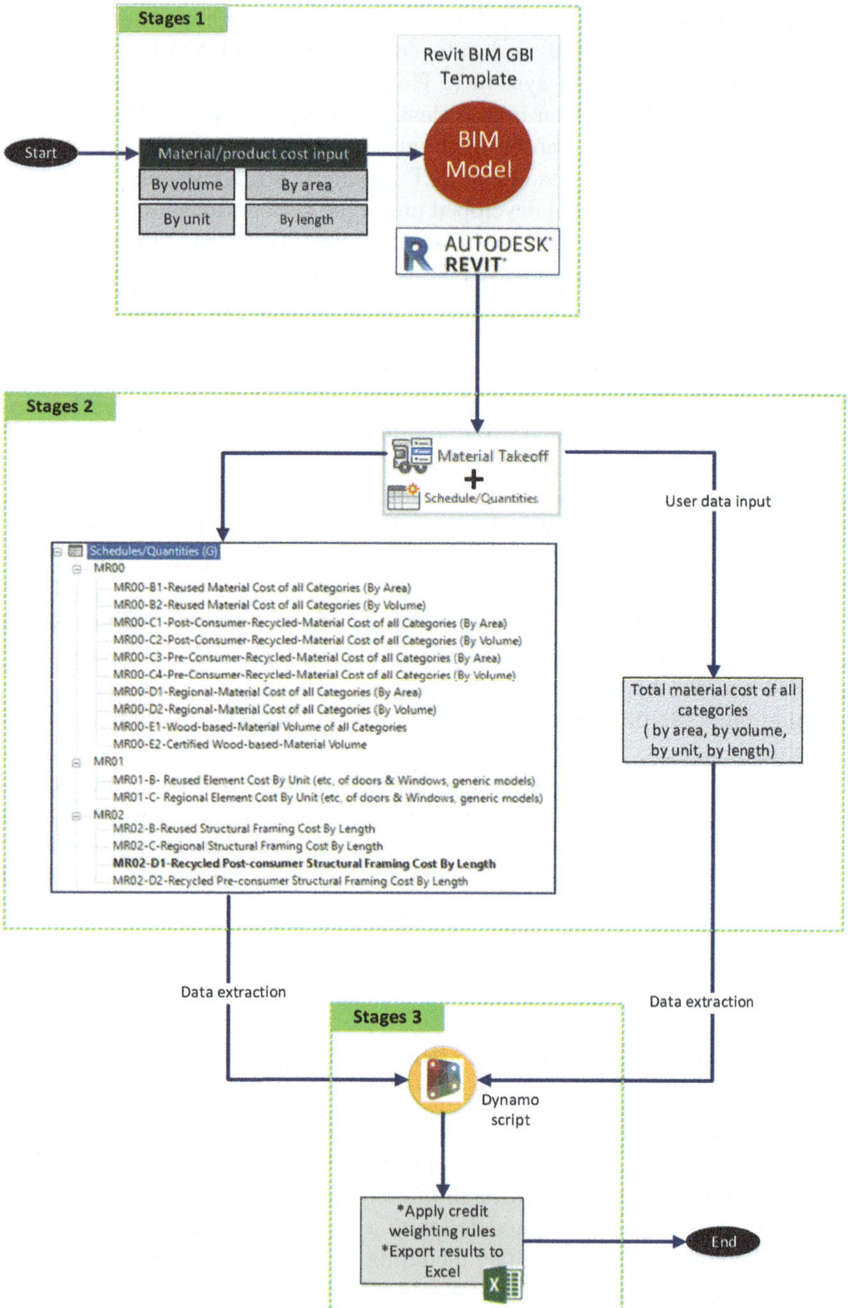

Fig. 1 Workflow of the BIM-MR integration system

3.1 BIM Model and Cost Data

The BIM model consists of a database where all the necessary data for the assessment of MR criteria can be extracted from. For this reason, the modeling accuracy of the BIM model in Revit is highly important to ensure accurate data extraction. The calculation of MR credits depends on the quantity and cost of the building materials and products in the project. Quantity parameters such as area, volume, or linear length of building components depend on the building component geometries available in the BIM model. However, cost data should be inputted in the model by the user based on estimated prices.

3.2 Revit Template

The Revit template is a part of the BIM-MR integration system. It is developed using Autodesk Revit 2021. This template contains twelve material takeoff (MTO) schedules and ten Revit schedules/quantities. These schedules are created to cover all types of material cost quantification (by volume, by area, by length, and by unit). The template also contains new shared Boolean parameters (yes/no) created to facilitate the identification of sustainable material/product contributing to the MR criteria (e.g., reused, recycled, regional, etc.)

3.3 Dynamo Scripting and Excel Template

To ensure an automatic process of data extraction under the BIM-MR integration method, a Dynamo script (see Fig. 2) has been developed using Dynamo 2.5.0.7460. The script was developed using a combination of Dynamo built-in nodes and other nodes available in downloadable packages such as Archi-lab, Rhythm, Bimorphnodes, etc. In general, the script performs the required calculation using the total cost of MTO schedules generated using Revit. Then this data is processed to verify the number of obtained credits based on GBI requirements available in the user guide. Once all calculation is performed, the script exports the summary of the assessment results to an Excel template.

As shown in Fig. 2, the script processes the extracted data from the BIM model in four major phases. Phase A consists of calculating the total material cost of all building elements categories in the BIM model while considering all the different costs (cost by area, by volume, by meter, and by unit). In this phase, the script relies on the cost calculation generated by Revit Material take-off schedules which are already available in the BIM-MR template (see Fig. 3). Next, in phase B, only the cost of specific materials (e.g., reused materials) in the project is calculated based on the material takeoff schedules available in the template. Phase C focuses on

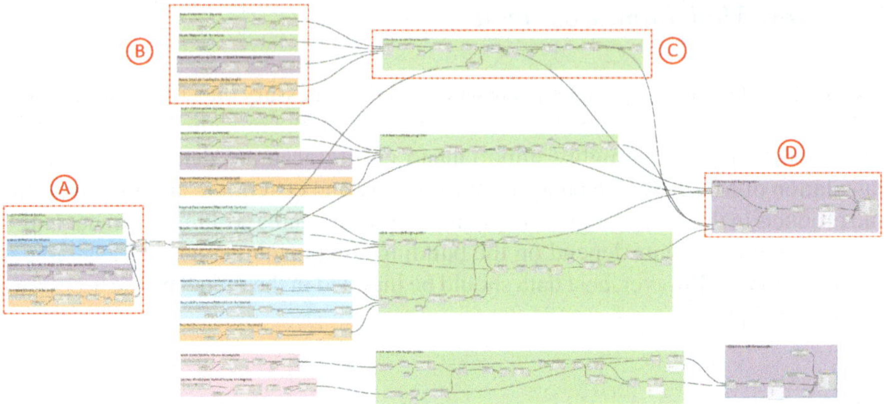

Fig. 2 Overview of developed dynamo script for BIM-MR assessment method

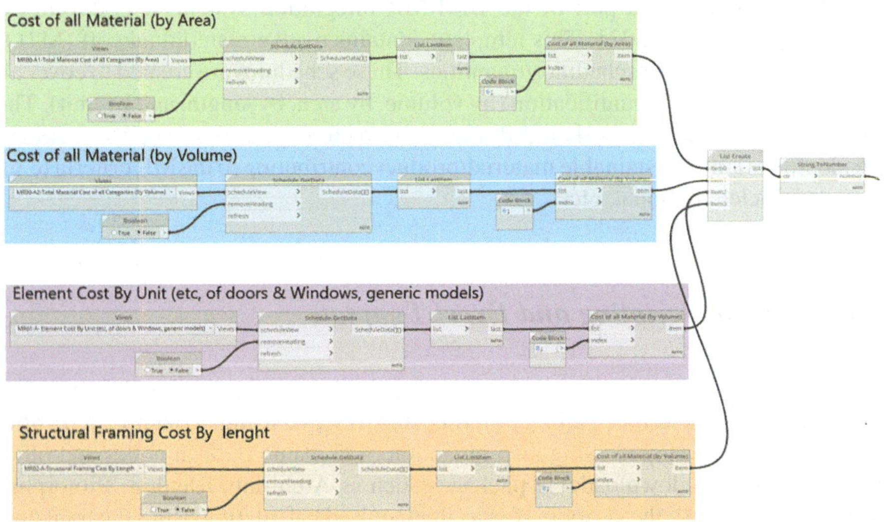

Fig. 3 Phase A of the developed dynamo script for the BIM-MR assessment method

managing the cost data calculated in the previous phases (A and B). In this phase, the script performs the necessary calculation and allocates the achieved credits based on GBI requirements. Finally, in phase D, all the calculation results are exported to the BIMM-MR Excel template for visualization purposes.

Case study building	Component	Material and resources (MR) criteria			
		MR1	MR2	MR3	MR4
	Walls			O	
	Floor		O	O	
	Ceiling				
	Doors				
	Windows	O			
	Structural Columns		O		
	Structural framing		O		
	Roof	O	O	O	
O: The building components contain materials/products contributing to a specific MR criterion under GBI					

Fig. 4 Case study building and applied scenario for MR criteria under GBI

4 Case Study

The BIM-MR integration system is tested on a case study (LOD 300) which consists of a typical office building with four levels, four facades, and 3,473 m^2 of gross floor area (see Fig. 4). Material layers and cost data of each component are inputted based on the Malaysian current practice. For testing purposes, one scenario regarding material usage is applied to the case study. As shown in Fig. 3, windows and roof elements contain reused materials/products (MR1). Whereas, floor, structure components, and roof elements contain recycled materials (MR2). Floors, walls, and roof elements also contain regional materials (MR3). However, none of the building materials or products contain sustainable wood (MR4).

5 Results and Discussion

This section discusses the obtained results of the BIM-MR integration system following its application to the case study. Based on the applied scenario (see Fig. 4), the case study acquired a total of 3 credits out of 6 for MR criteria under GBI. Two out of two credits for reused materials (MR1), one out of two credit for recycled materials (MR2), zero out of one credit for both regional material (MR3) and sustainable timber (MR4) (see Fig. 6). The total cost of materials/products quantified based on area and volume are 1,103,925.01 RM and 234,598.68 RM respectively (see Fig. 5).

Regional material (MR3) and sustainable wood (MR4) criteria obtained 0 credit. This occurred because the applied scenario did not achieve the minimum required quantity of regional materials in the project on one hand, and due to the unavailability of wood-based materials/products on the other hand. The Dynamo script took

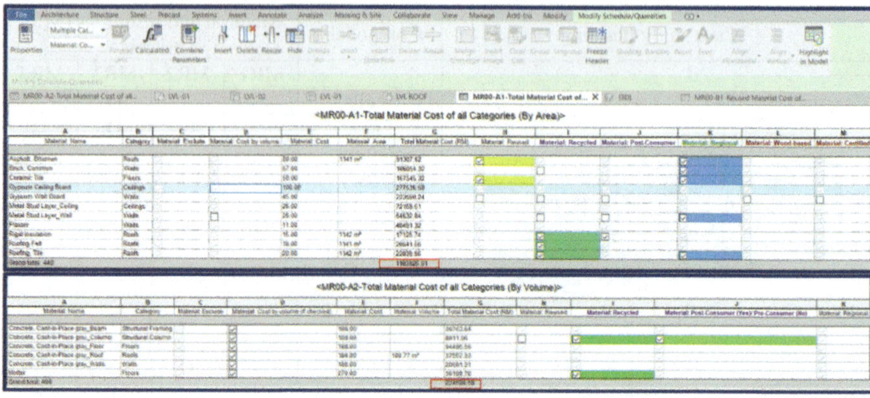

Fig. 5 Example of the BIM-MR integration template showing MTO schedules, shared parameters, and total material cost

Materials & Resources (MR) under GBI

Total cost of materials/products contributing to MR criteria (RM)
942689.43

		Material cost (RM)	Aquired GBI credits	Max credits
MR1	Materials Reuse	348997.89	2	2
MR2	Recycled Content Materials (Post & Pre-Consumer)	142371.72	1	2
MR3	Regional Materials	451319.82	0	1

		Total Volume (m³)	Aquired GBI Credits	Max credits
MR4 Sustainable Timber	Wood-based Material/Products	0.00	0	1
	Certified Wood-based Material/Products	0.00		

Fig. 6 Final assessment results for MR criteria under GBI using the BIM-MR integration method

only four seconds to extract material data, perform the necessary calculations, and export the results to the excel template. In contrast to the traditional method, the proposed BIM-MR integration method eliminated manual calculations and manual data processing to generate the final results of MR criteria. The results accuracy in this study is subjected to the LOD of the BIM model.

6 Conclusion

This paper presented the development of a BIM-based method (BIM-MR integration method) for automated assessment of MR criteria under the GBI, which is the green building rating system. The developed method relies on Autodesk Revit MTO functionalities and a Dynamo script developed to perform automated data extraction and calculation. The BIM-MR integration method is still under development; however, this study consists of a proof of concept that a hybrid combination of BIM MTO and visual scripting could automate the MR assessment under GBI and eliminate manual calculation and manual data processing.

Acknowledgements Authors wishing to acknowledge the financial support provided by UCSI University. Additionally, they would like to acknowledge the research funding from IJM Construction Sdn. Bhd. and Technology Depository Agency (TDA) Berhad through the Contract Research Grant, project no. 4C270, titled "Building Information Modelling (BIM) for Green Building Assessment".

References

1. GBI Organisation (2009) Green building index assessment criteria for non-residential new construction (NRNC)
2. National Institute of Building Sciences (2007) United States national building information modeling standard version 1—part 1: overview, principles, and methodologies
3. Sacks R, Eastman CM, Lee G, Teicholz PM (2018) BIM handbook: a guide to building information modeling for owners, designers, engineers, contractors, and facility managers, 3rd edn. Wiley, New York
4. Lu Y, Wu Z, Chang R, Li Y (2017) Building Information Modeling (BIM) for green buildings: a critical review and future directions. Autom Constr 83:134–148. https://doi.org/10.1016/j.autcon.2017.08.024
5. Bečvarovská R, Matějka P (2014) Comparative analysis of creating traditional quantity takeoff method and using a BIM tool. In: Construction maeconomics conference
6. Sattineni A, Bradford R (2011) Estimating with BIM: a survey of US construction companies. In: Proceedings of the 28th international symposium on automation and robotics in construction (ISARC), Seoul, Korea, pp 564–569. https://doi.org/10.22260/ISARC2011/0103
7. Khosakitchalert C, Yabuki N, Fukuda T (2019) Improving the accuracy of BIM-based quantity takeoff for compound elements. Autom Constr 106:102891. https://doi.org/10.1016/j.autcon.2019.102891
8. Khosakitchalert C, Yabuki N, Fukuda T (2020) Automated modification of compound elements for accurate BIM-based quantity takeoff. Autom Constr 113:103142. https://doi.org/10.1016/j.autcon.2020.103142
9. Azhar S, Carlton WA, Olsen D, Ahmad I (2011) Building information modeling for sustainable design and LEED ® rating analysis. Autom Constr 20:217–224. https://doi.org/10.1016/j.autcon.2010.09.019
10. Thai Green Building Institute (TGBI) (2014) TREES-NC/CS Thai's rating of energy and environmental sustainability for new construction and major renovation and core and shell building. Bangkok, Thailand
11. Wong JKW, Kuan KL (2014) Implementing "BEAM Plus" for BIM-based sustainability analysis. Autom Constr 44:163–175. https://doi.org/10.1016/j.autcon.2014.04.003
12. Barnes S, Castro-Lacouture D (2009) BIM-enabled integrated optimization tool for LEED decisions. In: Proceedings of the 2009 ASCE international workshop on computing in civil engineering, vol 346. American Society of Civil Engineers, pp 258–268. https://doi.org/10.1061/41052(346)26
13. Biswas T, Wang T-H, Krishnamurti R (2008) Integrating sustainable building rating systems with building information models. In: Proceedings of the 13th international conference on computer aided architectural design research in Asia (CAADRIA 2008), Chiangmai, Thailand, pp 193–200
14. Khoshdelnezamiha G, Liew SC, Bong VNS, Ong DEL (2020) A BIM-based automated assessment tool for green building index. IOP Conf Ser: Mater Sci Eng 943:012059. https://doi.org/10.1088/1757-899X/943/1/012059

15. Seghier TE, Wah LY, Ahmad MH, Samuel WO (2017) Building envelope thermal performance assessment using visual programming and BIM, based on ETTV requirement of Green Mark and GreenRE. Int J Built Environ Sustain 4:227–235. https://doi.org/10.11113/ijbes.v4.n3.216
16. Seghier TE, Ahmad MH, Lim Y (2019) Automation of concrete usage index (CUI) assessment using computational BIM. Int J BES 6:23–30. https://doi.org/10.11113/ijbes.v6.n1.319
17. Lim Y-W, Shahsavari F, Sediadi E, Fazlenawati N, Azli MN (2016) Building information modelling for building energy efficiency evaluation integration with green building index (GBI) in Malaysia. In: The 4th annual international conference on architecture and civil engineering (ACE 2016), Singapore, pp 42–48
18. Ilhan B, Yaman H (2016) Green building assessment tool (GBAT) for integrated BIM-based design decisions. Autom Constr 70:26–37. https://doi.org/10.1016/j.autcon.2016.05.001
19. Nik-Bakht M, Lee J, Dehkordi SH (2021) Bim-based reverberation time analysis. J Inf Technol Constr 26:28–38. https://doi.org/10.36680/j.itcon.2021.003
20. Abbasi S, Noorzai E (2021) The BIM-based multi-optimization approach in order to determine the trade-off between embodied and operation energy focused on renewable energy use. J Clean Prod 281:125359. https://doi.org/10.1016/j.jclepro.2020.125359

Formulating Ph.D. Construction Management Research Question or Problem for Mixed Method (Qualitative and Quantitative)

Samuel Quashie

Abstract Demand to formulate the research question or problem for the Ph.D. Construction Management started with literature reviews and experience of practical problems. The research question's construction stands on three classifications originality, relevance, and rigour; furthermore, Ph.D. Construction Management Research Design framework formed the links and contributed to developing the research question. The research question's development would be falsifiable, observation concern by the researcher, service users, and participants involvement within Built Environment. The chosen research method is a mixed-methods (qualitative and quantitative), which is primarily qualitatively driven. This research method reinforced that descriptive theory would also support mixed methods in answering the question, which has two parts. A well-stated research question is essential to guide the thesis and pinpoints what the research is to find out, giving the inquiry a clear focus and purpose; this will be achieved. In two parts Initial Part A and Final Part B, Junctures for the research design: a review of literature in line with the research area to identify the problem or phenomenon by studying the development of an Integrated Management System Post-Disaster Reconstruction. The research question is formulated using Part A. Mixed methodologists and methods embrace all to generate an integrated management system.

Keywords Ph.D. Construction management research question · Methodology and mixed methods · Service users-participants and research design

1 Introduction

The research question or problem construction of the research question stands on three classifications originality, relevance, and rigour. These classifications would contribute to improving the quality of the statement of the research question. For example, Watt's [1] explains that an excellent research question has three categories:

S. Quashie (✉)
Navajo Technical University, Crownpoint, NM 87313, USA
e-mail: squashie@navajotech.edu

T. Kang and Y. Lee (eds.), *Proceedings of 2021 4th International Conference on Civil Engineering and Architecture*, Lecture Notes in Civil Engineering 201,
https://doi.org/10.1007/978-981-16-6932-3_47

relevance, originality, rigour, and all interrelated and inspired by the U.K. Research Excellence Framework (REF).

The Ph.D. Construction Management Research Design framework in Fig. 2 formed the links and contributed to developing the research question. UNIVERSITY LIBRARIES, (2) emphasizes that research design aims to answer the research question and ensure that the data obtained help answer the research question more effectively. The research design framework consists of methods (quantitative, qualitative, or mixed), the methodology, etc. Good research design ensures that the data received will answer the research question more effectively. Gavin [2], wrote that the research question aligns with methods (quantitative, qualitative, or mixed) and a specific research design supported by the methodology. Creswell [3], gave an example that a study begins with qualitative, and later in the study, the quantitative phase is addressed. The mixed methods research question is to investigate and relies on neither quantitative or qualitative research alone. The qualitative (qual) is the core component, and quantitative (quant) is the supplementary component study answer the question to develop the integrated management system for the post-disaster reconstruction management would be conducted sequentially during the data collection and analyses. Morgan [4], and Morse and Niehaus [5], suggested that the approach is essential to combining qualitative and quantitative research design methods, as confirmed by Morgan. And a few combinations of the two provides adequate information for the mixed method research question. Qualitative and quantitative research questions need to be advanced in the mixed-method study to narrow and focus on the statement.

1.1 Studies and Reports that Addressed the Problem or Research Question

Global Facility for Disaster Reduction and Recovery (GFDRR) [6], addressed the importance and relevance of the practice of Post-Disaster Reconstruction more broadly.

In addition, United Nations Development Program (UNDP)—Bureau for Crisis Prevention and Recovery (BCPR), 2011expressed the opinion that after the 2008 Wenchuan earthquake in China, countries incorporated strategically mainstream disaster risk management into their national development strategies, Global Facility for Disaster Reduction and Recovery (GFDRR) [6]. The deficiencies in the studies of post-disaster reconstruction management research are because it is a new area. Ismaila et al. [7], agreed that it is necessary for the Governments; International Nongovernmental Organisations (NGOs), and international organizations to research the issues and challenges are within working in Post Disaster Reconstruction environments and come out with a concrete solution to overcome the complexity and uncertainties of post-disaster reconstruction. Witt et al. [8], typical post-disaster reconstruction project failure cases often refer to problems in integration, financial, inappropriate assessment, communication and coordination, inadequacies of resource procurement, ineffective design, transportation, corruption, and delay.

1.2 Significance of the Research to Particular Audiences

The research question's importance is to develop an integrated management system for post-disaster reconstruction that is significant to these particular audiences: Academic Scholars, Governments; International Nongovernmental Organisations (NGOs), Researchers, and International Organizations.

1.3 Recognize the Assumption

Recognize the assumption is an essential step used to formulating the research question. Gorvine et al. [9], proposed a need to recognize the assumptions as an indispensable step for preparing the research question. The researcher assumes that service users' (participants) involvement in the research case studies, observations, interviews, and questionnaires will be straightforward, thus reducing the workload between researcher participants and service users influencing the research question construction. As Agee [10], explained, the qualitative study has moved toward involving the researcher, participants in the investigation process. Participants are occasionally invited to collaborate on formulating research questions, especially in this participatory action research, to create an integrated management system framework for post-disaster reconstruction [10].

1.4 Methodology

The methodology part presents the stages used for the creation of the Structure of Ph.D. Construction Management Research Question or Problem—Mixed Method Qualitative and Quantitative. The methodology section is critical for other researchers to redo the Construction Management Research Question or Problem stages from literature review to conclusion, helping researchers know the steps used better, as illustrated in Fig. 1. Bhattacharyya [11], Enago.com Academy [12], and Kallet [13] highlights that the methods section of a research paper is a roadmap leading to the research's core to guide readers, an essential aspect of a research paper. It provides how a study's validity is ultimately judged. Replicate the experiment(s) or the non-experimental of the research.

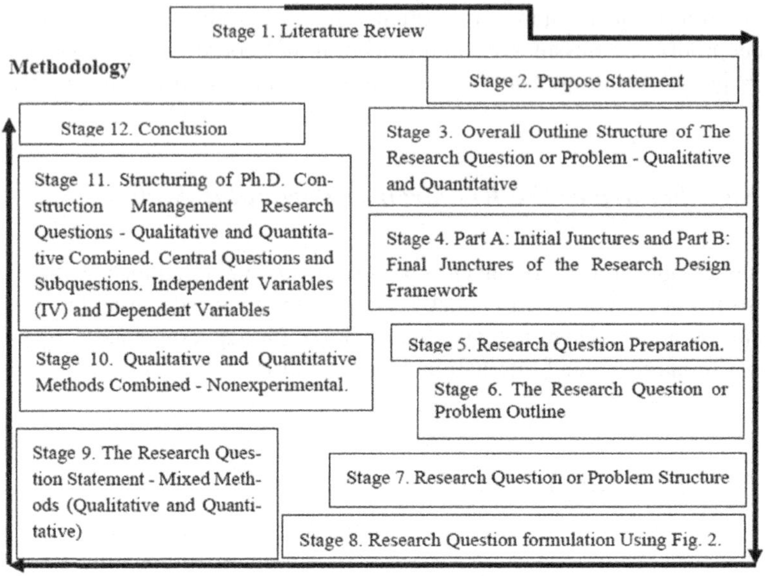

Fig. 1 Methodology

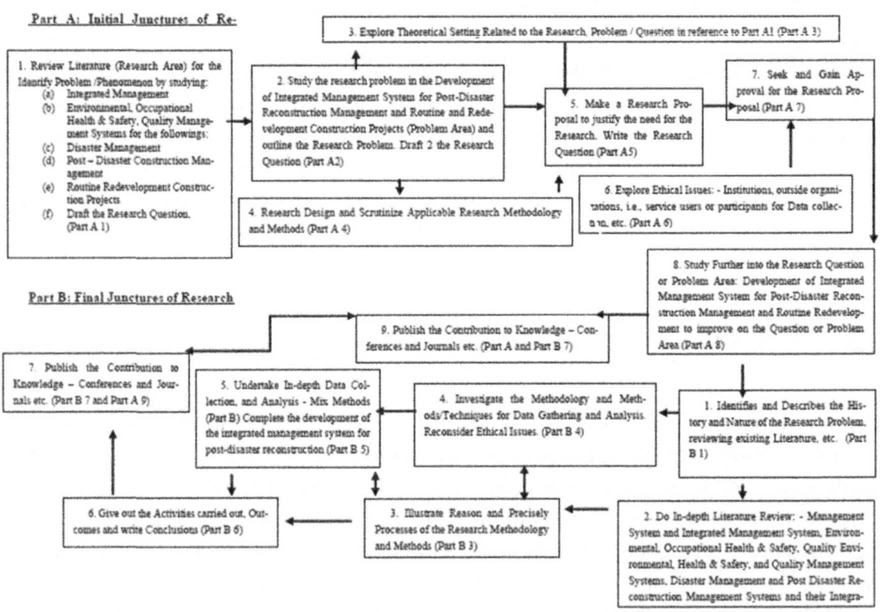

Fig. 2 The two parts A and B junctures for Ph.D. Construction management research design

2 Literature Review

Problem Formulation: The formation of the Ph.D. Construction Management research question to generate an integrated management system for the post-disaster reconstruction management phase is after identifying the research topic. My understanding of the problem formulation draws from my work in the Hatti aftermath of 2010, from a comprehensive literature review and research within the post-disaster reconstruction construction phase management. Owen and Koh [14], chapter two research problem and literature review of their book suggested that research problems will vary according to researcher experience contemplating an investigation. Booth et al. [15], Passi and Barocas [16], and Shoket [17], pointed out that once a topic is stated, there will be a question to answer, identification of research problem leads to conducting research, the problem must be made into question(s) that study can answer. Furthermore, Passi and Barocas [16], and Shoket [17], the research problem should depict what is to be determined, and the study's cope and identification of the research problem lead to conducting research. Ph.D. Construction Management research within the Built Environment is closely related to social science. As Kehily [18], pointed out, it is an inter-discipline, which makes an excellent point, as explained by STUDY AND EXAM.com [19] formation of a social research question (problem) follows these steps: identify a broad research area of the interest, dissect the broad area into sub-areas, select one of the sub-areas, raise research questions, and formulate the objectives, these would reflect and use on the formation of the research question.

Highlights of Research Question. Besides CENTER [20], and Kross and Giust [21], highlight that the research question statement would be falsifiably capable of being tested (verified or falsified) by observation concern. Therefore, the research question's development would be falsifiable, observation concern by the researcher participants and service users involvement within Built Environment. Formation of the mixed-method research question involves considering the following: Relevance, originality, rigour, and rigour of the research question. The research question provides enough specifies that the researcher can understand its purpose without needing additional explanation. Narrow enough can be answered in the given time frame; furthermore, potential answers are open to debate instead of accepted facts and make people see things differently or motivate them to learn more through discussion.

3 Relevance Originality and Rigour of Research Question

Watt's [1], explains that an excellent research question has three categories: relevance, originality, rigour, interrelated, and inspired by the U.K. Research Excellence Framework (REF). The U.K. Research Excellence Framework. is a peer assessment of U.K. universities' research quality in all disciplines. It has replaced the Research Assessment Exercise (RAE). More importantly, Mattick et al. [22], and Watt [1], stated writing up the research question, start with one of the three categories, and revisit each multiple times (this means changing the research question slightly or entirely). The construction of the research question stands on three classifications originality, relevance, and rigour. These classifications would contribute to improving the quality of the statement of the research question. The time used to review each type would be well investing time to make up relevance, originality, and rigour research question.

Relevance: Home [23], McCombes [24], and Newman et al. [4, 25], suggested the question focused on the specific, appropriately complex, and relevant to science, as in the sciences of the built environment—post-disaster reconstruction phase management. The question's relevance is that post-disaster reconstruction management research is a new area in the Built Environment. Ismaila et al. [26], emphasizes that governments are necessary; International Nongovernmental Organisations (NGOs) and international organizations research the issues and challenges working in Post Disaster Reconstruction environments. And come out with a firm solution to overcome the difficulty and uncertainties of post-disaster reconstruction.

Originality: McCombes [24], stated that a good research question is essential to guide a research thesis. Pinpoints what the researcher wants to find out and gives your work a clear focus and purpose, focused on a single problem or issue. The research question's originality focuses on developing an integrated framework for planning and managing post-disaster reconstruction projects.

Rigour: Davies and Dodd [27], referred to rigour as the reliability and validity of the research to answer the research question, to develop an integrated management system for post-disaster reconstruction, with the contribution of service users (participants). The research question's riguor is to generate an integrated management system that will be structured, measurable, systemized, ordered, uniform, neutral, and credible. Mattick et al. [22], explain that well-written research questions will include rigour and guide the research on a search to identify and collect data that can be analyzed and interpreted. Thus, it will provide new insights to help develop Post-Disaster Reconstruction Integrated Management System.

Recognize Background Assumptions of the Research Question. Gorvine et al. [8] claims that recognizing the assumptions is essential to formulating the research question. For instance, it is assumed that service users' (participants) involvement in the research case studies, observations, interviews, and questionnaires will be straightforward, thus reducing the workload between the researcher and service users (participants). These influence the research question formulation. Appropriate qualitative research question drives both the research method and findings. The issue presents a formal and concise statement of the question (problem) under-investigated. That is the development of an integrated management system for post-disaster reconstruction Gorvine et al. [9], McCombes [24], and Newman et al. [25]. The research question focused on a single problem or issue to produce Post-Disaster Reconstruction Integrated Management System Framework—(Integrating Disaster Management, ISO 9000 Quality Management, ISO 14000 Environmental Management, ISO 45001 Occupational Health and Safety and Post-Disaster Management).

4 Purpose Statement

Business and I.P. Centre [28], identified that the purpose statement sets out the qualitative research's core objective or intent. Also, stating the purpose statement researcher needs to use qualitative words such as discovering, understanding, and exploring to help set the tone of the question and demonstrate your desire to delve deeper. A mixed-method research question purpose statement combines strands of qualitative and quantitative purpose statements, which requires a little of each method, as suggested by Watt [1]. Using sequential mixed methods (qualitative and quantitative) involves case studies, working with service users (participants), observations, interviews, questioners in the construction industry and related industries. The drive or the purpose of forming the mixed methods research question is for Ph.D.'s Construction Management research thesis to solve the study issue for post-disaster reconstruction phase management. And at the end, to generate an integrated management system for post-disaster reconstruction management in the construction industry and associated sectors. More importantly, Global Facility for Disaster Reduction and Recovery (GFDRR), 2009, the World Bank- Global Facility for Disaster Reduction and Recovery (GFDRR) report for 2009 maintained that countries incorporated strategically mainstream disaster risk management into their national development strategies [26].

Furthermore, in their program, the United Nations Development Program (UNDP)—Bureau for Crisis Prevention and Recovery (BCPR), 2011 made a statement after the 2008 Wenchuan earthquake in China [2]. The GEDRR then developed a strategy that integrated disaster prevention and risk reduction in the post-disaster reconstruction program [26]. Secondary Sources—offer an analysis or restatement of primary sources, such as the International Organization for Standardization (ISO) International Standardization covering almost all aspects of technology and manufacturing—ISO 9000 Family of Quality Management. The research question is feasible to answer within the timeframe for the Ph.D. research and practical constraints specific enough to answer the research question thoroughly. Complex enough to develop the answer over the space of the Ph.D. research thesis, as highlighted by Home [3], and Newman [4].

Overall Outline Structure of The Research Question or Problem.

The chosen research method is a mixed-methods (qualitative and quantitative), which is primarily qualitatively driven. Christensen [29], Morse and Niehaus [5], and Rouse [30], reinforced that descriptive theory would also support the mixed method in answering the research question or solving the problem, which has two parts. First, an inductive (qualitative) process will predict and arrive at a specific conclusion. Second, a deductive (quantitative) is to confirm findings and arrive at a generalized conclusion. Qualitative driven is the core component and is inductive. The qualitative research method with this mixed-method design is also referred to as a naturalistic inquiry. It is a distinct research field with its research philosophy, theory, and methodology, as explained by Rucke [31]. More importantly, Morse and Niehaus [5], defined the mixed-method as combining more methodological approaches or practices drawn from a second method (quantitative supplementary component) into a single research study. To access some part of the phenomena of interest that cannot be accessed using the first method alone (qualitatively (qual.) driven core component). The research is discovery and exploration to contribute to the body knowledge or revelation, which helps integrate management for post-disaster reconstruction framework development, as confirmed by Hughes [32], and Morse and Niehaus [5].

Rucke [31], explained that the research question guides the inquiry and reflects the researcher's stance. Therefore, the vocabulary of the items (questions) will usually suggest the researcher intends to explore a specific phenomenon in its natural context. Business and I.P. Centre [28], stated that well-articulated qualitative wording of a research question would include the following qualitative words to engineer the interviewees' responses: how, what, generate, identify, describe, meaning, outline, and experience.

5 Part A: Initial Junctures and Part B: Final Junctures of the Research Design Framework

A well-stated research question is essential to guide the thesis and pinpoints what the research is to find out, giving the inquiry a clear focus and purpose; this will be achieved. As illustrated in Fig. 2, the Initial and Final Junctures for the research design are the undertaking: a literature review in line with the research area to identify the problem or phenomenon for developing an Integrated Management System for use in Post-Disaster Reconstruction Management Phase. Teddie and Tashkkori [33], highlights that literature review is used to create or refine the research questions and hypotheses. More importantly, Teddie and Tashkkori [4], also mentioned that several researchers use the literature review to develop or refine their research questions.

Parts A and B Junctures for Ph.D. Construction Management Research Design Framework—Fig. 2. illustrates Parts A and B Junctures for Ph.D. Construction Management Research Design, Parts A Junctures formed the links for developing the research question. UNIVERSITY LIBRARIES, (2018) says research design aims to answer the research question and ensure that the data obtained help answer the research question more effectively. Good research design ensures that the data received will answer the research question more effectively. Gavin [2], states that the research question aligns with methods (quantitative, qualitative, or mixed) and a specific design supported by the methodology.

The research design comprises process steps Part A Initial Junctures and Part B Final Junctures of the research design framework. The research design framework contents would strengthen the overall research and the research question's formation. Iedunote.com [34], highlights that all research endeavours share a common goal of furthering understanding the problem, and thus, all traverse through certain primary stages, forming a process called the research process. Again, Iedunote.com [34], stated that understanding the research process is necessary to carry out research and sequence the process's steps effectively. Part A—Initial Junctures: 1. Review Literature of research area for the identified problem or phenomenon by studying (a) Integrated Management, (b) Environmental, Occupational Health and Safety, Quality Management Systems for the followings: Disaster Management, Post—Disaster Construction Management, Routine Redevelopment Construction Projects and Draft 1 the Research Question (Part A 1). As explained by Kebritchi, the 2017 research problem is the core of an investigation and provides the framework for understanding the research findings. Kebritchi [35], and Snyder [36], stated that the researcher would need to know where to look for the research problem and evaluate its success. Also, Kebritchi [35], and Snyder [36], again pointed out that literature reviews help provide an overview of a particular issue or research problem. The research problems could be identified by reviewing recent literature, reports, and databases in the research field, Kebritchi [2].

1. Study the research problem in the Development of Integrated Management System for Post-Disaster Reconstruction Management and Routine and Redevelopment Construction Projects (Problem Area) and outline the Research Problem. Draft 2 the Research Question (Part A2)
2. Explore Theoretical Setting Related to the Research, Problem/Question regarding Part A1 (Part A 3). Studying theoretical background concerning the research problem contributes to the research's theory building and testing theory. Khan [37], suggested that grounded theory is a data collection approach in qualitative research methods based on data rather than emerging theory. Furthermore, the research questions and literature review lead and support conceptual thinking and theory building. And rather than empirical testing of the theory and the study follows an inductive theory-building approach. Kebritchi [35], maintained that looking at what theories need to be tested is another opportunity to develop a research problem.
3. Research Design and Scrutinise Applicable Research Methodology and Methods (Part A 4). This part A 4 process would identify and elaborate the essential factors of a research methodology and methods. Khan [37] supported these by laying out that the literature reviews the efforts to identify and further elaborate the necessary components of a research methods chapter. Those are an integral part of any research paper while conducting a grounded theory approach [5].
4. Make a Research Proposal to justify the need for the research. Then, write the Research Question (Part A5).
5. Explore Ethical Issues: Institutions, outside organizations, i.e., service users or participants for Data collection, etc. (Part A6)
6. Seek and Gain Approval for the Research Proposal (Part A7)
7. Study Further into the Research Question or Problem Area: Development of Integrated Management System for Post-Disaster Reconstruction Management and Routine Redevelopment to improve on the Question or Problem Area (Part A8)

6 Research Question Preparation

CENTER, THE WRITING [20], suggested that research questions are essential to the research process. The research question formulated is outside the realm of scientific inquiry. It can be confirmed or refuted by observations, or service users' (participants) experiences can be verified easily empirically. Developing or making a research question statement is to select a topic that the researcher is interested in as it is essential to the research progression. Then, the researcher does adequate studies in journals, publications, conference papers, etc., covering the chosen topic. Help the researcher focus on the research by providing a path through the research and writing and fieldwork observation processes.

Furthermore, CENTER, THE WRITING [20], also stated that researchers do some background reading in reference sources, current periodicals, and journals to see what's already been done and help narrow the researcher focus. Besides, Creswell [26], highlights that a robust mixed-methods study should start with a mixed methods research question to shape the overall research design methods. Furthermore, a research problem is a statement: about an area of concern, a condition to be improved, a difficulty to be eliminated, or a troubling question in scholarly literature, in theory, or practice (such as the concern to develop an integrated management system of post-disaster reconstruction) that points to the need for meaningful understanding and deliberate investigation. Construction Management research within the Built Environment is closely related to Social Science, which does not lend itself to hypothesis testing and statistical analysis. California [38], maintained four general processes of forming (conceptualizations) of a research problem or question in social science. The four available processes question in social science are, Casuist (Method of Reasoning) Research Problem; this problem relates to the determination of right and wrong in conduct or conscience questions. Second, difference Research Problem, this type of problem statement is used when the researcher compares or contrasts two or more phenomena. Descriptive Research Question the underlying purpose is to describe the significance of a situation, state, or existence of a specific phenomenon related to this research. This problem is often associated with revealing hidden or understudied issues. Finally, a relational Research Problem proposes a relationship between two or more variables to be investigated to develop an integrated management system. The underlying purpose is to investigate specific qualities or characteristics that may be connected in some way. The research question to be formulated is a Descriptive Research Problem to describe the situation, state, or existence of any particular phenomenon and reveal hidden or understudied issues. The researcher uses the help of participants and service users to formulate the question and develop an integrated management system, a Descriptive Research Problem. Agee [10], claims that Flick (2006), Lassiter (2005), and Maxwell (2005) pointed out that qualitative inquiry has moved toward involving the researcher and participants in the study process.

6.1 The Research Question or Problem Outline

The research question or problem statement would establish: clarity, demonstrate a researchable issue, what would be studied, identify an overarching question or set of questions, supplemented by critical concepts and terms, and present the conceptual study and boundaries of parameters or limitations. As Kross and Giust [21], supported, the research question must not be an unambiguous statement that openly pronounces the phenomenon you plan to investigate. California, n.d.; Dean's Office—Student Academic Affairs [39], Royal Essaywriters.co.uk [40], and Sacred Heart University Library [41], identifies that in some social science disciplines, the problem is posed in the form of a question. Again, construction management science within the built environment is closed related to social science as it is an inter-discipline, as pointed out by Kehily [18]. California, n.d.; Dean's Office—Student Academic Affairs [39], Royal Essaywriters.co.uk [40], and Sacred Heart University Library [41], maintained that in some social science disciplines, the problem is posed in the form of a question.

This research question is based on the researcher doing literature review studies and observations to understand the title and the research question. Again, construction management science within the built environment is closed related to social science as it is an inter-discipline, as Kehily [18], pointed out. Dainty [3], claims that those engaged in social science research in the built environment could usefully embrace the emerging multi-strategy or multi methodology research design principles. To better understand the complex network of relationships that shapes the construction industry practice. And guide the research, why something is happening, and answer the question or solve the problem. Sherratt and Leicht [42], make a case that multi-methodological perspectives on a problem (question) should be adopted to generate more holistic understandings and more comprehensive illuminations of the phenomena in practice. Dainty [43], suggested using multimethodological views to support developing and establishing a good construction engineering management research discipline, both in academic scholarship and relevant practice. These will help embrace the emerging principles of multi-strategy and multi methodology in the construction management research design. Library, n.d. stated that the research problem is typically posed in a question in some social science disciplines. In support of this, Zou et al. [44], explained that standard research methodologies are embraced in social science research: quantitative, qualitative, and mixed methods, which contribute to knowledge or increase knowledge. Stringer [45], argued that stakeholders—that is, the Researcher, Participants, and Service Users, those whose lives are affected by the problem under study, should be engaged in the investigation processes. Bothello and Roulet [26], explained that by makeup, management research is interdisciplinary, drawing from multiple humanities and social sciences fields and often bringing them together to generate novel contributions to research.

6.2 Research Question or Problem Structure

Creswell [3], and Royal Essaywriters.co.uk [40], emphasizes that selecting a suitable topic for the research project requires deciding whether to use research questions or hypothesis methods. To make up the research question, that is a mixed-methods research (qualitative and quantitative) question. Renwick [26], suggested using qualitative and quantitative methods to answer a single research question means building a complete understanding of the topic or title the research is investigating. Creswell [3], similarly narrows the focus to specific questions to be answered or predictions based on the hypothesis to be tested. Renwick [29], also stated that quantitative data would tell what is happening and help understand the magnitude, while qualitative data (/qualitative-research-methods) can tell why something is happening. The study's research title/topic and the question would have been conceivable to the Ph.D. supervisors, others, and the researcher and contribute to knowledge for post-disaster reconstruction management. Lead to produce professional research studies to be polished during and after the work is completed. Verba [7], relates that the topic selected allows creating a professional, polished piece of work within a limited amount of time, making an original contribution to the field.

The topic is—**Development of Integrated Management System**: Environmental, Occupational Health, and Safety and Quality Management Systems into Post-Disaster Reconstruction Management Phase in the Construction Industry and Related Industries. The topic came about from a broad statement, then a narrower one as stated above; from the topic comes the question to ask. The stated topic is in two parts knowledge—Development of Integrated Management System: Environmental, Occupational Health, and Safety and Quality Management Systems) and consequence of knowledge—into Post-Disaster Reconstruction Management Phase in the Construction Industry and Related Industries. Booth et al. [15], confirmed these principles; first part knowledge or flawed to the understanding, and second part consequence knowledge or flawed to the understanding. Also, Booth et al. [5], said once the researcher has a topic, it should find the question(s) or problem(s) to answer. This study called for to be conducted because post-disaster reconstruction management phase systems need to be improved.

In the Global Facility for Disaster Reduction and Recovery (GFDRR), 2009, the World Bank- Global Facility for Disaster Reduction and Recovery (GFDRR) report in 2009 maintained that countries incorporated strategically mainstream disaster risk management into their national development strategies. In addition, the United Nations Development Program (UNDP)—Bureau for Crisis Prevention and Recovery (BCPR) [46], made a statement after the 2008 Wenchuan earthquake in China. As a result, the Global Facility for Disaster Reduction and Recovery (GFDRR) [6], formulated a post-disaster recovery and reconstruction master plan. The GEDRR then developed a strategy that integrated disaster prevention and risk reduction in the post-disaster reconstruction program. The above statement statements demonstrate the supports to undertake that study. First, the research question is formulated using Fig. 3. Part A Junctures, assembling background material for Formation of Ph.D.

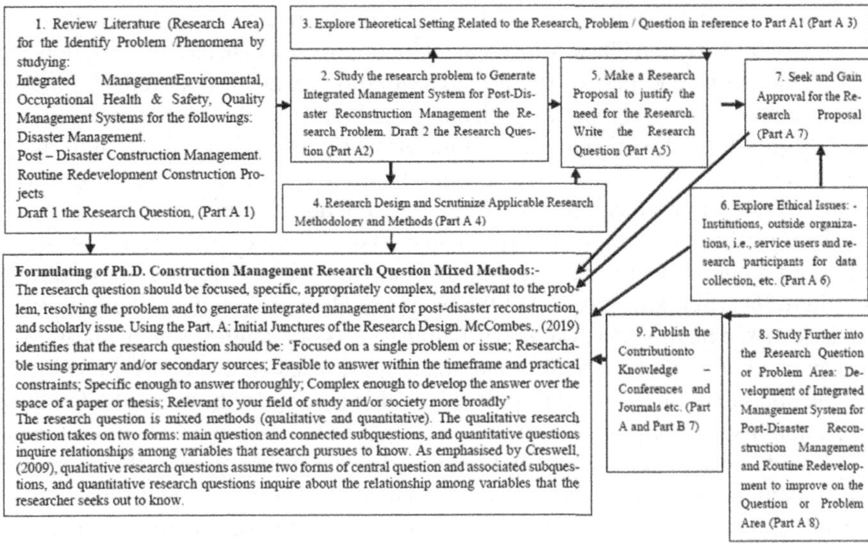

Fig. 3 Part A junctures-1, 2, 3, 4, 5, 6, 7, 8, and 9

Construction Management Research Question. Libraries.indiana.edu [47], picked a topic and collected the background information to focus on the research question. Next is the research question's forming using Parts A: 1, 2, 3, 4, 5, 6, 7, 8, and 9 and revisiting the topic and purpose statement.

The Research Question or Problem Statement—Mixed Methods (Qualitative and Quantitative) is

Generate an Integrated Management System Framework for Post-Disaster Reconstruction—Integrating Disaster Management, Environmental, Occupational Health and Safety, Quality Management Systems into Post-Disaster Reconstruction Phase Management for Improvement of Post-Disaster Reconstruction Phase Management in the Construction Industry, and Related Industries. The research question statement came from the topic stated above; Booth, et al. [15], pointed out that there will be a question to answer once the topic is formed. Furthermore, Passi and Barocas [26], stated problem must be made into question(s) that science (research) can answer, and the problem is realistic to the original research objectives. Finally, Shoket [6], agreed that identifying the research problem leads to conducting research.

The formulation of the research problem should depict what is to be determined and the scope of the study. Thus, generate is one of the sound words to be used in qualitative research questions. Business and I.P. Centre [28], pointed out suitable types of qualitative wording for research questions, including the following kinds of qualitative words to engineer the research's variety of responses from the interviewees: How, What, Generate, Identify, Meaning, Outline, and Experience. As is

Generate is one of the sound words to be used in qualitative research questions. Post-Disaster Reconstruction Phase Management for improvement in the Construction and Related Industries end part of the research question makes it researchable, makes up questions, and answers the research question. MESHGuides [26], confirmed that research questions should be researchable, allowing researchers to research them. This research question statement is a single research question; the mixed methods would explore different aspects of a single research question.

The researcher and service users (participants) would use separate but related qualitative and quantitative research questions to answer the research questions during the field studies. Schoonenboom and Johnson [48], also confirmed that the researcher could use mixed methods to examine different aspects of a single research question or use separate but related qualitative and quantitative research questions. In addition, Schoonenboom and Johnson [48], pointed out that mixing methods, methodologies, and paradigms will help answer the research questions and improve a more fundamental study design. As a result, and fuller and richer information will be obtained in the mixed methods study. For example, construction management science within the Built Environment is closed related to social science. Sacred Heart University Library [41], and Royal Essaywriters.co.uk [40], suggested that the researcher using observations to understand the title and the research question and guide the research, why something is happening.

Stating the research question statement is followed by considering which qualitative, quantitative, and mixed-method methods match the stated research question. The three approaches have value, but each process differs in its characteristics and the effects and contributions to answering the research question(s). The appropriate method to the above-stated research question is a mixed-method process, combining qualitative and quantitative methods. MESHGuides [49], again stated that research questions should be researchable, allowing researchers to research them, and this is the use of mixed methods. The end part of the research question makes it researchable, makes up questions, and answers the research question for Post-Disaster Reconstruction Phase Management for improvement in the Construction Industry and Related Industries. The researcher and service users (participants) would use separate but related qualitative and quantitative research questions to answer the research questions during the field studies. Schoonenboom and Johnson [48], confirmed that the researcher could use mixed methods to examine different aspects of a single research question or use separate but related qualitative and quantitative research questions. In addition, Schoonenboom and Johnson [48], relates that mixing methods, methodologies, and paradigms will help answer the research questions and improve a more fundamental study design. As a result, and fuller and richer information will be obtained in the mixed methods study. Also, in some social science disciplines, the problem is posed in the form of a question. Therefore, the appropriate method to the above-stated research question is a mixed-method process, combining qualitative and quantitative methods. Stating the research question statement is followed by considering which qualitative, quantitative, and mixed-method methods match the stated research question. The three approaches have value, but each process differs in its characteristics and the effects and contributions to answering the research question(s).

6.3 Assessment of Research Question's Formation

The above insight has contributed to formating the research question. Using the topic as the basis to come about with the research question did make it very simple. Booth et al. [15], pointed out that there will be a question to answer once a topic is formed. The formulation of the research question using Part A Junctures is excellent. The research question's forms using Parts A: 1, 2, 3, 4, 5, 6, 7, 8, and 9 and revisiting the topic and purpose statement reinforced the expression by, Libraries.indiana.edu [47].

7 Conclusion

The formation of the above question proved to be relevant as post-disaster reconstruction management research is new in the Built Environment. Therefore, it is necessary for Governments; International Nongovernmental Organisations (NGOs), and international organizations to research the issues and challenges working in Post Disaster Reconstruction environments. And come out with a firm solution to overcome the difficulty and uncertainties of post-disaster reconstruction. The research question's originality focuses on developing an integrated framework for planning and managing post-disaster reconstruction projects. The well-written research question includes rigour and guides the research on a search to identify and collect data that can be analyzed and interpreted. It will provide new insights to help develop Post-Disaster Reconstruction Integrated Management System. Part of making up the research question is recognizing background assumptions are essential to formulating the research question. The word Generate used in the research question's statement mixed methods (qualitative and quantitative) to answer the problem: to produce an Integrated Management System to improve post-disaster reconstruction management. The research question statement is a single research question; the mixed methods would explore different aspects of a single research question.

References

1. Creswell JW (2009) The use of theory. In: Knight V (ed) Research design. Thousand Oak, SAGE, pp 49–72
2. United Nations Development Program (UNDP)—Bureau for Crisis Prevention and Recovery (BCPR) (2011) Methodological guide for pre-disaster recovery planning processes guidelines and actions for national, regional and local governments. Cia Ltda, Ecuador

3. Home, Library. Primary, Secondary, and Tertiary Sources [Online] [Cited: Jan 9, 2019]. https://www.crk.umn.edu/library/primary-secondary-and-tertiary-sources
4. Newman I et al (1997) Theses and dissertations. University Press of America, Lanham
5. Khan SN (2014) Qualitative research method: grounded theory. Int J Bus Manage 9(11):224–233
6. Shoket M (2014) Research problem: identification and formulation. Int J Res 1(4):512–518
7. Agee J (2009) Developing qualitative research questions: a reflective process. Int J Qual Stud Educ 22:431–447
8. Gorvine B et al (2017) Research methods from theory to practice, 1st edn. s.l. Oxford University Press
9. McCombes S, Scribbr.com. [Online] 2019. [Cited: May, July, 2019]. https://www.scribbr.com/research-process/research-question-examples/
10. Kebritchi M (2018) How to identify an appropriate research problem [Online] 2017. [Cited: Sept 21, 2018]. https://research.phoenix.edu/blog/how-identify-appropriate-research-problem
11. Royal Essaywriters.co.uk. What is the difference between research questions and hypothesis? [Online] 2020. [Cited: Feb 24, 2020]. https://royalessaywriters.co.uk/research-questions-and-hypothesis/
12. Verba C (2019) How to pick a dissertation topic [Online] 2012. [Cited: Oct 23, 2019]. https://gsas.harvard.edu/news/stories/how-pick-dissertation-topic
13. MESHGuides. Meshguides.org [Online] [Cited: Feb 10, 2019]. https://www.meshguides.org/guides/node/469#:~:text=%20Wording%20your%20research%20questions%20%201%20Comparing,research%20questions%20should%20be%20written%20as...%20More%20
14. Thomas L, Scribbr.Com. [Online] 2020. [Cited: Aug 10, 2020]. https://www.scribbr.com/methodology/independent-and-dependent-variables/
15. Thomas L, Scribbr.Com [Online] 2020. [Cited: July 9, 2020]. https://www.scribbr.com/methodology/independent-and-dependent-variables/
16. Libraries.indiana.edu. (2019) Narrowing a topic and developing a research question [Online] 2008. [Cited: May 5, 2019]. https://libraries.indiana.edu/sites/default/files/Develop_a_Research_Question.pdf
17. Christensen CM (2006) The ongoing process of building a theory of disruption. Prod Innov Manage 23:39–55
18. Rouse M, TECHTARGET NETWORK [Online] 2013. [Cited: May 15, 2019]. https://whatis.techtarget.com/definition/deductive-reasoning
19. Morse J, Niehaus L (2011) Mixed method design: principles mixed method design: principles. Qualit Soc Res 12(1)
20. Homenick K, lifepersona.com [Online] 2017. [Cited: May 12, 2018]. https://www.lifepersona.com/5-examples-of-theoretical-framework-for-understanding-the-concept
21. USCLibraries. Research guide [Online] 2018. [Cited: July 20, 2018]. http://libguides.usc.edu/c.php?g=235034&p=1561763
22. Statistics solutions. Theoretical framework [Online] 2018. [Cited: June 18, 2018]. https://www.statisticssolutions.com/theoretical-framework/
23. Clavero J (2018) What is integrated project management? [Online] 2018. [Cited: Nov 6, 2018]. https://esub.com/what-is-integrated-project-management/
24. Standards store. Integrated standards [Online] [Cited: Dec 12, 2017]. http://integrated-standards.com/articles/what-is-integrated-management-system/
25. Rucke M (2018) Qualitative research question examples [Online] 2016. [Cited: Mar 16, 2018]. https://unstick.me/qualitative-research-question-examples/
26. Global Facility for Disaster Reduction and Recovery (GFDRR) (2009) Integrating disaster risk reduction into the fight against poverty. GFDRR Secretariat, Washington
27. Morse JM, Niehaus L (2009) Qualitatively-driven mixed method designs. In: Morse J (ed) Mixed method design principles and procedures. Left Coast Press Inc., Walnut Crerk California, pp 85–115
28. Morgan DL (1998) Practical strategies for combining qualitative and quantitative methods: applications to health research. Qualit Health Res 8(3):362–376

29. Kross J, Giust A (2019) Elements of research questions in relation to qualitative inquiry. Qualit Rep 24(1):24–30
30. Wikipedia. Built environment [Online] 12 Mar 2020. [Cited: Mar23, 2020]. https://en.wikipedia.org/wiki/Built_environment
31. Sacred Heart University Library. Organizing academic research papers: the research problem/question. [Online] [Cited: June 18, 2019]. https://library.sacredheart.edu/c.php?g=29803&p=185918
32. Watt D (2018) Scotland's Rural College [Online] 2014. [Cited: Mar 12, 2018]. https://www.sruc.ac.uk/info/120447/class_8_commercial_publications/848/research_excellence_framework
33. California, University of Southern. Research Guides [Online] [Cited: Dec 2, 2019]. https://libguides.usc.edu/writingguide/introduction/researchproblem
34. Sherratt F, Leicht R (2019) Unpacking ontological perspectives in CEM research: everything is biased. J Constr Eng Manage 146(2):1–9
35. Dean's Office—Student Academic Affairs. Ph.D. Handbook. The New School for Social Research, New York, NY
36. Davies D, Dodd J (2002) Qualitative research and the question of rigor. Qualit Health Res 12(2):279–289
37. Scribbr.Com [Online] 2019. [Cited: June 4, 2019]. https://www.scribbr.com/research-process/research-questions/
38. Niemoller J, Perillon.com [Online] 2019. [Cited: Nov 6, 2019]. http://www.perillon.com/blog/12-benefits-of-iso-45001
39. EHS MANAGEMENT BLOG [Online] 2019. [Cited: Nov 6, 2019]. http://www.perillon.com/blog/12-benefits-of-iso-45001
40. Gavin D (2018) Constructing a study design: aligning research question with methodology, design, and degree program [Online] 2016. [Cited: May 15, 2018]. https://research.phoenix.edu/blog/constructing-study-design-aligning-research-question-methodology-design-and-degree-program
41. Ismaila D et al (2014) Project management success for post-disaster reconstruction projects: international NGOs perspectives. Elsevier B.V, Salford Quays, United Kingdom
42. CENTER, THE WRITING. George Mason University [Online] Aug 12, 2018. [Cited: Oct 7, 2018]. https://writingcenter.gmu.edu/guides/how-to-write-a-research-question#:~:text=Steps%20to%20developing%20a%20research%20question%3A%201%20Choose,and%20%E2%80%9Cwhy%E2%80%9D%20questions%20about%20your%20...%20More%20items
43. Witt E, Sharma K, Lill I (2014) Mapping construction industry roles to the disaster management. Sciencedirect.Com, Salford Quays U.K.
44. Bothello J, Roulet T (2019) Social sciences pace [Online] 2018. [Cited: Sept 23, 2019]. https://www.socialsciencespace.com/2018/06/management-is-a-social-science/
45. Dawson C (2002) How to decide upon methodology. In: Brueton D (ed) Practical research methods. How To Books Ltd., Oxford U.K., pp 14–23
46. Business and I.P. Centre (2019) What makes a good qualitative research question? [Online] 2019. [Cited: June 20, 2019]. https://www.bl.uk/business-and-ip-centre/articles/what-makes-a-good-qualitative-research-question
47. Integrity, The Office of Research. U.S. Department of Health and Human Service [Online] [Cited: Oct 12, 2020]. https://ori.hhs.gov/content/module-3-elements-research-section-1
48. Renwick D (2019) What is mixed methods research? [Online] 2019. [Cited: Sept 10, 2019]. https://blog.optimalworkshop.com/what-is-mixed-methods-research/
49. Tomorrow's professor postings. Choosing a dissertation topic [Online] [Cited: Mar 6, 2020]. https://tomprof.stanford.edu/posting/1482

Use and Application of Symbolic Solar Lighting in Mayan and Inca Architecture: A Literary Review

José Ramón Leyva Guzmán and Marcelo Villacis Ormaza

Abstract This literary review examines and analyzes the use of sunlight for symbolic purposes in the architecture of the main pre-hispanic cultures, analyzing how this knowledge has been used since the beginning of human civilization. The first civilizations that studied the apparent solar movement discovered that this knowledge meant an advantage to be able to survive and evolve, so they began to incorporate this knowledge into their architectural works; initially as a tribute, as it was one of the first divinities; then as lighting and later with symbolic or phenomenological purposes by using the effect that light generates when it travels through the architectural elements. With this data collection it is intended to manifest the basic knowledge of applied solar astronomy, to understand the knowledge used by the Mayans and Incas thanks to their empirical and experimental methods. From this review it can be concluded that pre-hispanic cultures were able to study the apparent solar movement and, with it, build their calendars based on their survival needs and at the same time, build religious buildings in which the Sun, through its illumination, plays a leading role for symbolic purposes.

Keywords Inca architecture · Mayan architecture · Symbolic lighting

1 Introduction

The ancient peoples did not understand the existence of the Sun and before this position they considered it as a divinity and worshiped it as such, since they sensed that life, death and scarcity could derived from it, as happened in times of droughts. In early civilizations it was granted as sovereign status because from above, it sees all things and in daily journey to the subsoil it is linked to the afterlife [1].

J. R. L. Guzmán (✉) · M. V. Ormaza
Faculty of Architecture, Arts and Design (FAAD), Universidad Tecnológica Indoamérica, Quito, Ecuador
e-mail: joseleyva@uti.edu.ec

M. V. Ormaza
e-mail: marcelovillacis@uti.edu.ec

T. Kang and Y. Lee (eds.), *Proceedings of 2021 4th International Conference on Civil Engineering and Architecture*, Lecture Notes in Civil Engineering 201,
https://doi.org/10.1007/978-981-16-6932-3_48

Some civilizations, such as Mesopotamia, used sunlight for planning their cities [2], the context dominated by a strong summer with intense sun, compromised life in those areas and consequently determined the location and disposition of settlements.

The pre-hispanic Mayan and Inca cultures also felt the need to be able to explain the presence of the Sun and its power to condition their development. Both peoples studied and understood its journey, impact on the climate and the seasons; getting to develop their calendars and thereby identify which were the ideal dates to carry out each agricultural activity. In the same way, they used solar lighting for symbolic purposes in their architecture, generally religious [3].

2 Methodology

As a large amount of information was found on the object of analysis of this article, it was necessary to find a way to synthesize the information for a clear, holistic and replicable understanding. As a method of approach to the research, a systematic review of the literature was applied to know and compare empirical evidence that can help to draw conclusions [4].

This review can be described as a systematic method of collecting files or documents that present relevant evidence, which later will be evaluated through a comparative qualitative methodology [5].

Two stages in the investigation were defined; the first was made up the analysis of the solar evidences found in each culture, while in the second stage comparative tables are made tabulating the data from previous stage.

The archaeological or primary sources from which this work emerges come from both written sources—historical and textual archeology—and material sources, latter being understood as the structures—architecture—made by these cultures, which were analyzed from the lens of archeoastronomy [1].

3 Mayan Solar Architecture

The Mayan culture stands out among all pre-hispanic civilizations for its development in general, standing out in different aspects such as writing, mathematics, agriculture, astronomy and architecture; right with these last ones this analysis emerges, the domain of knowledge of solar movement, sunrise and sunset, was used for the orientation of their civic and ceremonial buildings in different settlements [6].

In the analysis carried out by Sánchez and Šprajc in 87 sites in the Mayan lowlands, it is concluded that the orientations of civic and religious architecture were governed by the east–west disposition according to the sunrise and sunset on certain dates, predominantly sunrise from February 12th to October 30th, an interval that corresponds to the ritual calendar cycle and with a division of 260 days between these [6].

Fig. 1 View of the north facade of Kukulkán's temple or 'El Castillo' in Chichén Itzá. There are several elements that are significant in the pyramid, for instance, the number of steps corresponds to the number of days in each quarter of a solar year: 91 × 4 is 364 where the solar year is 364.2422 days. *Source* Authors' creation

There is also evidence that the location of their monumental buildings facilitated the agricultural cycle activities schedule [7].

Chichén Itzá is undoubtedly one of the most important Mayan cities, built at the height of the empire between the late classic or early post-classic periods (800–1100 AC). The Kukulkán temple or "El Castillo" is the most important construction. This enclosure pays tribute to the Mayan entity that gives the temple its name and that represents, according to the Mayan language, the Feathered Snake. According to some sources studied, on the days of the spring and autumn equinoxes, sunlight and shadow take a leading role in the body of the pyramid shortly before sunset, simulating the descent of the Mayan deity [8]. It is pertinent to clarify that there are other criteria that suggest that the days of such solar effects do not correspond to equinoxes, since this solar effect is also seen on nearby dates [9] (Figs. 1 and 2).

In recent studies it was determined that this event has its reason in the called "days of the quarter" [7], which are the days in which the year is divided into four equal parts (91 days × 4 = 364), which, according to Alfred Tozzer, are also the number of steps on each of the four steps of the temple [10].

In the same way, both approaches take the Sun and its light as a symbolic element in this hierophany, showing the Mayans' knowledge of astronomy and their desire to use the sun king to pay tribute to their Gods.

4 Inca Solar Architecture

The Incas considered the sun, the moon and the stars as divine beings that influenced people, animals and plants. The astronomical observation together with the precise maintenance of its calendar; made possible to forecast the behavior of the climate,

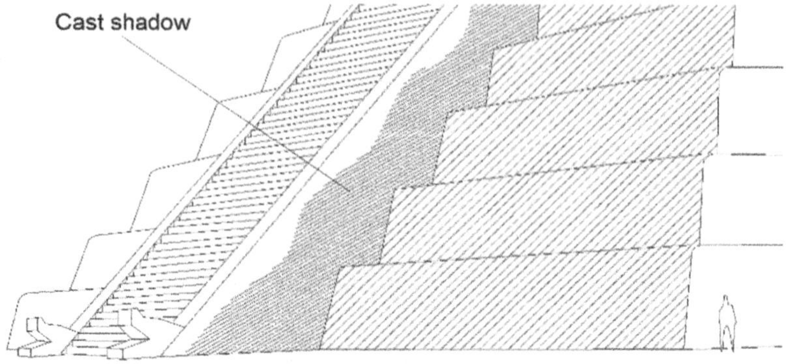

Fig. 2 The image shows the solar phenomenon that occurs on the "days of the quarter" which corresponds to the dates March 23rd and September 21st, located two days after and before the spring and autumn equinoxes, respectively. The shadow projected on the north staircase when the sun sets, corresponds to the stepped edges of the pyramid. *Source* Authors' creation

and thus determine the times of sowing and harvesting; as well as to determine exactly the celebration dates for their ceremonies [3].

It is known from chroniclers' stories that the Incas used architectural or landscape elements to coordinate their calendar, and to make their religious ceremonies coincide with key solar days such as the solstices and equinoxes [11]. There is documentary evidence, but not physical, that in Cusco were constructions as observation columns, both in Juan de Betanzos' text "Suma y narración de los Incas", and in "Comentarios Reales de Garcilaso de la Vega" and in the description of Pedro Sarmiento de Gamboa, who was a cosmographer and mathematician [1].

In Machu Picchu, unlike Cusco, there is a carved stone at the top of a rocky settlement that allowed them to make observations of the Sun with precision, this construction or "Intihuatana", which in quechua means: Inti—Sun—and huatana—tie up-, provided them with a precise follow-up of the sunrise and sunset and the passage through the zenith, essential knowledge to make their calendar, or as Herrmann assures in his research: Machu Picchu was an astronomical religious pilgrimage place [12].

A clear evidence of the knowledge of the Sun is its calendar, the celebrations and activities in general of the Inca people were determined by the climate conditions imposed by the Sun in its apparent annual movement, for example, we have Raymi—December in the Gregorian calendar—it means the great Easter sun, Jatuncusqui—May—harvest month and Aucaycusqui—June—estival in honor of the sun God (Inti Raymi), coincides with the winter solstice for the Southern hemisphere [12] (Fig. 3).

Fig. 3 The image shows the carved stone known as Intihuatana, used by the Incas in Machu Picchu to study the Sun. Source: "Intihuatana in Machu Picchu", by Alesegura is licensed under CC BY4.0

5 Solar Knowledge Applied by the Mayan and Inca Cultures. Brief Explanation

Evidences found in Inca towns, both in Cusco and in Machu Picchu, indicate that they applied the experimental empirical method, which is based on monitoring daily and annual solar trajectories, for this they had to use gnomes such as the Intihuatana. This patient and prolonged analysis allowed them to determine all the singular dates such as solstices, equinoxes and days of the zenith Sun [3]. Although it is unknown how the Mayans understand the movement of the Sun, it is assumed that they applied experimental methods not very different from Incas.

To understand the movement of the Sun at any latitude, it is necessary to apply a series of concepts such as: solstices, equinoxes, zenith Sun, homologous days, sunrise and sunset. These concepts are intended to be explained briefly, to understand the astronomical knowledge of pre-hispanic cultures.

The apparent movement of the Sun is understood as the daily movement of the star king through the celestial vault (BC) for each latitude, in which the solar trajectories seen from the Earth are observed as arcs of circumferences above the horizon (Fig. 4). This movement over the course of a year is determined by the rising and setting of the Sun on the horizon at points that range roughly from Northeast (NE) to Southeast (SE) for the rising, and from Northeast (NW) to the Southwest (SW) for setting, corresponding to half a year for each group of trajectories [7].

The summer and winter solstices occur when the sun rises and sets at the ends of its trajectory, only once a year and occurs on dates close to June 21st and December 21st, respectively (Fig. 4). The equinoxes are the days when the sun rises in the East and sets in the West, approximately March 21st and September 21st (Fig. 4).

The Zenith Sun is when the Sun rises through the zenith (Fig. 5), the day when the shadow of the gnome at noon disappears when the sun is at 90° or perpendicular to

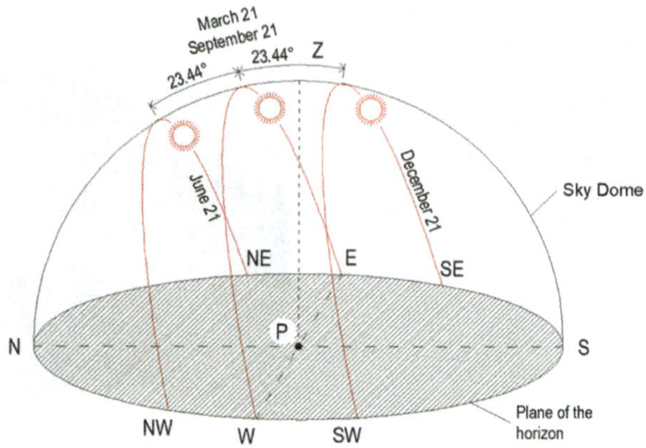

Fig. 4 Solstices solar path diagram (June 21st and December 21st) and equinoxes (March 21st and September 21st). For an observer (P) located approximately 15° South latitude (Machu Picchu). *Source* Authors' creation

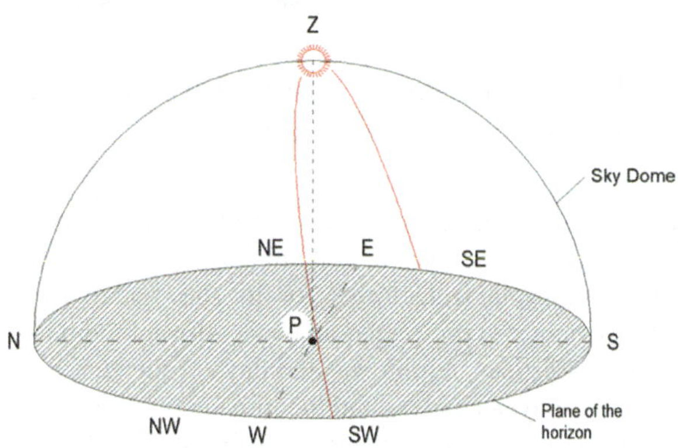

Fig. 5 Zenith Sun solar path diagram (Z), it occurs two days a year depending on the latitude of the place. For an observer (P) located approximately 15° South latitude (Machu Picchu). *Source* Authors' creation

the plane of the horizon. This phenomenon occurs twice a year for observers located between Cancer and Capricorn tropics.

Due to the above, it is pertinent to clarify that astronomy as a science began to gain strength in the XVII century with the contributions of Galileo, Kepler and Newton. For this reason, it is logical to think that the dominance of the Mayan and Inca empires was elemental. Enough to be able to predict the arrival of each season and thus to be able to anticipate times of droughts and rains; allowing them to plan the harvests and their religious rituals [1].

6 Conclusions

The dominance of solar astronomy in the Mayan and Inca empires is evident, although each people used solar illumination to satisfy their religious and subsistence perspectives, established according to their geographical conditions and their individuality.

The Mayans generally used solar lighting for the location of their buildings and to pay tribute to their deities. The concepts they applied were the sunrise and sunset on significant days linked to their ritual and agricultural calendar, the solstices and equinoxes, and the days of the fourth (Fig. 6) in architectural works such as Chichén Itzá.

The rise of the Sun through the zenith, the sunrise and sunset on the days of the equinoxes and solstices were the concepts applied by the Incas, in addition to the creation of an exact calendar that contained all their ceremonies, in which they stand out rites to the sun and the distribution of all agricultural and subsistence activities by months.

The knowledge acquired by pre-hispanic cultures, which is translated in this article through graphics of the apparent movement of the sun; conveys the concern that existed in ancient peoples regarding the sun king and its impact on climate, the sensitivity in the use of lighting for symbolic purposes and motivates us to continue researching the use of solar lighting and its application in architecture (Table 1).

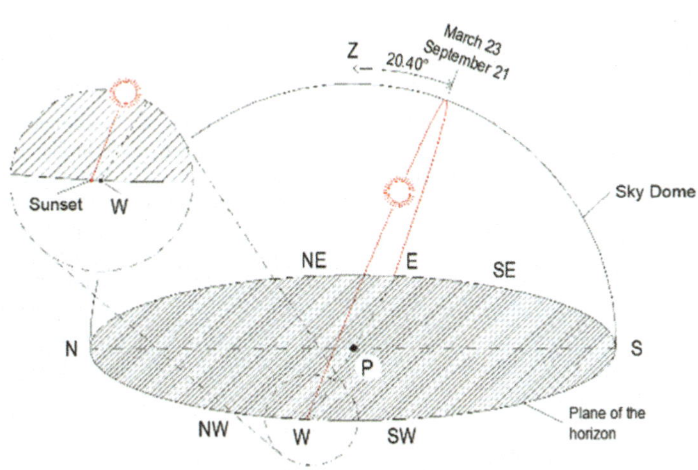

Fig. 6 The figure represents the sun path on the "quarter days" corresponding to the dates of March 23rd and September 21st, located two days after and before the spring and autumn equinoxes, respectively. *Source* Authors' creation

Table 1 Mayan and Incan solar parameters records

	Settlements	Solar effect or phenomenon used	Source	Description
Mayan	Chichen Itza: Kukulkán temple or «El Castillo», temple of warriors and upper temple of the Jaguars	Equinoxes. Days of the fourth sunset	Material sources: archaeological sources (archeoastronomy)	It is presumed that there was some Gnomes or pillar to carry out readings of the sunrise and sunset and thus make the calendar, religious schedule and agricultural activities. As well as being able to accurately locate their buildings
	"Siete Muñecas" temple in Dzibilchaltún [13]	Days of the fourth sunset	Material sources: archaeological sources (archeoastronomy)	
	Becán structure I, Calakmul structure I, Comalcalco structure I, Edzná "Casa de la Luna", "Pava del Mirador" pyramid, "Kinich Kak Moo de Izamal" building, la Pomoná structure IV, and others [7]	Sunrise Sunset	Material sources: archaeological sources (archeoastronomy)	
Inca	Cusco	Equinoxes Solstices Zenith Sun Sunrise Sunset	Historical sources: written	According to the texts, there was a pillar or gnome to carry out readings of the sunrise and sunset and thus make the calendar, religious schedule and agricultural activities
	Machu Picchu	Equinoxes Solstices Zenith Sun Sunrise Sunset	Material sources: archaeological sources (archeoastronomy)	Gnomes or Intihuatana construction to study the Sun and thus make the calendar and be able to schedule religious and agricultural activities

References

1. Ortiz E (2012) Una aproximación a los estudios arqueoastronómicos en los andes precolombinos [An approach to archaeoastronomical studies in the pre-Columbian Andes] Diálogo Andino, pp 57–72
2. Shepperson M (2009) Urban forms create complex spatial environments which form, and are formed by, the lives of their. World Archaeol:378
3. Ortiz E (2012) Los incas y el Sol: métodos de observación solar y calendario incaicos [The Incas and the Sun: methods of solar observation and Inca calendar] Revista Española de Antropología Americana, pp 127–143
4. Davis J, Mengersen K, Bennett S et al (2014) Viewing systematic reviews and meta-analysis in social research through different lenses. SpringerPlus 3:511. https://doi.org/10.1186/2193-1801-3-511
5. Snyder H (2019) Literature review as a research methodology: an overview and guidelines. J Bus Res 104:333–339
6. Nava PFS (2015) Orientaciones astronómicas en la arquitectura maya de las tierras bajas, México: Instituto nacional de antropología e historia [Astronomical Orientations in Lowland Maya Architecture, Mexico: National Institute of Anthropology and History]
7. Šprajc I, Sánchez P (2012) Astronomía en la arquitectura de Chichén Itzá: una reevaluación [Astronomy in the architecture of Chichén Itzá: a reappraisal]. Estudio de Cultura Maya XLI:60
8. Carlson JB (1999) Pilgrimage and the Equinox 'Serpent of Light and Shadow' Phenomenon at the Castillo, Chichén Itzá, Yucatán Archaeoastronomy. J Astron Culture:136–152
9. Aveni AF, Milbrath S, Peraza C (2004) Chichen Itza's legacy in the astronomically oriented architecture of Mayapán. Res Anthropol Aesthetics 45(1):123–143
10. Tozzer AM (1941) Relación de Las Cosas de Yucatán de Landa: una traducción [Relationship of the Things of Yucatán de Landa: a translation]. Edited with notes by Tozzer Alfred M Papers of the Peabody Museum of American Archaeology and Ethnology, Harvard University, vol XVIII, pp 394
11. Acosta J (1894) Historia natural y moral de las Indias [Natural and moral history of the Indies], Madrid Press by Ramón Angés
12. Herrmann DB (2020) Sobre la inclinación del Intihuatana de Machu Picchu [On the inclination of the Intihuatana of Machu Picchu] Filosofía de la educación en astronomía. Conferencias y ensayos seleccionados 1995–2020, p 244
13. Šprajc I, Sánchez P (2018) El Sol en Chichén Itzá y Dzibilchaltún: la supuesta importancia de los equinoccios en Mesoamérica [The Sun in Chichén Itzá and Dzibilchaltún: the supposed importance of the equinoxes in Mesoamerica] Arqueología Mexicana, vol 149, pp 26

The Performance of Multinational Foreign Labours at Construction Project in Klang Valley, Malaysia

Mohd Faisal Mat Ali, Mazlina Zaira Mohammad, and Nor Syamimi Samsudin

Abstract Malaysia is experiencing a huge influx of foreigners worldwide, especially of low-skilled workers. Malaysia has been suffering labour shortages in many key sectors of the country such as construction, plantation, industrial and service. Foreign labour is another country's worker who comes to work in Malaysia for a period that is also known as economic labour. This study was conducted to study the performance of multinational foreign labours on a construction project in Klang Valley. This research focuses more in identifying the most significant factors and effects on the performances of multinational foreign labours at the construction project. Besides that, this research will determine the strategy to improve the project performances of multinational foreign labours at the construction project. This study utilised a survey instrument by using a questionnaire and observations conducted on 80 respondents at construction companies. Data was collected and analysed by using SPSS version 22 and Microsoft Excel. Descriptive statistics such as mean, percentage and standard deviation were used to analyse the data. The research findings show that there are positive and negative factors and effects of the achievement of a diversity of multinational foreign Labours in construction projects. Besides that, the strategies also were identified to improve the performance work involving the multinational foreign labours at construction projects in Klang Valley.

Keywords Multinational labours · Foreign labours performance · Project performance

M. F. M. Ali
Institut Kemahiran Belia Negara (IKBN) Jitra, Kementerian Belia dan Sukan (KBS), KM 23 Jalan Changloon, 06000 Jitra, Kedah, Malaysia

M. Z. Mohammad (✉)
School of Civil Engineering, College of Engineering, Universiti Teknologi MARA Shah Alam, 40450 Shah Alam, Selangor, Malaysia
e-mail: mazlinazaira@uitm.edu.my

N. S. Samsudin
Department of Built Environment Studies and Technology, Faculty of Architecture, Planning and Surveying, Universiti Teknologi MARA, Perak Branch, Seri Iskandar Campus, 32610 Seri Iskandar, Perak, Malaysia

T. Kang and Y. Lee (eds.), *Proceedings of 2021 4th International Conference on Civil Engineering and Architecture*, Lecture Notes in Civil Engineering 201,
https://doi.org/10.1007/978-981-16-6932-3_49

1 Introduction

The globalisation has created a new phenomenon and giving varying impacts to economic development between countries, especially developed and developing countries. However, most developing countries are developing tend to stay behind in the current flow of globalisation compared to developed countries. The process of globalisation has resulted in wide imbalance in terms of technology, distribution source, level of production and access to finance worldwide and further contributes to income distribution global imbalance. International economic insecurity international will continue to follow the pattern the development of globalisation. Some important impacts globalisation of global economic growth are differences in income gap, differences in foreign direct investment gaps, ICT development gaps, international migration either by foreigners or refugees, economic globalisation, environmental issues as well implications of policy in a country [1].

The impact of globalisation will be felt by all nations and societies all over the world. Quality management in the construction industry is vital for sustainability development hence the efficient construction management handling a multinational foreign labour also important [2]. This research discusses the impacts of multinational foreign labours at construction project in Klang Valley against the local labours. The issue of cross-border immigration is a global issue. This also happens in developing countries with a wide range attractive factors of attractiveness such as economic opportunities, employment, stability, security as well as gain protection. Australia, Canada, Malaysia and Saudi Arabia are among the countries chosen by illegal immigrants or invalid workers [3]. For the construction companies that use foreign labours, it will cause some problems to their companies. Among the problems are unsatisfactory productivity level and bad quality of work. Most construction companies take advantage of hiring foreign labours because their salaries are low to pay them. Nevertheless, the company which hire the foreign labours must bear the consequences of such factors as lack of experience in field, weaknesses in communication between construction workers and misinterpretation in understanding the instruction manuals provided by employers.

1.1 Construction Sector Contribution

There have been many theories and studies on the factors affecting the arrival of foreign labours in the construction sector and other sectors in Malaysia. The study has collected 20 journals related to the current study. According to [4], it was stated that foreign worker is consider temporarily employed in other countries. Foreign labour has been employed by recruitment agencies or contracted by temporary contracts by certain companies to meet labour requirements in certain sectors of the countries. In addition, foreign labours can be categorised into two types which are workers with valid work permit and illegal permits. Foreign labours with valid work permits will usually be employed by employers who need their services. While the illegal labours are called the unauthorized labours [5].

According to [6], it was mentioned that Malaysia experienced a large influx of foreigners in the world, especially in low-skilled workers. Malaysia has experienced labour shortages in major sectors of the country especially in construction industry. With that, with the entry of foreign labours it can help the construction development growth in Malaysia. For this study, the researchers intend to identify the most significance factors and effects on the performance of the multinational foreign labour at construction project. Apart from that, the study was conducted to determine the strategy to improve the project performance with multinational foreign labours at construction site.

1.2 Research Purpose

This research studies about the action with power of authority especially Construction Industry Development Board (CIDB) and Ministry of Human Resources (MOHR) to monitor, analyse and give right trainings to the foreign labours to enhance their job skills from lower skills to the competent skills labour. The null hypothesis is multinational foreign labours have a significance relationship to the site performance of construction project. The research purposes are to identify the main factors of performance multinational foreign labour in a project, to determine the effect of these labours to the project performance and to identify the strategies to improve the construction project performance associated to multinational foreign labours manpower. Figure 1 shows the theoretical framework focus on two parts which is the multinational foreign labours impacts towards construction industry.

The following research objectives (RO) have been identified in order to achieve this research study.

RO1: To identify the most significance influence factors on the performances of multinational foreign labours at construction project.

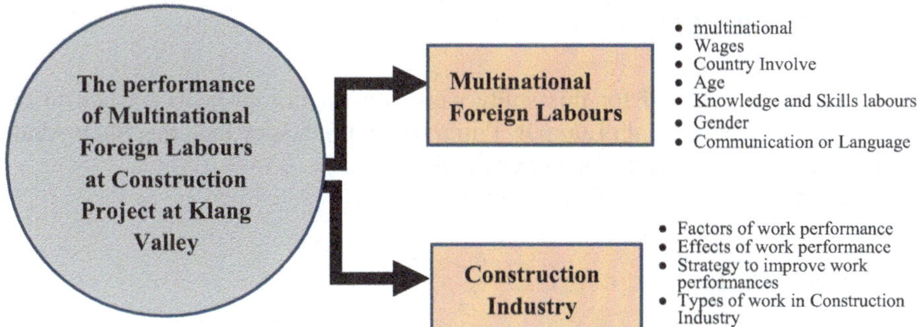

Fig. 1 Theoretical framework

RO2: To identify the most significance of effects of multinational foreign labours intake on the performance of construction project.

RO3: To determine the strategy to improve the performances of construction project related to the intake of multinational foreign labours.

2 Research Methodology

This study utilised quantitative research method which is a survey-based instrument by using a cross-sectional survey design. The quantitative data can be measured through a process such as a questionnaire and data were analysed using the Statistical Package for the Social Sciences (SPSS). According to [7], cross-sectional design involves the method of collecting data on a single sample from a single study population based on the existing respondents' attributes cited by [8]. Through this design, information and descriptions by each category of respondents can be carried out comprehensively.

2.1 Questionnaire Development

The questionnaire was divided into the five (5) sections. The first section which is Section A involves demographic of respondents. Section B involves questions in identifying the most significance influencing factors on the performances of multinational foreign labours at construction project. Section C serves to identify significance of effects of multinational foreign labours intake on the performance of construction project. Section D serves to determine the strategy to improve the performances of construction project. Last but not least is Section E that requests the comments and suggestions from respondents for improvement about this research.

2.2 Data Sampling

The populations selected for this study were construction projects located in the Klang Valley namely Kuala Lumpur, Putrajaya and Selangor includes Shah Alam, Petaling Jaya, Klang, Puchong, and Gombak.

2.3 Pilot Study

The pilot study aims to ensure that the methods used have high validity and reliability and are able to collect data as needed. Furthermore, the findings from the pilot study will enable the researcher to identify the weaknesses of the instrument and to determine the strength of the study. In addition, this pilot study was conducted to address the difficulty of these questions whether realistic and practical and appropriate and meet the needs of this study. Therefore, any question that is inappropriate or difficult to understand or can be misleading, can be improved or eliminated. The questions are composed of closed-form questions that include age, gender, place of residence, status, occupation type, and so on. Prior to this study, a pilot study was conducted with 20 respondents. The respondents involved among 10 officers at the Ministry of Human Resources (MOHR) Putrajaya, 5 officers of Vocational Training at CIAST Shah Alam and 5 students of the Faculty of Civil Engineering at UiTM Shah Alam. The choice of respondents to make is based on the fact that they have characteristics similar to those of the respondents required in this study.

2.4 Data Collection

Respondents were given 10 min to respond and complete the questionnaire. The questionnaire was more focused on the information or background data of the respondents and their perceptions of foreign labours. In English version questionnaire was prepared with 100 sets through face-to-face handover the survey. However, eighty respondents' feedback was obtained.

3 Findings

Their construction company has a diverse workforce of foreign labours, especially from the third developing countries such as Indonesia, Bangladesh, India, Nepal, Cambodia, Myanmar and there is one other country from China. Foreign labours who are willing to come to Malaysia because the Malaysian construction industry has given foreign laborers the freedom to work in high-risk job, dirty, dangerous and complex sectors such as farming, mining and others including the construction sector. They are willing to work with a low wage even with a daily wage. The findings shown the highest foreign labours are came from Indonesia 47.5%. The second highest is Bangladeshi 43.8% and followed by China 3.8%, India, Nepalese and Pakistani. Most of construction company at Klang Valley had more than 5 nationalities of foreign labour in one project.

In the construction sector in the Klang Valley, foreign labours are involved in the construction of MRTs, LRTs, residential houses, high rise building, road and highways. Tables 1, 2 and 3 tabulated the percentage of respondents' feedback on the survey associated this research. Most of the respondents were employees working as Project Engineers, Project Manager, Site Manager, Site Engineer, Safety Officer, Safety Supervisor and Clerk of Work at construction site company in Kelang Valley. A total of 61 respondents were from Selangor, 16 respondents from Kuala Lumpur

Table 1 Influence factors on the performance of multinational foreign

Element	Strongly disagree (%)	Disagree (%)	Neutral (%)	Agree (%)	Strongly agree (%)
Lack of knowledge to use the standard operating procedure (SOP)	0	3.8	30	41.2	25
Educational issues	2.4	10	31.3	36.3	20
Lacks of adequate awareness on safety and health	0	8.7	30	33.8	27.5

Table 2 Effects of multinational foreign labour intake on the project performance

Element	Strongly disagree (%)	Disagree (%)	Neutral (%)	Agree (%)	Strongly agree (%)
Lacks of skills at work field	2.5	12.5	25	36.3	23.7
Lacks of confident level	0	11.3	48.8	30	10
Lacks of communication to speak in Malay, English or their mother tongue language	0	10	32.5	27.5	30

Table 3 The strategies to improve the performances of construction project related to the intake of multinational foreign labours

Element	Strongly disagree (%)	Disagree (%)	Neutral (%)	Agree (%)	Strongly agree (%)
Educate the foreign labour	2.5	1.3	30	52.5	13.8
Company commitment and involvement to improve the foreign labour skills	0	2.5	25	51.3	21.2
Communicate, inform and discuss between employer and employee	0	3	24	51	2

and 3 respondents from Putrajaya. This analysis uses ranking by following the high percentage (%) which respondents have answered the related questions about the research topic.

3.1 Influence Factors on the Performance of Multinational Foreign Labour

Table 1 shown the influence factors on the performance of multinational foreign labour in construction works project. The findings indicate the most influence factor is lack of knowledge to use the standard operating procedure (SOP), followed by lacks of adequate awareness on safety and health and educational issues. The foreign labour had the limited understanding regarding the instruction or SOP in construction work activities. If the foreign labour unable to follow the SOP, it may cause unproductivity to the project performance. The second factor shows lack of awareness on safety and health is matter on the performance of multinational foreign labour because safety is priority in construction project. Education also contributes to their performance factor. However, it is not the main concern due to their job scope is more on practical skills rather than critical thinking as a professional job.

3.2 Effects of Multinational Foreign Labour Intake on the Project Performance

Based on result shown in Table 2 listed the effects of multinational foreign labour working together in one construction project. The first highest effect is lacks of skills at work field because no knowledge and skills at construction activities before they leave their country. The second highest influence effect for taking the multinational foreign labours is lacks of communication or language barrier among foreign labours specially to speak in Malay and English. This will be affected to the construction project planning because they work in team and difficult to understand the work instruction from management to run the construction activities. They will understand on their own mother tongue language to communicate each other with similar nationality. Hence, the conflict may always occur among the foreign labours. The third effect is lacks of confident level at work field. Foreign labours in Malaysian construction are mixing with at least five different nationalities may lower the confident level of each nationality.

3.3 The Strategies to Improve the Performances of Construction Project Related to the Intake of Multinational Foreign Labour

Table 3 indicates the strategies to improve the performances of construction project related to the intake of multinational foreign labour. The highest result is consider based on the percentage summation of agree and strongly agree. Company commitment and involvement to improve the foreign labour skills is vital in order to improve the performance of construction project. Followed by the other strategy which is to educate the foreign labour either through on-site training or daily toolbox meeting. The construction industry in Malaysia also must set a requirement for the selection of foreign labour to work in Malaysia. Each country that allows their people to work in Malaysia is require to educate their citizen with the construction skills and knowledge and also basic Malay language and culture. The purpose is before they come to Malaysia, it will be ease for them to adapt the nature of work. The company management have to pay attention on this issue to avoid the production or progress of construction project had the long-term problem. Other important strategy is communication between employee and employer to well inform and discuss must be achieved at the same wavelength to maintain and improve the project productivity. Therefore, the construction companies can carry their project smoothly.

4 Discussion

The construction employers require the best solution to provide a good training for foreign labours before they can start the job at construction areas. This finding supports the research done by [9]. It actually gives the positive impacts when the foreign labours do the real task and it is the strategy to improve their knowledge and skills in construction industry. In ensuring the success for this planning, this activity needs involvement from employers and companies to handle the training sessions. In saving the government's or company's allocation to train new foreign labours that comes to work in Malaysia, the suggestion is that the foreign labours origin countries need to handle their new labours before coming to enter Malaysia construction industry. Florina [10] highlighted the involvement of construction foreign labour in Malaysian is crucial and requires action by authority.

The involvement of government that dictates only the skilled labours or experienced labours can be given visa and permits. The government needs to states the requirements to all embassies if has their citizens wish to work in Malaysia country especially in construction industry. The government must to ensure the foreign labours to be legally certified. References [11, 12] were explains Malaysia is depending on foreign labour from various country only to support the operation works in construction project. Hence, this is very important to control our country, for safety and security and ensuring the best health condition to our Malaysian citizen from contagious disease for example the pandemic which is currently being faced worldwide now which is Covid-19.

Construction companies need to employ highly skilled labours to handle new technologies in construction sites. The construction industry should bring in new technologies to limit their dependency on manual labours especially foreign labours. Employing foreign workers usually lead to reducing construction cost, but if the employers simply employ labours just for the sake of reducing cost without taking into consideration in terms of their knowledge, skills, and experience, it will lead to low project quality. Hence, this will affect the country's economic growth due to lack of trust in Malaysian construction projects. According to [13] also asserted the impact of hiring outside labour will contribute either direct or indirectly to positive and negative impact for the country. It is a vital for the construction key players to take an initiative concern on how to handling the multinational foreign labour in the most effective way.

5 Conclusion

This research study achieved three research objectives which are first is to identify the most significance influence factors on the performances of multinational foreign labours at construction project, second is to identify the most significance of effects of multinational foreign labours intake on the performance of construction project. Lastly is to determine the strategy to improve the performances of construction project related to the intake of multinational foreign labours.

The first objective identified the most influential factor is lack of knowledge to use the SOP, followed by lacks of awareness on safety and health and educational issues that affect the performance of foreign labor at the construction project. The second objective identified the most significant effect by taking multinational foreign labour working together in one project is lacks skills at work field, followed by lacks of communication to speak Malay language and lastly is lacks confident level. The third objective determined the strategy to improve the performances of construction project related to the intake of multinational foreign labours which are by prioritizing the company commitment and involvement to improve the foreign labour skills, then educate the foreign labour in many ways and lastly is to communicate, inform and discuss between employer and employee with suitable method to ensure the foreign labour understand the instruction very well especially there is a language barrier.

In the year of 2020, the world has been hit by pandemic Covid-19. Malaysia has been facing many difficulties during restriction movement order and difficult for the researcher to get sufficient data from questionnaires survey on site. This limitation in getting the respondents is consider insufficient feedback. Therefore, for the future studies the following scopes of studies are recommended. The researcher suggests that future researchers should use the interviewing face-to-face approach to elicit data from respondents at construction sites and take more time to get more accurate and effective answers from respondents. Question items in questionnaires must be written in short sentences, clearly constructed and precise because the respondents do not have enough time to read and answer the question items.

Acknowledgements This work was received financial support from internal special research grant (Geran Penyelidikan Khas) by Universiti Teknologi MARA (UiTM)—600-RMC/GPK 5/3 (261/2020).

References

1. Aini N, Koon GL (2009) The impact of globalization on poverty. Dalam, Ekonomi Malaysia towards balanced development. Universiti Kebangsaan, Malaysia
2. Mohammad MZ, Viana M, Raghib AT (2019) Benefits and implications of the different types of quality management in the Malaysian construction industry. IOP Conf Ser Mater Sci Eng 650(1):012008
3. Manu P, Emuze F, Saurin TA, Hadikusumo BH (eds) (2019) Construction health and safety in developing countries. Routledge
4. Ramesh KM, Charles SPR, Peter Y (2012) The impact of employment of foreign workers, local employability and trade union roles in Malaysia
5. Abdul Rahim AH, Bachan S, Wan Zulkifli WY, Aminah MY, Norzamzila M (2012) Problems faced by contractors in managing foreign workers on construction sites
6. Intan S, Sity D (2016) Implications for foreign worker inclusion in Malaysia: a preliminary review
7. Malhotra NK, Agarwal J, Peterson M (1996) Methodological issues in cross-cultural marketing research: a state-of-the-art review. Int Marketing Rev
8. Jumaat AK, Yasiran SS, Malek AA, Rahman WEZW, Badrin N, Osman SH, Mahmud R (2014) Performance comparison of Canny and Sobel edge detectors on balloon snake in segmenting masses. In: 2014 international conference on computer and information sciences (ICCOINS). IEEE, pp 1–5
9. Mohammad MZ, Hadikusumo BH (2015) A model of integrated multi-level safety intervention practices in construction industry. Proc CIB W099:49–61
10. Florina A, Norhanim MAM, Hamimah A, Har Einur B, Mohd Reza E, Ahmad (2012) Involvement of foreign labour in Malaysian construction Industry
11. Arzlee H (2012) Dependency of foreign workers in Malaysian construction industry
12. Wen Y, Albert PC, Chan (2014) Critical review of labor productivity research in construction journal
13. Hamzah AR, Chen W, Lincoln CW, Shu FL (2012) Negative impact induced by foreign workers: evidence in Malaysian construction sector

Investigating the Causes of Disputes from the Perspective of Project Characteristics

Mathusha Francis and Thanuja Ramachandra

Abstract Disputes have become an inherent feature of construction projects and often, lead to unsuccessful projects at the end. This encourages the construction industry to step towards dispute prevention which could begin at the early stage of a project where salient project characteristics are visible and responsible for the prevalence of disputes. For example, the causes which contribute to disputes such as complexity of design, unrealistic time target and inappropriate selection of procurement method can be addressed by the project characteristics of degree of deign complexity, time constraint and procurement method respectively. This research therefore aims to explore the causes of disputes from the perspective of project characteristics towards managing disputes effectively. A questionnaire survey was attended seventy-eight (78) professionals who involved in construction disputes, drawn through snowball sampling technique. A total of 52 causes and 14 project characteristics were identified through literature review and carried for the questionnaire survey. Amongst 30 causes and 10 project characteristics found to be significantly contributing to disputes. The significant causes further analysed from the perspective of project characteristics. Amongst 14 causes are correlated with project characteristics conceptually. Therefore, this study concludes that the careful selection of project characteristics would contribute to mitigate the disputes in construction projects.

Keywords Causes · Disputes · Project characteristics

1 Introduction

Disputes are common in the construction industry due to its complexity and multiparty involvement in projects. In general, each stakeholder has different expectations from the project and this derives to conflicts in early stage and eventually develops into disputes. Such disputes result in several ill effects. The growing construction disputes can lead to disruption of construction schedules, increased project costs,

---

M. Francis (✉) · T. Ramachandra
Department of Building Economics, University of Moratuwa, Moratuwa, Sri Lanka
e-mail: mathushaf@uom.lk

T. Kang and Y. Lee (eds.), *Proceedings of 2021 4th International Conference on Civil Engineering and Architecture*, Lecture Notes in Civil Engineering 201,
https://doi.org/10.1007/978-981-16-6932-3_50

and even adversely affect the relationships between project participants [1]. When construction disputes are not resolved in a timely manner, it becomes very expensive in terms of finances, personnel, time, and opportunity costs [2]. High cost are resolved in lawsuits arising out of disputes and conflicts in the United States construction projects.

These situations encourage the construction industry to step towards dispute avoidance as per the proverb "prevention is better than cure" [3]. To this note, Gerber [4] explains that dispute avoidance provides a refreshing new way of looking at and responding to the underlying causes of disputes. Further the dispute avoidance strategies such as allocating fair contract risk, drafting dispute clauses, team building, provision of a neutral arbitrator and binding arbitration as the strategies would enable preventing conflicts from escalating into disputes [4, 5]. On the other hand, a relationship could be seen between procurement methods and the frequency of disputes. The traditionally procured projects experience more disputes while partnering and alliancing provide less prospects for disputes. On a different note, the complexity of project causes disputes in construction projects while degree of complexity is one of the project characteristics which could directly influencing dispute occurrences [3]. Similarly, few authors stated unrealistic time targets lead to dispute whereas time constrain is a project characteristic [6, 7]. The causes of unclear and unfair risk allocation, inappropriate selection of procurement method and inadequate brief can be identified in the inception stage of project. These causes could be originated from procurement method. However, the studies fail to address the causes in terms of project characteristics. This makes clear that the disputes could be mitigated where possible by giving due consideration to project characteristics appropriately. Therefore, the current study investigates the causes of disputes and the project characteristics which influence on disputes and thereby tends to mitigate disputes from the perspective of project characteristics.

2 Literature Review

The causes of disputes have been the focus of several researches over the decades. Many researchers have investigated the causes of disputes from different perspectives across nations over decades. For example, a study on Australian infrastructure projects concluded with ten (10) key problematic areas in disputes such as variations to scope, contract interpretation, EOT claims, site conditions, late or incomplete information, obtaining approvals, site access, quality of design, and availability of resources [8]. Out of which, contract interpretation and site conditions could be attributed to project characteristics of type of standard contract document and site factor. Recently, an author clustered 29 direct and 32 indirect causes of disputes. The analysis indicated that the top five severe direct causes are: delay in progress payment by owner, unrealistic contract duration, change orders, poor quality of completed works, and labor inefficiencies respectively. While the top five indirect causes are: inadequate contractor's experience, lack of communication between construction parties, ineffective planning and scheduling of project by contractor, cash problems during construction and poor estimation practices. Amongst, the direct causes such as unrealistic contract duration and change orders can be related to the project characteristics of time constraints, and scope of work. A researcher clustered the causes of disputes in terms of contract and specification, financial and environmental and concluded that delay in progress payment by owner was the most critical factor [7]. Similarly, another research classified the causes from the perspective of owner related, contractor related, design related, contract related, human behavior related, project related and external factor related. The authors found that contractor related causes are significantly contributing to disputes [9]. In another archival study of ligation cases conducted to ascertain the causes of disputes in the UK construction industry, it was indicated that most contributing causes are defects, variations, design changes, non-completion, extension of time and poor workmanship.

On a different view to previous researchers, the project characteristics could influence on claims and disputes. The project characteristics of complexity, size of project and duration of project are the top three influencing factors. Authors further found that correlations exit between the size of project and duration of project along with the amount claimed [10]. Claims and disputes are frequent in large projects and can cause budgetary difficulties to employers and loss of liquidity to contractors. Construction projects are often described as being complex. In support of this, the authors established that the construction projects are complex in terms of organisational complexity (people involved and their relationships), uncertainty and overlap of construction elements. Further, the construction industry is highly fragmented as the projects are divided into parts that are subcontracted to individual enterprises. On the other hand, fragmented structure of construction industry causes disputes in projects [9]. Scope changes and rework can lead to acceleration of work activity. These scope changes may remain as ambiguous and create disagreement between parties and turns into disputes [11]. These are usually the result of inadequate planning at the project definition stage or because of lack of involvement of the owner in the design phase. These evidenced that the root causes of disputes originated at the early stages of construction projects.

Certain projects are subjected to tight time schedules. Timing will influence whether subsequent activity can occur as planned and, in many cases, may severely affect those factors identified as critical or high priority in the business case [12]. Further nature of the projects also decides the duration of project however most clients are expecting project in shorter period. The unexpected weather conditions also cause time constraints in projects. The time constrains also arise as a need of the client to finish on a specific date or time period. However, many authors suggested that acceleration due to the unrealistic expectation of client could lead to disputes [2, 6, 7, 9]. On a different note, the constraints in a multi-party working situation bring complications in project management which then further develops into conflicts and disputes, results in cost consequences to clients and contractors. From the funding perspective, the contractual disputes is one of the major problems affecting the prosperity of public funded projects in Taiwan [13].

The foregoing review evidences that there is a relationship exists between project characteristics and causes of disputes. In line with that, literature suggests that dispute in a construction project can be addressed from the causes of dispute and the characteristics of the project. Thereafter there seems no research investigated disputes from the perspective of project characteristics. Therefore, the current research further investigates the causes of disputes and project characteristics.

3 Methodology

Initially, a comprehensive literature review was carried out to revisit the causes of disputes and project characteristics. Accordingly, fifty-two (52) causes of disputes and fourteen (14) project characteristics were identified and incorporated into a questionnaire. The questionnaire survey was responded by a sample of seventy-eight (78) experienced professionals who are experts in managing disputes. The required sample was drawn using snowballing, as it was difficult to quantify the targeted population. A minimum number of 30 sample for statistical analyses provides a useful rule of thumb for the smallest number in each category within your overall sample. Thus three categories of respondents such as adjudicators/arbitrators, senior quantity surveyors and project managers are employed. Respectively, numbers of 12, 34 and 32 were engaged in the current research. It was approachable only 12 adjudicators/arbitrators due to lack of practicing group. Finally, seventy-eight (78) questionnaires were carried to the analysis.

The questionnaire required the respondents to record their opinions using a 5 point Likert scale where the scale of 1 to 5 refers to very low, low, moderate, high and very high respectively. The reliability of mean and standard deviation are inadequate to the overall ranking [14]. Therefore, Relative Importance Index (RII) was computed using to rank the significant project characteristics and causes of disputes.

$$\text{Relative Important Index(RII)} = \frac{\Sigma w}{A x N}$$

where 'w' refers to Weighting given to each factor by each respondents, 'A'—Highest Weight and 'N'—Total number of respondents.

According to Akadiri [15], five important levels are transformed from RI values: high (H) ($0.8 \le \text{RI} \le 1$), high medium (H–M) ($0.6 \le \text{RI} \le 0.8$), medium (M) ($0.4 \le \text{RI} \le 0.6$), medium–low (M-L) ($0.2 \le \text{RI} \le 0.4$) and low (L) ($0 \le \text{RI} \le 0.2$).

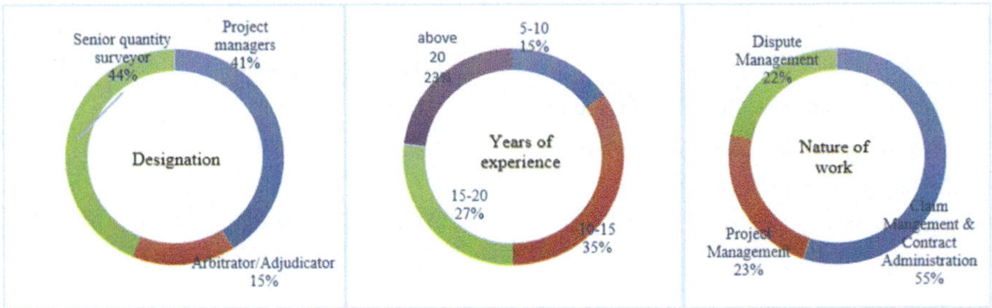

Fig. 1 Profile of participants

4 Research Analysis and Findings

4.1 Profile of Respondents

In order to ensure the reliability of the survey findings, due consideration was given to designation, years of experience and the nature of work in selection of participants. The professionals who have involved in disputes and related matters are involved in the questionnaire survey. Figure 1 depicts the profile of survey participants.

4.2 Causes of Disputes

The list of causes identified through literature review was presented to the research participants and requested to indicate the level of contribution of those causes to disputes in Sri Lankan construction projects. The level of contribution was recorded on the scale of 1–5 where '1' represents 'very low' while '5' represents 'very high'. Based on the responses, the RII values were calculated. Table 1 presents the causes of disputes, the respective RII values and their significance level.

A total of 30 causes which have above 0.800 RII values were considered as highly contributing causes to disputes. The top three causes of disputes are errors, contradiction and ambiguities of documents; client's slow response to decision making and discrepancies in contract provisions. This finding is in support of previous study [7] that slow response to decision making is one the top most leading causes of disputes. In line with the current findings, several researchers also identified discrepancies in contract provisions is one of the significant causes leads to disputes [2, 6, 9]. The findings revealed that payment delays as the significant cause leading to disputes with the RII value of 0.892. On a similar note, another researcher [7] also identified payment delays as a critical factor contributing to disputes. Contrarily, delay in payment is one of the least treacherous causes of disputes in Pakistan construction industry [2]. As shown in Table 1, major accidents, unexpected social event and environmental pollution are the least contributing causes of disputes. In line with

Table 1 Causes of disputes

Causes of disputes	RII	Rank	Sig. level
Errors, contradictory, ambiguities of documents	0.946	1	H
Client-slow response to decision making	0.933	2	H
Discrepancies in contract provisions	0.931	3	H
Design complexity	0.908	4	H
Frequent changes to scope	0.903	5	H
Payment delay	0.892	6	H
Unrealistic requirement (i.e. time)	0.890	7	H
Cost overruns	0.885	8	H
Inadequate contract administration	0.869	9	H
Inadequate/errors in soil investigation	0.867	10	H
Inadequate brief	0.862	11	H
Delayed drawings	0.862	11	H
Late instructions	0.856	13	H
Incomplete source of fund	0.849	14	H
Inappropriate procurement method	0.849	14	H
Unstable financial status of client	0.846	14	H
Unfavourable risk allocation	0.846	14	H
Time overruns	0.846	14	H
Delay in progress of work	0.846	14	H
Fragmentation of the system	0.841	20	H
Inappropriate selection of contractor	0.838	21	H
Poor communication	0.836	22	H
Poor documentation of contractor	0.823	23	H
Delay in evaluation of completed works	0.823	23	H
Unrealistic tender price	0.821	25	H
Inadequate site investigation	0.821	25	H
Unjust and untimely presentation of claims	0.815	27	H
Delay in approvals	0.813	28	H
Error in estimation (over or under)	0.810	29	H
Design errors and poor quality design	0.810	29	H
Unpredictable site/ground conditions	0.797	31	H-M
Opportunistic behaviour	0.792	32	H-M
Poor planning and scheduling	0.790	33	H-M
Lack of team spirit	0.790	33	H-M
Inappropriate contract type	0.790	33	H-M
Negligence/lack of professionalism	0.782	36	H-M

(continued)

Table 1 (continued)

Causes of disputes	RII	Rank	Sig. level
Acceleration of work	0.782	36	H-M
No contractor involvement in design stage	0.772	38	H-M
Poor supervision	0.767	39	H-M
Different interests of project parties	0.762	40	H-M
Lack of risk management	0.759	41	H-M
Competitive tendering and price competition	0.751	42	H-M
Inflation	0.733	43	H-M
Excessive and exaggerated claims	0.728	44	H-M
Unclear milestones	0.723	45	H-M
Changes in government regulations	0.723	45	H-M
Poor labour productivity	0.721	47	H-M
Weather changes	0.649	48	H-M
Material shortage	0.644	49	H-M
Major accidents	0.569	50	H-M
Unexpected social event	0.554	51	H-M
Environmental pollution	0.551	52	H-M
Unpredictable site/ground conditions	0.797	31	H-M
Opportunistic behaviour	0.792	32	H-M
Poor planning and scheduling	0.790	33	H-M
Lack of team spirit	0.790	33	H-M
Inappropriate contract type	0.790	33	H-M
Negligence/lack of professionalism	0.782	36	H-M
Acceleration of work	0.782	36	H-M
No contractor involvement in design stage	0.772	38	H-M
Poor supervision	0.767	39	H-M
Different interests of project parties	0.762	40	H-M
Lack of risk management	0.759	41	H-M
Competitive tendering and price competition	0.751	42	H-M
Inflation	0.733	43	H-M
Excessive and exaggerated claims	0.728	44	H-M
Unclear milestones	0.723	45	H-M
Changes in government regulations	0.723	45	H-M
Poor labour productivity	0.721	47	H-M
Weather changes	0.649	48	H-M
Material shortage	0.644	49	H-M
Major accidents	0.569	50	M
Unexpected social event	0.554	51	M

(continued)

Table 1 (continued)

Causes of disputes	RII	Rank	Sig. level
Environmental pollution	0.551	52	M

the current findings, the previous study also identified noise pollution as the least contributing cause to disputes [7].

4.3 Project Characteristics Which Influence on Disputes

The participants' views on the degree of influence of project characteristics on disputes were sought on a scale of 1–5 where '1' represents very low while '5' represents very high. Based on the responses, RII values were calculated and Table 2 represents project characteristics, their RII values and the respective significance level.

The findings show that degree of design complexity (0.949) is the top most influencing project characteristic on disputes in construction projects while other characteristics such as time constraints, procurement method used, type of client, size of project, site factors, standard form of contract and source of funding next level influencing project characteristics with more 0.800 RII values which referred as highly significant by Akadiri [15]. In line with the current study, another related study [7]

Table 2 Project characteristics influencing disputes

Project characteristics	RII	Rank	Sig. level
Degree of design complexity	0.949	1	H
Time constrains	0.915	2	H
Procurement method	0.887	3	H
Type of client	0.879	4	H
Site factors	0.813	5	H
Standard form of contract	0.810	6	H
Source of funding	0.805	7	H
Project cost	0.800	8	H
Payment method	0.779	9	H-M
Project scope	0.754	10	H-M
Duration of project	0.741	11	H-M
Type of project (i.e. building, road, bridge…)	0.677	12	H-M
Project location	0.674	13	H-M
Construction method and technological advancement	0.662	14	H-M

identified complexity of project as the top most project characteristic which influencing claims and disputes in Nigeria. On the other hand, many researchers have claimed complexity as an inherent feature of construction project. The literature suggested that size and duration of the project significantly influence on disputes. In line with, current study indicated that size of the project significantly effects on dispute however, duration found to be less significant [7].

4.4 Relationship Between Causes of Disputes and Project Characteristics

This section of the current study tends to investigate the relationship between the significant causes of disputes and the project characteristics which influences on dispute occurrence, were described in the Sects. 4.2 and 4.3 respectively. Table 3 shows the conceptual relationship between the significant causes of disputes and the project characteristics.

As per Table 3, a total of fourteen (14) significant causes (out of 30) found to be conceptually related with project characteristics. For example, the top significant cause; errors, contradictory, ambiguities of documents can be addressed from the perspective of type of standard document; Client-slow response to decision making is directly related to type of client; Different interpretation of contract provisions which marked as third most significant cause which would be managed by the standard form of contract; Design complexity is the 4th ranked cause can be directly

Table 3 Relationship between causes of disputes and project characteristics

Causes of disputes	Rank	Project characteristics
Errors, contradictory, ambiguities of documents	1	Standard form of contract
Client-slow response to decision making	2	Type of client
Discrepancies in contract provisions	3	Standard form of contract
Design complexity	4	Degree of design complexity
Frequent changes to scope	5	Scope of project
Payment delays	6	Type of client, source of funding
Unrealistic requirement (i.e. time)	7	Time constrains, type of client
Inadequate/errors in soil investigation	10	Site condition
Incomplete source of fund	14	Type of client, source of funding
Inappropriate procurement method	14	Procurement method
Unstable financial status of client	14	Type of client
Unfavourable risk allocation	14	Procurement method
Fragmentation of the system	20	Procurement method
Delay in approvals	28	Type of client

linked to degree of complexity which is the top most project characteristics which influences on disputes. Further explaining the cause, Client-slow response to decision making, the different type of clients such as government, corporate company, private company and individuals create a different pattern in decision making and follow different procedures as well. For example, government client would practice a lengthy procedure including cabinet approval for any decision making. Thus, slow response to decision making formulate a relationship with the type of client. Thus, certain causes are identified to create correlation with project characteristics as depicted in Table 3.

5 Conclusions

The research aims to formulate the relationship between causes of disputes and project characteristics. Initially, a comprehensive literature review was conducted on causes of disputes and project characteristics. The review identified a total of fifty-two (52) causes of disputes and fourteen (14) project characteristics which are then carried to a questionnaire survey in order to identify the significant causes which contributing to disputes and project characteristics which influence on disputes. Thereby the study tends to formulate a conceptual relationship among causes and project characteristics.

The survey findings revealed that errors, contradictory, ambiguities of documents; slow response to decision making and discrepancies in contract provisions are the top most significant causes contributing to disputes in Sri Lankan construction projects. The research found that degree of design complexity, time constrains and procurement methods are the top three significant project characteristics which influence on disputes. The study further analysed the significant causes from the perspective of project characteristics and indicated a conceptual relation exist between the causes and project characteristics. For example, the causes of errors, contradictory, ambiguities of documents and discrepancies in contract provisions can be addressed by standard form of contract whereas client-slow response to decision making can be related to type of client. Therefore, the research recommends the industrial professionals to address the disputes beforehand from the perspective of project characteristics.

References

1. Yiu K, Cheung SO (2004) Significant dispute sources of construction mediation. In: Proceedings of the 1st international conference on the world of construction project management. 27th–28th May, Toronto, Canada
2. Farooqui RU, Azhar S (2014) Key causes of disputes in the Pakistani construction industry—assessment of trends from the viewpoint of contractors. In: 50th ASC annual international conference proceedings. Associated Schools of Construction, Pakistan
3. Cheung S, Yiu T (2006) Are construction disputes inevitable? IEE Trans Eng Manage 53(3):456–470

4. Gerber PA (2013) Dispute avoidance procedures ("DAPs")—the changing face of construction dispute management. Int Constr Law Rev 2001(1):123–129
5. Jannadia MO, Assaf S, Bubshait AA, Naji A (2000) Contractual methods for dispute avoidance and resolution (DAR). Int J Project Manage 18(2000):41–49
6. Love P, Davis P, Ellis J, Cheung SO (2010) Dispute causation: identification of pathogenic influences in construction. Eng Constr Archit Manage 17(4):404–423
7. Na Ayudhya BI (2011) Common disputes related to public work projects in Thailand. J Sci Technol 33(5):565–573
8. Austrailian Constructors Association (2006) Scope for improvement; a survey of pressure points in Australian construction and infrastructure projects. Blake Dawson Waldron, Australia
9. Cakmak E, Cakmak PI (2013) An analysis of causes of disputes in the construction industry using analytical network process. Social Behavioral Sci 109(2014):183–187
10. Bajere PA, Galadima UN, Durodola OD (2017) Impact of selected project characteristics on construction claims in Niger State, Nigeria. Covenant J Res Built Environ (CJRBE) 5(1)
11. Love PE, Davis P, London K, Jasper T (2008) Causal modelling of construction disputes. In: Proceeding of the twenty-fourth annual association of researchers in construction management conference. 1st–3rd Sept 2008. ARCOM, England, pp 869–878
12. Kelly J, Morledge R, Wilkinson S (2002) Best value in construction. Blackwell Publishing in Conjunction with the RICS Foundation, Oxford
13. Dada MO (2013) Conflicts in construction projects procured under traditional and integrated systems: a correlation analysis. Int J Constr Supply Chain Manage 3(1):1–15
14. Doloi H (2013) Cost overruns and failure in project management: understanding the roles of key stakeholders in construction projects. J Constr Manage Eng Manage:267–279
15. Akadiri OP (2011) Development of a multi-criteria approach for the selection of sustainable materials for building projects. PhD Thesis. University of Wolverhampton, UK

Post-pandemic Project Change Management Model: An Adaptable Framework Utilizing Levenberg–Marquardt Algorithm and Dynamic Causal Loop Diagram for Construction Innovation

Dante Silva, Kevin Lawrence de Jesus, Bernard Villaverde, Rachel Grace Dela Torre, Nicolich Espero, Kezia Justine Fermin, and Raymond Robie Ramirez Jr.

Abstract The coronavirus pandemic (Covid-19) had a detrimental effect on the majority of sectors, including the construction industry. The industry's operations, productivity, expenditures, and profit have all been impacted. With the present state of the economy, it's natural that certain methods and frameworks are no longer as useful, efficient, or effective as they once were. This study examines the various relationships between delays, profit, productivity rate, and project acquisition, as well as their impact on the project-change management practices of companies in the Philippine construction industry, with the goal of developing a new project-change management framework that is flexible and adaptable to future changes. This research employs a variety of statistical techniques to ascertain the significance of each variable and its contribution to project change management. Through the use of artificial neural network modeling and causal loop diagramming, the different connections between the variables were graphically interpreted. The artificial neural network modeling of the variables demonstrated a high degree of accuracy, with a R all value of 0.99492; scatter plots also revealed a high degree of positive correlation. The causal loop diagram illustrates the many connections between the input variables and their contributing components. Together, the models and statistical tests summarize the criteria employed by the researchers to create a project-change management framework that is focused on its capacity to be used for both short- and long-term objectives, is applicable to the new normal, and is adaptable to any future change.

Keywords Dynamic simulation model · Project-change management · Construction project changes · Productivity rate · Levenberg–Marquardt algorithm

---

D. Silva (✉) · B. Villaverde · R. G. D. Torre · N. Espero · K. J. Fermin · R. R. Ramirez Jr.
School of Civil, Environmental, and Geological Engineering, Mapua University, Muralla St., 1002 Intramuros, Manila, Philippines

K. L. de Jesus
School of Graduate Studies, Mapua University, Muralla St., 1002 Intramuros, Manila, Philippines

T. Kang and Y. Lee (eds.), *Proceedings of 2021 4th International Conference on Civil Engineering and Architecture*, Lecture Notes in Civil Engineering 201,
https://doi.org/10.1007/978-981-16-6932-3_51

1 Introduction

A worldwide epidemic has started, causing chaos across the globe, including the Philippines [1]. Several regions throughout the nation implemented community lockdowns in March 2020. The construction sector suffered significant losses as a result of community rules established by the Philippine government in the second quarter of 2020, which also resulted in downsizing and liquidation of certain businesses. There was a decrease in labor productivity and increased difficulties transporting goods [2]. The building sector in the Philippines had collapsed suddenly as the epidemic worsened. The industry had not anticipated that limitations would be implemented in the majority of the country's industries, affecting businesses' productivity rates and project change management [3].

It is certain that the epidemic has impacted both small and large construction businesses. Prior to the epidemic, the Philippines was well-known for its thriving and rapidly expanding building sector. Unfortunately, the government's funding for construction-related projects has been cut by 15.8 percent as a result of budget reallocation to combat this epidemic [4]. The Philippine Statistics Authority (PSA) noted an increase in the unemployment rate as a result of the community lockdowns [5].

2 Methodology

This study used a series of evaluations to determine the different factors that should be taken into account while developing a new project-change management framework. Figure 1 depicts the overall process of this study.

2.1 *Description of the Data*

A statistical analysis was performed of the survey data using IBM SPSS software [6–8] on the data from the "Effective project-change management model for construction

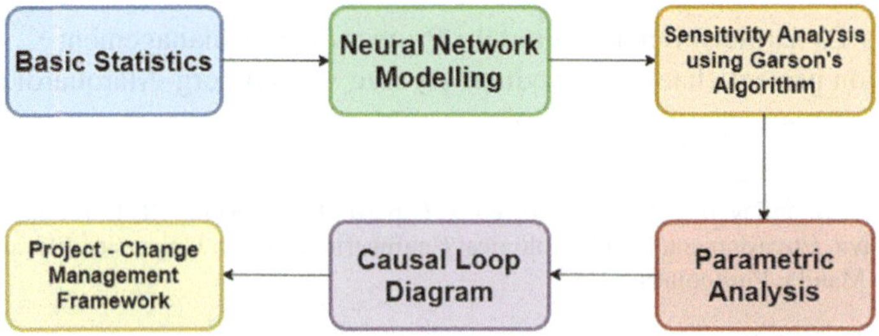

Fig. 1 Overall process of the study

Table 1 Descriptive statistics of the dataset

Parameter	Delays	Profit	Productivity rate	Project acquisition	Project-change management rating
Mean	4.142	4.000	3.745	3.760	3.888
Std. dev.	0.392	0.600	0.575	0.508	0.347
Skewness	− 0.131	− 1.807	− 0.131	− 0.450	− 0.039
Kurtosis	0.186	5.539	− 0.667	0.576	− 1.031
Minimum	3.10	1.40	2.50	2.30	3.30
Maximum	4.90	4.90	4.80	3.70	4.60

companies in the Philippines impacted by the pandemic" survey. Table 1 outlines the dataset's features and description for this investigation.

2.2 *Neural Network Modelling*

Using the data, develop an Artificial Neural Network Model [9, 10] that corresponds to the prediction of the project-change management rating for Philippines-based construction firms affected by pandemic.

The project-change management rating model for construction firms in the Philippines affected by the factors caused by the pandemic has internal characteristics presented in Table 2. Levenberg–Marquardt algorithm [11] was the training algorithm employed in the development of the model while the transfer function used is the hyperbolic tangent sigmoid (tansig) function [12, 13]. The number of hidden neurons used for the project-change management rating model development is equal to the "m" number of input parameters in reference to the study of [14].

The design and architecture of the developed project-change management rating model for construction firms in the Philippines affected by the factors caused by the pandemic are shown in Figs. 2 and 3.

Table 2 Internal characteristic of the project-change management rating model

Internal characteristics	Value
Training algorithm	Levenberg–Marquardt
Transfer function	Hyperbolic tangent sigmoid
Number of hidden neurons	4
Number of hidden layers	2 (1 hidden layer and 1 output layer)
Performance indicators	Pearson's correlation coefficient (R) and mean squared error

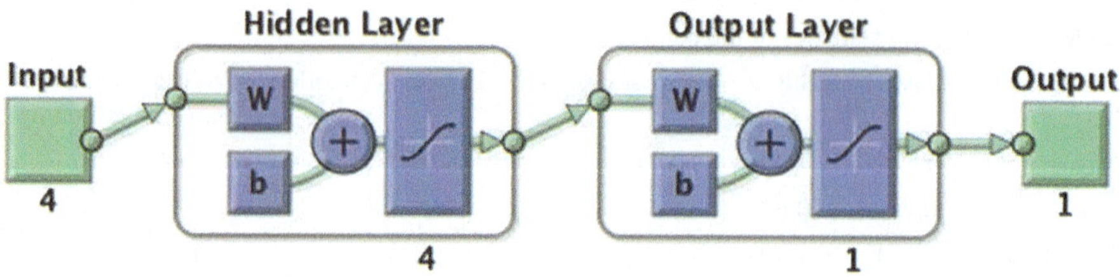

Fig. 2 Design of the developed project-change management rating model

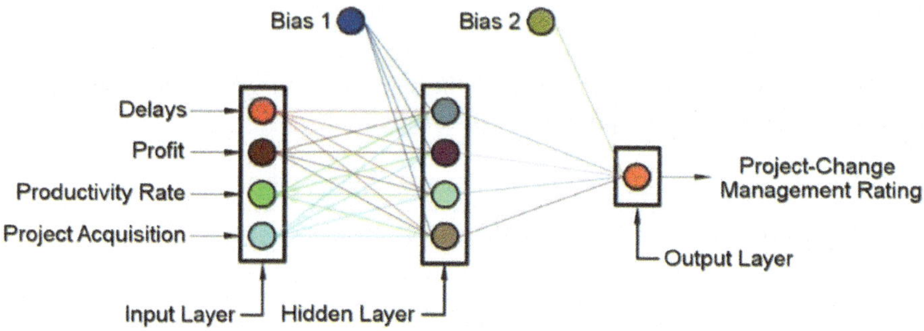

Fig. 3 Architecture of the developed project-change management rating model

2.3 Relative Importance Using Garson's Algorithm

The project-change management rating model utilized four factors as input parameters. It includes delays [15], profit [16], productivity, and project acquisition [17]. The relative importance of the factors contributing to the project-change management rating for construction firms in the Philippines affected by the factors caused by the pandemic was calculated using Garson's Algorithm [18].

2.4 Parametric Analysis

In identifying the behavior of the project-change management rating considering varying values of delay, profit, productivity, and project acquisition A parametric analysis of the Factors contributing to the project-change management rating Philippines-based construction firms affected by the factors caused by the pandemic was performed [19].

2.5 Causal Loop Diagram and Project-Management Framework

A Causal Loop Diagram of the input variables (delays, profit, productivity rate, and project acquisition) [20] employing the results from the calculation of basic statistics and relative importance was constructed to develop the project-management framework for construction firms in the Philippines in the post-pandemic era.

2.6 Respondent's Feedback

The respondents for the previously gathered data were presented as the final output. This is to measure if the achieved goal can successfully be applied in real life working companies that qualify in the affected population. The respondents were asked if the revised framework could be applied in their construction company and if it is more efficient than the existing ones during this pandemic [21].

3 Results and Discussion

3.1 Description of the Data

Based on the descriptive statistics of the instrument responses shown in Table 1, the highest mean responses were observed in the delay indicators while the least mean responses were observed in the productivity rate indicators. All parameters have returned a negative skewness which suggests that the tail is longer on the left side of the distribution. Only the profit indicators have a skewness less than -1 which implies that response in this indicator were highly skewed. Among the project-change management rating factors only the productivity rate has a negative kurtosis which denotes that distribution is flatter than a normal curve with the same mean and standard deviation.

3.2 Neural Network Modelling

The accuracy of the project-change management rating model was measured based on its Pearson's Correlation Coefficient, Mean Squared Error and Mean Absolute Percentage Error. The developed project-change management rating model has a very high R_{all} value of 0.99492 and extremely low MSE and MAPE of 0.0014274 and 0.553%, respectively. The correlation and performance plots of the project-change management rating model were displayed in Figs. 4 and 5.

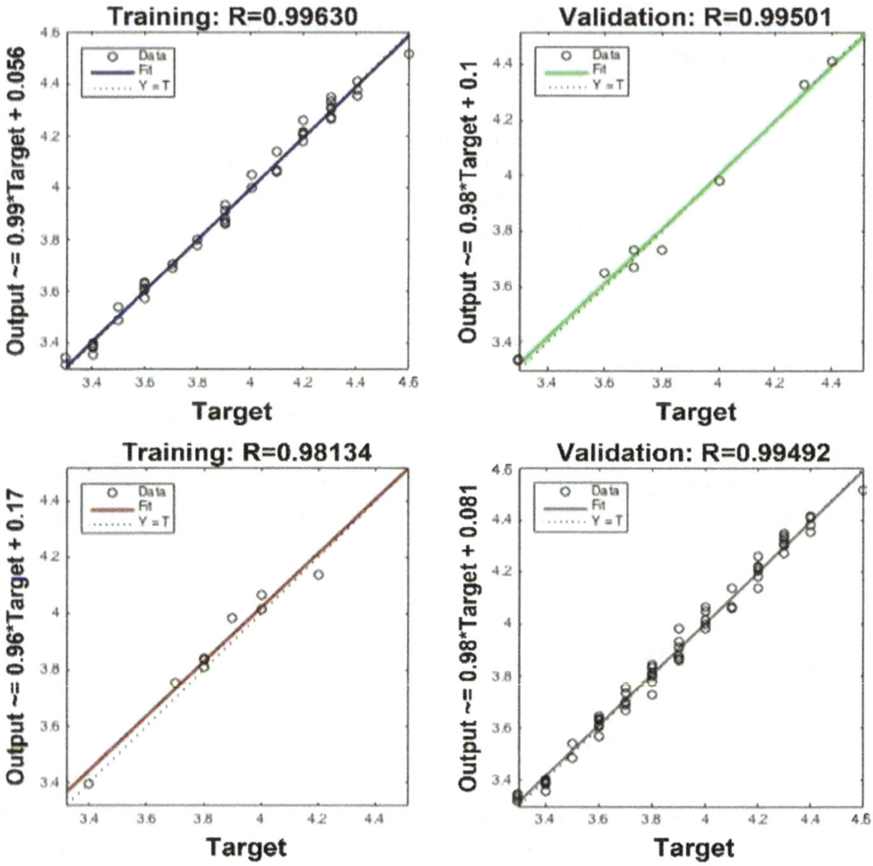

Fig. 4 Correlation plots of the project-change management rating model

3.3 Relative Importance Using Garson's Algorithm

Utilizing the connection weights and applying the Garson's Algorithm, the ranking of influence of these factors is described as Productivity < Profit < Delay < Project Acquisition wherein Productivity is the least influential factor with relative importance (R.I.) = 20.388% and Project Acquisition is the most influential factor with RI = 28.918%. Table 3 presents the product weights and the R.I. of the factors contributing to the project-change management rating Philippines-based construction firms affected by the factors caused by the pandemic. Figure 6 illustrates the summary of the relative importance of the factors contributing to project-change management rating.

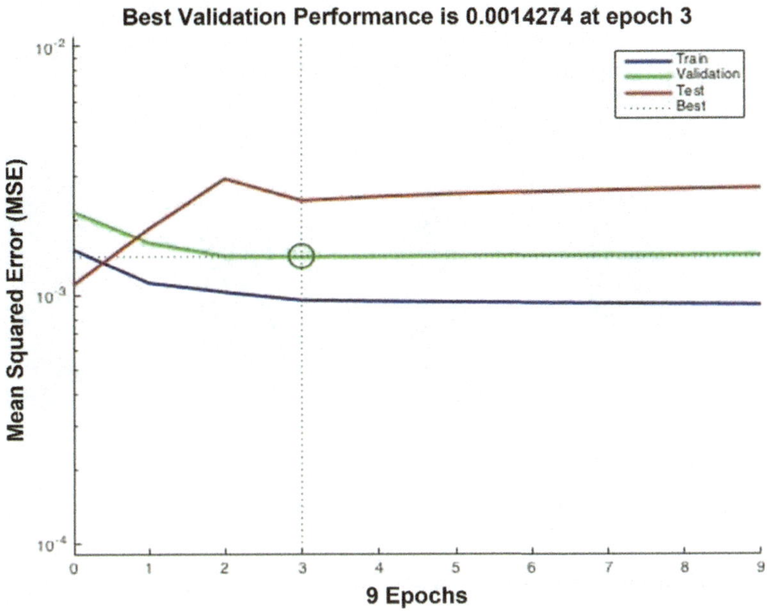

Fig. 5 Performance plot of the project-change management rating model

Table 3 Product weights and the R.I. of the factors contributing to the project-change management rating Philippines-based construction firms affected by the factors caused by the pandemic

	Delay	Profit	Productivity rate	Project acquisition	Sum
Hidden 1	0.37145	0.23275	0.13323	1.21585	1.95328
Hidden 2	1.73466	2.75610	1.50244	2.25864	8.25184
Hidden 3	0.03064	0.01157	0.02693	0.00232	0.07146
Hidden 4	0.61371	0.82970	0.46607	0.56435	2.47383
R.I. %	26.9325	23.7620	20.3875	28.9180	

3.4 Parametric Analysis

The behavior of the project-change rating considering varying values of the factors' Ratings while maintaining a fixed mean value for the remaining factors was depicted in Figs. 7, 8, 9 and 10.

Based on the parametric analysis plots shown in Figs. 6, 7, 8 and 9, all parameters considered in this study have a positive impact to the project-change management rating. The value of the project-change management rating increase as the value of the factors increases.

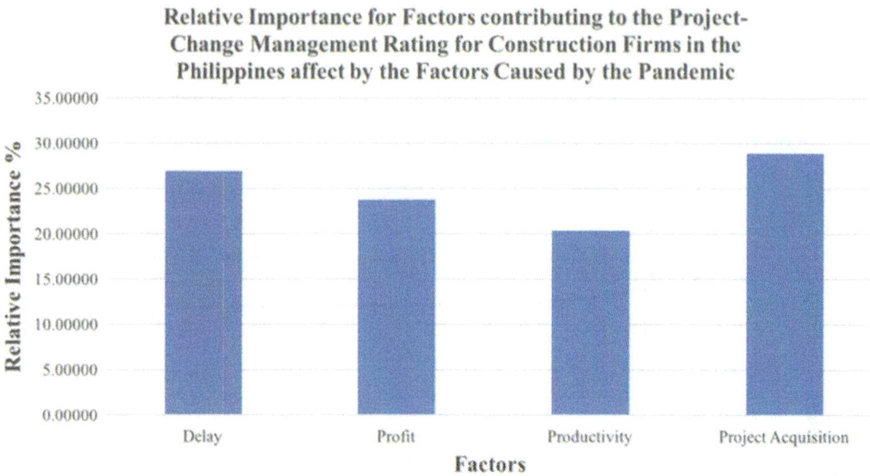

Fig. 6 Relative importance of the factors contributing to the project-change management rating

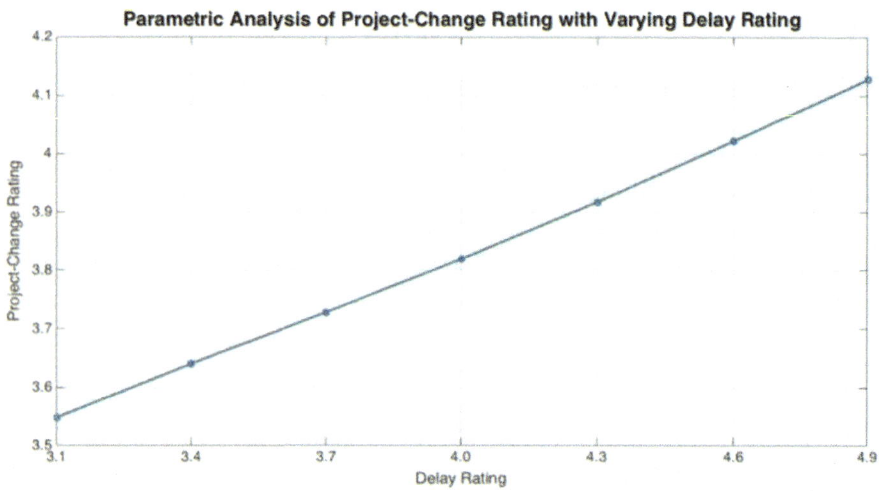

Fig. 7 Parametric analysis of project-change management rating with varying delay rating

3.5 *Causal Loop Diagram and Project-Management Framework*

The interrelationships between the input variables and their contributory factors are visualized in Fig. 6, these are shown through positive (blue loops) and negative (red loops) relationships. When two variables change in the same direction such as increasing or decreasing together it is considered as a positive relationship. On the other hand, variables that change in different directions such as one increasing

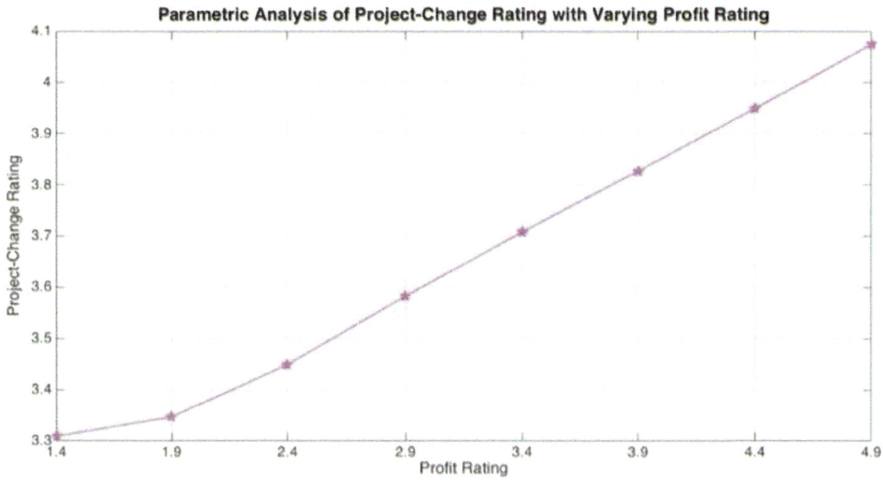

Fig. 8 Parametric analysis of project-change management rating with varying profit rating

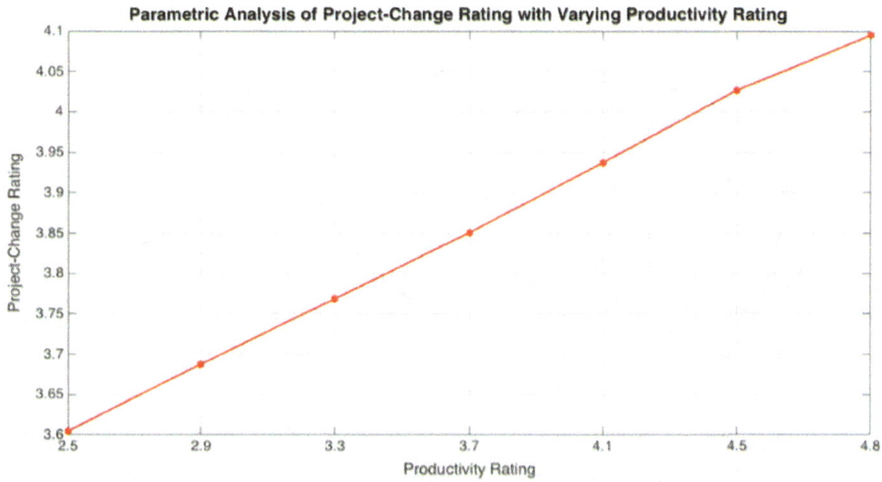

Fig. 9 Parametric analysis of project-change management rating with varying productivity rating

while the other decreases or vice versa, are considered a negative relationship. From Fig. 11, it could be observed that there are more positive relationships than negative relationships, but the number of negative relationships is an odd number. The odd number of negative relationships is essential in classifying the causal loop that is present in this study, it is a balancing loop. A balancing loop shows a feedback process that characterizes the growth-limiting behavior of the variables in the study.

In comparison to the existing project-change management framework in the new normal, it is visibly clear that the proposed project-change management is

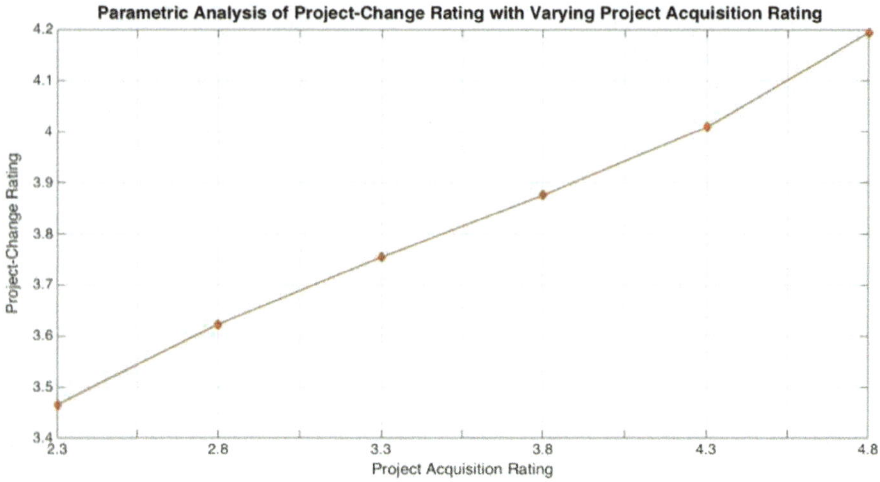

Fig. 10 Parametric analysis of project-change management rating with varying project acquisition rating

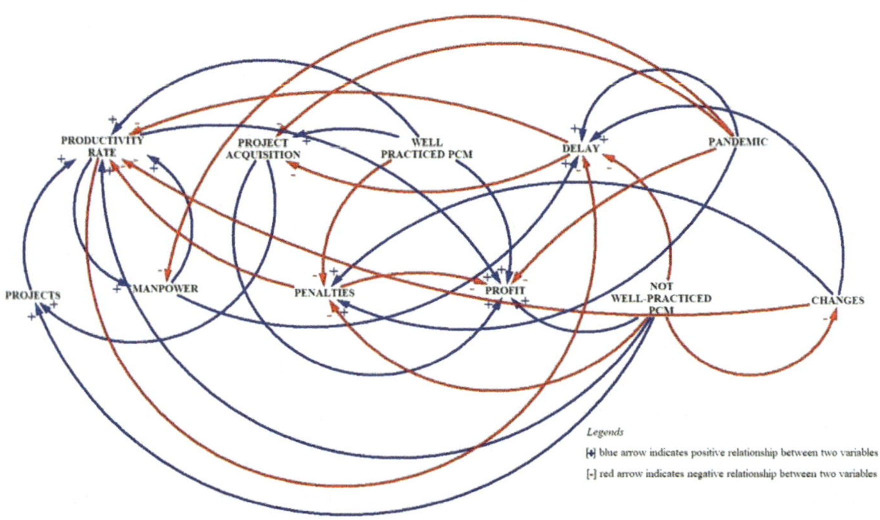

Fig. 11 Causal loop diagram of input variables

more extensive and has a wider scope. The proposed post-pandemic project change management framework begins with being aware of the problem. The better understanding one has of the change and its causes and effects, the more we can assess the situation efficiently. Awareness includes being knowledgeable of its effects. As the first part of the framework, the awareness and assessment satisfy the recognition of the change and what needs to be addressed.

The design of the protocol plan starts with understanding the aims and goals of the company and knowing which are needed to be resolved first. The longevity of the effectiveness of the action plan is one of the most considering factors because the proposed framework also aims to create a project-change management framework that is efficient and can be applied in the future struggles faced by the company. Another consideration encouraged by the framework is the company's reaction to the implementation method of the action plan. The framework made sure to include health and safety considerations in creating and implementing the action plan.

The application of the implemented action plan will then be monitored and continuously revised according to the changes encountered over time. Other existing frameworks encourage the implementations to be rigid and not be recalibrated. Recalibration of the action plan every time there is a change in construction projects makes the project-change management framework more efficient. After monitoring, reinforcing the plan will be the last step to success. Previous action plans that were inefficient or unsuccessful are encouraged to be reviewed and be taken note of for further reference of what and how to avoid inefficiency. The proposed project-change management framework in the new normal is presented in Fig. 12.

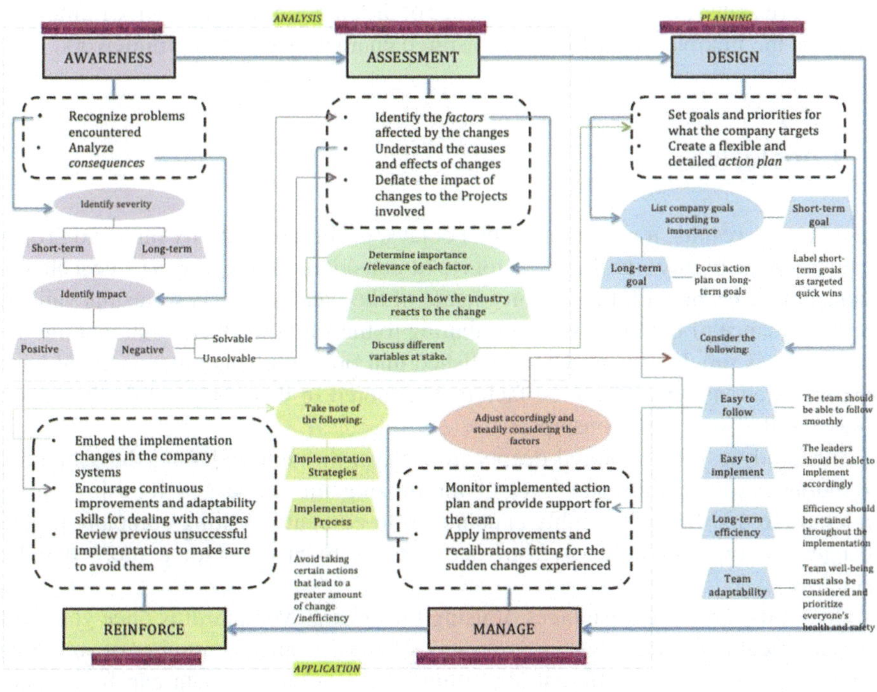

Fig. 12 Proposed project-change management framework in the new normal

3.6 Respondent's Feedback

Most of the respondents reacted positively towards the revised project change management framework and approved the framework and are willing to try the framework and implement it within their company. The approvals also came with a few different revisions from different companies. The revisions and suggestions they would apply to the framework are for the framework to be well fit and applied in their own company setting. Such suggestions and little adjustments they would change in the framework includes the additional considerations for designing an action plan for dealing with changes. Most of the respondents suggested that the designing considerations should include the economic status of the construction industry. This was a note taken by the research seriously since the pandemic has brought a fluctuating economic performance to the country.

On the other hand, some of the respondents commended the thoroughness of recognizing the problem. A few respondents appreciated the work put out in revising a new framework, but respectfully declined its implementation in their company. Some reasons that were also given are that the framework is too elaborate and requires more effort than the existing project-change management they have. The researchers noted the reasons; however, the significance of the elaborate and extensive framework serves as the flexible trait which is an aspired characteristic of the framework.

4 Conclusion

Based on the results, it was inferred that delays, profit, productivity rate, and project acquisition are highly significant factors to be considered in creating a project change management framework that is considerate of the pandemic. These results validate the significance of the factors in contributing to the framework and that making this framework is timely. It was also established that project-change management has a positive relationship with all the factors considered in this study, while the factors coexist with diverse relationships.

This study can be explored deeper in 2 ways, through widening the scope and extending it to a continental level or minimizing the scope and focusing on the regional factors. It is also highly encouraged to compare existing company project change management policies from one another to arrive with a more detailed framework.

One of the greatest advantages of using Levenberg–Marquardt is once you input the numerical data in the system, it provides an instantaneous number of solutions. If one was to utilize a different algorithm where numerical data can be inputted manually, perhaps it can result in a more specific and precise analysis, but it would require more data.

Awareness of how the company is keeping up with the new normal system. (e.g., what are the problems that the company experienced during the pandemic, or what are the possible consequences?) Assess the issue to formulate new kinds of hypotheses for a better implementation. (e.g., what are the factors involved in those disruptive changes, what are the causes and effects of changes, or what are the impacts of changes to the projects involved?) Plan, (e.g., how are we going to set the goal and priorities for what the company targets or how are we going to create a flexible and detailed action plan that considers long-term effectiveness where everyone can easily follow?) Manage, (e.g., monitoring of implemented action plan continuously and provide support for the team or recalibrating and applying new alternative action plans for the sudden change.) Reinforce, (e.g., who is the best fit in determining the effect of change in important variables of the company or who are going to apply improvement and risk assessment plans for sudden changes?).

Normalizing the importance of critical path milestones. A lot of the construction companies today are not as focused on determining the critical path milestone. Critical path milestone helps to determine where the project is lacking in different activities. It is much easier to monitor the project schedule and determine the possible bottlenecks. Considering the critical path will help to eliminate project errors, making sure that all the deliverables are being done in the most efficient way.

Moreover, building a healthy relationship with the stakeholders is vital. Stakeholders will help the company to remain stable. Frameworks will initially give clarity to the stakeholders by letting them involve in all decision-making of the framework.

References

1. Vithana NDI, Bandara KPSPK, Jayasooriya SD (2016) Impact of Covid-19 pandemic to construction industry in Sri Lanka
2. Ogunnusi M, Hamma-Adama M, Salman H, Kouider T (2020) COVID-19 pandemic: the effects and prospects in the construction industry. Int J Real Estate Stud 14 (Special Issue 2)
3. Nicola M, Alsafi Z, Sohrabi C, Kerwan A, Al-Jabir A, Iosifidis C, Agha M, Agha R (2020) The socio-economic implications of the coronavirus pandemic (COVID-19): a review. Int J Surg 78:185–193
4. Lim JA (2020) The Philippine economy during the COVID pandemic. Ateneo Center Econ Res Dev 16:1–28
5. Baticulon RE, Sy JJ, Alberto NRI, Baron MBC, Mabulay REC, Rizada LGT, Tiu CJS, Clarion CA, Reyes JCB (2021) Barriers to online learning in the time of COVID-19: a national survey of medical students in the Philippines. Med Sci Educ 31(2):615–626
6. Everitt BS, Skrondal A (2010) The Cambridge dictionary of statistics
7. Ajayi SO, Oyedele LO, Bilal M, Akinade OO, Alaka HA, Owolabi HA (2017) Critical management practices influencing on-site waste minimization in construction projects. Waste Manage 59:330–339
8. Vogt WP, Johnson B (2011) Dictionary of statistics and methodology: a nontechnical guide for the social sciences. Sage
9. Silva DL, de Jesus KLM, Villaverde BS, Cahilig CJC, Cruz JPLD, Sario JMS (2020) Sensitivity analysis and strength prediction of fly ash—based geopolymer concrete with polyethylene terephtalate using artificial neural network. In: 2020 IEEE 12th international conference on humanoid, nanotechnology, information technology, communication and control, environment, and management (HNICEM). IEEE, pp 1–6

10. Park M, Nepal MP, Dulaimi MF (2004) Dynamic modeling for construction innovation. J Manage Eng 20(4):170–177
11. Zayani R, Bouallegue R, Roviras D (2008) Adaptive predistortions based on neural networks associated with Levenberg–Marquardt algorithm for satellite down links. EURASIP J Wirel Commun Netw 2008:1–15
12. Monjardin CEF, de Jesus KLM, Claro KSE, Paz DAM, Aguilar KL (2020) Projection of water demand and sensitivity analysis of predictors affecting household usage in urban areas using artificial neural network. In: 2020 IEEE 12th international conference on humanoid, nanotechnology, information technology, communication and control, environment, and management (HNICEM). IEEE, pp 1–6
13. Lan TS, Lan YH, Chen KL, Chen PC, Lin WC (2013) A study of developing a system dynamics model for the learning effectiveness evaluation. Math Probl Eng 2013
14. El-Sawalhi NI, Shehatto O (2014) A neural network model for building construction projects cost estimating. J Constr Eng Project Manage 4(4):9–16
15. Macariola RN, Silva DL (2019) Coping with the information age: development of A data flow diagram-based knowledge management system for mitigating delays for construction. IOP Conf Ser Mater Sci Eng 652(1):012070
16. Tansey P, Meng X, Cleland D (2013) A critical review of response strategies adopted by construction companies during an economic recession. In: Proceedings of the 29th annual ARCOM conference. Association of Researchers in Construction Management, Reading, pp 679–689
17. Wang N, Gong Z, Xu Z, Liu Z, Han Y (2021) A quantitative investigation of the technological innovation in large construction companies. Technol Soc 65:101533
18. Laroza Silva DANTE, Marcelo De Jesus KL (2020) Backpropagation neural network with feature sensitivity analysis: pothole prediction model for flexible pavements using traffic and climate associated factors. In: 2020 the 3rd international conference on computing and big data, pp 60–67
19. Rashidi MM, Bég OA, Parsa AB, Nazari F (2011) Analysis and optimization of a transcritical power cycle with regenerator using artificial neural networks and genetic algorithms. Proc Inst Mech Eng Part A J Power Energy 225(6):701–717
20. Choi YG, Cho KT (2020) A cause analysis of the construction incident using causal loop diagram: safety culture perspective. J Korean Soc Safety 35(2):34–46
21. Wong JM, Chan AP, Chiang YH (2007) Forecasting construction manpower demand: a vector error correction model. Build Environ 42(8):3030–3041

Application of BIM Technology in the Formulation of Building Projects in Peru—2020

Jorge Vásquez Calderón, Doris Esenarro, Ciro Rodriguez, and Elizabeth Segovia

Abstract The objective of this research is to explain how it significantly influences the application of BIM technology in the formulation of building projects, given the construction deficiencies that currently exist with specialties, as a methodology, a survey of builders is elaborated and also shows the processes of the transition from the 2D model to the new 3D model (BIM). which allows the articulation of specialties during the project formulation process and minimizes failures, errors in the process of execution of the work, as the results were obtained from those surveyed to the builders, that 80% agree that the use of BIM technology due to significantly improved performance and quality of the work, while 20% consider that BIM technology is complicated.

Keywords Bim technologies · Project formulation · Projects

1 Introduction

The research-based on the Application of BIM Technology in the Formulation of Building Projects in Peru—2020, consists of using this new BIM technology [1]. in the formulation of building projects in the three-dimensional form [2], with the insertion of Software of Architecture, structures, electrical, sanitary, electromechanical installations and online texts in an assistive and collaborative way in real-time, guaranteeing a multidisciplinary quality work [3].

J. V. Calderón (✉) · D. Esenarro · C. Rodriguez
EUPG Graduate School Federico Villarreal University—UNFV, San Miguel de Allende, Peru

D. Esenarro
e-mail: desenarro@unfv.edu.pe

C. Rodriguez
e-mail: crodriguez@unfv.edu.pe

D. Esenarro · E. Segovia
Specialized Institute for Ecosystems and Natural Resources Research (INERN), San Miguel de Allende, Peru
e-mail: esegovia@unfv.edu.pe

T. Kang and Y. Lee (eds.), *Proceedings of 2021 4th International Conference on Civil Engineering and Architecture*, Lecture Notes in Civil Engineering 201,
https://doi.org/10.1007/978-981-16-6932-3_52

Currently, we have many technical deficiencies in the elaboration of building projects in Peru [4], by continuing to work traditionally and disjointed from the different technical specialties of a project, this generates problems in the evaluations, more time invested in correcting and raising observations, on the other hand, the technical issues and professional responsibilities that a poorly prepared project entails at the time of the execution of the work. This has a direct impact on the loss of time and cost overruns. Therefore, the purpose of the research is the use of BIM technology in the formulation of building projects; through the results of this research, we will be able to have the tools and technical support to insert this technology into our work environment [5], which will allow us to optimize time and present quality work. Therefore, the formulation of the statement "What is the result of using BIM technology application in the formulation of building projects in Peru?" is made, with the primary objective of showing the characteristics and advantages of using BIM technology. Likewise, BIM can be applied in any entity corresponding to the three levels of government, subject to the National System of Multiannual Programming and Investment Management, and executing investment projects or investments of optimization, marginal expansion, replacement, and rehabilitation—IOARR, in any of the phases of the investment cycle. International BIM standards exist in countries where the methodology has already been implemented at the government level. These include Australia, South Korea, Finland, the Netherlands, Norway, New Zealand, Singapore, and the United States. At the regional level, significant efforts are being made to improve the quality of public spending. For this reason, methodologies and technologies have been incorporated at all stages of the investment cycle to achieve this objective. Peru is no stranger to this effort. To participate in the development of collaboration activities and the definition of regional and international standards, it is part of the BIM Network of Latin American Governments [6].

BIM roles: There are four levels of BIM users; these levels are: BIM Review, BIM Modeling, BIM Coordination, BIM Management, and BIM Direction, and they are divided according to the competencies and attitudes that will be described [7].

BIM review: Reviews geometric and non-geometric information of the models, depending on the stage in which it is. Manages normative and technical knowledge of the project. It can carry out validation, audit, control, development, inspection and execution tasks.

BIM modeling: Create the 3d models, according to your specialty, setting the geometric information content as the respective non-geometric. Master format swapping. Manages normative and technical knowledge of the project.

BIM coordination: It is in charge of guaranteeing the correct flow of information between those involved according to the stage of the project. Validate, integrate, coordinate, collect information and follow up on a correct modeling. Coordinate with the modelers the respective advances and improvements. Manages normative and technical knowledge of the project. Possesses leadership skills.

BIM management: He spearheads the scheduling of activities, personnel management, and the resources required for implementation. It shapes the standards and guidelines that will be used. Leadership skills.

Address in BIM: Implement BIM in an organization, responding to the revised strategies. Experience in project and organization strategy management.

2 Method

2.1 Type of Investigation

To obtain the results, this study cited BIM y CAD [8] regarding the 2D workflow and BIM workflow for the comparison of time savings and data loss, as shown in Figs. 1 and 2. In this sense, it is described using independent traditional Software [9] that allows us to generate good works according to each specialty. Still, the

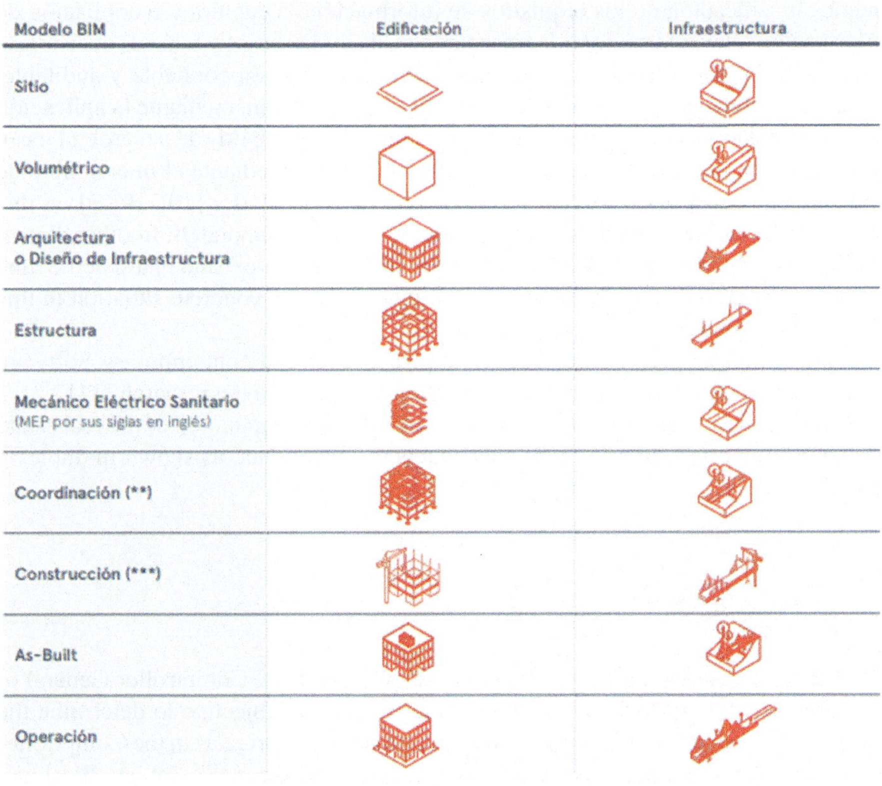

Fig. 1 BIM standard for public projects [4]

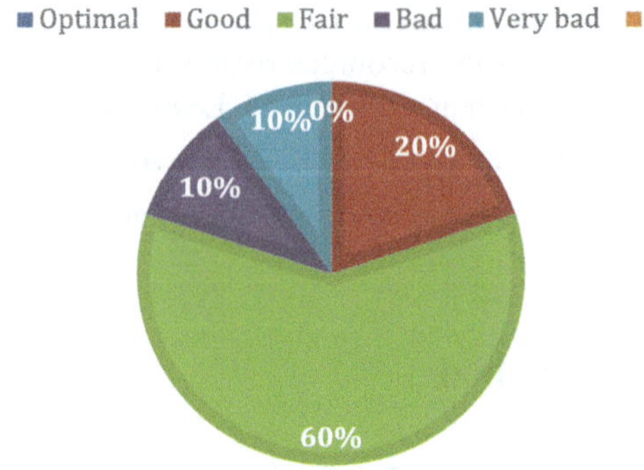

Fig. 2 Result of the modeling process

investment time is higher; once a project is finished, it has to be evaluated, observed, corrected again, observed, and corrected for approval. During all this time, the costs have often changed and what is coming is a budget update and modification. We continue to invest more time than expected, leading to more significant investment and general discomfort. BIM nos enmarca objetivos especificos que permiten mejorar la calidad de los expedientes técnicos o documentos equivalentes, a través de una mejor especificación de los requisitos de información, la captura y recopilación de información relevante, mejorar la transparencia de los procesos y toma de decisiones, a través de la disponibilidad de información de la inversión confiable y auditable, mejorar la calidad y la eficiencia de ejecución de la inversión, mediante la aplicación de los estándares, herramientas y usos de la metodología BIM, de acuerdo al caso, mejorar las capacidades digitales de la cadena de valor mediante el intercambio de conocimientos, espacios comunes y desarrollo de capacidades [10]. Based on this information, a diagnosis of the current situation will be elaborated. In other words, the facts are observed as they are presented in their real context and in a specific time frame for their analysis. Therefore, we do not construct a concrete situation in this design, but we observe the existing ones.

Likewise, bibliographic information on traditional and contemporary Software will be compiled to provide theoretical technical support to the research [11].

A survey-based on questionnaires was conducted in a population of the real estate sector in Lima, in a sample of 10 consultants to obtain data, as shown in Tables 1 and 2.

2.2 Procedures

BIM, in the constructability of infrastructure projects in the Comptroller General of the Republic, Jesús María, 2016" which had as general objective to determine the level of knowledge of the constructability of infrastructure projects in the Comptroller General of the Republic, Jesús María, 2016, and use of the methodology "Building Information Modeling", this study was carried out with a quantitative approach of

Table 1 Valuation of the use of BIM in design and construction

How do you rate the use of BIM technology in the design process and construction process in building projects?		
Experiences	No. 30 of respondents	Percentage (%)
Optimal	24	80
Good	3	10
Fair	3	10
Bad	0	0
Very bad	0	0
Total	30	100

Table 2 Rating of the use of traditional systems

How do you rate the use of BIM technology in the design process and construction process in building projects?		
Experiences	No. 30 of respondents	Percentage (%)
Optimal	0	0
Good	6	20
Fair	18	60
Bad	3	10
Very bad	3	10
Total	30	100

descriptive character, since statistical methods are applied to analyze survey data. It was concluded that there is a 95% of respondents have a level of knowledge of constructability of infrastructure projects, through the use of BIM [12].

BIM allows achieving better results in terms of cost, time and quality, it is efficient, because it allows generating savings in the use of public and private funds throughout the investment cycle, since it allows having a better control of the technical and management information of the investment. This is reflected in a reduction of cost overruns and delays during the execution of the investments, as well as in a rational use of resources destined to their operation and maintenance, it is of quality because it allows to have a better quality control of the investments in buildings or infrastructure, through the collaborative work and the management of the information, it makes possible the analysis and control of the quality standards, as well as the verification of the fulfillment of norms applicable to the investment, it is collaborative because it allows to establish standards for the participation, communication and exchange of information between the diverse agents involved in the development of an investment, It is transparent because it allows to make explicit the different decisions taken during the development of the investment, through the management of the information regarding all the operators involved throughout the investment cycle, as well as the management of the information gathered, obtained or generated for this purpose and respect to the time allows to have a better control of the time in each phase of the investments in buildings or infrastructure, through collaborative work and information management, enables the timely identification of factors that have a direct impact on the programming during the planning and execution of the investment [13].

1. These surveys were conducted to 10 consultants dedicated to developing real estate projects in Peru, on their comparative experience in using BIM technology and traditional systems in the design process and construction process in building projects.
2. Exemplification of the application of BIM technology in the formulation of an infrastructure project.

3 Results

The following are the results obtained in the surveys conducted with the ten consultants on comparing traditional systems and BIM technology in building projects corresponding to the Plus dimension.

According to Table 1, 24 out of 30 consultants, representing 80%, consider optimal BIM technology in the design and construction process in building projects. While three consultants, representing 10%, considers the use of BIM technology in the design and construction process in building projects as good, and 3, representing 10%, considers the use of BIM technology in the design and construction process in building projects as regular, showing the poor mastery of BIM [14].

In Fig. 2, It is concluded that a majority representing 80%, agree that BIM technology significantly improves the performance and quality of work, while 20% consider using BIM technology is tricky.

According to Table 2, 6 out of 30 consultants, representing 20%, consider using traditional systems in the design and construction process in building projects.

In Fig. 3 show 18 out of 30 consultants, representing 60%, evaluate the use of conventional methods in the design and construction process in building projects as fair, 03, representing 10%, considers it as evil and 03, representing 10%, considers the use of traditional systems in the design and construction process in building projects as very bad.

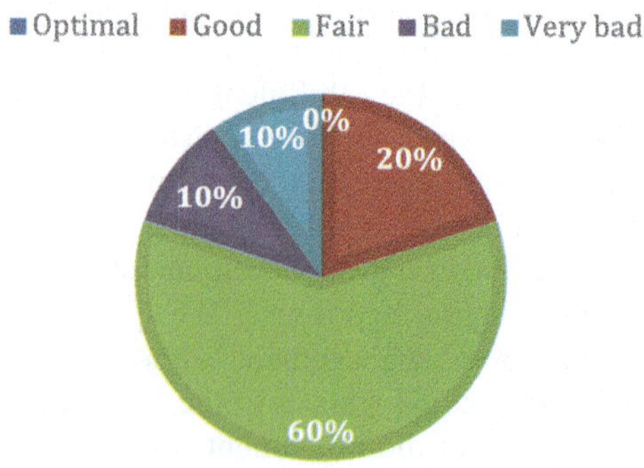

Fig. 3 Result of use of traditional systems

Fig. 4 01 housing modeling process [12]

It is concluded that a majority representing 60% of the respondents are convinced that traditional systems comply with the requirements to carry out the work, but it entails a more significant effort. A 20% consider traditional methods to be sound because they are secure and reliable Software, while 20% are convinced that conventional systems already belong to the past.

The results obtained in the application of BIM technology in the formulation of an infrastructure project are shown below the formulation of an infrastructure project.

The fundamental objective of BIM is to avoid losing information throughout the project cycle, with the traditional method, which forces a more significant production effort in the amount of information in the project's different phases. This result of the application of the BIM system [15]. In the value of the data can be seen in Fig. 4.

Figure 4 shows the formulation stage of the infrastructure project with the application of BIM technology; we can see the distribution plan of the first level plotted in 2D, as a starting point, in the following figure, we can see the elevation at 3D level, the height is given to the walls with the shades in color according to the cutting plans and the placement of doors and windows, in which you can define the type of material and finishes of these elements and their harmonic integration to the volumetry, the furniture is placed to determine the dimensions of each environment, correcting from the design process space problems, you can also see the direction of the sunlight to reconsider the location of doors and windows.

Figure 5 shown in the formulation stage of the infrastructure project with the application of BIM technology, the volumetric set of the second floor is observed, the transition from 2 to 3D design, by managing the volumetry we can visualize errors, for example with the placement of the bleachers allows us to have a better perception of the functionality, regarding the passageways that connect the various environments, we can raise with greater criteria the location and direction of the doors.

Figure 6 As shown in the formulation stage of the infrastructure project with the application of BIM technology, we observe the arrangement of the furniture and the placement of basic decoration accessories in a building to give us the functional spaces of each environment, we visualize the post of the cover showing the percentages of inclination and its intersection with the walls, we consider from this stage the quality of the material and the structural resources for its support, it is necessary

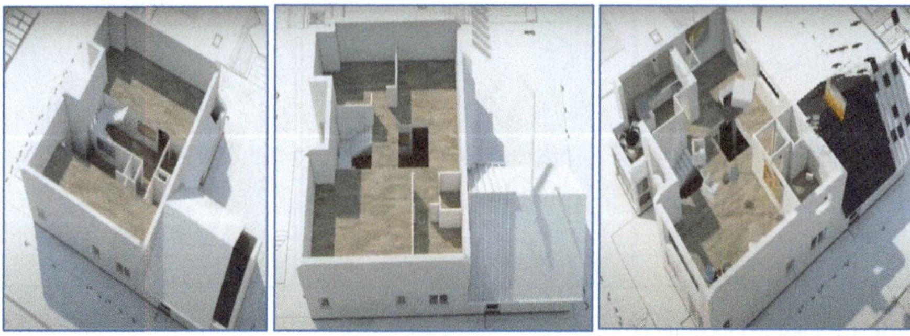

Fig. 5 Modeling process 02 housing [12]

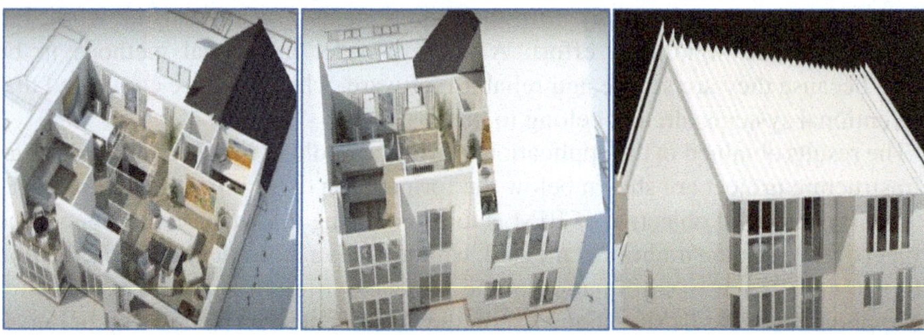

Fig. 6 03 housing modeling process [12]

to mention that any type of alteration, changes, and modifications to the volumetry automatically generates the modification level of plans, metrics, and budget. We are developing an optimization of time and resources

As shown in Fig. 7, through the BIM application, the house's design is much faster in all its stages, with significant savings compared to the traditional system.

As shown in Fig. 8, using the BIM application, the margin of error is reduced due to the advantage of 3D modeling in real-time.

As shown in Fig. 9, using the BIM application, 3D modeling house view results in real-time.

4 Discussion

It is observed that with the use of BIM technology, such as Archicad 24 software, it is possible to work with large amounts of data, it was also observed, the increase in productivity when making changes in the designs, since the 3D design objects, such as planimetry, walls, floors, equipment, furniture among others, are linked, allowing to automatically generate all project documentation, such as plans,

Fig. 7 04 housing modeling process [12]

Fig. 8 Modeling process 05 housing [12]

Fig. 9 Modeling process 06 housing

tables, schedules, the number of materials, 3D modeling, videos, etc. The research concludes that the effects—benefits found after implementing the project management model with BIM software, Archicad 24, increased efficiency, optimizing workflows, improved information management, and architectural design. This has been achieved by de-emphasizing the use of traditional systems in the formulation of infrastructure projects [16]. It is also important to highlight that the BIM methodology establishes procedures for delivering information between the operators of the National System of Programming and Investment Management and other operators involved in the development of investments. Its use is applicable in investments in buildings or infrastructure during Formulation and Evaluation, Execution, and Operation. It serves to improve and optimize the execution of public assets. The BIM methodology is a transparent, effective, and reliable decision-making tool since it ensures better efficiency and quality in delivering investment information [17].

5 Conclusions

It is concluded that a majority representing 80%, agree that the use of BIM technology significantly improves the performance and quality of the work. In comparison, 20% consider the use of BIM technology to be complicated [18].

It is concluded that a majority representing 60% of the respondents are convinced that traditional systems comply with the requirements to perform the works, but it entails a more significant effort. 20% consider traditional methods to be sound because they are safe and reliable Software, while 20% are convinced that conventional systems are a thing of the past.

Using BIM technology, using ARCHICAD 24 software, offers intelligent modeling tools for creating building elements, furniture, and customized and free-form elements. With the help of Twinmotion 2020 (a free and open-source tool for creating layouts, animations, or interactive visualizations from data) any object, or shape can be modeled and documented according to architectural standards. Likewise, the structural model in a simple, fast, and streamlined workflow while your engineering partner works with powerful structural analysis software such as SCIA Engineer. MEP 3D modeling (duct, piping, and cable tray systems) reduces costs and incident last-minute modifications during the construction phase.

References

1. Valdepeñas P, Esteban M, Henche C, Rodríguez R, Fernández G, López J (2020) Application of the BIM Method in the management of the maintenance in port infrastructures. J Mar Sci Eng 8:981. https://doi.org/10.3390/jmse8120981
2. Sun JJJ, Wang Z, Xu T (2017) The construction of BIM application value system for residential buildings' design stage in China based on traditional DBB mode. Proc Eng:851–858. https://doi.org/10.1016/j.proeng.2017.04.246

3. Jacques S, Bissey S, Martín A (2016) Multidisciplinary project based learning within a collaborative framework: a case study on urban drone conception, Revista internacional de tecnologías emergentes en el aprendizaje (iJET) 11(12):36. https://doi.org/10.3991/ijet.v11i12.5996
4. Koseoglu O, Keskin B, Ozorhon B (2019) Challenges and enablers in BIM-enabled digital transformation in mega projects: the Istanbul new airport project case study. Buildings 9:115. https://doi.org/10.3390/buildings9050115
5. Adom D, Hussein EK, Adu-Agyem J (2018) Theoretical and conceptual framework: mandatory ingredients of a quality research. Int J Sci Res 7(1):438–441
6. Succar B (2020) https://bimdictionary.com/en/common-data-environment/1/
7. ISO 19650-1(2018) 3.3.11 y 3.3.12, modificado
8. Ramos L, Esenarro D, Rodriguez C, Lagos J (2020) Recovery of public spaces for the conservation of green areas in Tablada Lurin. IOP Conf Ser Mater Sci Eng 910:012020. https://doi.org/10.1088/1757-899X/910/1/01
9. Tahir MM, Haron NA, Alias AH, Al-Jumaa AT, Muhammad IB (2017) Applications of building information model (BIM) in Malaysian construction industry. IOP Conf Ser Mater Sci Eng 291:012009. https://doi.org/10.1088/1757
10. Bohorquez F, Esenarro D, Rodriguez C, Tafur V, Tafur G (2021) Modular construction camp as an architectural model and its different construction systems—Moquegua, Perú 2020. Mater Today Proc. https://doi.org/10.1016/j.matpr.04.399
11. Esenarro D, Escate I, Anco L, Tassara C, Rodriguez C (2020) Proposal for an ecological research center for the recovery and revaluation of biodiversity in the town of Quichas-Lima, Peru. Int J Environ Sci Dev 11(4):212–216. ISSN 2010-0264. https://doi.org/10.18178/ijesd.2020.11.4.1253
12. Rodriguez C, Esenarro D, Alburquerque C, Vega M, Ramirez V (2020) Theme park of renewable energies for mitigation of CO_2 in the urban area of the district of Chorrillos, Peru, IOP Conf Ser Mater Sci Eng 910:012021. https://doi.org/10.1088/1757-899X/910/1/012021
13. Graphisoft: building together "Copyright © GRAPHISOFT. Todos los derechos reservados" (2020)
14. Namli E, Isikdag U, Kocakaya MN (2019) Building information management (BIM), a new approach to project management. J Sustain Constr Mater Technol 4(1):323–332. https://doi.org/10.29187/jscmt.2019.36
15. Namhyuk J, Min KY, Kim JH, Kim J (2020) A study on application of BIM (building information modeling) to pre-design in construction project. https://doi.org/10.1109/ICCIT.2008.190
16. Namli E, Isikdag U, Kocakaya MN (2019) Building information management (BIM), a new approach to project management. J Sustain Constr Mater Technol 4(1):323–332
17. Leśniak A, Górka M, Skrzypczak I (2021) Barriers to BIM implementation in architecture, construction, and engineering projects—the polish study. Energies 14:2090. https://doi.org/10.3390/en14082090
18. Badawy A (2020) Collaboration and BIM model maturity to produce green buildings as an organizational strategy. HBRC J 16(1):243–268. https://doi.org/10.1080/16874048.2020.1807815

Ancient Architecture and Architectural History

Expressions of Local Identity via Architectural Practice in UNESCO World Heritage Sites

Surasak Kangkhao, Chaturong Louhapensang, and Koompong Noobanjong

Abstract Via its investigations on three types of buildings, consisting of residential, religious, and educational structures, this research: (1) provides guidelines for the creation of local identity via architecture erected in the areas adjacent to UNESCO World Heritage sites; and (2) examines the mechanism in creating such an identity. As exhibited by the upcoming discussions, the economic, social, cultural and technological conditions are central not only for practical requirements and structural integrity of those buildings, but also for expressions of the local identity. In addition to maintaining a delicate balance between aesthetics, architectural realization, and local wisdom, the creations of the case studies engage all involving parties in the following five domains as well, which are: (1) generation of public consciousness; (2) design participation; (3) establishment of Heritage Academic Research Community; (4) recognitions for the values of the Thai identity; and (5), creative integration of stakeholders with community networks.

Keywords Expression · Local identity · Architectural · Practice · World heritage site

1 Introduction

Aside from being a cultural artifact, architecture is a part of the creative industry. On the one hand, historical buildings are tangible assets of societies as shown by numerous tourist attractions throughout the world. On the other hand, the same structures symbolize the images or identities to which those societies aspire to assume. Nevertheless, as societies evolve, historical places cannot always provide space for sheltering new utilitarian requirements effectively. For that reason, a critical challenge

S. Kangkhao · C. Louhapensang (✉) · K. Noobanjong
Department of Architectural Education and Design, Faculty of Industrial Education and Technology, King Mongkut's Institute of Technology Ladkrabang, Bangkok, Thailand
e-mail: chaturong.lo@kmitl.ac.th

S. Kangkhao
e-mail: surasak.ka@kmitl.ac.th

T. Kang and Y. Lee (eds.), *Proceedings of 2021 4th International Conference on Civil Engineering and Architecture*, Lecture Notes in Civil Engineering 201,
https://doi.org/10.1007/978-981-16-6932-3_53

dwells on how to devise a building that is not alienated from its own cultural roots, yet at the same time flexible enough to accommodate modern functions necessitated by our current way of living.

World heritage site area: Kamphaeng Phet.

In this research, the objects of inquiries rest in the specialized monumental areas of the Historic Town of Sukhothai and Associated Historic Towns. Standing approximately 450 km. away from Bangkok in the lower northern region of Thailand, the locations are in the vicinity of Kamphaeng Phet historical park, which is a UNESCO World Heritage site in Kamphaeng Phet province.

1.1 Cultural Heritage, Identities, and Creative Livelihoods

Cultural capital and cultural heritage are something that reflects the way of life. and expresses the cultural wisdom of the area. According to Hall [1], cultural identity representative the livelihoods and wisdom of a group of people. In any case, culture and identity define the boundaries of community culture [2]. In addition, as remarked by Jensen et al. [3], heritage is an important heirloom that should be valued and preserved.

In this regard, cultural heritage is the tangible and intangible attributes of society that are linked from the past and maintained by the present to pass on to the next generation [4].

Concerning creative livelihood, the concept stands for the way of living that offers an equally accessible means of socio-economic advancement to all members of the community. According to Blaikie et al. [5], the term of expanding concepts towards conservation and respect for cultural rights.

Heritage is our asset in the past, what we need to maintain and develop into a sustainable future in an appropriate context.

Cultural heritage is the tangible and intangible legacy of a group or society that is handed down from the previous generations, preserved in the present, and given the benefit for future generations.

Tangible heritage Including historical buildings and places, artifacts, etc., which are considered worthy of conservation in the future. This includes objects that are important to the archeology, architecture, science, or technology of a particular culture. Ancient objects are important to the study of human history. because it shows concrete evidence for the idea and can check Their preservation represents a necessity acceptance of the past and the things that tell the story.

The reality of this is that cultural heritage is constantly changing and deteriorating. Thus, what is considered conservative is truly transformed, focusing on values connected to the past.

Culture and cultural heritage expressing the identity and history of the holder of cultural rights can be instrumental in creating identity and differentiation, while at the same time creating a universal shared experience characterized by the contact and integration of culture. Destinations depend on both the physical assets and the experience set created from them. which generally extends to "Culture of living" and atmosphere of the place using local heritage and identity in cultural tourism strategies We can differentiate our travel products from competitors. but at the same time, we also support developments that benefit local communities in the protection of local values, livelihoods, and economies.

Enhancing the cultural identity of the local community and community involvement is essential. It must benefit and take pride in being a better host. The goal of conservation in the development of cultural heritage is to learn how to preserve, develop and use prominence. by maintaining authenticity and local identity It is the only way that tourism activities can be culturally sustainable. (and economically).

The manifestation of local wisdom could be customs, life habits, styles or ways of life, or various cultural traditions, such as religious rites, life cycle rituals, and traditional arts. In other words, local wisdom was the embodiment of a local culture inherited for generations to guide life towards the common good.

Similar to cultural heritage, local wisdom could be both abstract (tangible) and concrete (intangible), but the important characteristics were that it originated from experiences or truth gained from life. By integrating the body, the spirit, and the environment, local wisdom was accumulated through real experiences, focusing on the attainment of inner satisfaction rather than material wealth.

In most rural Thai societies, local wisdom constituted a simple way of living that could benefit people in every status. By placing a great value on the interactions between family and community, local wisdom provided a strategy for people to work together on a task that could not be accomplished by an individual acting alone, thus representing the essence of helpfulness, harmony, and unity that created the feelings of belonging and solidarity that became a basis for forming a collective identity among community members.

2 Research Methodology

See Table 1.

Table 1 Research procedures

Input	Operating Procedures	Expected outcome
The creation of architectural identity, consisting of residential, religious, and educational structures, in the areas adjacent to UNESCO World Heritage sites in Kamphaeng Phet	Creation of guidelines for expressions of local identity via architectural designs for buildings near UNESCO World Heritage sites	1. Analysis of the factors contributing to the conceptual guideline for expressions of local identity 2. Proposition of the guideline via architectural designs

Source The Authors

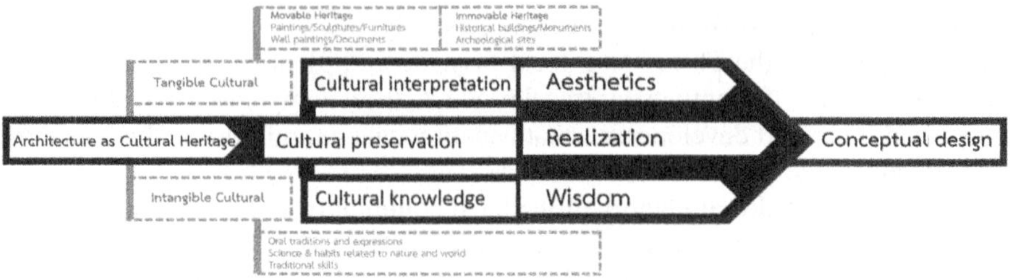

Fig. 1 Conceptual context diagrammatic. *Source* The Authors

Table 2 Architectural typology and composition of design conceptual awareness guideline

Architectural typology	Composition of design conceptual awareness guideline
Residential structures	LA, CC, CI, PC, PD
Religious structures	CI, CC, PC, PD, LA
Academic places	PD, CC, CI, PC, LA

* Remark the attribute factor develop from reference source

3 Result of Research

3.1 Architecture as Cultural Heritage

The creation of architectural identity, consisting of residential, religious, and educational structures, in the areas adjacent to UNESCO World Heritage sites in Kamphaeng Phet province. Structural and relationship components models are present in Fig. 1 and Table 2. Explore Architectural Typology via Composition of Design Conceptual awareness Guideline including LA = livelihood assets (Human capital, Social capital, Natural capital, Physical capital, Financial capital); CC = cultural capital (Tangible, Intangible); CI = cultural interpretation (Appearance, Aesthetic and Local wisdom); PC = public consciousness; PD = participation design.

3.2 Expressions of Local Identity Guidelines

Creation of guidelines for expressions of local identity via architectural designs for residential, religious, and educational structures near UNESCO World Heritage sites in Kamphaeng Phet.

3.3 *Realization of Local Identity via Architectural Designs*

Realization of local identity via architectural designs for residential, religious, and educational structures in the vicinity of UNESCO World Heritage sites in Kamphaeng Phet.

Architectural Typology: Residential Structures. Baan Fai Rim Ping: The local balcony is welcome for everybody this area is show all activities that owner can do such as sit on the floor, lay down, relax with bare foot. That able to touch Teak wood floor. Every rooms face on the different angle beautiful river views with the balcony. Character of the house still respectful of the mountain skyline (Fig. 2).

Kamphaeng Phet Air Residence: It's located in a prime limited area of Kamphaeng Phet city, in front of the Ping river. Expressions from local textile to make bricks pattern facade. The building is conveying as the biggest laterite keystone (Fig. 3).

Architectural Typology: Academic places Meditation Hall at Kamphaeng Phet Rajabhat University: It's located inside the lake which represents the symbolic icon of the university. Expressions from an archaeological site in Kamphaeng Phet Historical Park (UNESCO World Heritage site) (Fig. 4).

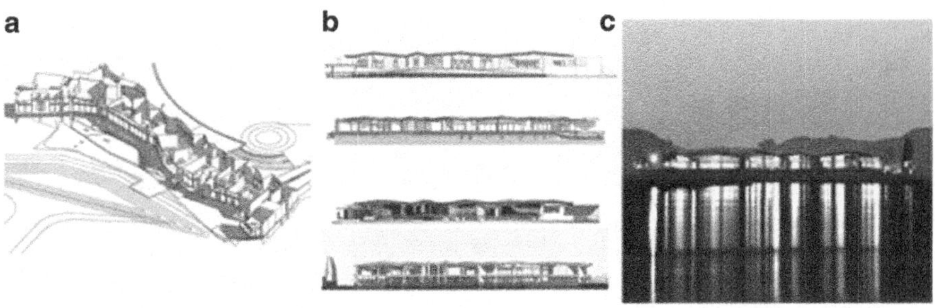

Fig. 2 **a** Perspective, **b** elevation, **c** construction

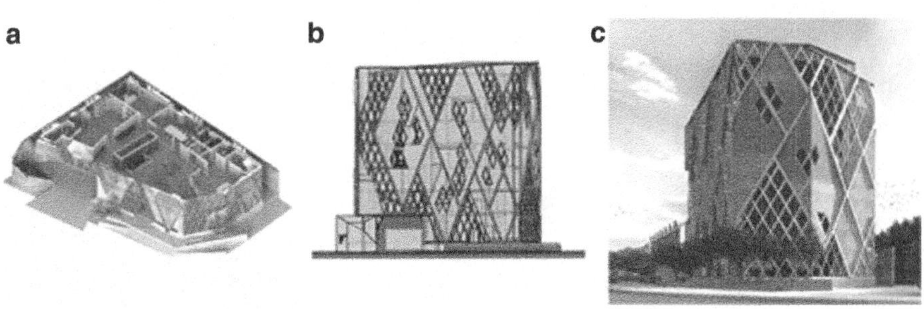

Fig. 3 **a** Perspective, **b** elevation, **c** construction

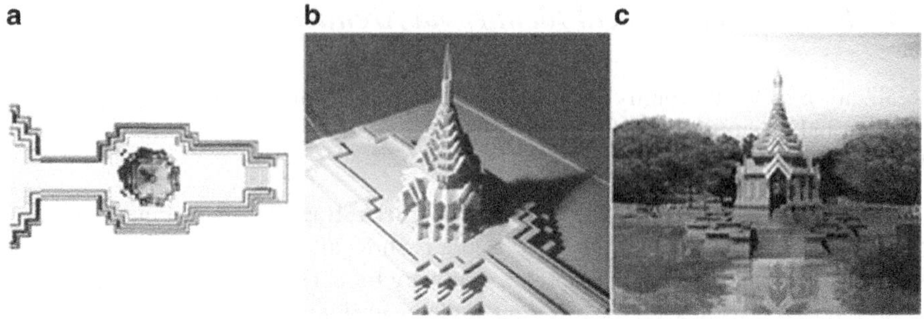

Fig. 4 **a** Plan, **b** perspective, **c** construction

Art Gallery at Kamphaeng Phet Rajabhat University: It represents the public area for arts and performance events and exhibitions of the Kamphaeng Phet city. Therefore the building distributes directly acknowledgement to the public (Fig. 5).

Multi-purpose Building at Kamphaeng Phet Rajabhat University: This building serves international students and faculty members which is suitable for the local environment (Green design) (Fig. 6).

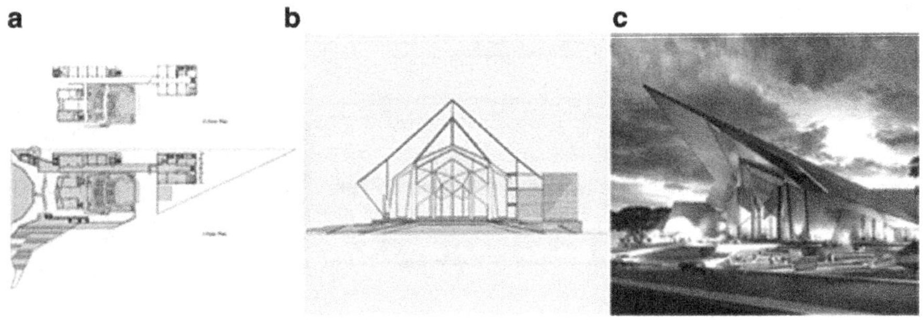

Fig. 5 **a** Plan, **b** elevation, **c** construction

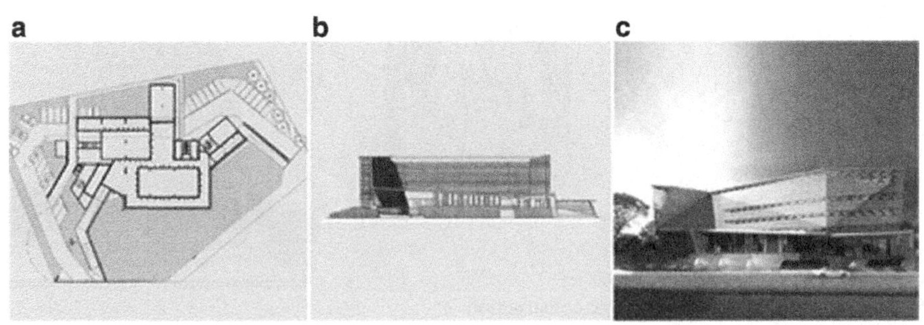

Fig. 6 **a** Plan, **b** elevation, **c** construction

Architectural Typology: Religious Structures. Multi-purpose Building at Ku Yang Buddhist Temple: The building serves for divinity examinations and activities of local northern regional district monks (Fig. 7).

Grand Stupa at Pak Ang Buddhist Temple (Fig. 8).

Grand Stupa at Sai Ngam Buddhist Forest Monastery (Fig. 9).

Grand Stupa at Chai Rangsi Buddhist Forest Monastery (Fig. 10).

The buildings represent from Philosophy of Buddhism 7 elements; Three majors combination including morality, concentration, and wisdom. Expressions to Three

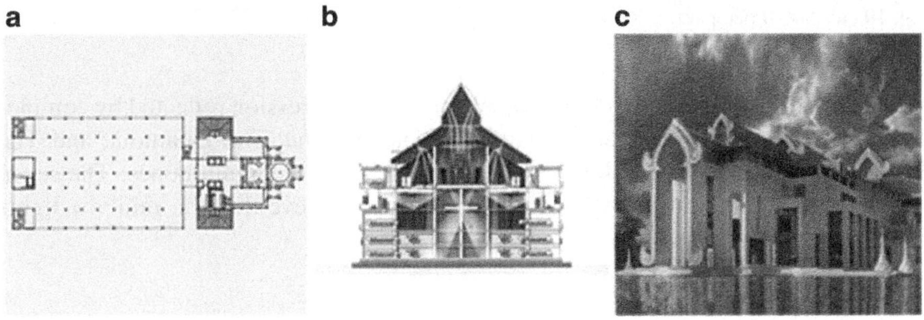

Fig. 7 **a** Plan, **b** elevation, **c** construction

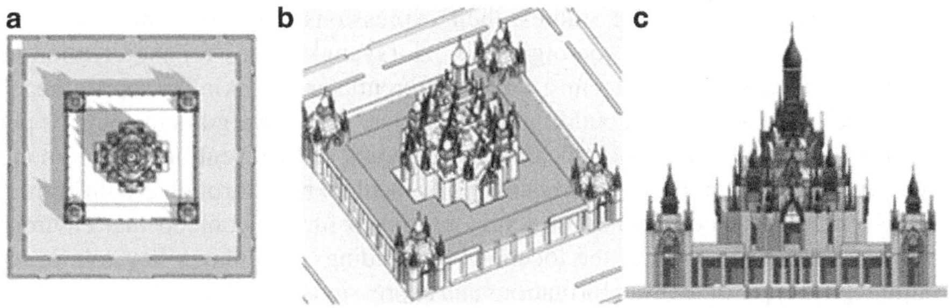

Fig. 8 **a** Plan, **b** perspective, **c** construction

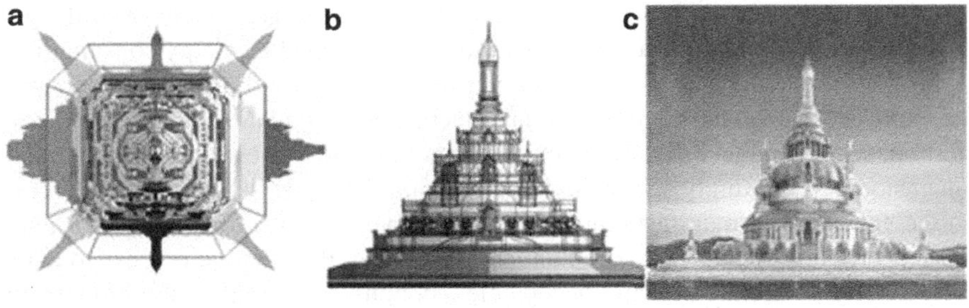

Fig. 9 **a** Plan, **b** elevation, **c** construction

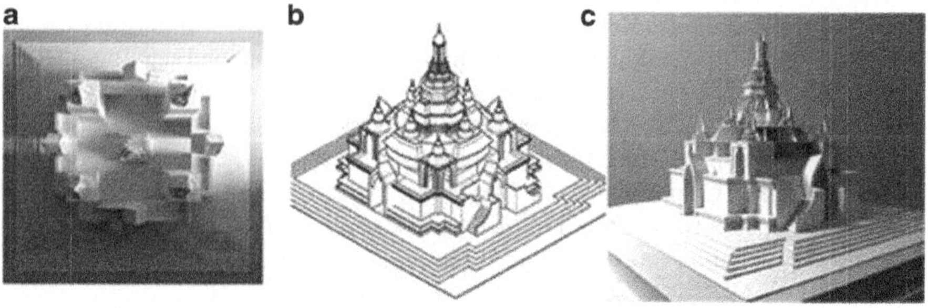

Fig. 10 a Plan, b perspective, c construction

majors color including red, yellow, and blue another expression reflected by complementary color via Triple Gem/Rattana Tri; The Buddha, The Dhamma, and The Sangha. Which the white color is the seventh element to contain bones. The seven colors of elements appear in the stupas. Buddhist believe that the rainbow lights colors as a bridge to heaven.

4 Conclusion and Discussion

As demonstrated by the case studies, their expressions of local identity seem to resonate with the approach promulgated by Horayangkura [6] for the creation of identity by modern Thai architecture, which differentiates the "skin" or "appearance" from "core" or "essence" of a building. In a corollary view, an argument could be put forward that a combination of appearance, aesthetics and conventional wisdom are three contributing factors for manifestations of local identity through architecture.

The aforementioned observation lends ground to further contend that environmental conditions specific to the locations of buildings also play a key role in re-examining the topic of identity formations and expressions in built forms. This remark is particularly important when architects seek to engage the globalizing world in a meaningful way while maintaining their cultural roots, resulting in what Nagashima [7] called "glocal" architecture (Glocal = Global + Local). Such a phenomenon encompasses a variety of concerns that affect humankind as a whole regardless of their places of origins, such as global warming and sustainable energy consumption.

In a nutshell, the case studies collectively testify that the design guidelines for expressions of local identity via architecture are stipulated by the economic, cultural, social, and technological conditions both in the domestic and international dimensions. Accordingly, a concluding note could be made that there is no prêt-à-porter "formula" for manifestations of identity through architectural designs. Rather, those expressions result from constant adjustments in the creative process to come up the designs that are socially acceptable, economically viable, environmentally sustainable, technologically innovative, aesthetically pleasing, and culturally unique.

Acknowledgements The authors would like to extend our warm and sincere gratitude to the governor of Kamphaeng Phet province, local community leaders, as well as the owners of those residential, religious and educational buildings.

References

1. Hall S (1997) Representation: cultural representations and signifying practices. Sage in Aassociation with the Open University, London Thousand Oaks, California. ISBN 9780761954323
2. Jelincic DA (2009) Splintering of tourism market: new appearing forms of cultural tourism as a consequence of changes in everyday lives. CollAntropol 33(1)
3. Jensen AL, Arnett JJ, McKenzie J (2011) Globalization and cultural identity. In: Schwartz SJ et al (eds) Handbook of identity and theory research. Springer, New York, pp 285–301
4. Richards G, Wilson J (2006) Developing creativity in tourist experiences: a solution to the serial reproduction of culture? Tour Manage 27:1209–1223
5. Wisner B, Blaikie P, Cannon T, Davis I (2004) At risk: natural hazards, people's vulnerability and disasters, 2nd edn. Routledge, London. ISBN/ISSN9780415252164
6. Horayangkura V (2011) The creation of cultural heritage towards the development of a modern thai architectural identity. J Arch/Plann Res Stud 8(1):1–19. ISSN: 1905-2022
7. Nagashima K (1996) Asian architecture of the future: towards glocal architecture. In: Roots, a pamphlet distributed in exhibitions in Tokyo and other cities in Japan
8. Bourdieu P (1990) In other words: essays towards a reflexive sociology. Adamson, M. (Trans). Stanford University Press, Stanford, California
9. Department for International Development (DFID) (1999) Sustainable livelihoods guidance sheets. Department for international development, London, UK
10. Bhandari PB (2013) 2013 Rural livelihood change? Household capital, community resources and livelihood transition. J Rural Stud 32:126–136
11. Denny JM, Case PM, Metzger A, Ivanova M, Asfaw A (2018) Power in participatory processes: reflectionsfrom multi-stakeholder workshops in the horn of Africa. Sustain Sci 2018(13):879–893

Sustainable Construction of Wetland Stilt House in Indonesia

Amelia Rizyan Nyssa, Dalhar Susanto, and Toga H. Panjaitan

Abstract The stilt house is one of Indonesia's traditional architectures renowned for its ability to withstand wetland environments since ancient times. This paper then aims to discuss the uniqueness of stilt house construction on each wetlands area in Indonesia and draw conclusions about which construction elements can affecting the stilt house's durability. The method used in this research is a literature study. The literature on wetland and wooden house construction will be discussed at the beginning and used to classify the construction used on stilt houses in each wetland area in Indonesia. From each wetland area, it is found that stilt houses use several types of constructions. The stilt house with pedestal foundation uses rigid pillar type called continued pillars. In contrast, the stilt house with pile foundation uses a more flexible type of pillars such as discontinued pillars or a combination of continued and discontinued pillars. The choice of pillar types and foundations compatible with their wetlands context can make the stilt house durability last longer. This is one of the evidence of local people's knowledge turns out bringing sustainability values that can make them survive in a long time.

Keywords Sustainable construction · Vernacular · Wetlands · Stilt house

1 Introduction

Indonesia is one of the largest wetland contributors in the world. According to the 2016–2020 Peatland Restoration Strategic Plan [1], 47% of the world's wetlands are located in Indonesia. Wetlands are an essential component of an ecosystem that must be maintained stable. People in Indonesia have been used to living side by side with wetlands since ancient times. One form of adaptation made by wetland communities is how they make their dwelling place in the form of a stilt house in an environment that is always watery all year round. This wetland stilt house was built with the local wisdom of the local community. Each region in Indonesia has

A. R. Nyssa (✉) · D. Susanto · T. H. Panjaitan
Departement of Architecture, Universitas Indonesia, Kampus UI Depok, Jakarta, West Java 16424, Indonesia
e-mail: amelia.rizyan@ui.ac.id

T. Kang and Y. Lee (eds.), *Proceedings of 2021 4th International Conference on Civil Engineering and Architecture*, Lecture Notes in Civil Engineering 201,
https://doi.org/10.1007/978-981-16-6932-3_54

its uniqueness, both from natural materials and the construction techniques they use. With local wisdom, the stilt house can sustainably survive in wetlands for hundreds of years. Knowledge of local wisdom is essential to learn. This knowledge can later be developed to build stilt houses with good durability while paying attention to the natural environment. There is quite a lot of literature that discusses the peculiarities of stilt house construction in each wetland area, but until now, no one has elaborated the literature in each area and analyzed it so that conclusions can be drawn in the form of a typology regarding stilt house construction in wetlands. This paper then aims to discuss the uniqueness of the stilt house construction of each wetland area in Indonesia and draw conclusions about what elements of construction can make the stilt house last in the long term.

2 Methods

This study used a qualitative method through literature studies on construction systems and materials on stilt houses in Indonesia's wetlands. First, we will study the literature on definitions and categories of wetlands and then explain wetland stilt house characteristics in Indonesia. This literature will then be used to classify the constructions used on stilt houses in each wetland area and how this construction affects the building's durability and condition. This classification will then be synthesized to obtain the overall construction typology of each stilt house and which construction has the better durability. Finally, it will be followed by concluding the discussions as the closing part of this paper.

3 Literature Review

3.1 Wetlands Definition and Classification

The term wetland was first adopted after signing the 1971 Ramsar Convention on Wetlands of International Importance. The area which can be described as the wetland is an area that includes brackish, swampy, peat, or aquatic areas. Whether natural or artificial, permanent or temporary, with flowing or stationary (stagnant), fresh, brackish, or salty water, including areas with seawater whose depth at low tide (low tide) does not exceed six meters [2]. Based on river currents and high tide from the sea, wetlands are divided into three zones. Each of these zones has a different physiographic form and the distribution and characteristics of each wetland zone, including the soil bearing capacity and water content [3]: Zone I is a saltwater/brackish tidal marsh zone. This zone has stronger ocean currents than river currents. This area also includes the area of islands or deltas located at the mouth of large rivers. Zone II is a freshwater tidal swamp zone. This zone is a swamp with slightly larger water

Table 1 Indonesia wetland stilt house constructions

Construction	Types		References
Substructure	Pedestal foundation 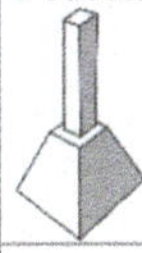	Pile foundation 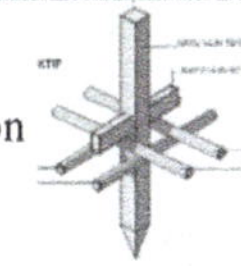	[4, 5]
Wood Joints	Fixed wood joints 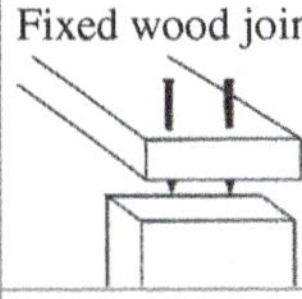	Detachable wood joints 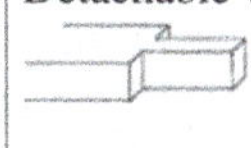	[6, 7]
Superstructure	Continued pillar 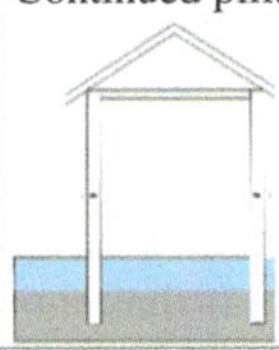	Discontinued pillar 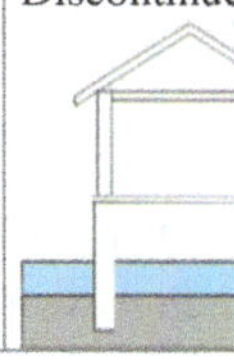	[7–9]

currents, such as currents in rivers. Swamplands in this zone are influenced by daily tides but are not influenced by salt/brackish water. The tides that occur do not cause overflows in the surrounding area. Zone III is a non-tidal swamp or swamp zone. In this zone, the daily tides that occur in river water are not visible, but the water level still depends on river water but does not change significantly in a short time. Each zone has a different soil bearing capacity and water content. This will then affect the material resistance and construction of each type of stilt house.

3.2 Wetland Stilt House Construction

Construction on stilt houses is generally divided into sub-structure and superstructure construction. Table 1 of construction types found in Indonesia's wetland stilt houses, including the wood joint used by the local people.

3.3 Indonesia Wetlands Stilt Houses

Below is an illustration of each stilt house's location's will be mention in the literature synthesis section. These locations are located in South Sulawesi, Central Sulawesi, East Kalimantan, Central Kalimantan, South Kalimantan, Aceh, and South Sumatra.

Fig. 1 Wetland stilt house locations. The location taken is the location of wetlands that are spread across almost all islands in Indonesia. These locations are also the location taken by previous researchers who examined the stilt house and Indonesia's traditional houses. *Image sources* [11] edited by author

Sulawesi and Sumatra are classified as earthquake-prone areas. Each stilt house uses local wood that belongs to the mangrove type of plant that can resist the wetland's damp soil. The locals' selection of wood material in vernacular houses can support the building's durability [10] (Fig. 1).

4 Synthesis

Table 2 of synthesis results from previous research.

Based on the table above, it can be seen that most of the stilt houses can last until 100 years, and some of them can reach more than 100 years. Each area has its combination of substructure, superstructure, and wood joints based on the wetland conditions. Wetland Stilt house that stands on an earthquake-prone area will use a pedestal foundation as the substructure construction. A pedestal foundation is more suitable to handle the wetland's wetness and the horizontal loads that come from the earthquake. For the wetland areas that don't affect by the earthquake, the pile foundations are more suitable for handling the soft texture soils and the water content in the wetland itself. In the superstructure part, the pedestal foundation requires the continued pillar to make the structure more rigid so that the earthquake loads can be distributed equally. In contrast, the pile foundation provides more flexibility for the community to choose pillars adjusting to their maintenance or repair easiness needs. Some of the stilt house building conditions are poor because the old woods start to deteriorate after many years of use. This deterioration mostly happens in the building envelope part, the structural part of all stilt houses still quite good and durable to handle the loads and the environment. If we see the synthesis as a whole, we can see the most durable constructions in the wetland are found on stilt houses

Table 2 Wetland stilt house literature synthesis

	Location	Wet-land zone	Sub-structure construction	Superstructure construction	Wood joints	Building age	Building conditions	Picture
[12]	Cambaya coastal area, makassar	I	Umpak (pedestal foundation)	Continued pillar	Detachable	< 100 years	Poor	
[13]	Bajo stilt house, parigi moutong coastal area	I	Cerucuk (pile foundation)	Mixed	Fixed	< 100 years	Poor	
[14]	Martapura riverside, banjarmasin	II	*Galam* wood cerucuk (pile foundation)	Discontinued pillars	Fixed	10–100 years	Good	
[15]	Kutai stilt house, tenggarong	II	*Galam* wood cerucuk (pile foundation)	Mixed	Detachable	> 80 years	Good	
[16–18]	Kahayan riverside stilt house and *Huma Loendjoe,* Palangkaraya	II	*Galam* wood cerucuk (pile foundation)	Discontinued pillars	Detachable	Around 111 years	Good	
[19]	Traditional house of kampung tenun samarinda	III	Cerucuk (pile foundation)	Mixed	Detachable	> 100 years	Poor	

(continued)

Table 2 (continued)

	Location	Wet-land zone	Sub-structure construction	Superstructure construction	Wood joints	Building age	Building conditions	Picture
[20]	*Rumoh Santeut,* west coast of aceh	III	Umpak (pedestal foundation)	Continued pillar	Detachable	> 100 years	Poor	
[21]	Situs air sugihan settlements, south sumatera	III	Umpak (pedestal foundation)	Continued pillar	Detachable	> 100 years	Poor	

located in non-seismic wetland zone II in Kalimantan. This stilt house can reach the age of 111 years with good overall conditions. The construction system used in Kalimantan's stilt house uses pile foundations of *Galam* wood material, discontinued and combination pillars, and fixed and detachable wood joints systems.

5 Conclusions

From the literature study that has been carried out, we can see some unique characteristics possessed by durable stilt houses in the wetland. First, the material used is natural material in local wood, which is already suitable with the wetland's original conditions. Second, Generally, the stilt house is intended to continue to function for a long period. Most homes can be used for 2–3 generations. Third, Wetland stilt houses tend to use a construction system considering the ease of repair/maintenance process. Fourth, The construction installation techniques used vary according to local wisdom and each region's natural context. And the last, stilt house placement is very friendly to the environment. Wetland stilt houses communities generally avoid landfilling in the area around their house. Overall, it can be concluded that the local architectural approach in Indonesia, in this context, the stilt house of wetlands, can fulfill two aspects of sustainability, both ecological and durable. This proves that local people's life habits in ancient times kept values that we can still learn to develop sustainable architecture. There are still limited studies that can analyze the uniqueness of construction that traditional architecture has from a more scientific and measurable perspective. Criteria or requirements to build a stilt house that fits Indonesia's local context haven't been defined. For further research, this research can become one of the references for other research that wants to learn more about the uniqueness of Indonesia's traditional architectural constructions in a more comprehensive way.

References

1. Rencana Strategis Badan Restorasi Gambut Tahun 2016–2020, Nov 2016, p 72 (2016)
2. Soendjoto MA (2015) Potensi, Peluang, dan Tantangan Pengelolaan Lingkungan Lahan Basah Secara Berkelanjutan. Pros. Semin. Univ. Lambung Mangkurat 2015, Sept 2015, p 3
3. Subagyo H (1997) Potensi pengembangan dan tata ruang lahan rawa untuk pertanian. In: Prosiding simposium Nasional dan Kongres VI PERAGI, pp 17–55
4. Nursaniah C, Machdar I, Azmeri AM, Irwansyah M, Sawab H (2019) Transformation of stilt houses: a way to respond to the environment to be sustainable. IOP Conf Ser Earth Environ Sci 365(1)
5. Heldiansyah JC, Aufa N, Wastuty PW (2017) KONSEP KONSERVASI KAWASAN PUSAKA LAHAN BASAH UNTUK MELESTARIKAN RUMAH BUBUNGAN TINGGI TELOK SELONG (The concept of conservation on wetlands heritage areas to conserve). Rumah Bubungan Tinggi Teluk Selong. Pros. Semin. Nas. Lahan Basah, pp 983–993
6. Zwerger K (2011) Wood and wood joints, vol 4, no 1. The German National Library, New York

7. Gunawan Y, Oktarina S, Fauzan MAF (2016) Discontinued pillars : the adaptability of water Stilt-House at Kuin River, Banjarmasin, South Borneo, Indonesia. In: International symposium on architectural interchange Asia (ISAIA 2016), Sendai, Japan, May, pp 1–5
8. Seman S, Irhamna (2011) Arsitektur Tradisional Banjar Kalimantan Selatan, Cetakan Ke. Banjarmasin: Lembaga Pengkajian dan Pelestarian Budaya Banjar Kalimantan Selatan
9. Idham NC (2019) Indonesian architecture and earthquake vulnerability: the development of building safety through the civilization. MATEC Web Conf 280:01004
10. Saskia CS, Susanto D, Suganda E (2020) Acehnese vernacular house: the study of constructions and materials resilience. IOP Conf Ser Earth Environ Sci 452(1)
11. Map of Indonesia (2019) Online. Available https://www.sejarah-negara.com/en/25345/peta-buta-indonesia/
12. Sastrawati I (2009) The characteristics of the self-support stilt-houses towards the disaster potentiality at the cambaya coastal area, Makassar. Dimens. (Jurnal Tek. Arsitektur) 37(1):33–40
13. Rifai AJ (2010) Perkembangan Struktur Dan Konstruksi Rumah Tradisional Suku Bajo Di Pesisir Pantai Parigi Moutong. J ruang 2(1)
14. Afdholy AR (2017) Tipomorfologi Permukiman Tepian Sungai Martapura Kota Banjarmasin. Local Wisdom J Ilm Kaji Kearifan Lokal 9(1)
15. Hidayati Z, Cisyulia O (2013) Studi Adaptasi Rumah Vernakular Kutai Terhadap Lingkungan Rawan Banjir di Tenggarong. Dimens (J Archit Built Environ) 40(2):89–97
16. Hamidah MAMN, Rijanta R, Setiawan B (2014) Urban riverside settlement model case: Kahayan riverside settlement, Palangkaraya. J Pemukim 9(1):17–27
17. Murti NK (2020) Tipologi Bangunan di Permukiman Bantaran Sungai Berdasarkan Lokasi dan Jenis Konstruksi (Studi Kasus: Permukiman Bantaran Sungai Kahayan, Palangkaraya) 7(1):14–24
18. Widyastuti M, Putranto D (2018) Sistem Konstruksi Rumah Panggung Pada Tanah Gambut Kalimantan Tengah (Studi Kasus: Huma Loendjoe). J Mhs Jur Arsitektu
19. Hayuni N, Syahbana JA (2014) Upaya Revitalisasi Permukiman Tradisional Tepi Sungai (Studi Kasus : Permukiman Kampung Tenun Samarinda). J Pembang Wil Kota 10(4):400
20. Nursaniah C, Izziah I, Qadri L (2017) Konsep Kearifan Lokal dari Konstruksi Rumah Vernakular di Pesisir Barat Aceh untuk Perancangan Arsitektur Modern (Studi Kasus: Wilayah DAS Krueng Tripa, kabupaten Nagan Raya). Tesa Arsitektur J Archit Discourses 14(2):55–63
21. Vita V (2016) Adaptasi Masyarakat Pra-Sriwijaya di Lahan Basah Situs Air Sugihan, Sumatera Selatan. Kalpataru 25(1):1

The Architecture of Dutch East Indies Colonial Quarantine Station in Onrust Island (1911–1933)

Ikrar Raksaperdana and Kemas Ridwan Kurniawan

Abstract In 1911–1933, Onrust Island, an island located 14 km off the coast of Batavia (Jakarta), Dutch East Indies (Indonesia), went from being a military base to a quarantine station for Hajj pilgrims. Narratives about quarantine stations, especially in colonial site, often described by inadequate spatial conditions, while also possessed a disciplinary mechanisms. In context of Dutch East Indies' Hajj quarantine, there were narratives beyond a pathogenic-quarantine discourse, such as political efforts to supervise the pilgrims' political behaviour. This issue raises a question: How was, architecturally, the Onrust Island Quarantine Station's condition? Was it indecent? Was it possessed a disciplinary mechanism? This paper is a research on architecture history, which will use interpretive research methods on spatial conditions, by collecting data related to historical archives from the Onrust Island Quarantine Station and put it on the context of quarantine architecture.

Keywords Quarantine · Architecture · Onrust Island · Disciplinary mechanism · Dutch East Indies

1 Introduction

Since the fourteenth century, the application of architectural settings as a tool for quarantine has been a strategy of preventing disease outbreaks. Quarantine derived from Italian word "*quaranta*", which means 40 (forty) and since its first application in the fourteenth century, quarantine has been used as a protocol to isolate humans, animals and objects exposed to infectious diseases for 40 days [1]. In Rosen's (1958) paper about public health history, since its early development, quarantine was a form of public health practice, which translates medical procedures using physical factors such as environmental configuration to suppress pathogens [2]. Rosen [2] also stated that in a quarantine site, we must first avoid corrupted air which may come from muddy places, from stagnant water and ditches, from burial places, and from stables of animal [2]. Theoretically, in modern medical and spatial science, quarantine in

I. Raksaperdana (✉) · K. R. Kurniawan
Department of Architecture, Universitas Indonesia, Beji, Depok 16424, Indonesia

T. Kang and Y. Lee (eds.), *Proceedings of 2021 4th International Conference on Civil Engineering and Architecture*, Lecture Notes in Civil Engineering 201,
https://doi.org/10.1007/978-981-16-6932-3_55

context of place, is a form of "therapeutic space" which is essentially a place with an attempt to treat or heal [3]. It then takes us to a broader question, does a quarantine site always have a medical effects? Or the role of quarantine station is the same as a prison, where there is a practice of "isolating and separating people". This question refers to Foucault's [4] thoughts, which notes that on the bubonic plague, quarantine were a form of "panopticon". Foucault [4] narrated that a quarantine on the bubonic plague were implementing a "disciplinary mechanism", and also demonstrated the use of power in order to divide people [4]. Foucault's [4] thoughts takes us to how was the condition of a quarantine station in context of colonial site? Where a quarantine station, especially in the Dutch East Indies, was specifically aimed for separating indigenous people in context of immense possibility of transmitting pathogens in their bodies [5] while also a political effort to control the debarkation of Hajj pilgrims from the influence of anti-western propaganda and politics in the Middle East [6]. Speaking of which, Onrust Island Quarantine Station is one example of a quarantine station that operated in the colonial site of Dutch East Indies. The questions lead us to, how was the actual spatial condition of the Onrust Island Quarantine Station? Was there any "hidden agenda" such as "disciplinary mechanism" in the creation of this quarantine station? Was it architecturally indecent? Or in reality, it was a quarantine product that was architecturally decent as a quarantine station?

After the publication of Hajj pilgrims sailing regulation in 1898 and 1922, as well as regulations for quarantine in 1911, three new quarantine station were formed in the Dutch East Indies, including Rubiah Island, Sabang Island, and Onrust Island (including Kuipers Island as a satellite island of Onrust) [7]. In the process, Onrust Island was used for the arrival of pilgrims [5]. Onrust Island itself is an island located in the Thousand Islands cluster, Dutch East Indies (Indonesia), with an area of 3.5 km^2 and located 14 km from the Batavia (Jakarta) coast. The long history of Onrust Island dates back to the seventeenth century when the *Verenigde Oost Indische Compagnie* (VOC) was granted permission to set up a shipyard there, which in following years developed into a fortress and military base [8]. In 1800, all buildings and forts on Onrust Island were destroyed by the British fleet and reconstructed into a shipyard [8]. In 1806, the whole building was again abandoned by the British until 1827 Governor General Van der Capellen restored the function of Onrust Island [8]. Since the relocation of shipping function to Tanjung Priok, Onrust Island lost its role again, until 1911, Onrust Island changed its function once again to became a quarantine station as an implication of quarantine and Hajj regulations [8]. The utilization of Onrust Island as a quarantine station lasted until 1933, in 1940 Onrust Island was temporarily transformed into an internment camp for European rebels in the "Seven Ship Incident" (*Zeven Provinciën*) and now become a historical site [9].

1.1 Methods

As this research aims to examine spatial systems in a historical object, Onrust Island Quarantine Station, this research will be conducted using an interpretative method with a spatial analysis approach to historical sources regarding the Onrust Island Quarantine Station, including photos, maps, site plans and historical literature. Interpretation will be carried out referring to architecture history research methods which described in Groat and Wang [10] (Architectural Research Methods), by conducting investigations on physical and social phenomena occured in the Onrust Island Quarantine Station. Interpretive methods in historical and architectural contexts, according to Groat and Wang [10] requires a justification of data or historical evidence using a specific approach that is in accordance with the context of research' object. In addition, Groat and Wang [10] also explained that historical research requires a deep understanding of the socio-political context to gain a perspective on the interaction between humans and objects of study (architecture) [10]. Therefore, a spatial analysis approach will be carried out, by collecting data related to the Onrust Island Quarantine Station and analyzed it from the perspective of technical architecture aspects gathered from quarantine station precedents and social aspects examined from Foucault's [4] thoughts about quarantine.

2 Onrust Island Quarantine Station Architecture

Onrust Island Quarantine Station has an area of 3.5 km^2 or 350 ha. The main access to Onrust Island could be reached from harbor which located on the south. In general, the Onrust Island Quarantine Station function were divided into, quarantine functions, such as quarantine barracks, health support functions, such as hospital, doctors' house, kitchen and patient warehouse, as well as operational functions that oversee management and support for general activities in the area, such as docks, guard post, electrical, reservoir, warehouse, island administrator's house and European's house. In addition, there were also ruins of a former prison building and a former VOC tomb in the Onrust Island Quarantine Station (Fig. 1). Figure 1 also shows a model on Onrust Island's circulation, interpretated from narrative of Onrust Island archeologist, Attahiyat [11], and archives of the *Department van Burgerlijke Openbare Werken* (B.O.W) (1922) gathered from National Archives of Republic of Indonesia. After the pilgrims went through the debarkation border control process on Kuiper Island, the pilgrims headed to Onrust Island through a jetty for health checks and arrangement, to be divided into groups for each quarantine barrack's unit [11]. The management and/or medical personnel produce a circulation that connected the service and administrative zones, hospital, and specified zone for medical personnel (doctor's house, European house). The pilgrims' circulation and medical personnel was not completely separated, there was still a circulation that overlaps with one another (Fig. 1), although the main access to the island was separated, as the pilgrims' arrived from jetty that connected Onrust and Kuiper and medical personnel from the harbor.

All buildings other than quarantine barracks at Onrust Island Quarantine Station reused old buildings, such as the hospital building which used former barracks and military director's dwelling, then the doctor's house, which was previously a home for naval officer, as well as supporting functions that have already been existed

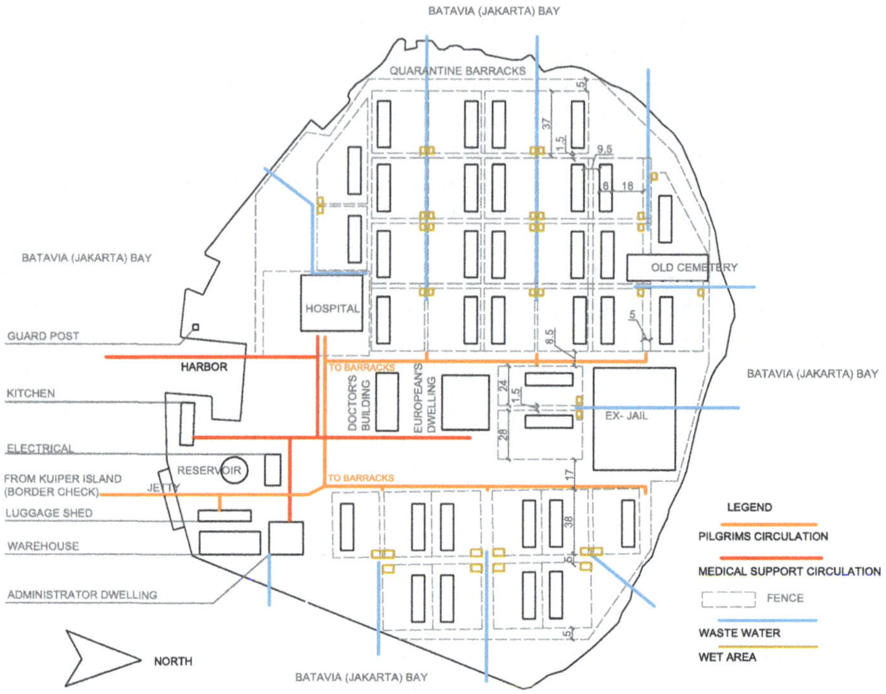

Fig. 1 Onrust Island quarantine station map (1922). *Source* National Archives of Republic of Indonesia (ANRI)

before (see Dutch Colonial Maps (no. 04523)—Universiteit Leiden). The quarantine function at the Onrust Island Quarantine Station was consisted with 35 barracks, Each barracks has a mass orientation towards north–south, with placement carried out referring to the Onrust Island grid. Each quarantine barracks surrounded by rat-proof wire fences which were supported by poles (Fig. 2) [9]. This fences was created

Fig. 2 Quarantine barrack (1940). *Source* KITLV 19,059

as a barrier for each unit also there was a larger fences that surrounded the island. Each zone of the quarantine barracks has a communal function in the quarantine barracks yard (Fig. 4) while the interior provided for resting area for 100 peoples [5], it also had a specified space for washing area and wet area (toilet) (Fig. 4). Every wet areas in the island was connected by a waste water system that drained the water straight to the Batavia (Jakarta) Bay. In terms of building, each unit was levelled approximately 5 cm from the ground and it can be seen that placement of the structural brick walls uses the one-brick bonding technique with a wall thickness of 20 cm (data gathered from observations on quarantine ruins in Onrust Archeological Park). Each barracks has a linear shape with an arch-shaped tin roof, with roof overhangs on each side of the building (Fig. 2) There were also openings formed from a space given between barrack's roof and wall (Fig. 4). The barrack's roof supported by a frame structure that can be seen from interior and exterior. There was also a bracing that binds the columns. The quarantine barracks' wall that faced north–south have a lower height than the wall faced east–west (Fig. 4) and all outer wall materials were unfinished (Fig. 2). In the inner space, we could saw the frame structure and an arch roof which was visually exposed, then the floor was formed from plastered cement and sand (Fig. 3), and there was grass throughout the quarantine barracks unit zone exterior, with shade trees in some of quarantine barracks units.

The first interpretation carried out at a site scale, where the Onrust Island Quarantine Station has a site placement characteristics similar to Howard's [12] narratives

Fig. 3 Quarantine barrack interior (1940). *Source* KITLV 19,072

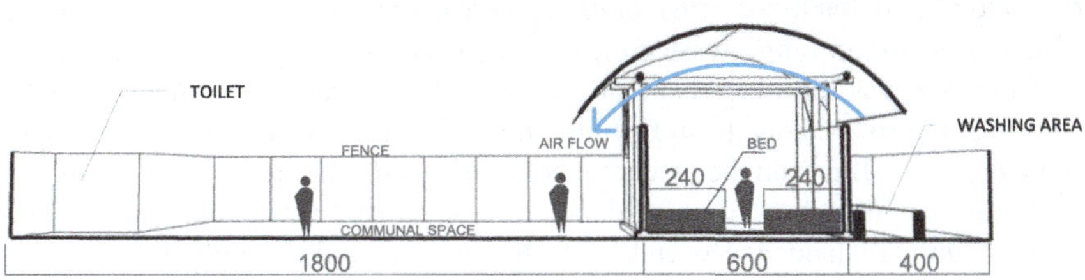

Fig. 4 Quarantine barrack section. *Source* National Archives of Republic of Indonesia (ANRI)

on medieval European quarantine station at the beginning of its development, which was located on a small island and has a single access from the pier for each user [12]. Most of Onrust Island's archives and literatures [11] discusses that disinfectant, isolation and border check activities were carried out on Kuipers Island, àn island that originally served as an island for a weather tower, which located not far from Onrust Island (400 m between piers), however, it is unfortunate that the core activities from a quarantine station, disinfectant of the users, did not have a specified space on Onrust Island. Then from user's circulation, it was found that there still a circulation from each users that intersects each other, this is quite different from other earlier quarantine station architectural models from Howard's [12] note that prioritize separation of circulation and accessibility for each user [12]. The existence of old cemetery adjacent to the quarantine zone in the northern part of Onrust Island also contradicts to Rosen's [2] statement which suggests to avoid burial places [2]. Another unfortunate thing was there were only one mass afforded for pilgrim's luggage, which is quite concerning to see that while in one quarantine session, the Onrust Island Quarantine Station could facilitate up to 3500 people [9]. On the buildings, it appears that the quarantine barracks on Onrust Island looks quite feasible, if we compare it to the models on Howard's [12] note, which explains that a quarantine station mostly has a linear mass shape with vents at both ends to facilitate air flow, a roof shape that can protect indoor elements from weather such as rain to prevent development of disease due to wetness of the interior, and having open space on each patient mass [12], we could see that Onrust Island's quarantine barracks had achieved that narration by the application of a linear shaped building with a space created between roof and vertical planes which would provide a cross ventilation. The thick vertical planes, where the east–west facing vertical plane was built higher than the north–south facing plane, also provides maximum protection from the sun, while the thickness ensure that no moisture would penetrated to interior. The arched roof's overhang also provides protection for the inner space from rain. There was also an open space on the terrace of each barracks unit.

One unfortunate thing was the unfinished materials of vertical planes and floors, and direct ground after barrack's exit access, which is quite risky in long term condition, to cause weathering, puddles and microbes to nest in surfaces, a condition that Rosen [2] suggest to avoid from a quarantine station [2]. However, from the unfinished materials' use, which were also composed by simple materials and constructed with simple structural compositions, and considering that the construction of this quarantine station coincided with the publication of the quarantine ordinance in 1911 [5], it can be interpreted that these considerations were made in order to achieve maximum productivity with minimum wasted effort or expense. The important thing that must be highlighted was the placement of a wet area (toilet and cleaning area) which was located approximately 15 m from the pilgrim's barrack, a direct water disposal system to Batavia Bay, and the placement of a specified space for pilgrim's waste and washing area, which showed an evidence of a developed waste and water disposal system on Onrust Island, as Rosen [2] emphasizes, on modern public health system, the disposal of sewage and waste are some aspects that are more fundamental than a sophisticated scientific knowledge of disease causation and transmission [13].

In terms of disciplinary mechanism aspect carried out at the Onrust Island Quarantine Station, the map on Fig. 1 shows that there was only one guard post located at the pier, then the mass organization system did not manifest a "panopticon", with the presence of doctor's and European's people house as a site supervisor that were located in the middle of the grid-arranged site, which lowered the effectivity of surveillance act from the quarantine supervisor, where Foucault [4] explains that a panopticon formed when the supervisor can observe all people's movements from an ideal angle (watchtower; commonly radial spatial organization form). However, as Foucault [4] stated on Discipline and Punish, this "decent or useful space" could be identified as disciplinary mechanism's "rule of functional site", coupled with the wire fences application in each barrack's units, and a larger fence that surrounded the island, this was a staggering issue that could be interpreted as an application of disciplinary mechanism at the Onrust Island Quarantine Station. This fences were also reminiscent to barriers' application at the El Tor pilgrims Quarantine Station, in Egypt, which functioned to prevent quarantined people from entering and leaving easily [14], the only difference from the El Tor was there was no specified guard post for a military personnel. There was no explanation of how each pilgrims divided into a group, but if assumed that there was a separation based on age or gender groups as were done in several quarantine stations [12], the application of barriers on each unit was considered a normal protocols in quarantine in terms of segregating groups of people to prevent disease transmission. Even so, one thing that needs to be emphasized was the wire material use on the barrier, and with a dimensions approximately taller than indigenous peoples height, which could build an interpretation of disciplinary mechanism usage in the island, to control the pilgrims.

3 Conclusions

Architecturally, the Onrust Island Quarantine Station, in terms of quarantine barrack design was quite decent with a composition of architectural elements that can protect from the weather and the application of vents which created from spaces between roof and vertical elements at both ends to facilitate air flow. The unfinished material use which combined with simple structural compositions is tolerable, concerning the facts that the construction of this quarantine station coincided with the publication of the quarantine ordinance in 1911, it can be interpreted that these considerations were made in order to achieve maximum productivity with minimum wasted effort or expense. The evidence of a developed waste and water disposal system also emphasized that this quarantine station was well designed. However, on a site scale there were still several shortcomings, such as the absence of a specified mass for disinfectant, evidence of a burial near the northern barrack's unit, also there was still an evidence of a circulation from each users that were not separated, and there was only one mass for luggage afforded on this quarantine station. Last but not least, there were other aspect that would lead to a broader discussion, particularly the disciplinary mechanisms, shown from wire fences with a dimensions approximately taller than indigenous peoples height that enclosed each barracks and the island. Although things such as Foucault's [4] "panopticon" was not significantly showed, the wire fences application could lead an interpretation of disciplinary mechanism, and if so, the "decent useful space" of the Onrust Island would be what Foucault [4] stated as a strategy to create one of the disciplinary mechanism aspects, "the rule of functional site".

This research is financially supported by the Directorate of Research and Community Engagement (DRPM) Universitas Indonesia, under the international proceeding publication grant (PUTI Prosiding), contract number NKB-3785/UN2.RST/HKP.05.00/2020.

References

1. Tognotti E (2013) Lessons from the History of Quarantine, from Plague to Influenza. Emerg Infect Dis 19(2):254–259
2. Rosen G (1958) A history of public health. MD Publications, Inc., New York, pp 29–30
3. Collins J (2020) The architecture and landscape of health: a historical perspective on therapeutic places 1790–1940, Chap 1. Routledge, New York
4. Foucault M (1977) Discipline and punish: the birth of prison. In: Sheridan A (ed) Penguin Books, Great Britain, pp 195–198
5. Pols H (2016) Quarantine in the Dutch East Indies. In: Bashford A (ed) Macmillan, UK, pp 85–103
6. Alexanderson K (2019) Subversive seas: anticolonial networks across the twentieth-century Dutch empire. Cambridge University Press, Cambridge
7. Baihaqi AF (2016) Pelayaran Angkutan Jamaah Haji di Hindia Belanda (Tahun 1911–1930) [The Voyage of Hajj Pilgrims in Dutch East Indies (Year 1911–1930)]. Buletin Al-Turas: Mimbar Sejarah, Sastra, Budaya, dan Agama 22(1):1–24
8. Dharmaputra G (1985) Bangunan, Pemukiman, dan Penduduk di Pulau Onrust Tahun 1803 dan 1864: Sebuah Kajian Hipotesis [Buildings, Settlements, and Dwellers in Onrust Island Year 1803 and 1864: a hypothetical study]. Universitas Indonesia, Depok
9. Attahiyat C (2000) Onrust Island. Dinas Museum dan Pemugaran, Jakarta
10. Groat LN, Wang D (2013) Architectural research methods, vol 2, Chap 6. In: Aufl (ed) Wiley, New Jersey
11. Attahiyat C (2001) Laporan Penelitian/Pengumpulan Informasi tentang Bangunan di Pulau Onrust [Research report/information gathering on buildings in Onrust Island]. Dinas Kebudayaan dan Permuseuman, Jakarta
12. Howard J (1791) An account of the principal lazarettos in Europe. In: Johnson J, Dilly C, Cadell T (eds), London
13. Rosen G (1958) A history of public health. MD Publications, Inc., New York, pp xix–xx
14. Long J (1902) The quarantine camp at el tor. Publ Health Rep (1896–1970) 17(20):1156–1159

Author Index

T. Kang and Y. Lee (eds.), *Proceedings of 2021 4th International Conference on Civil Engineering and Architecture*, Lecture Notes in Civil Engineering 201,
https://doi.org/10.1007/978-981-16-6932-3

The manufacturer's authorised representative in the EU is Springer
Nature Customer Service Centre GmbH, Europaplatz 3, 69115 Heidelberg,
Germany. If you have any concerns regarding our products, please
contact ProductSafety@springernature.com

Printed and bound by CPI Group (UK) Ltd, Croydon, CR0 4YY
30/09/2026
02225332-0019